Sample Size Calculations in Clinical Research

Third Edition

Published Titles

Sample Size Calculations in Clinical Research

Third Edition

Shein-Chung Chow, Jun Shao,
Hansheng Wang, and Yuliya Lokhnygina

CRC Press
Taylor & Francis Group
Boca Raton London New York

CRC Press is an imprint of the
Taylor & Francis Group, an **informa** business

A CHAPMAN & HALL BOOK

CRC Press
Taylor & Francis Group
6000 Broken Sound Parkway NW, Suite 300
Boca Raton, FL 33487-2742

First issued in paperback 2020

ISBN-13: 978-1-138-74098-3 (hbk)
ISBN-13: 978-0-367-65776-5 (pbk)

Library of Congress Cataloging-in-Publication Data

Names: Chow, Shein-Chung, 1955- editor. | Shao, Jun (Statistician) editor. | Wang, Hansheng, 1977- editor. | Lokhnygina, Yuliya, editor.
Title: Sample size calculations in clinical research / [edited by] Shein-Chung Chow, Jun Shao, Hansheng Wang, Yuliya Lokhnygina.
Description: Third edition. | Boca Raton : Taylor & Francis, 2017. | Series: Chapman & Hall/CRC biostatistics series | "A CRC title, part of the Taylor & Francis imprint, a member of the Taylor & Francis Group, the academic division of T&F Informa plc." | Includes bibliographical references.
Identifiers: LCCN 2017011239 | ISBN 9781138740983 (hardback)
Subjects: LCSH: Clinical medicine--Research--Statistical methods. | Drug development--Statistical methods. | Sampling (Statistics)
Classification: LCC R853.S7 S33 2017 | DDC 610.72/7--dc23 LC record available at https://lccn.loc.gov/2017011239

Visit the Taylor & Francis Web site at
http://www.taylorandfrancis.com

and the CRC Press Web site at
http://www.crcpress.com

Contents

Preface

This third edition is expanded to 20 chapters with five completely new chapters and numerous updates and new sections of the existing chapters. These new chapters include cluster randomized trial design (Chapter 15), zero-inflated Poisson distribution (Chapter 16), sample size estimation based on clinical trial simulation (Chapter 17), two-stage seamless adaptive trial design (Chapter 18), and sample size estimation based on clinical trial simulation (Chapter 19). New sections in the existing chapters include negative binomial regression in Chapter 5 and sample size requirement for analytical similarity assessment in Chapter 10. Numerous updates have been included in several chapters especially the introduction chapter. Similar to the first two editions, this third edition also concentrates on concepts and implementation of methodology rather than technical details. We keep the mathematics and statistics covered in this edition as fundamental as possible and illustrate the concepts and implementation through real working examples whenever possible.

The first and second editions of this book have been well received by pharmaceutical and clinical scientists/researchers and biostatisticians. The first two editions are widely used as a reference source and a graduate textbook in pharmaceutical and clinical research and development. Following the tradition of the first two editions, the purpose of this third edition is also to provide a comprehensive and unified presentation of the principles and methodologies for sample size calculation under various designs and hypotheses for various clinical trials across different therapeutic areas. This revision is to give a well-balanced summary of current regulatory requirements and recently developed statistical methods in this area. It is our continuing goal to provide a complete, comprehensive, and updated reference and textbook in the area of sample size calculation in clinical research. In the past decade, tremendous progress has been made in statistical methodology to utilize innovative design and analysis in pharmaceutical and clinical research to improve the probability of success in pharmaceutical and clinical development. These methods include the use of adaptive design methods in clinical trials and others such as negative binomial regression for recurrent incidence rates, zero-inflated Poisson for the number of lesions, randomized cluster randomized trials, clinical trials with extremely low incidence rates, and sample size requirement for analytical similarity assessment for biosimilar products.

We have received much positive feedback and constructive suggestions from scientists and researchers in the biopharmaceutical industry, regulatory agencies such as the U.S. FDA, and academia. This third edition provides a comprehensive coverage of the current state-of-the-art methodology for sample size estimation of clinical trials. Therefore, we strongly believe that this new and expanded third edition is not only an extremely useful reference book for scientists and researchers, regulatory reviewers, clinicians, and biostatisticians in biopharmaceutical industry, academia, and regulatory agencies, but may also serve as a textbook for graduate students in clinical trials and biostatistics-related courses. In addition, this third edition can serve as a bridge among the biopharmaceutical industry, regulatory agencies, and academia.

We would like to thank David Grubbs at Taylor & Francis for his outstanding administrative assistance and support. We also express our sincere appreciation to the many clinicians, scientists, researchers, and biostatisticians from the FDA, academia, and pharmaceutical industry for their valuable feedback, support, and encouragement. Finally, the views expressed are those of the authors and not necessarily those of Duke University School of Medicine, Durham, North Carolina, USA; University of Wisconsin, Madison, Wisconsin, USA; and Peking University, Beijing, China. We are solely responsible for the contents and errors of this edition. Any comments and suggestions will be very much appreciated.

1

Introduction

In clinical research, during the planning stage of a clinical study, the following questions are of particular interest to the investigators: (i) How many subjects are needed to have a desired power for detecting a clinically meaningful difference (e.g., an 80% chance of correctly detecting a clinically meaningful difference)? (ii) What is the *trade off* between cost-effectiveness and power if only a small number of subjects are available for the study due to limited budget and/or certain medical considerations. To address these questions, a statistical evaluation for sample size calculation is often performed based on certain statistical inference (e.g., power or confidence interval) of the primary study endpoint with certain assurance. In clinical research, sample size calculation plays an important role for assuring validity, accuracy, reliability, and integrity of the intended clinical study.

For a given study, sample size calculation is usually performed based on some statistical criteria controlling type I error (e.g., a desired confidence level) and/or type II error (i.e., a desired power). For example, we may choose sample size in such a way that there is a desired precision at a fixed confidence level (i.e., fixed type I error). This approach is referred to as *precision analysis* for sample size calculation. The method of precision analysis is simple and easy to perform and yet it may have a small chance of correctly detecting a true difference. As an alternative, the method of prestudy *power analysis* is usually conducted to estimate sample size. The concept of the prestudy power analysis is to select the required sample size for achieving the desired power for detecting a clinically or scientifically meaningful difference at a fixed type I error rate. In clinical research, the prestudy power analysis is probably the most commonly used method for sample size calculation. In this book, we will focus on sample size calculation based on power analysis for various situations in clinical research.

In clinical research, to provide an accurate and reliable sample size calculation, an appropriate statistical test for the hypotheses of interest is necessarily derived under the study design. The hypotheses should be established to reflect the study objectives under the study design. In practice, it is not uncommon to observe discrepancies among study objectives (hypotheses), study design, statistical analysis (test statistic), and sample size calculation. These discrepancies can certainly distort the validity and integrity of the intended clinical trial.

In Section 1.1, regulatory requirement regarding the role of sample size calculation in clinical research is discussed. In Section 1.2, we provide some basic considerations for sample size calculation. These basic considerations include study objectives, design, hypotheses, primary study endpoint, and clinically meaningful difference. The concepts of type I and type II errors and procedures for sample size calculation based on precision analysis, power analysis, probability assessment, and reproducibility probability are given in Section 1.3. The aim and structure of this book is given in Section 1.4.2.

1.1 Regulatory Requirement

As indicated in Chow and Liu (1998, 2003, 2013), the process of drug research and development is a lengthy and costly process. This lengthy and costly process is necessary not only to demonstrate the efficacy and safety of the drug product under investigation, but also to ensure that the study drug product possesses good drug characteristics such as identity, strength, quality, purity, and stability after it is approved by the regulatory authority. This lengthy process includes drug discovery, formulation, animal study, laboratory development, clinical development, and regulatory submission. As a result, clinical development plays an important role in the process of drug research and development because all of the tests are conducted on humans. For approval of a drug product under investigation, the United States Food and Drug Administration (FDA) requires that at least two adequate and well-controlled clinical studies be conducted for providing substantial evidence regarding the efficacy and safety of the drug product (FDA, 1988). However, the following scientific/statistical questions are raised: (i) What is the definition of an adequate and well-controlled clinical study? (ii) What evidence is considered substantial? (iii) Why do we need at least two studies? (iv) Will a single large trial be sufficient to provide substantial evidence for approval? (v) If a single large trial can provide substantial evidence for approval, how large is considered large? In what follows, we will address these questions.

1.1.1 Adequate and Well-Controlled Clinical Trials

Section 314.126 of 21 CFR (Code of Federal Regulation) provides the definition of an adequate and well-controlled study, which is summarized in Table 1.1.

As can be seen from Table 1.1, an adequate and well-controlled study is judged by eight characteristics specified in the CFR. These characteristics include study objectives, methods of analysis, design, selection of subjects, assignment of subjects, participants of studies, assessment of responses, and assessment of the effect. For study objectives, it is required that the study objectives be clearly stated in the study protocol such that they can be formulated into statistical hypotheses. Under the hypotheses, appropriate statistical methods should be described in the study protocol. A clinical study is not considered adequate and well controlled if the employed study design is not valid. A valid study design allows a quantitative assessment of drug effect with a valid comparison with a

TABLE 1.1

Characteristics of an Adequate and Well-Controlled Study

Criteria	Characteristics
Objectives	Clear statement of investigation's purpose
Methods of analysis	Summary of proposed or actual methods of analysis
Design	Valid comparison with a control to provide a quantitative assessment of drug effect
Selection of subjects	Adequate assurance of the disease or conditions under study
Assignment of subjects	Minimization of bias and assurance of comparability of groups
Participants of studies	Minimization of bias on the part of subjects, observers, and analysis
Assessment of responses	Well defined and reliable
Assessment of the effect	Requirement of appropriate statistical methods

control. The selection of a *sufficient* number of subjects with the disease or conditions under study is one of the keys to the integrity of an adequate and well-controlled study. In an adequate and well-controlled clinical study, subjects should be randomly assigned to treatment groups to minimize potential bias by ensuring comparability between treatment groups with respect to demographic variables such as age, gender, race, height and weight, and other patient characteristics or prognostic factors such as medical history and disease severity. An adequate and well-controlled study requires that the primary study endpoint or response variable be well defined and assessed with a certain degree of accuracy and reliability. To achieve this goal, statistical inferences on the drug effect should be obtained based on the responses of the primary study endpoint observed from the sufficient number of subjects using appropriate statistical methods derived under the study design and objectives.

1.1.2 Substantial Evidence

The substantial evidence as required in the *Kefauver–Harris* amendments to the *Food and Drug and Cosmetics Act* in 1962 is defined as the evidence consisting of adequate and well-controlled investigations, including clinical investigations, by experts qualified by scientific training and experience to evaluate the effectiveness of the drug involved, on the basis of which it could fairly and responsibly be concluded by such experts that the drug will have the effect it purports to have under the conditions of use prescribed, recommended, or suggested in the labeling or proposed labeling thereof. Based on this amendment, the FDA requests that reports of adequate and well-controlled investigations provide the primary basis for determining whether there is substantial evidence to support the claims of new drugs and antibiotics.

1.1.3 Why At Least Two Studies?

As indicated earlier, the FDA requires that at least two adequate and well-controlled clinical trials be conducted for providing substantial evidence regarding the effectiveness and safety of the test drug under investigation for regulatory review and approval. In practice, it is prudent to plan for more than one trial in the phase III study because of any or combination of the following reasons: (i) lack of pharmacological rationale, (ii) a new pharmacological principle, (iii) phase I and phase II data are limited or unconvincing, (iv) a therapeutic area with a history of failed studies or failures to confirm seemingly convincing results, (v) a need to demonstrate efficacy and/or tolerability in different subpopulations, with different comedication or other interventions, relative to different competitors, and (vi) any other needs to address additional questions in the phase III program.

Shao and Chow (2002) and Chow, Shao, and Hu (2002) pointed out that the purpose of requiring at least two clinical studies is not only to assure the *reproducibility* but also to provide valuable information regarding *generalizability*. Reproducibility is referred to as whether the clinical results are reproducible from location (e.g., study site) to location within the same region or from region to region, while generalizability is referred to as whether the clinical results can be generalized to other similar patient populations within the same region or from region to region. When the sponsor of a newly developed or approved drug product is interested in getting the drug product into the marketplace from one region (e.g., where the drug product is developed and approved) to another region, it is a concern that differences in ethnic factors could alter the efficacy and safety of the drug product in the new region. As a result, it is recommended that a bridging study be conducted to generate a limited amount of clinical data in the new region to extrapolate the clinical data between the two regions (ICH, 1998a).

In practice, it is often of interest to determine whether a clinical trial that produced positive clinical results provides substantial evidence to assure reproducibility and generalizability of the clinical results. In this chapter, the reproducibility of a positive clinical result is studied by evaluating the probability of observing a positive result in a future clinical study with the same study protocol, given that a positive clinical result has been observed. The generalizability of clinical results observed from a clinical trial will be evaluated by means of a sensitivity analysis with respect to changes in mean and standard deviation of the primary clinical endpoints of the study.

1.1.4 Substantial Evidence with a Single Trial

Although the FDA requires that at least two adequate and well-controlled clinical trials be conducted for providing substantial evidence regarding the effectiveness of the drug product under investigation, a single trial may be accepted for regulatory approval under certain circumstances. In 1997, FDA published the *Modernization Act* (FDAMA), which includes a provision (Section 115 of FDAMA) to allow data from one adequate and well-controlled clinical trial investigation and confirmatory evidence to establish effectiveness for risk/benefit assessment of drug and biological candidates for approval under certain circumstances. This provision essentially codified an FDA policy that had existed for several years but whose application had been limited to some biological products approved by the Center for Biologic Evaluation and Research (CBER) of the FDA and a few pharmaceuticals, especially orphan drugs such as zidovudine and lamotrigine. As can be seen from Table 1.2, a relatively strong significant result observed from a single clinical trial (say, p-value is less than 0.001) would have about 90% chance of reproducing the result in future clinical trials.

Consequently, a single clinical trial is sufficient to provide substantial evidence for demonstration of efficacy and safety of the medication under study. However, in 1998, FDA published a guidance that shed light on this approach despite the FDA having recognized that advances in sciences and practice of drug development may permit an expanded role for the single controlled trial in contemporary clinical development (FDA, 1998).

TABLE 1.2

Estimated Reproducibility Probability Based on Results from a Single Trial

t-Statistic	p-Value	Reproducibility
1.96	0.050	0.500
2.05	0.040	0.536
2.17	0.030	0.583
2.33	0.020	0.644
2.58	0.010	0.732
2.81	0.005	0.802
3.30	0.001	0.901

1.1.5 Sample Size

As the primary objective of most clinical trials is to demonstrate the effectiveness and safety of drug products under investigation, sample size calculation plays an important role in the planning stage to ensure that there are sufficient subjects for providing accurate and reliable assessment of the drug products with certain statistical assurance. In practice, hypotheses regarding medical or scientific questions of the study drug are usually formulated based on the primary study objectives. The hypotheses are then evaluated using appropriate statistical tests under a valid study design to ensure that the test results are accurate and reliable with certain statistical assurance. It should be noted that a valid sample size calculation can only be done based on appropriate statistical tests for the hypotheses that can reflect the study objectives under a valid study design. It is then suggested that the hypotheses be clearly stated when performing a sample size calculation. Each of the above hypotheses has different requirements for sample size to achieve a desired statistical assurance (e.g., 80% power or 95% assurance in precision).

Basically, sample size calculation can be classified into sample size estimation/determination, sample size justification, sample size adjustment, and sample size reestimation. Sample size estimation/determination is referred to the calculation of the required sample size for achieving some desired statistical assurance of accuracy and reliability such as 80% power, while sample size justification is to provide statistical justification for a *selected* sample size, which is often a small number due to budget constraints and/or some medical considerations. In most clinical trials, sample size is necessarily adjusted for some factors such as dropouts or covariates to yield sufficient number of evaluable subjects for a valid statistical assessment of the study medicine. This type of sample size calculation is known as sample size adjustment. In many clinical trials, it may be desirable to conduct interim analyses (planned or unplanned) during the conduct of the trial. For clinical trials with planned or unplanned interim analyses, it is suggested that sample size be adjusted for controlling an overall type I error rate at the nominal significance level (e.g., 5%). In addition, when performing interim analyses, it is also desirable to perform sample size reestimation based on cumulative information observed up to a specific time point to determine whether the selected sample size is sufficient to achieve a desired power at the end of the study. Sample size reestimation may be performed in a blinded or unblinded fashion depending upon whether the process of sample size reestimation will introduce bias to clinical evaluation of subjects beyond the time point at which the interim analysis or sample size reestimation is performed. In this book, however, our emphasis will be placed on sample size estimation/determination. The concept can be easily applied to (i) sample size justification for a selected sample size, (ii) sample size adjustment with respect to some factors such as dropouts or covariates, and (iii) sample size reestimation in clinical trials with planned or unplanned interim analyses.

1.2 Basic Considerations

In clinical research, sample size calculation may be performed based on precision analysis, power analysis, probability assessment, or other statistical inferences. To provide an accurate and reliable sample size calculation, it is suggested that an appropriate statistical test for the hypotheses of interest be derived under the study design. The hypotheses should be established to reflect the study objectives and should be able to address statistical/medical questions of interest under the study design. As a result, a typical procedure for sample size calculation is to determine or estimate sample size based on an appropriate statistical method or test, which is derived under the hypotheses and the study design, for testing the hypotheses to achieve a certain degree of statistical inference (e.g., 95% assurance or 80% power) on the effect of the test drug under investigation. As indicated earlier, in practice, it is not uncommon to observe discrepancies among study objectives (hypotheses), study design, statistical analysis (test statistic), and sample size calculation. These discrepancies certainly have an impact on sample size calculation in clinical research. Therefore, it is suggested that the following be carefully considered when performing sample size calculation: (i) the study objectives or the hypotheses of interest be clearly stated, (ii) a valid design with appropriate statistical tests be used, (iii) sample size be determined based on the test for the hypotheses of interest, and (iv) sample size be determined based on the primary study endpoint, and (v) sample size be determined based on the clinically meaningful difference of the primary study endpoint that the clinical study is intended to detect.

1.2.1 Study Objectives

In clinical research, it is important to clearly state the study objectives of the intended clinical trials. The objectives of clinical studies may include one or more of the following four objectives: (i) demonstrate/confirm efficacy, (ii) establish a safety profile, (iii) provide an adequate basis for assessing the benefit/risk relationship to support labeling, and (iv) establish the dose–response relationship (ICH, 1998b). Since most clinical studies are conducted for the clinical evaluation of efficacy and safety of drug products under investigation, it is suggested that the following study objectives related to efficacy and safety be clarified before choosing an appropriate design strategy for the intended trial.

		Safety		
		Equivalence	Noninferiority	Superiority
	Equivalence	E/E	E/N	E/S
Efficacy	Noninferiority	N/E	N/N	N/S
	Superiority	S/E	S/N	S/S

For example, if the intent of the planned clinical study is to develop an alternative therapy to the standard therapy that is quite toxic, then we may consider the strategy of E/S, which is to show that the test drug has equal efficacy but less toxicity (superior safety). The study objectives will certainly have an impact on the sample size calculation. Sample size calculation provides the required sample size for achieving the study objectives.

1.2.2 Study Design

In clinical trials, different designs may be employed to achieve the study objectives. A valid study design is necessarily chosen to collect relevant clinical information for achieving the study objectives by addressing some statistical/medical hypotheses of interest, which are formulated to reflect the study objectives.

In clinical research, commonly employed study designs include parallel-group design, crossover design, enrichment design, and titration design (see, e.g., Chow and Liu, 1998). The design strategy can certainly affect sample size calculation because statistical methods or tests are usually derived under the hypotheses and study design. As an example, Fleming (1990) discussed the following design strategies that are commonly used in clinical therapeutic equivalence/noninferiority and superiority trials.

Design	Description
Classical	STD + TEST versus STD
Active control	TEST versus STD
Dual purpose	TEST versus STD versus STD + TEST

The classical design is to compare the combination of a test drug (TEST) and a standard therapy (STD) (i.e., STD + TEST) against STD to determine whether STD + TEST yields superior efficacy. When the intent is to determine whether a test drug could be used as an alternative to a standard therapy, one may consider an active control design involving direct randomization to either TEST or STD. This occurs frequently when STD is quite toxic and the intent is to develop an alternative therapy that is less toxic, yet equally efficacious. To achieve both objectives, a dual-purpose design strategy is useful.

Note that in practice, a more complicated study design, which may consist of a combination of the above designs, may be chosen to address more complicated statistical/medical questions regarding the study drug. In this case, standard procedure for sample size calculation may not be directly applicable and a modification will be necessary.

1.2.3 Hypotheses

In most clinical trials, the primary study objective is usually related to the evaluation of the effectiveness and safety of a drug product. For example, it may be of interest to show that the study drug is effective and safe as compared to a placebo for some intended indications. In some cases, it may be of interest to show that the study drug is as effective as, superior to, or equivalent to an active control agent or a standard therapy. In practice, hypotheses regarding medical or scientific questions of the study drug are usually formulated based on the primary study objectives. The hypotheses are then evaluated using appropriate statistical tests under a valid study design.

In clinical trials, a hypothesis is usually referred to as a postulation, assumption, or statement that is made about the population regarding the effectiveness and safety of a drug under investigation. For example, the statement that there is a direct drug effect is a hypothesis regarding the treatment effect. For testing the hypotheses of interest, a random sample is usually drawn from the targeted population to evaluate hypotheses about the drug product. A statistical test is then performed to determine whether the null hypothesis would be rejected at a prespecified significance level. Based on the test result, conclusion(s) regarding the hypotheses can be drawn. The selection of the hypothesis depends upon the study objectives. In clinical research, commonly considered hypotheses include point hypotheses for testing equality and interval hypothesis for testing equivalence/noninferiority and superiority, which are described below. A typical approach for demonstration of the efficacy and safety of a test drug under investigation is to test the following hypotheses.

1.2.3.1 Test for Equality

$$H_0 : \mu_T = \mu_P \quad \text{versus} \quad H_a : \mu_T \neq \mu_P, \tag{1.1}$$

where μ_T and μ_P are the mean response of the outcome variable for the test drug and the placebo, respectively. We first show that there is a statistically significant difference between the test drug and the placebo by rejecting the null hypothesis, and then demonstrate that there is a high chance of correctly detecting a clinically meaningful difference if such difference truly exists.

1.2.3.2 Test for Noninferiority

In clinical trials, one may wish to show that the test drug is as effective as an active agent or a standard therapy. In this case, Blackwelder (1982) suggested testing the following hypotheses:

$$H_0 : \mu_S - \mu_T \geq \delta \quad \text{versus} \quad H_a : \mu_S - \mu_T < \delta, \tag{1.2}$$

where μ_S is the mean for a standard therapy and δ is a difference of clinical importance. The concept is to reject the null hypothesis and conclude that the difference between the test drug and the standard therapy is less than a clinically meaningful difference δ and hence the test drug is as effective as the standard therapy. This study objective is not uncommon in clinical trials, especially when the test drug is considered to be less toxic, easier to administer, or less expensive than the established standard therapy.

1.2.3.3 Test for Superiority

To show superiority of a test drug over an active control agent or a standard therapy, we may consider the following hypotheses:

$$H_0 : \mu_T - \mu_S \leq \delta \quad \text{versus} \quad H_a : \mu_T - \mu_S > \delta. \tag{1.3}$$

The rejection of the above null hypothesis suggests that the difference between the test drug and the standard therapy is greater than a clinically meaningful difference. Therefore, we may conclude that the test drug is superior to the standard therapy by rejecting the null hypothesis of Equation 1.3. Note that the above hypotheses are also known as hypotheses for testing *clinical* superiority. When $\delta = 0$, the above hypotheses are usually referred to as hypotheses for testing *statistical* superiority.

1.2.3.4 Test for Equivalence

In practice, unless there is some prior knowledge regarding the test drug, usually we do not know the performance of the test drug as compared to the standard therapy. Therefore, hypotheses (1.2) and (1.3) are not preferred because they have predetermined the performance of the test drug as compared to the standard therapy. As an alternative, the following hypotheses for therapeutic equivalence are usually considered:

$$H_0 : |\mu_T - \mu_S| \geq \delta \quad \text{versus} \quad H_a : |\mu_T - \mu_S| < \delta. \tag{1.4}$$

We then conclude that the difference between the test drug and the standard therapy is of no clinical importance if the null hypothesis of Equation 1.4 is rejected.

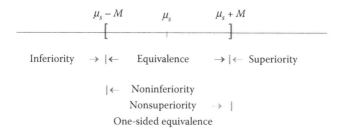

FIGURE 1.1
Relationship among noninferiority, superiority, and equivalence.

1.2.3.5 Relationship among Noninferiority, Superiority, and Equivalence

To study the relationship among noninferiority, equivalence, and superiority, we first assume that the noninferiority margin, equivalence limit, and superiority margin are the same. Let M denote the noninferiority margin (also equivalence limit and superiority margin). Also, let μ_T and μ_S be the mean responses of the test treatment and standard therapy (active control agent), respectively. If we assume that an observed mean response on the right-hand side of μ_S is an indication of improvement, then the relationship among noninferiority, equivalence, and superiority is illustrated in Figure 1.1.

It should be noted that a valid sample size calculation can only be done based on appropriate statistical tests for the hypotheses that can reflect the study objectives under a valid study design. It is then suggested that the hypotheses be clearly stated when performing a sample size calculation. Each of the above hypotheses has different requirement for sample size to achieve a desired power or precision of the corresponding tests.

1.2.4 Primary Study Endpoint

A study objective (hypotheses) will define what study variable is to be considered as the primary clinical endpoint and what comparisons or investigations are deemed most clinically relevant. The primary clinical endpoints depend upon therapeutic areas and the indications that the test drugs sought for. For example, for coronary artery disease/angina in cardiovascular system, patient mortality is the most important clinical endpoint in clinical trials assessing the beneficial effects of drugs on coronary artery disease. For congestive heart failure, patient mortality, exercise tolerance, the number of hospitalizations, and cardiovascular morbidity are common endpoints in trials assessing the effects of drugs in congestive heart failure, while mean change from baseline in systolic and diastolic blood pressure and cardiovascular mortality and morbidity are commonly used in hypertension trials. Other examples include change in forced expiratory volume in one second (FEV_1) for asthma in respiratory system, cognitive and functional scales specially designed to assess Alzheimer's disease and Parkinson's disease in central nervous system, tender joints and pain-function endpoints (e.g., Western Ontario and McMaster University Osteoarithritis Index) for osteoarthritis in musculoskeletal system, and the incidence of bone fracture for osteoporosis in the endocrine system.

It can be seen that the efficacy of a test drug in the treatment of a certain disease may be characterized through multiple clinical endpoints. Capizzi and Zhang (1996) classify the clinical endpoints into primary, secondary, and tertiary endpoints. Endpoints that satisfy the following criteria are considered primary endpoints: (i) should be of biological and/or clinical importance, (ii) should form the basis of the objectives of the trial, (iii) should not be highly correlated, (iv) should have sufficient power for the statistical hypotheses formulated from the objectives of the trial, and (v) should be relatively few (e.g., at most 4). Sample size calculation based on detecting a difference in some or all primary clinical endpoints may result in a high chance of false-positive and false-negative results for the evaluation of the test drug. Thus, it is suggested that sample size calculation should be performed based on a single primary study endpoint under certain assumption of the single primary endpoint. More discussion regarding the issue of false-positive and false-negative rates caused by multiple primary endpoints can be found in Chow and Liu (1998).

1.2.5 Clinically Meaningful Difference

In clinical research, the determination of a clinically meaningful difference, denoted by δ, is critical in clinical trials such as equivalence/noninferiority trials. In therapeutic equivalence trials, δ is known as the equivalence limit, while δ is referred to as the noninferiority margin in noninferiority trials. The noninferiority margin reflects the degree of inferiority of the test drug under investigation as compared to the standard therapy that the trials attempt to exclude.

A different choice of δ may affect the sample size calculation and may alter the conclusion of clinical results. Thus, the choice of δ is critical at the planning stage of a clinical study. In practice, there is no golden rule for the determination of δ in clinical trials. As indicated in the ICH E10 Draft Guideline, the noninferiority margin cannot be chosen to be greater than the smallest effect size that the active drug would be reliably expected to have compared with placebo in the setting of the planned trial, but may be smaller based on clinical judgment (ICH, 1999). The ICH E10 Guideline suggests that the noninferiority margin be identified based on past experience in placebo-controlled trials of adequate design under conditions similar to those planned for the new trial. In addition, the ICH E10 Guideline emphasizes that the determination of δ should be based on both statistical reasoning and clinical judgment, which should not only reflect uncertainties in the evidence on which the choice is based, but also be suitably conservative.

In some cases, regulatory agencies do provide clear guidelines for the selection of an appropriate δ for clinical trials. For example, as indicated by Huque and Dubey (1990), the FDA proposed some noninferiority margins for some clinical endpoints (binary responses) such as cure rate for anti-infective drug products (e.g., topical antifungals or vaginal antifungals). These limits are given in Table 1.3. For example, if the cure rate is between 80% and 90%, it is suggested that the noninferiority margin or a clinically meaningful difference be chosen as $\delta = 15\%$.

TABLE 1.3

Noninferiority Margins for Binary Responses

δ (%)	Response Rate for the Active Control (%)
20	50–80
15	80–90
10	90–95
5	>95

Source: FDA Anti-Infectives Drug Guideline (FDA, 1977).

On the other hand, for bioequivalence trials with healthy volunteers, the margin of $\delta = \log(1.25)$ for mean difference on log-transformed data such as area under the blood or plasma concentration–time curve (AUC) or maximum concentration C_{max} is considered (FDA, 2001).

Most recently, FDA published a draft guidance on *Non-Inferiority Clinical Trials*, which recommends that two noninferiority margins, namely, M_1 and M_2, should be considered (FDA, 2010b). The 2010 FDA draft guidance indicated that M_1 is based on (i) the treatment effect estimated from the historical experience with the active control drug, (ii) assessment of the likelihood that the current effect of the active control is similar to the past effect (the constancy assumption), and (iii) assessment of the quality of the noninferiority trial, particularly looking for defects that could reduce a difference between the active control and the new drug. Thus, M_1 is defined as the entire effect of the active control assumed to be present in the noninferiority study. On the other hand, FDA indicates that M_2 is selected based on a clinical judgment, which is never greater than M_1 even if for active control drugs with small effects. It should be noted that a clinical judgment might argue that a larger difference is not clinically important. Ruling out that a difference between the active control and test treatment that is larger than M_1 is a critical finding that supports the conclusion of effectiveness.

In clinical trials, the choice of δ may depend upon absolute change, percent change, or effect size of the primary study endpoint. In practice, a standard effect size (i.e., effect size adjusted for standard deviation) between 0.25 and 0.5 is usually chosen as δ, if no prior knowledge regarding clinical performance of the test drug is available. This recommendation is made based on the fact that the standard effect size of clinical importance observed from most clinical trials is within the range of 0.25 and 0.5.

1.3 Procedures for Sample Size Calculation

In practice, sample size may be determined based on either precision analysis or power analysis. Precision analysis and power analysis for sample size determination are usually performed by controlling type I error (or confidence level) and type II error (or power), respectively. In what follows, we will first introduce the concepts of type I and type II errors.

1.3.1 Type I and Type II Errors

In practice, two kinds of errors occur when testing hypotheses. If the null hypothesis is rejected when it is true, then a type I error has occurred. If the null hypothesis is not rejected when it is false, then a type II error has been made. The probabilities of making type I and type II errors, denoted by α and β, respectively, are given below:

$$\alpha = P\{\text{type I error}\}$$
$$= P\{\text{reject } H_0 \text{ when } H_0 \text{ is true}\},$$
$$\beta = P\{\text{type II error}\}$$
$$= P\{\text{fail to reject } H_0 \text{ when } H_0 \text{ is false}\}.$$

An upper bound for α is a significance level of the test procedure. Power of the test is defined as the probability of correctly rejecting the null hypothesis when the null hypothesis is false, that is,

$$\text{Power} = 1 - \beta$$
$$= P\{\text{reject } H_0 \text{ when } H_0 \text{ is false}\}.$$

As an example, suppose one wishes to test the following hypotheses:

$$H_0 : \text{The drug is ineffective} \quad \text{versus} \quad H_a : \text{The drug is effective}.$$

Then, a type I error occurs if we conclude that the drug is effective when in fact it is not. On the other hand, a type II error occurs if we claim that the drug is ineffective when in fact it is effective. In clinical trials, none of these errors is desirable. With a fixed sample size, a typical approach is to avoid a type I error but at the same time to decrease a type II error so that there is a high chance of correctly detecting a drug effect when the drug is indeed effective. Typically, when the sample size is fixed, α decreases as β increases and α increases as β decreases. The only approach to decrease both α and β is to increase the sample size. Sample size is usually determined by controlling both type I error (or confidence level) and type II error (or power).

In what follows, we will introduce the concepts of precision analysis and power analysis for sample size determination based on type I error and type II error, respectively.

1.3.2 Precision Analysis

In practice, the maximum probability of committing a type I error that one can tolerate is usually considered as the level of significance. The confidence level, $1 - \alpha$, then reflects the probability or confidence of not rejecting the true null hypothesis. Since the confidence interval approach is equivalent to the method of hypotheses testing, we may determine sample size required based on type I error rate using the confidence interval approach. For a $(1 - \alpha)100\%$ confidence interval, the precision of the interval depends on its width. The narrower the interval is, the more precise the inference. Therefore, the precision analysis for sample size determination is to consider the maximum half width of the $(1 - \alpha)100\%$ confidence interval of the unknown parameter that one is willing to accept. Note that the maximum half width of the confidence interval is usually referred to as the *maximum error* of an estimate of the unknown parameter. For example, let $y_1, y_2, ..., y_n$ be independent and identically distributed normal random variables with mean μ and variance σ^2. When σ^2 is known, a $(1 - \alpha)100\%$ confidence interval for μ can be obtained as

$$\bar{y} \pm z_{\alpha/2} \frac{\sigma}{\sqrt{n}},$$

where $z_{\alpha/2}$ is the upper $(\alpha/2)$th quantile of the standard normal distribution. The maximum error, denoted by E, in estimating the value of μ that one is willing to accept is then defined as

$$E = |\bar{y} - \mu| = z_{\alpha/2} \frac{\sigma}{\sqrt{n}}.$$

Thus, the sample size required can be chosen as

$$n = \frac{z_{\alpha/2}^2 \sigma^2}{E^2}. \tag{1.5}$$

Note that the maximum error approach for choosing n is to attain a specified precision while estimating μ, which is derived based only on the interest of type I error. A nonparametric approach can be obtained by using the following Chebyshev's inequality

$$P\{|\bar{y} - \mu| \le E\} \ge 1 - \frac{\sigma^2}{nE^2},$$

and hence

$$n = \frac{\sigma^2}{\alpha E^2}. \tag{1.6}$$

Note that the precision analysis for sample size determination is very easy to apply either based on Equation 1.5 or 1.6. For example, suppose we wish to have a 95% assurance that the error in the estimated mean is less than 10% of the standard deviation (i.e., 0.1σ). Thus,

$$z_{\alpha/2} \frac{\sigma}{\sqrt{n}} = 0.1\sigma.$$

Hence,

$$n = \frac{z_{\alpha/2}^2 \sigma^2}{E^2} = \frac{(1.96)^2 \sigma^2}{(0.1\sigma)^2} = 384.2 \approx 385.$$

The above concept can be applied to binary data (or proportions). In addition, it can be easily implemented for sample size determination when comparing two treatments.

1.3.3 Power Analysis

Since a type I error is usually considered to be a more important and/or serious error that one would like to avoid, a typical approach in hypothesis testing is to control α at an acceptable level and try to minimize β by choosing an appropriate sample size. In other words, the null hypothesis can be tested at predetermined level (or nominal level) of significance with a desired power. This concept for the determination of sample size is usually referred to as *power analysis* for sample size determination.

For the determination of sample size based on power analysis, the investigator is required to specify the following information. First of all, select a significance level at which the chance of wrongly concluding that a difference exists when in fact there is no real difference (type I error) one is willing to tolerate. Typically, a 5% level of significance is chosen to reflect a 95% confidence regarding the unknown parameter. Second, select a desired power at which the chance of correctly detecting a difference when the difference one wishes to achieve truly exists. A conventional choice of power is either 90% or 80%. Third, specify a clinically meaningful difference. In most clinical trials, the objective is to demonstrate effectiveness and safety of a drug under study as compared to a placebo. Therefore, it is important to specify what difference, in terms of the primary endpoint, is considered to be of clinical or scientific importance. Denote such a difference by Δ. If the investigator will settle for detecting only a large difference, then fewer subjects will be needed. If the difference is relatively small, a larger study group (i.e., a larger number of subjects) will be needed. Finally, the knowledge regarding the standard deviation (i.e., σ) of the primary endpoint considered in the study is also required for sample size determination. A very precise method of measurement (i.e., a small σ) will permit detection of any given difference with a much smaller sample size than would be required with a less precise measurement.

Suppose there are two group of observations, namely, x_i, $i = 1$, ..., n_1 (treatment) and y_i, $i = 1$, ..., n_2 (control). Assume that x_i and y_i are independent and normally distributed with means μ_1 and μ_2 and variances σ_1^2 and σ_2^2, respectively. Suppose the hypotheses of interest are

$$H_0 : \mu_1 = \mu_2 \quad \text{versus} \quad H_1 : \mu_1 \neq \mu_2.$$

For simplicity and illustration purpose, we assume (i) σ_1^2 and σ_2^2 are known, which may be estimated from pilot studies or historical data, and (ii) $n_1 = n_2 = n$. Under these assumptions, a Z-statistic can be used to test the mean difference. The Z-test is given by

$$Z = \frac{\bar{x} - \bar{y}}{\sqrt{\left(\dfrac{\sigma_1^2}{n}\right) + \left(\dfrac{\sigma_2^2}{n}\right)}}.$$

Under the null hypothesis of no treatment difference, Z is distributed as $N(0, 1)$. Hence, we reject the null hypothesis when

$$|Z| > z_{\alpha/2}.$$

Under the alternative hypothesis that $\mu_1 = \mu_2 + \delta$ (without loss of generality, we assume $\delta > 0$), a clinically meaningful difference, Z is distributed as $N(\mu^*, 1)$, where

$$\mu^* = \frac{\delta}{\sqrt{\left(\dfrac{\sigma_1^2}{n}\right) + \left(\dfrac{\sigma_2^2}{n}\right)}} > 0.$$

The corresponding power is then given by

$$P\{|N(\mu^*,1)| > z_{\alpha/2}\} \approx P\{N(\mu^*,1) > z_{\alpha/2}\}$$
$$= P\{N(0,1) > z_{\alpha/2} - \mu^*\}.$$

To achieve the desired power of $(1 - \beta)100\%$, we set

$$z_{\alpha/2} - \mu^* = -z_\beta.$$

This leads to

$$n = \frac{\left(\sigma_1^2 + \sigma_2^2\right)\left(z_{\alpha/2} + z_\beta\right)^2}{\delta^2}. \tag{1.7}$$

To apply the above formula for sample size calculation, consider a double-blind, placebo-controlled clinical trial. Suppose the objective of the study is to compare a test drug with a control and the standard deviation for the treatment group is 1 (i.e., $\sigma_1 = 1$) and the standard deviation of the control group is 2 (i.e., $\sigma_2 = 2$). Then, by choosing $\alpha = 5\%$ and $\beta = 10\%$, we have

$$n = \frac{\left(\sigma_1^2 + \sigma_2^2\right)\left(z_{\alpha/2} + z_\beta\right)^2}{\delta^2} = \frac{(1^2 + 2^2)(1.96 + 1.28)^2}{1^2} \approx 53.$$

Thus, a total of 106 subjects is required for achieving a 90% power for the detection of a clinically meaningful difference of $\delta = 1$ at the 5% level of significance.

1.3.4 Probability Assessment

In practice, sample size calculation based on power analysis for detecting a small difference in the incidence rate of rare events (e.g., 3 per 10,000) may not be appropriate. In this case, a very large sample size is required to observe a single event, which is not practical. In addition, small difference in the incidence rate (e.g., 2 per 10,000 versus 3 per 10,000) may not be of practical/clinical interest. Alternatively, it may be of interest to justify the sample size based on a probability statement, for example, there is a certain assurance (say $(1 - \varepsilon)100\%$) that the mean incidence rate of the treatment group is less than that of the control group with probability $(1 - \alpha)100\%$.

Suppose there are two groups of observations, namely, $x_i, i = 1, \ldots, n$ (treatment) and $y_i, i = 1, \ldots, n$ (control). Assume that x_i and y_i are independent and identically distributed as Bernoulli random variables with mean p_1 and p_2, that is, $B(1, p_1)$ and $B(1, p_2)$, respectively, and that

$$P(\bar{x} \leq \bar{y} \,|\, n) + P(\bar{x} > \bar{y} \,|\, n) = 1$$

for large n. Then

$$P(\bar{x} < \bar{y} \,|\, n) = p.$$

The hypotheses of interest are

$$H_0 : p \notin (\varepsilon, 1) \quad \text{versus} \quad H_1 : p \in (\varepsilon, 1),$$

for some $\varepsilon > 0$, where $p = P(\bar{x} = \bar{y} \,|\, n)$ for some n. A test for the hypothesis that $p \in (\varepsilon, 1)$ is

$$\phi(x, y) = I(\bar{x} < \bar{y}).$$

We then reject the null hypothesis if $\phi(x, y) = 1$. Then, given $p_1 < p_2$, the power is given by

$$\text{Power} = P(\bar{x} < \bar{y})$$

$$= P\left(\frac{\bar{x} - \bar{y} - (p_1 - p_2)}{\sqrt{\dfrac{p_1(1 - p_1) + p_2(1 - p_2)}{n}}} < \frac{p_2 - p_1}{\sqrt{\dfrac{p_1(1 - p_1) + p_2(1 - p_2)}{n}}} \right)$$

$$\approx \Phi\left(\frac{p_2 - p_1}{\sqrt{\dfrac{p_1(1 - p_1) + p_2(1 - p_2)}{n}}} \right).$$

Therefore, for a given power $1 - \beta$, the sample size, n, can be estimated by letting

$$\frac{(p_2 - p_1)}{\sqrt{\dfrac{p_1(1 - p_1) + p_2(1 - p_2)}{n}}} = z_\beta,$$

which gives

$$n = \frac{z_\beta^2[p_1(1 - p_1) + p_2(1 - p_2)]}{(p_2 - p_1)^2}.$$

To illustrate the above procedure, consider a double-blind, active-control trial. The objective of this trial is to compare a test drug with a reference drug (active control). Suppose the event rate of the reference drug is 0.075 and the event rate of the test drug is 0.030. Then, with $\beta = 10\%$, we have

$$n = \frac{1.28^2(0.075 \times (1 - 0.075) + 0.030 \times (1 - 0.030))}{(0.075 - 0.030)^2} \approx 80.$$

Thus, a total of 160 subjects is needed to achieve a 90% power for observing less accident rate in test drug group.

1.3.5 Reproducibility Probability

As indicated, current regulation for approval of a test drug under investigation requires at least two adequate and well-controlled clinical trials be conducted for proving substantial evidence regarding the effectiveness and safety of the drug product. Shao and Chow (2002) investigated the probability of reproducibility of the second trial and developed an estimated power approach. As a result, sample size calculation of the second clinical trial can be performed based on the concept of reproducibility probability. Suppose these are two groups of observations obtained in the first trial, namely, x_{1i}, $i = 1, ..., n_1$

(treatment) and x_{2i}, $i = 1, \ldots, n_2$ (control). Assume that x_{1i} and x_{2i} are independent and normally distributed with means μ_1 and μ_2 and common variances σ^2, respectively. The hypotheses of interest are

$$H_0 : \mu_1 = \mu_2 \quad \text{versus} \quad H_1 : \mu_1 \neq \mu_2.$$

When σ^2 is known, we reject H_0 at the 5% level of significance if and only if $|T| > t_{n-2}$, where t_{n-2} is the $(1 - \alpha/2)$th percentile of the t-distribution with $n - 2$ degrees of freedom, $n = n_1 + n_2$,

$$T = \frac{\bar{x}_1 - \bar{x}_2}{\sqrt{\dfrac{(n_1 - 1)s_1^2 + (n_2 - 1)s_2^2}{n - 2}} \sqrt{\dfrac{1}{n_1} + \dfrac{1}{n_2}}},$$

and $\bar{x}_i$ and s_i^2 are the sample means and variances calculated based on data from the ith group, respectively. Thus, the power of T is given by

$$\begin{aligned}
p(\theta) &= P(|T(x)| > t_{n-2}) \\
&= 1 - T_{n-2}(t_{n-2} \mid \theta) + T_{n-2}(-t_{n-2} \mid \theta),
\end{aligned} \tag{1.8}$$

where

$$\theta = \frac{\mu_1 - \mu_2}{\sigma \sqrt{\dfrac{1}{n_1} + \dfrac{1}{n_2}}},$$

and $T_{n-2}(\cdot \mid \theta)$ denotes the distribution function of the t-distribution with $n - 2$ degrees of freedom and the noncentrality parameter θ. Let x be the observed data from the first trial and $T(x)$ be the value of T based on x. Replacing θ in the power in Equation 1.8 by its estimate $T(x)$, the estimated power

$$\hat{P} = p(T(x)) = 1 - T_{n-2}(T_{n-2} \mid T(x)) + T_{n-2}(-t_{n-2} \mid T(x)),$$

is defined by Shao and Chow (2002) as a reproducibility probability for the second trial. Based on this concept, sample size calculation for the second trial can be obtained as

$$n^* = \frac{(T^*/\Delta T)^2}{\dfrac{1}{4n_1} + \dfrac{1}{4n_2}},$$

where T^* is the value obtained such that a desired reproducibility probability is attained and Δ is given by

$$\Delta = \frac{1 + \in /(\mu_1 - \mu_2)}{C},$$

where ε and C reflect the population mean and variance changes in the second trial. In other words, in the second trial, it is assumed that the population mean difference is changed from $\mu_1 - \mu_2$ to $\mu_1 - \mu_2 + \varepsilon$ and the population variance is changed from σ^2 to $C^2\sigma^2$, where $C > 0$.

1.3.6 Sample Size Reestimation without Unblinding

In clinical trials, it is desirable to perform a sample size reestimation based on clinical data accumulated up to the time point. If the reestimated sample size is bigger than the originally planned sample size, then it is necessary to increase the sample size to achieve the desired power at the end of the trial. On the other hand, if the reestimated sample size is smaller than the originally planned sample size, a sample size reduction is justifiable. Basically, sample size reestimation involves either unblinding or without unblinding of the treatment codes. In practice, it is undesirable to perform a sample size reestimation with unblinding of the treatment codes as even the significance level will be adjusted for potential statistical penalty for the unblinding. Thus, sample size reestimation without unblinding the treatment codes has become very attractive. Shih (1993) and Shih and Zhao (1997) proposed some procedures without unblinding for sample size reestimation within interim data for double-blinded clinical trials with binary outcomes. The detailed procedure for sample size reestimation with unblinding will be given in Chapter 8.

In practice, it is suggested that a procedure for sample size reestimation be specified in the study protocol and should be performed by an external statistician who is independent of the project team. It is also recommended that a data monitoring committee (DMC) be considered to maintain the scientific validity and integrity of the clinical trial when performing sample size reestimation at the interim stage of the trial. More details regarding sample size reestimation are provided in Chapter 8.

1.4 Aims and Structure of this Book

1.4.1 Aim of this Book

As indicated earlier, sample size calculation plays an important role in clinical research. Sample size calculation is usually performed using an appropriate statistical test for the hypotheses of interest to achieve a desired power for the detection of a clinically meaningful difference. The hypotheses should be established to reflect the study objectives for clinical investigation under the study design. In practice, however, it is not uncommon to observe discrepancies among study objectives (or hypotheses), study design, statistical analysis (or test statistic), and sample size calculation. These inconsistencies often result in (i) wrong test for right hypotheses, (ii) right test for wrong hypotheses, (iii) wrong test for wrong hypotheses, or (iv) right test for right hypotheses with insufficient power. Therefore, the aim of this book is to provide a comprehensive and unified presentation of statistical concepts and methods for sample size calculation in various situations in clinical research. Moreover, this book will focus on the interactions between clinicians and biostatisticians that often occur during various phases of clinical research and development. This book is also intended to provide a well-balanced summarization of current and emerging clinical issues and recently developed statistical methodologies in the area of sample size calculation in clinical research. Although this book is written from the viewpoint of clinical research and development, the principles and concepts presented in this book can also be applied to a nonclinical setting.

1.4.2 Structure of this Book

It is our goal to provide a comprehensive reference book for clinical researchers, pharmaceutical scientists, clinical or medical research associates, clinical programmers or data coordinators, and biostatisticians in the areas of clinical research and development, regulatory agencies, and academia. The scope of this book covers sample size calculation for studies that may be conducted during various phases of clinical research and development. Basically, this book consists of 20 chapters, which are outlined below.

Chapter 1 provides a brief introduction and a review of regulatory requirement regarding sample size calculation in clinical research for drug development. Also included in this chapter are statistical procedures for sample size calculation based on precision analysis, power analysis, probability assessment, and reproducibility probability. Chapter 2 covers some statistical considerations such as the concept of confounding and interaction, a one-sided test versus or a two-sided test in clinical research, a crossover design versus a parallel design, subgroup/interim analysis, and data transformation. Also included in this chapter are unequal treatment allocation, adjustment for dropouts or covariates, the effect of mixed-up treatment codes, treatment or center imbalance, multiplicity, multiple-stage design for early stopping, and sample size calculation based on rare incidence rate.

Chapter 3 focuses on sample size calculation for comparing means with one sample, two samples, and multiple samples. Formulas are derived under different hypotheses testing for equality, superiority, noninferiority, and equivalence with equal or unequal treatment allocation. In addition, sample size calculation based on Bayesian approach is also considered in this chapter. Chapter 4 deals with sample size calculation for comparing proportions based on large sample tests. Formulas for sample size calculation are derived under different hypotheses testing for equality, superiority, noninferiority, and equivalence with equal or unequal treatment allocation. In addition, issues in sample size calculation based on the confidence interval of the relative risk and/or odds ratio between treatments are also examined.

Chapter 5 considers sample size calculation for binary responses based on exact tests such as the binomial test and Fisher's exact test. Also included in this chapter are optimal and flexible multiple-stage designs that are commonly employed in phase II cancer trials. The emphasis of Chapter 6 is placed on tests for contingency tables such as the goodness-of-fit test and test for independence. Procedures for sample size calculation are derived under different hypotheses for testing equality, superiority, noninferiority, and equivalence with equal or unequal treatment allocation.

Chapter 7 provides sample size calculation for comparing time-to-event data using Cox's proportional hazards model and weighted log-rank test. Formulas are derived under different hypotheses testing for equality, superiority, noninferiority, and equivalence with equal or unequal treatment allocation. Chapter 8 considers the problems of sample size estimation and reestimation in group sequential trials with various alpha spending functions. Also included in this chapter are the study of conditional power for the assessment of futility and a proposed procedure for sample size reestimation without unblinding.

Chapter 9 discusses statistical methods and the corresponding sample size calculation for comparing intrasubject variabilities, intrasubject coefficient of variations (CV), intersubject variabilities, and total variabilities under replicated crossover designs and parallel-group designs with replicates. Chapter 10 summarizes sample size calculation for the assessment of population bioequivalence, individual bioequivalence, and *in vitro* bioequivalence under replicated crossover designs as suggested in the FDA 2001 guidance (FDA, 2001).

Chapter 11 summarizes sample size calculation for dose-ranging studies, including the determination of minimum effective dose (MED) and maximum tolerable dose (MTD). Chapter 12 considers sample size calculation for microarray studies controlling false discovery rate (FDR) and family-wise error rate (FWER). Bayesian sample size calculation is discussed in Chapter 13. Sample size calculation based on nonparametrics for comparing means with one or two samples is discussed in Chapter 14. Chapter 15 discusses statistical methods for data analysis under a cluster randomized trial. Sample size requirements for analysis at the cluster level and at individual subject level (within the cluster) are examined.

Sample size requirements for comparing clinical data (e.g., number of lesions) that follow zero-inflated Poisson distribution are studied in Chapter 16. Also included in this chapter are formulas for sample size calculation for testing noninferiority/superiority and equivalence. Chapter 17 provides a compromised approach (compromise between precision analysis and power analysis in conjunction with probability assessment) for sample size calculation for clinical studies with extremely low incidence rate. Chapter 18 provides a comprehensive summarization of sample size requirements for two-stage seamless adaptive trial designs with/without different study objectives and study endpoints at different stages.

Chapter 19 provides an alternative method for sample size calculation using clinical trial simulation when there exists no closed form for test statistics and/or formulas or procedures for sample size calculation for complex clinical studies. Chapter 20 includes sample size calculations in other areas of clinical research such as QT/QTc studies, where the QT interval is a measure of the time between the start of the Q wave and the end of T wave in the heart's electrical cycle and QTc is corrected QT interval in cardiology, the use of propensity score analysis in nonrandomized or observational studies, analysis of variance with repeated measurements, quality-of-life assessment, bridging studies, and vaccine clinical trials.

For each chapter, whenever possible, real examples concerning clinical studies of various therapeutic areas are included to demonstrate the clinical and statistical concepts, interpretations, and their relationships and interactions. Comparisons regarding the relative merits and disadvantages of statistical methods for sample size calculation in various therapeutic areas are discussed whenever deemed appropriate. In addition, if applicable, topics for future research development are provided.

2

Considerations Prior to Sample Size Calculation

As indicated in Chapter 1, sample size calculation should be performed using appropriate statistical methods or tests for hypotheses, which can reflect the study objectives under the study design based on the primary study endpoint of the intended trial. As a result, some information including study design, hypotheses, mean response and the associated variability of the primary study endpoint, and the desired power at a specified α level of significance are required when performing sample size calculation. For good statistics practice, some statistical considerations such as stratification with respect to possible confounding/interaction factors, the use of a one-sided test or a two-sided test, the choice of a parallel design or a crossover design, subgroup/interim analyses, and data transformation are important for performing an accurate and reliable sample size calculation. In addition, some practical issues that are commonly encountered in clinical trials, which may have an impact on sample size calculation, should also be taken into consideration when performing sample size calculation. These practical issues include unequal treatment allocation, adjustment for dropouts or covariates, mixed-up treatment codes, treatment study center imbalance, multiplicity, multiple-stage design for early stopping, and sample size calculation based on rare or extremely low incidence rate.

In Section 2.1, we introduce the concepts of confounding and interaction effects in clinical trials. Section 2.2 discusses the controversial issues between the use of a one-sided test and a two-sided test in clinical research. In Section 2.3, we summarize the difference in sample size calculation between a crossover design and a parallel design. The concepts of group sequential boundaries and alpha spending function in subgroup/interim analyses in clinical trials are discussed in Section 2.4. Section 2.5 clarifies some issues that are commonly seen in sample size calculation based on transformed data such as log-transformed data under a parallel design or a crossover design. Section 2.6 provides a discussion regarding some practical issues that have impact on sample size calculation in clinical trials. These issues include unequal treatment allocation in randomization, sample size adjustment for dropouts or covariates, the effect of mixed-up treatment codes during the conduct of clinical trials, the loss in power for treatment and/or center imbalance, the issue of multiplicity in multiple primary endpoints and/or multiple comparisons, multiple-stage design for early stopping, and sample size calculation based on rare or extremely low incidence rate in safety assessment.

2.1 Confounding and Interaction

2.1.1 Confounding

Confounding effects are defined as effects contributed by various factors that cannot be separated by the design under study (Chow and Liu, 1998, 2003, 2013). Confounding is an important concept in clinical research. When confounding effects are observed in a clinical trial, the treatment effect cannot be assessed because it is contaminated by other effects contributed by various factors.

In clinical trials, there are many sources of variation that have an impact on the primary clinical endpoints for clinical evaluation of a test drug under investigation. If some of these variations are not identified and properly controlled, they can become mixed with the treatment effect that the trial is designed to demonstrate, in which case the treatment effect is said to be confounded by effects due to these variations. In clinical trials, there are many subtle, unrecognizable, and seemingly innocent confounding factors that can cause ruinous results of clinical trials. Moses (1992) gives the example of the devastating result in the confounder being the personal choice of a patient. The example concerns a polio-vaccine trial that was conducted on 2 million children worldwide to investigate the effect of Salk poliomyelitis vaccine. This trial reported that the incidence rate of polio was lower in children whose parents refused injection than in those who received placebo after their parents gave permission (Meier, 1989). After an exhaustive examination of the data, it was found that susceptibility to poliomyelitis was related to the differences between families who gave the permission and those who did not. Therefore, it is not clear whether the effect of the incidence rate is due to the effect of Salk poliomyelitis vaccine or due to the difference between families giving permission.

2.1.2 Interaction

An interaction effect between factors is defined as a joint effect with one or more contributing factors (Chow and Liu, 1998, 2003, 2013). *Interaction* is also an important concept in clinical research. The objective of a statistical interaction investigation is to conclude whether the joint contribution of two or more factors is the same as the sum of the contributions from each factor when considered alone. When interactions among factors are observed, an overall assessment on the treatment effect is not appropriate. In this case, it is suggested that the treatment must be carefully evaluated for those effects contributed by the factors.

In clinical research, almost all adequate and well-controlled clinical trials are multicenter trials. For multicenter trials, the FDA requires that the treatment-by-center interaction be examined to evaluate whether the treatment effect is consistent across all centers. As a result, it is suggested that statistical tests for homogeneity across centers (i.e., for detecting treatment-by-center interaction) be provided. The significant level used to declare the significance of a given test for a treatment-by-center interaction should be considered in light of the sample size involved. Gail and Simon (1985) classify the nature of interaction as either quantitative or qualitative. A quantitative interaction between treatment and center indicates that the treatment differences are in the same direction across centers but the magnitude differs from center to center, while a qualitative interaction reveals that substantial treatment differences occur in different directions in different centers. More discussion regarding treatment-by-center interaction can be found in Chow and Shao (2002).

2.1.3 Remark

In clinical trials, a stratified randomization is usually employed with respect to some prognostic factors or covariates, which may have confounding and interaction effects on the evaluation of the test drug under investigation. Confounding or interaction effects may alter the conclusion of the evaluation of the test drug under investigation. Thus, a stratified randomization is desirable if the presence of the confounding and/or interaction effects of some factors is doubtful. In practice, although sample size calculation according to some stratification factors can be similarly performed within each combination of the stratification factors, it is not desirable to have too many stratification factors. Therefore, it is suggested that the possible confounding and/or interaction effects of the stratification factors should be carefully evaluated before a sample size calculation is performed and a stratified randomization is carried out.

2.2 One-Sided Test versus Two-Sided Test

In clinical research, there has been a long discussion on whether a one-sided test or a two-sided test should be used for clinical evaluation of a test drug under investigation. Sample size calculations based on a one-sided test and a two-sided test are different at a fixed α level of significance. As it can be seen from Equation 1.3, the sample size for comparing the two means can be obtained as

$$n = \frac{\left(\sigma_1^2 + \sigma_2^2\right)\left(z_{\alpha/2} + z_\beta\right)^2}{\delta^2}.$$

When $\sigma_1^2 = \sigma_2^2 = \sigma^2$, the above formula reduces to

$$n = \frac{2\sigma^2\left(z_{\alpha/2} + z_\beta\right)^2}{\delta^2}.$$

If $\delta = c\sigma$, then the sample size formula can be rewritten as

$$n = \frac{2\left(z_{\alpha/2} + z_\beta\right)^2}{c^2}.$$

Table 2.1 provides a comparison for sample sizes obtained based on a one-sided test or a two-sided test at the α level of significance. The results indicate that sample size may be reduced by about 21% when switching from a two-sided test to a one-sided test for testing at the 5% level of significance with an 80% power for detection of a difference of 0.5 standard deviation.

The pharmaceutical industry prefers a one-sided test for demonstration of clinical superiority based on the argument that they will not run a study if the test drug would

TABLE 2.1

Sample Sizes Based on One-Sided Test and Two-Sided Test at α Level of Significance

α	δ^*	One-Sided Test		Two-Sided Test	
		80%	90%	80%	90%
0.05	0.25σ	198	275	252	337
	0.50σ	50	69	63	85
	1.00σ	13	18	16	22
0.01	0.25σ	322	417	374	477
	0.50σ	81	105	94	120
	1.00σ	21	27	24	30

TABLE 2.2

Comparison between One-Sided Test and Two-Sided Test at α Level of Significance

Characteristic	One-Sided Test	Two-Sided Test
Hypotheses	Noninferiority/superiority	Equality/equivalence
One trial	1/20	1/40
Two trials	1/400	1/1600

be worse. In practice, however, many drug products such as drug products in the central nervous system may show a superior placebo effect as compared to the drug effect. This certainly argues against the use of a one-sided test. Besides, a one-sided test allows more *bad* drug products to be approved because of chances as compared to a two-sided test.

As indicated earlier, the FDA requires that at least two adequate and well-controlled clinical studies be conducted to provide substantial evidence regarding the effectiveness and safety of a test drug under investigation. For each of the two adequate and well-controlled clinical trials, suppose the test drug is evaluated at the 5% level of significance. Table 2.2 provides a summary of comparison between one-sided test and two-sided test in clinical research. For the one-sided test procedure, the false-positive rate is one out of 400 trials (i.e., 0.25%) for the two trials, while the false-positive rate is one out of 1600 trials (i.e., 0.0625%) for two trials when applying a two-sided test.

Some researchers from the academia and the pharmaceutical industry consider this false-positive rate as acceptable and the evidence provided by the two clinical trials using the one-sided test procedure as rather substantial. Hence, the one-sided test procedure should be recommended. However, in practice, a two-sided test may be preferred because placebo effect may be substantial in many drug products such as drug products regarding diseases in the central nervous system.

2.2.1 Remark

Dubey (1991) indicated that the FDA prefers a two-sided test over a one-sided test procedure in clinical research and development of drug products. In situations where (i) there is concern with outcomes in only one tail and (ii) it is completely inconceivable that results could go in the opposite direction, one-sided test procedure may be appropriate (Dubey, 1991). Dubey (1991) provided situations where one-sided test procedure may be justified. These situations include (i) toxicity studies, (ii) safety evaluation, (iii) the analysis of occurrences of adverse drug reaction data, (iv) risk evaluation, and (v) laboratory data.

2.3 Crossover Design versus Parallel Design

As indicated earlier, an adequate and well-controlled clinical trial requires that a valid study design be employed for a valid assessment of the effect of the test drug under investigation. As indicated in Chow and Liu (1998), commonly used study designs in clinical research include parallel design, crossover design, enrichment design, titration design, or a combination of these designs. Among these designs, crossover and parallel designs are probably the two most commonly employed study designs.

2.3.1 Intersubject and Intrasubject Variabilities

Chow and Liu (1998, 2003, 2013) suggested that relative merits and disadvantages of candidate designs should be carefully evaluated before an appropriate design is chosen for the intended trial. The clarification of the intrasubject and intersubject variabilities is essential for sample size calculation in clinical research when a crossover design or a parallel design is employed.

Intrasubject variability is the variability that could be observed by repeating experiments on the same subject under the same experimental condition. The source of intrasubject variability could be multifold. One important source is biological variability. Exactly the same results may not be obtained even if they are from the same subject under the same experimental condition. Another important source is measurement or calculation error. For example, in a bioequivalence study with healthy subjects, it could be (i) the error when measuring the blood or plasma concentration–time curve, (ii) the error when calculating AUC (area under the curve), and/or (iii) the error of rounding after log-transformation. Intrasubject variability could be eliminated if we could repeat the experiment infinitely many times (in practice, this just means a large number of times) on the same subject under the same experimental condition and then take the average. The reason is that intrasubject variability tends to cancel each other on average in a large scale. If we repeat the experiment on different subjects infinitely many times, it is possible that we may still see that the averages of the responses from different subjects are different from each other even if the experiments are carried out under exactly the same conditions. Then, what causes this difference or variation? It is not due to intrasubject variability, which has been eliminated by averaging infinitely repeated experiments; it is not due to experimental conditions, which are exactly the same for different subjects. Therefore, this difference or variation can only be due to the unexplained difference between the two subjects.

It should be pointed out that sometimes people may call the variation observed from different subjects under the same experimental condition intersubject variability, which is different from the intersubject variability defined here. The reason is that the variability observed from different subjects under the same experimental condition could be due to unexplained difference among subjects (pure intersubject variability); it also could be due to the biological variability, or measurement error associated with different experiments on different subjects (intrasubject variability). Therefore, it is clear that the observed variability from different subjects incorporates two components. They are, namely, pure intersubject variability and intrasubject variability. We refer to it as the total intersubject variability. For simplicity, it is also called total variability, which is the variability one would observe from a parallel design.

In practice, no experiment can be carried out infinitely many times. It is also not always true that the experiment can be repeatedly carried out on the same subject under the same experimental condition. But, we can still assess these two variability components (intra- and inter-) under certain statistical models, for example, a mixed effects model.

2.3.2 Crossover Design

A crossover design is a modified randomized block design in which each block receives more than one treatment at different dosing periods. In a crossover design, subjects are randomly assigned to receive a sequence of treatments, which contains all the treatments in the study. For example, for a standard two-sequence, two-period 2×2 crossover design, subjects are randomly assigned to receive one of the two sequences of treatments (say, *RT* and *TR*), where *T* and *R* represent the test drug and the reference drug, respectively. For subjects who are randomly assigned to the sequence of *RT*, they receive the reference drug first and then crossover to receive the test drug after a sufficient length of washout. The major advantage of a crossover design is that it allows a within-subject (or intrasubject) comparison between treatments (each subject serves as its own control) by removing the between-subject (or intersubject) variability from the comparison. Let μ_T and μ_R be the mean of the responses of the study endpoint of interest. Also, let σ_S^2 and σ_e^2 be the intersubject variance and intrasubject variance, respectively. Define $\theta = (\mu_T - \mu_R)/\mu_R$ and assume that the equivalence limit is $\delta = 0.2\mu_R$. Then, under a two-sequence, two-period crossover design, the formula for sample size calculation is given by (see, also Chow and Wang, 2001)

$$n \geq \frac{CV^2(t_{\alpha,2n-2} + t_{\beta/2,2n-2})^2}{(0.2 - |\theta|)^2},$$

where $CV = \sigma_e/\mu_R$.

2.3.3 Parallel Design

A parallel design is a complete randomized design in which each subject receives one and only one treatment in a random fashion. The parallel design does not provide independent estimates for the intrasubject variability for each treatment. As a result, the assessment of treatment effect is made based on the total variability, which includes the intersubject variability and the intrasubject variability.

Under a parallel design, assuming that the equivalence limit $\delta = 0.2\mu_R$, the following formula is useful for sample size calculation (Chow and Wang, 2001):

$$n \geq \frac{2CV^2(t_{\alpha,2n-2} + t_{\beta/2,2n-2})^2}{(0.2 - |\theta|)^2},$$

where $CV = \sigma/\mu_R$ and $\sigma^2 = \sigma_S^2 + \sigma_e^2$.

2.3.4 Remark

In summary, in a parallel design, the comparison is made based on the intersubject variation, while in a crossover design the comparison is made based on the intrasubject variation. As a result, sample size calculation under a parallel design or a crossover design is similar and yet different. Note that the above formulas for sample size calculation are obtained based on raw data. Sample size formulas based on log-transformation data under a parallel design or a crossover design can be similarly obtained (Chow and Wang, 2001). More discussion regarding data transformation such as a log-transformation is given in Section 2.5.

2.4 Subgroup/Interim Analyses

In clinical research, subgroup analyses are commonly performed in clinical trials. Subgroup analyses may be performed with respect to subject prognostic or confounding factors such as demographics or subject characteristics at baseline. The purpose of this type of subgroup analysis is to isolate the variability due to the prognostic or confounding factors for an unbiased and reliable assessment of the efficacy and safety of the test drug under investigation. In addition, many clinical trial protocols may call for an interim analysis or a number of interim analyses during the conduct of the trials for the purpose of establishing early efficacy and/or safety monitoring. The rationale for interim analyses of accumulating data in clinical trials has been well established in the literature. See, for example, Armitage et al. (1969), Haybittle (1971), Peto et al. (1976), Pocock (1977), O'Brien and Fleming (1979), Lan and DeMets (1983), PMA (1993), and DeMets and Lan (1994).

2.4.1 Group Sequential Boundaries

For interim analyses in clinical trials, it is suggested that the number of planned interim analyses should be specified in the study protocol. Let N be the total planned sample size with equal allocation to the two treatments. Suppose that K interim analyses is planned with equal increment of accumulating data. Then we can divide the duration of the clinical trial into K intervals. Within each stage, the data of $n = N/K$ patients are accumulated. At the end of each interval, an interim analysis can be performed using the Z-statistic, denoted by Z_i, with the data accumulated up to that point. Two decisions will be made based on the result of each interim analysis. First, the trial will continue if

$$|Z_i| \leq z_i, i = 1, \ldots, K-1, \tag{2.1}$$

where the z_i are some critical values that are known as the *group sequential boundaries*. We fail to reject the null hypothesis if

$$|Z_i| \leq z_i, \quad \text{for all } i = 1, \ldots, K. \tag{2.2}$$

Note that we may terminate the trial if the null hypothesis is rejected at any of the K interim analyses ($|Z_i| > z_i, i = 1, \ldots, K$). For example, at the end of the first interval, an interim analysis is carried out with data from n subjects. If we fail to reject the null hypothesis, we continue the trial to the second planned interim analysis. Otherwise, we reject the null hypothesis and may stop the trial. The trial may be terminated at the final analysis if we fail to reject the null hypothesis at the final analysis. Then we declare that the data from the trial provide sufficient evidence to doubt the validity of the null hypothesis. Otherwise, the null hypothesis is rejected and we conclude that there is statistically significant difference in change from baseline between the test drug and the control.

In contrast to the fixed sample where only one final analysis is performed, K analyses are carried out for the K-stage group sequential procedure. Suppose that the nominal significance level for each of the K interim analyses is still 5%. Then, because of repeated testing based on the accumulated data, the overall significance level is inflated. In other words, the probability of declaring at least one significance result increases due to K interim analyses. Various methods have been proposed to maintain the overall significance level at the prespecified nominal level. One of the early methods was proposed by Haybittle (1971) and Peto et al. (1976). They proposed to use 3.0 as group sequential boundaries for all interim analyses except for the final analysis for which they suggested 1.96. In other words,

$$z_i = \begin{cases} 3.0, & \text{if } i = 1, \ldots, K-1, \\ 1.96, & \text{if } i = K. \end{cases}$$

Therefore, their method can be summarized as follows:

Step 1: At each of the K interim analyses, compute Z_i, $i = 1, \ldots, K - 1$.

Step 2: If the absolute value of Z_i crosses 3.0, then reject the null hypothesis and recommend a possible early termination of the trial; otherwise, continue the trial to the next planned interim analysis and repeat Steps 1 and 2.

Step 3: For the final analysis, use 1.96 for the boundary. Trial stops here regardless of whether the null hypothesis is rejected.

Haybittle and Peto's method is very simple. However, it is a procedure with ad hoc boundaries that are independent of the number of planned interim analyses and stage of interim analyses. Pocock (1977) proposed different group sequential boundaries, which depend upon the number of planned interim analyses. However, his boundaries are constant at each stage of interim analyses. Since limited information is included in the early stages of interim analyses, O'Brien and Fleming (1979) suggested posing conservative boundaries for interim analyses scheduled to be carried out during an early phase of the trial. Their boundaries not only depend upon the number of interim analyses but are also a function of stages of interim analysis. As a result, the O'Brien–Fleming boundaries can be calculated as follows:

$$z_{ik} = \frac{c_k \sqrt{k}}{i}, \quad 1 \leq i \leq k \leq K, \tag{2.3}$$

where c_k is the critical value for a total of k planned interim analyses. As an example, suppose that five planned interim analyses are scheduled. Then, $c_5 = 2.04$ and boundaries for each stage of these five interim analyses are given as

$$z_{i5} = \frac{2.04\sqrt{5}}{i}, \quad 1 \leq i \leq 5.$$

Thus, O'Brien–Fleming boundary for the first interim analysis is equal to $(2.04)(\sqrt{5}) = 4.561$. The O'Brien–Fleming boundaries for the other four interim analyses can be similarly computed as 3.225, 2.633, 2.280, and 2.040, respectively. The O'Brien–Fleming boundaries are very conservative so that the early trial results must be extreme for any prudent and justified decision-making in recommendation of a possible early termination when very limited information is available. On the other hand, for the late phase of the trial when the accumulated information approaches the required maximum information, their boundaries also become quite close to the critical value when no interim analysis had been planned. As a result, the O'Brien–Fleming method does not require a significant increase in the sample size for what has already been planned. Therefore, the O'Brien–Fleming group sequential boundaries have become one of the most popular procedures for the planned interim analyses of clinical trials.

2.4.2 Alpha Spending Function

The idea of the alpha spending function proposed by Lan and DeMets (1983) is to spend (i.e., distribute) the total probability of false-positive risk as a continuous function of the information time. The implementation of the alpha spending function requires the selection and specification of the spending function in advance in the protocol. One cannot change and choose another spending function in the middle of the trial. Geller (1994) suggested that the spending function should be convex and have the property that the same value of a test statistic is more compelling as the sample sizes increase. Because of its flexibility and lack of requirement for total information and equal increment of information, there is a potential to abuse the alpha spending function by increasing the frequency of interim analyses as the results approach the boundary. However, DeMets and Lan (1994) reported that alteration of the frequency of interim analyses has very little impact on the overall significance level if the O'Brien–Fleming-type or Pocock-type continuous spending function is used.

Pawitan and Hallstrom (1990) studied the alpha spending function with the use of the permutation test. The permutation test is conceptually simple and provides an exact test for small sample sizes. In addition, it is valid for complicated stratified analysis in which the exact sampling distribution is, in general, unknown and large-sample approximation may not be adequate. Consider the one-sided alternative. For the kth interim analyses, under the assumption of no treatment effect, the null joint permutation distribution of test statistics $(Z_1, \ldots, Z_K)$ can be obtained by random permutation of treatment assignments on the actual data. Let $(Z^*_{1b}, \ldots, Z^*_{Kb})$, $b = 1, \ldots, B$, be the statistics computed from B treatment assignments and B be the total number of possible permutations. Given $\alpha(s_1)$, $\alpha(s_2) - \alpha(s_1)$, $\ldots$, $\alpha(s_K) - \alpha(s_{K-1})$, the probabilities of type I error allowed to spend at successive interim analyses, the one-sided boundaries $z_1, \ldots, z_K$ can be determined by

$$\frac{\text{Number of}\left(Z^*_1 > z_1\right)}{B} = \alpha(s_1),$$

and

$$\frac{\text{Number of}\left(Z^*_1 > z_1 \text{ or } Z^*_2 > z_2, \ldots, \text{ or } Z^*_k > z_k\right)}{B} = \alpha(s_k) - \alpha(s_{k-1}),$$

$k = 1, \ldots, K$. If B is very large, then the above method can be executed with a random sample with replacement of size B. The α spending function for an overall significance level of 2.5% for one-sided alternative is given by

$$\alpha(s) = \begin{cases} \dfrac{\alpha}{2} s, & \text{if } s < 1, \\ \alpha, & \text{if } s = 1. \end{cases}$$

In the interest of controlling the overall type I error rate at the α level of significance, sample size is necessarily adjusted according to the α spending function to account for the planned interim analyses. In some cases, sample size reestimation without unblinding may be performed according to the procedure described in Section 1.3 of Chapter 1. More details can be found in Chapter 8.

2.5 Data Transformation

In clinical research, data transformation on clinical response of the primary study endpoint may be necessarily performed before statistical analysis for a more accurate and reliable assessment of the treatment effect. For example, for bioavailability and bioequivalence studies with healthy human subjects, the FDA requires that a log-transformation be performed before data analysis. Two drug products are claimed bioequivalent in terms of drug absorption if the 90% confidence interval of the ratio of means of the primary pharmacokinetic (PK) parameters, such as area under the blood or plasma concentration–time curve (AUC) and maximum concentration (C_{max}), is entirely within the bioequivalence limits of (80%, 125%). Let μ_T and μ_R be the population means of the test drug and the reference drug, respectively. Also, let X and Y be the PK responses for the test drug and the reference drug. After log-transformation, we assume that log X and log Y follow normal distributions with means μ_X^* and μ_Y^* and variance σ^2. Then,

$$\mu_T = E(X) = e^{\mu_X^* + (\sigma^2/2)} \quad \text{and} \quad \mu_R = E(Y) = e^{\mu_Y^* + (\sigma^2/2)},$$

which implies

$$\log\left(\frac{\mu_T}{\mu_R}\right) = \log\left(e^{\mu_X^* + \mu_Y^*}\right) = \mu_X^* - \mu_Y^*.$$

Under both the crossover and parallel design, an exact $(1 - \alpha)100\%$ confidence interval for $\mu_X^* - \mu_Y^*$ can be obtained based on the log-transformed data. Hence, an exact $(1 - \alpha)100\%$ confidence interval for μ_T/μ_R can be obtained after the back transformation.

Chow and Wang (2001) provided sample size formulas under a parallel design or a crossover design with and without log-transformation. These formulas are different but very similar. In practice, scientists often confuse them with one another. The following discussion may be helpful for clarification.

We note that the sample size derivation is based on normality assumption for the raw data and log-normality assumption for the transformed data. Thus, it is of interest to study the distribution of log X when X is normally distributed with mean μ and variance σ^2. Note that

$$\text{Var}\left(\frac{X - \mu}{\mu}\right) = \frac{\sigma^2}{\mu^2} = CV^2.$$

If CV is sufficiently small, $(X - \mu)/\mu$ is close to 0. As a result, by Taylor's expansion,

$$\log X - \log \mu = \log\left(1 + \frac{X - \mu}{\mu}\right) \approx \frac{X - \mu}{\mu}.$$

Then,

$$\log X \approx \log \mu + \frac{X - \mu}{\mu} \sim N(\log \mu, CV^2).$$

This indicates that when CV is small, log X is still approximately normally distributed, even if X is from a normal population. Therefore, the procedure based on log-transformed

TABLE 2.3

Posterior Power Evaluation under a Crossover Design

Data Type	Power
Raw data	$1 - 2\Phi\left(\dfrac{0.2}{CV\sqrt{(1/n_1) + (1/n_2)}} - t_{\alpha, n_1 + n_2 - 2}\right)$
Log-transformed data	$1 - 2\Phi\left(\dfrac{0.223}{\sigma_e\sqrt{(1/n_1) + (1/n_2)}} - t_{\alpha, n_1 + n_2 - 2}\right)$

data is robust in some sense. In addition, the CV observed from the raw data is very similar to the variance obtained from the log-transformed data.

Traditionally, for the example regarding bioavailability and bioequivalence (BE) with raw data, BE can be established if the 90% confidence interval for $\mu_T - \mu_R$ is entirely within the interval of $(-0.2\mu_R, 0.2\mu_R)$ (Chow and Liu, 1992). This is the reason why 0.2 appears in the formula for raw data. However, both the 1992 FDA and the 2000 FDA guidances recommended that a log-transformation be performed before bioequivalence assessment is made. For log-transformed data, the BE can be established if the 90% confidence interval for μ_T/μ_R is entirely located in the interval (80%, 125%). That is why log 1.25 appears in the formula for log-transformed data. It should be noted that log 1.25 = −log 0.8 = 0.2231. In other words, the BE limit for the raw data is symmetric about 0 (i.e., $\pm 0.2\mu_R$), while the BE limit for the log-transformed data is also symmetric about 0 after log-transformation.

2.5.1 Remark

For the crossover design, since each subject serves as its own control, the intersubject variation is removed from comparison. As a result, the formula for sample size calculation derived under a crossover design only involves the intrasubject variability. On the other hand, for the parallel design, formula for sample size calculation under a parallel design includes both the inter- and intrasubject variabilities. In practice, it is easy to get confused with the sample size calculation and/or evaluation of posterior power based on either raw data or log-transformed data under either a crossover design or a parallel design (Chow and Wang, 2001). As an example, posterior powers based on raw data and log-transformed data under a crossover design when the true mean difference is 0 are given in Table 2.3.

2.6 Practical Issues

2.6.1 Unequal Treatment Allocation

In a parallel design or a crossover design comparing two or more than two treatments, sample sizes in each treatment group (for parallel design) or in each sequence of treatments (for crossover design) may not be the same. For example, when conducting a placebo-controlled clinical trial with very ill patients or patients with severe or life-threatening diseases, it may not be ethical to put too many patients in the placebo arm. In this case, the investigator may prefer to put fewer patients in the placebo (if the placebo arm is considered necessary to demonstrate the effectiveness and safety of the drug under investigation). A typical ratio of patient allocation for situations of this kind is 1:2, that is, each patient will have a one-third chance to be assigned to the placebo group and two-third chance to receive the active drug. For different ratios of patient allocation, the sample size formulas discussed can be directly applied with appropriate modification of the corresponding degrees of freedom in the formulas.

When there is unequal treatment allocation, say κ to 1 ratio, sample size for comparing two means can be obtained as

$$n = \frac{\left(\sigma_1^2/\kappa + \sigma_2^2\right)(z_{\alpha/2} + z_\beta)^2}{\delta^2}.$$

When $\kappa = 1$, the above formula reduces to Equation 1.3. When $\sigma_1^2 = \sigma_2^2 = \sigma^2$, we have

$$n = \frac{(\kappa + 1)\sigma^2(z_{\alpha/2} + z_\beta)^2}{\kappa\delta^2}.$$

Note that unequal treatment allocation will have an impact on randomization in clinical trials, especially in multicenter trials. To maintain the integrity of blinding of an intended trial, a blocking size of 2 or 4 in randomization is usually employed. A blocking size of 2 guarantees that one of the subjects in the block will be randomly assigned to the treatment group and the other one will be randomly assigned to the control group. In a multicenter trial comparing two treatments, if we consider a 2 to 1 allocation, the size of each block has to be a multiple of 3, that is, 3, 6, or 9. In the treatment of having a minimum of two blocks in each center, each center is required to enroll a minimum of six subjects. As a result, this may have an impact on the selection of the number of centers. As indicated in Chow and Liu (1998), as a rule of thumb, it is not desirable to have the number of subjects in each center to be less than the number of centers. As a result, it is suggested that the use of a κ to 1 treatment allocation in multicenter trials should take into consideration the blocking size in randomization and the number of centers selected.

2.6.2 Adjustment for Dropouts or Covariates

At the planning stage of a clinical study, sample size calculation provides the number of *evaluable* subjects required for achieving a desired statistical assurance (e.g., an 80% power). In practice, we may have to enroll more subjects to account for potential dropouts. For example, if the sample size required for an intended clinical trial is n and the potential dropout rate is p, then we need to enroll $n/(1 - p)$ subjects to obtain n evaluable subjects at the completion of the trial. It should also be noted that the investigator may have to screen more patients to obtain $n/(1 - p)$ *qualified* subjects at the entry of the study based on inclusion/exclusion criteria of the trial.

Fleiss (1986) pointed out that a required sample size may be reduced if the response variable can be described by a covariate. Let n be the required sample size per group when the design does not call for the experimental control of a prognostic factor. Also, let n^* be the required sample size for the study with the factor controlled. The relative efficiency (RE) between the two designs is defined as

$$\text{RE} = \frac{n}{n^*}.$$

As indicated by Fleiss (1986), if the correlation between the prognostic factor (covariate) and the response variable is r, then RE can be expressed as

$$RE = \frac{100}{1-r^2}.$$

Hence, we have

$$n^* = n(1-r^2).$$

As a result, the required sample size per group can be reduced if the correlation exists. For example, a correlation of $r = 0.32$ could result in a 10% reduction in the sample size.

2.6.3 Mixed-Up Randomization Schedules

Randomization plays an important role in the conduct of clinical trials. Randomization not only generates comparable groups of patients who constitute representative samples from the intended patient population, but also enables valid statistical tests for clinical evaluation of the study drug. Randomization in clinical trials involves random recruitment of patients from the targeted patient population and random assignment of patients to the treatments. Under randomization, statistical inference can be drawn under some probability distribution assumption of the intended patient population. The probability distribution assumption depends on the method of randomization under a randomization model. A study without randomization results in the violation of the probability distribution assumption and consequently no accurate and reliable statistical inference on the evaluation of the safety and efficacy of the study drug can be drawn.

A problem commonly encountered during the conduct of a clinical trial is that a proportion of treatment codes are mixed up in randomization schedules. Mixing up treatment codes can distort the statistical analysis based on the population or randomization model. Chow and Shao (2002) quantitatively studied the effect of mixed-up treatment codes on the analysis based on the intention-to-treat (ITT) population, which are described below.

Consider a two-group parallel design for comparing a test drug and a control (placebo), where n_1 patients are randomly assigned to the treatment group and n_2 patients are randomly assigned to the control group. When randomization is properly applied, the population model holds and responses from patients are normally distributed. Consider first the simplest case where two patient populations (treatment and control) have the same variance σ^2 and σ^2. Let μ_1 and μ_2 be the population means for the treatment and the control, respectively. The null hypothesis that $\mu_1 = \mu_2$ (i.e., there is no treatment effect) is rejected at the α level of significance if

$$\frac{|\bar{x}_1 - \bar{x}_2|}{\sigma\sqrt{(1/n_1)+(1/n_2)}} > z_{\alpha/2}, \tag{2.4}$$

where $\bar{x}_1$ is the sample mean of responses from patients in the treatment group, $\bar{x}_2$ is the sample mean of responses from patients in the control group, and $z_{\alpha/2}$ is the upper ($\alpha/2$)th percentile of the standard normal distribution. Intuitively, mixing up treatment codes does not affect the significance level of the test.

The power of the test, that is, the probability of correctly detecting a treatment difference when $\mu_1 \neq \mu_2$, is

$$p(\theta) = P\left(\frac{|\bar{x}_1 - \bar{x}_2|}{\sigma\sqrt{(1/n_1)+(1/n_2)}} > z_{\alpha/2}\right) = \Phi(\theta - z_{\alpha/2}) + \Phi(-\theta - z_{\alpha/2}),$$

where Φ is the standard normal distribution function and

$$\theta = \frac{\mu_1 - \mu_2}{\sigma\sqrt{(1/n_1)+(1/n_2)}}. \tag{2.5}$$

This follows from the fact that under the randomization model, $\bar{x}_1 - \bar{x}_2$ has the normal distribution with mean $\mu_1 - \mu_2$ and variance $\sigma^2((1/n_1)+(1/n_2))$.

Suppose that there are m patients whose treatment codes are randomly mixed up. A straightforward calculation shows that $\bar{x}_1 - \bar{x}_2$ is still normally distributed with variance $\sigma^2((1/n_1)+(1/n_2))$, but the mean of $\bar{x}_1 - \bar{x}_2$ is equal to

$$\left[1 - m\left(\frac{1}{n_1} + \frac{1}{n_2}\right)\right](\mu_1 - \mu_2).$$

It turns out that the power for the test defined above is

$$p(\theta_m) = \Phi(\theta_m - z_{\alpha/2}) + \Phi(-\theta_m - z_{\alpha/2}),$$

where

$$\theta_m = \left[1 - m\left(\frac{1}{n_1} + \frac{1}{n_2}\right)\right]\frac{\mu_1 - \mu_2}{\sigma\sqrt{(1/n_1)+(1/n_2)}}. \tag{2.6}$$

Note that $\theta_m = \theta$ if $m = 0$, that is, there is no mix-up.

The effect of mixed-up treatment codes can be measured by comparing $p(\theta)$ with $p(\theta_m)$. Suppose that $n_1 = n_2$. Then $p(\theta_m)$ depends on m/n_1, the proportion of mixed-up treatment codes. For example, suppose that when there is no mix-up, $p(\theta) = 80\%$, which gives that $|\theta| = 2.81$. When 5% of treatment codes are mixed up, that is, $m/n_1 = 5\%$, $p(\theta_m) = 70.2\%$. When 10% of treatment codes are mixed up, $p(\theta_m) = 61.4\%$. Hence, a small proportion of mixed-up treatment codes may seriously affect the probability of detecting treatment effect when such an effect exists. In this simple case, we may plan ahead to ensure a desired power when the maximum proportion of mixed-up treatment codes is known. Assume that the maximum proportion of mixed-up treatment codes is p and that the original sample size is $n_1 = n_2 = n_0$. Then,

$$\theta_m = (1 - 2p)\theta = \frac{\mu_1 - \mu_2}{\sigma\sqrt{2}}\sqrt{(1 - 2p)^2 n_0}.$$

Thus, a new sample size $n_{\text{new}} = n_0/(1 - 2p)^2$ will maintain the desired power when the proportion of mixed-up treatment codes is no larger than p. For example, if $p = 5\%$, then

$n_{\text{new}} = 1.23 n_0$, that is, a 23% increase of the sample size will offset a 5% mix-up in randomization schedules.

The effect of mixed-up treatment codes is higher when the study design becomes more complicated. Consider the two-group parallel design with an unknown σ^2. The test statistic is necessarily modified by replacing $z_{\alpha/2}$ and σ^2 by $t_{\alpha/2;n_1+n_2-2}$ and

$$\hat{\sigma}^2 = \frac{(n_1-1)s_1^2 + (n_2-1)s_2^2}{n_1+n_2-2},$$

where s_1^2 is the sample variance based on responses from patients in the treatment group, s_2^2 is the sample variance based on responses from patients in the control group, and $t_{\alpha/2;n_1+n_2-2}$ is the upper $(\alpha/2)$th percentile of the t-distribution with $n_1 + n_2 - 2$ degrees of freedom. When randomization is properly applied without mix-up, the two-sample t-test has the α level of significance and the power is given by

$$1 - T_{n_1+n_2-2}(t_{\alpha/2;n_1+n_2-2} \mid \theta) + T_{n_1+n_2-2}(-t_{\alpha/2;n_1+n_2-2} \mid \theta),$$

where θ is defined by Equation 2.5 and $T_{n_1+n_2-2}(\cdot \mid \theta)$ is the noncentral t-distribution function with $n_1 + n_2 - 2$ degrees of freedom and the noncentrality parameter θ. When there are m patients with mixed-up treatment codes and $\mu_1 \neq \mu_2$, the effect on the distribution of $\bar{x}_1 - \bar{x}_2$ is the same as that in the case of known σ^2. In addition, the distribution of $\hat{\sigma}^2$ is also changed. A direct calculation shows that the expectation of $\hat{\sigma}^2$ is

$$E(\hat{\sigma}^2) = \sigma^2 + \frac{2(\mu_1-\mu_2)^2 m}{n_1+n_2-2}\left[2 - m\left(\frac{1}{n_1} + \frac{1}{n_2}\right)\right].$$

Hence, the actual power of the two-sample t-test is less than

$$1 - T_{n_1+n_2-2}(t_{0.975;n_1+n_2-2} \mid \theta_m) + T_{n_1+n_2-2}(-t_{0.975;n_1+n_2-2} \mid \theta_m),$$

where θ_m is given by Equation 2.6.

2.6.4 Treatment or Center Imbalance

In multicenter clinical trials, sample size calculation is usually performed under the assumption that there are equal numbers of subjects in each center. In practice, however, we may end up with an imbalance in sample sizes across centers. It is a concern (i) what the impact is of this imbalance on the power of the test and (ii) whether sample size calculation should be performed in such a way as to account for this imbalance. In this section, we examine this issue by studying the power with sample size imbalance across centers.

For a multicenter trial, the following model is usually considered:

$$y_{ijk} = \mu + T_i + C_j + (TC)_{ij} + \varepsilon_{ijk},$$

where $i = 1, 2$ (treatment), $j = 1, \ldots, J$ (center), $k = 1, \ldots, n_{ij}$, T_i is the ith treatment effect, C_j is the effect due to the jth center, $(TC)_{ij}$ is the effect due to the interaction between the ith treatment in the jth center, and ε_{ijk} are random errors, which are normally distributed with mean 0 and variance σ^2. Under the above model, a test statistic for

$$\mu_1 - \mu_2 = (\mu + T_1) - (\mu + T_2) = T_1 - T_2$$

is given by

$$T^* = \frac{1}{J} \sum_{j=1}^{J} (\bar{y}_{1j} - \bar{y}_{2j})$$

with $E(T^*) = T_1 - T_2$ and

$$\text{Var}(T^*) = \frac{\sigma^2}{J^2} \sum_{j=1}^{J} \left(\frac{1}{n_{1j}} + \frac{1}{n_{2j}} \right).$$

If we assume that $n_{1j} = n_{2j} = n_j$ for all $j = 1, \dots, J$, then

$$\text{Var}(T^*) = \frac{\sigma^2}{J^2} \sum_{j=1}^{J} \frac{2}{n_j}.$$

In this case, the power of the test is approximately given by

$$\text{Power} = 1 - \Phi \left(z_{\alpha/2} - \frac{\delta}{\sigma / J \sqrt{\sum_{j=1}^{J} 2/n_j}} \right).$$

When $n_j = n$ for all j,

$$\text{Var}(T^*) = \frac{2\sigma^2}{Jn}$$

and the power of the test becomes approximately

$$\text{Power} = 1 - \Phi \left(z_{\alpha/2} - \frac{\delta}{\sigma \sqrt{2/(Jn)}} \right).$$

As it can be seen that

$$1 - \Phi \left(z_{\alpha/2} - \frac{\delta}{\sigma / J \sqrt{\sum_{j=1}^{J} 2/n_j}} \right) \leq 1 - \Phi \left(z_{\alpha/2} - \frac{\delta}{\sigma \sqrt{2/(Jn)}} \right).$$

To achieve the same power, the only choice is to increase the sample size if we assume that the variance remains the same. In this situation, the total sample size $N = \sum_{j=1}^{J} n_j$ should satisfy

$$\frac{\delta}{\sigma/\sqrt{\sum_{j=1}^{J} 2/n_j}} = \frac{\delta}{\sigma\sqrt{2/n}}.$$

The difficulty is that $n_j, j = 1, \ldots, J$, are not fixed and we are unable to predict how many subjects will be in each center at the end of the trial although we may start with the same number of subjects in each center. The loss in power due to treatment and/or center imbalance may be substantial in practice.

2.6.5 Multiplicity

In many clinical trials, multiple comparisons may be performed. In the interest of controlling the overall type I error rate at the α level, an adjustment for multiple comparisons such as the Bonferroni adjustment is necessary. The formulas for sample size calculation can still be applied by simply replacing the α level with an adjusted α level. In practice, it may be too conservative to adjust the α level when there are too many primary clinical endpoints or there are too many comparisons to be made. As a rule of thumb, Biswas, Chan, and Ghosh (2000) suggested that a multiplicity adjustment to the significance level be made when at least one significant result (e.g., one of several primary clinical endpoints or several pairwise comparison) is required to draw conclusion. On the other hand, a multiplicity adjustment is not needed when (i) all results are required to be significant to draw conclusion or (ii) the testing problem is closed. A test procedure is said to be *closed* if the rejection region of a particular univariate null hypothesis at a given significance α level implies the rejection of all higher-dimensional null hypotheses containing the univariate null hypothesis at the same significance α level (Marcus, Peritz, and Gabriel, 1976). When a multiplicity adjustment is required, it is recommended that either the method of Bonferroni or the procedures described in Hochberg and Tamhane (1987) be used.

2.6.6 Multiple-Stage Design for Early Stopping

In phase II cancer trials, it is undesirable to stop a study early when the treatment appears to be effective but desirable to terminate the trial when the treatment seems to be ineffective. For this purpose, a multiple-stage design is often employed to determine whether a test drug is promising enough to warrant further testing (Simon, 1989). The concept of a multiple-stage design is to permit early stopping when a moderately long sequence of initial failures occur. For example, in Simon's two-stage optimal design, n_1 subjects are treated and the trial terminates if all n_1 are treatment failures. If there are one or more successes in stage 1, then stage 2 is implemented by including the other n_2 subjects. A decision is then made based on the response rate of the $n_1 + n_2$ subjects. The drawback of Simon's design is that it does not allow early termination if there is a long run of failures at the start. To overcome this disadvantage, Ensign et al. (1994) proposed an optimal three-stage design that modifies the Simon's two-stage design. Let p_0 be the response rate that is not of interest for conducting further studies and p_1 be the response rate of definite interest ($p_1 > p_0$). The optimal three-stage design is implemented by testing the following hypotheses:

$H_0 : p \leq p_0$ versus $H_a : p \geq p_1$.

Rejection of H_0 indicates that further study of the test drug should be carried out. At stage 1, n_1 subjects are treated. We would reject H_a (i.e., the test drug is not promising) and stop the trial if there is no response. If there are one or more responses, then proceed to stage 2 by including additional n_2 subjects. We would then reject H_1 and stop the trial if the total number of responses is less than a prespecified number of r_2; otherwise continue to stage 3. At stage 3, n_3 more subjects are treated. We reject H_a if the total number of responses is less than a prespecified number of r_3. In this case, we conclude the test treatment is ineffective. Based on the three-stage design described above, Ensign et al. (1994) considered the following to determine the sample size. For each value of n_2 satisfying

$$(1-p_1)^{n_1} < \beta,$$

where

$$\beta = P(\text{reject } H_1 | p_1),$$

compute the values of r_2, n_2, r_3, and n_3 that minimize the null expected sample size $EN(p_0)$ subject to the error constraints α and β, where

$$EN(p) = n_1 + n_2\{1 - \beta_1(p)\} + n_3\{1 - \beta_1(p) - \beta_2(p)\},$$

and β_i are the probability of making type II error evaluated stage i. Ensign et al. (1994) use the value of

$$\beta - (1-p_1)^{n_1}$$

as the type II error rate in the optimization along with type I error

$$\alpha = P(\text{reject } H_0 | p_0)$$

to obtain r_2, n_2, r_3, and n_3. Repeating this, n_1 can be chosen to minimize the overall $EN(p_0)$.

2.6.7 Rare Incidence Rate

In most clinical trials, although the primary objectives are usually for the evaluation of the effectiveness and safety of the test drug under investigation, the assessment of drug safety has not received the same level of attention as the assessment of efficacy. Sample size calculations are usually performed on the basis of a prestudy power analysis based on the primary efficacy variable. If sample size is determined based on the primary safety variable such as adverse event rate, a large sample size may be required especially when the incidence rate is extremely rare. For example, if the incidence rate is one per 10,000, then we will need to include 10,000 subjects to observe a single incidence. In this case, we may justify a selected sample size based on the concept of probability statement as described in Section 1.3.5. O'Neill (1988) indicated that the magnitude of rates that can be feasibly studied in most clinical trials is about 0.01 and higher. However, observational cohort studies usually can assess rates on the order of 0.001 and higher. O'Neill (1988) also indicated that it is informative to examine the sample sizes that would be needed to estimate a rate or to detect or estimate differences of specified amounts between rates for two different treatment groups. Sample size requirement for clinical studies with extremely low incidence rate will be further investigated in Chapter 17.

3

Comparing Means

In clinical research, clinical trials are usually conducted for the evaluation of the efficacy and safety of a test drug under investigation as compared to a placebo-controlled or an active control agent (e.g., a standard of care or therapy) in terms of mean responses of some primary study endpoints. The objectives of the intended clinical trials usually include (i) the evaluation of the drug effect and (ii) the demonstration of therapeutic equivalence/non-inferiority, or (iii) the establishment of superiority. For the evaluation of the effect within a given treatment, the null hypothesis of interest is to test whether there is a significant difference in mean response between pre- and posttreatment or mean change from baseline to endpoint. We refer to this testing problem as a one-sample problem. For the establishment of the efficacy and safety of the test drug, a typical approach is to test whether there is a difference between the test drug and the placebo control and then evaluate the chance of correctly detecting a clinically meaningful difference if such a difference truly exists. Thus, it is of interest to first test the null hypothesis of equality and then evaluate the power under the alternative hypothesis to determine whether the evidence is substantial for regulatory approval. For the demonstration of therapeutic equivalence/noninferiority and/or superiority as compared to an active control or standard therapy, it is of interest to test hypotheses for equivalence/noninferiority and/or superiority as described in Chapter 1. In this chapter, under a valid design (e.g., a parallel design or a crossover design), methods for sample size calculation are derived for achieving a desired power of statistical tests for appropriate hypotheses.

In Section 3.1, testing in one-sample problems is considered. Sections 3.2 and 3.3 summarize procedures for sample size calculation in two-sample problems under a parallel design and a crossover design, respectively. Sections 3.4 and 3.5 present procedures in multiple-sample problems under a parallel design (one-way analysis of variance [ANOVA]) and a crossover design (Williams design), respectively. Section 3.6 discusses some practical issues regarding sample size calculation for comparing means in clinical research, including sample size reduction when switching from a two-sided test to a one-sided test or from a parallel design to a crossover design, sensitivity analysis with respect to change in variability, and a brief discussion regarding Bayesian approach.

3.1 One-Sample Design

Let x_i be the response from the ith sampled subject, $i = 1, \ldots, n$. In clinical research, x_i could be the difference between matched pairs such as the pretreatment and posttreatment responses or changes from baseline to endpoint within a treatment group. It is assumed that x_i's are independent and identically distributed (i.i.d.) normal random variables with mean 0 and variance σ^2. Let

$$\bar{x} = \frac{1}{n}\sum_{i=1}^{n} x_i \quad \text{and} \quad s^2 = \frac{1}{n-1}\sum_{i=1}^{n}(x_i - \bar{x})^2 \tag{3.1}$$

be the sample mean and sample variance of x_i's, respectively. Also, let $\varepsilon = \mu - \mu_0$ be the difference between the true mean response of a test drug (μ) and a reference value (μ_0). Without loss of generality, consider $\varepsilon > 0$ ($\varepsilon < 0$) an indication of *improvement* (*worsening*) of the test drug as compared to the reference value.

3.1.1 Test for Equality

To test whether there is a difference between the mean response of the test drug and the reference value, the following hypotheses are usually considered:

$$H_0 : \varepsilon = 0 \quad \text{versus} \quad H_a : \varepsilon \neq 0.$$

When σ^2 is known, we reject the null hypothesis at the α level of significance if

$$\left| \frac{\bar{x} - \mu_0}{\sigma/\sqrt{n}} \right| > z_{\alpha/2},$$

where z_a is the upper αth quantile of the standard normal distribution. Under the alternative hypothesis that $\varepsilon \neq 0$, the power of the above test is given by

$$\Phi\left(\frac{\sqrt{n\varepsilon^2}}{\sigma} - z_{\alpha/2} \right) + \Phi\left(-\frac{\sqrt{n\varepsilon^2}}{\sigma} - z_{\alpha/2} \right),$$

where Φ is the cumulative standard normal distribution function. By ignoring a small value $\leq \alpha/2$, the power is approximately

$$\Phi\left(\frac{\sqrt{n}|\varepsilon|}{\sigma} - z_{\alpha/2} \right).$$

As a result, the sample size needed to achieve power $1 - \beta$ can be obtained by solving the following equation:

$$\frac{\sqrt{n}|\varepsilon|}{\sigma} - z_{\alpha/2} = z_\beta.$$

This leads to

$$n = \frac{(z_{\alpha/2} + z_\beta)^2 \sigma^2}{\varepsilon^2} \tag{3.2}$$

(if the solution of Equation 3.2 is not an integer, then the smallest integer that is larger than the solution of Equation 3.2 should be taken as the required sample size). An initial value of ε (or ε/σ) is needed to calculate the sample size according to Equation 3.2. A lower bound of ε/σ, usually obtained from a pilot study or historical data, can be used as the initial value. A lower bound of ε/σ can also be defined as the clinically meaningful difference between the response means relative to the standard deviation (SD) σ.

When σ^2 is unknown, it can be replaced by the sample variance s^2 given in Equation 3.1, which results in the usual one-sample t-test, that is, we reject the null hypothesis H_0 if

$$\left|\frac{\bar{x} - \mu_0}{s/\sqrt{n}}\right| > t_{\alpha/2, n-1},$$

where $t_{a,n-1}$ is the upper αth quantile of a t-distribution with $n-1$ degrees of freedom. Under the alternative hypothesis that $\varepsilon \neq 0$, the power of the one-sample t-test is given by

$$1 - T_{n-1}\left(t_{\alpha/2,n-1}\left|\frac{\sqrt{n\varepsilon^2}}{\sigma}\right.\right) + T_{n-1}\left(-t_{\alpha/2,n-1}\left|\frac{\sqrt{n\varepsilon^2}}{\sigma}\right.\right),$$

where $T_{n-1}(\cdot|\theta)$ is the cumulative distribution function of a noncentral t-distribution with $n-1$ degrees of freedom and the noncentrality parameter θ. When an initial value of ε/σ is given, the sample size needed to achieve power $1 - \beta$ can be obtained by solving

$$T_{n-1}\left(t_{\alpha/2,n-1}\left|\frac{\sqrt{n\varepsilon^2}}{\sigma}\right.\right) - T_{n-1}\left(-t_{\alpha/2,n-1}\left|\frac{\sqrt{n\varepsilon^2}}{\sigma}\right.\right) = \beta. \tag{3.3}$$

By ignoring a small value $\leq \alpha/2$, the power is approximately

$$1 - T_{n-1}\left(t_{\alpha/2,n-1}\left|\frac{\sqrt{n}\,|\varepsilon|}{\sigma}\right.\right).$$

Hence, the required sample size can also be obtained by solving

$$T_{n-1}\left(t_{\alpha/2,n-1}\left|\frac{\sqrt{n}\,|\varepsilon|}{\sigma}\right.\right) = \beta. \tag{3.4}$$

Table 3.1 lists the solutions of this equation for some values of α, β, and $\theta = |\varepsilon|/\sigma$. When n is sufficiently large, $t_{\alpha/2,n-1} \approx z_{\alpha/2}$, $t_{\beta,n-1} \approx z_\beta$, and

$$T_{n-1}\left(t_{\alpha/2,n-1}\left|\frac{\sqrt{n}\varepsilon}{\sigma}\right.\right) \approx \Phi\left(z_{\alpha/2} - \frac{\sqrt{n}\varepsilon}{\sigma}\right). \tag{3.5}$$

TABLE 3.1

Smallest n with $T_{n-1}(t_{\alpha,n-1} \mid \sqrt{n}\theta) \leq \beta$

| | $\alpha = 2.5\%$ | | $\alpha = 5\%$ | | | $\alpha = 2.5\%$ | | $\alpha = 5\%$ | |
| | $1 - \beta =$ | | $1 - \beta =$ | | | $1 - \beta =$ | | $1 - \beta =$ | |
θ	80%	90%	80%	90%	θ	80%	90%	80%	90%
0.10	787	1053	620	858	0.54	29	39	23	31
0.11	651	871	513	710	0.56	28	36	22	29
0.12	547	732	431	597	0.58	26	34	20	27
0.13	467	624	368	509	0.60	24	32	19	26
0.14	403	539	317	439	0.62	23	30	18	24
0.15	351	469	277	382	0.64	22	28	17	23
0.16	309	413	243	336	0.66	21	27	16	22
0.17	274	366	216	298	0.68	19	25	15	20
0.18	245	327	193	266	0.70	19	24	15	19
0.19	220	293	173	239	0.72	18	23	14	18
0.20	199	265	156	216	0.74	17	22	13	18
0.21	180	241	142	196	0.76	16	21	13	17
0.22	165	220	130	179	0.78	15	20	12	16
0.23	151	201	119	164	0.80	15	19	12	15
0.24	139	185	109	151	0.82	14	18	11	15
0.25	128	171	101	139	0.84	14	17	11	14
0.26	119	158	93	129	0.86	13	17	10	14
0.27	110	147	87	119	0.88	13	16	10	13
0.28	103	136	81	111	0.90	12	16	10	13
0.29	96	127	75	104	0.92	12	15	9	12
0.30	90	119	71	97	0.94	11	14	9	12
0.32	79	105	62	86	0.96	11	14	9	11
0.34	70	93	55	76	0.98	11	14	8	11
0.36	63	84	50	68	1.00	10	13	8	11
0.38	57	75	45	61	1.04	10	12	8	10
0.40	52	68	41	55	1.08	9	12	7	9
0.42	47	62	37	50	1.12	9	11	7	9
0.44	43	57	34	46	1.16	8	10	7	8
0.46	40	52	31	42	1.20	8	10	6	8
0.48	37	48	29	39	1.30	7	9	6	7
0.50	34	44	27	36	1.40	7	8	5	7
0.52	32	41	25	34	1.50	6	7	5	6

Hence, formula (3.2) may still be used in the case of unknown σ.

3.1.2 Test for Noninferiority/Superiority

The problem of testing noninferiority and superiority can be unified by the following hypotheses:

$$H_0 : \varepsilon \leq \delta \quad \text{versus} \quad H_a : \varepsilon > \delta,$$

where δ is the superiority or noninferiority margin. When $\delta > 0$, the rejection of the null hypothesis indicates superiority over the reference value. When $\delta < 0$, the rejection of the null hypothesis implies noninferiority against the reference value.

When σ^2 is known, we reject the null hypothesis H_0 at the α level of significance if

$$\frac{\bar{x} - \mu_0 - \delta}{\sigma/\sqrt{n}} > z_\alpha.$$

If $\varepsilon > \delta$, the power of the above test is

$$\Phi\left(\frac{\sqrt{n}(\varepsilon - \delta)}{\sigma} - z_\alpha\right).$$

As a result, the sample size needed to achieve power $1 - \beta$ can be obtained by solving

$$\frac{\sqrt{n}(\varepsilon - \delta)}{\sigma} - z_\alpha = z_\beta,$$

which leads to

$$n = \frac{(z_\alpha + z_\beta)^2 \sigma^2}{(\varepsilon - \delta)^2}. \tag{3.6}$$

When σ^2 is unknown, it can be replaced by s^2 given in Equation 3.1. The null hypothesis H_0 is rejected at the α level of significance if

$$\frac{\bar{x} - \mu_0 - \delta}{s/\sqrt{n}} > t_{\alpha, n-1}.$$

The power of this test is given by

$$1 - T_{n-1}\left(t_{\alpha, n-1} \left| \frac{\sqrt{n}(\varepsilon - \delta)}{\sigma} \right.\right).$$

The sample size needed to achieve power $1 - \beta$ can be obtained by solving

$$T_{n-1}\left(t_{\alpha, n-1} \left| \frac{\sqrt{n}(\varepsilon - \delta)}{\sigma} \right.\right) = \beta.$$

By letting $\theta = (\varepsilon - \delta)/\sigma$, Table 3.1 can be used to find the required sample size. From approximation (3.5), formula (3.6) can be used to calculate the required sample size when n is sufficiently large.

3.1.3 Test for Equivalence

The objective is to test the following hypotheses:

$$H_0 : |\varepsilon| \geq \delta \quad \text{versus} \quad H_a : |\varepsilon| < \delta.$$

The test drug is concluded to be equivalent to a gold standard on average if the null hypothesis is rejected at significance level α.

When σ^2 is known, the null hypothesis H_0 is rejected at significance level α if

$$\frac{\sqrt{n}(\bar{x} - \mu_0 - \delta)}{\sigma} < -z_\alpha \quad \text{and} \quad \frac{\sqrt{n}(\bar{x} - \mu_0 + \delta)}{\sigma} > z_\alpha.$$

The power of this test is given by

$$\Phi\left(\frac{\sqrt{n}(\delta - \varepsilon)}{\sigma} - z_\alpha\right) + \Phi\left(\frac{\sqrt{n}(\delta + \varepsilon)}{\sigma} - z_\alpha\right) - 1. \tag{3.7}$$

Although the sample size n can be obtained by setting the power in Equation 3.7 to $1 - \beta$, it is more convenient to use the following method. Note that the power is larger than

$$2\Phi\left(\frac{\sqrt{n}(\delta - |\varepsilon|)}{\sigma} - z_\alpha\right) - 1. \tag{3.8}$$

Hence, the sample size needed to achieve power $1 - \beta$ can be obtained by solving

$$\Phi\left(\frac{\sqrt{n}(\delta - |\varepsilon|)}{\sigma} - z_\alpha\right) = 1 - \frac{\beta}{2}.$$

This leads to

$$n = \frac{(z_\alpha + z_{\beta/2})^2 \sigma^2}{(\delta - |\varepsilon|)^2}. \tag{3.9}$$

Note that the quantity in Equation 3.8 is a conservative approximation to the power. Hence, the sample size calculated according to Equation 3.9 is conservative. A different approximation is given in Chow and Liu (1992, 2000), which leads to the following formula for sample size calculation:

$$n = \begin{cases} \dfrac{(z_\alpha + z_{\beta/2})^2 \sigma^2}{\delta^2} & \text{if } \varepsilon = 0 \\[2ex] \dfrac{(z_\alpha + z_\beta)^2 \sigma^2}{(\delta - |\varepsilon|)^2} & \text{if } \varepsilon \neq 0. \end{cases}$$

When σ^2 is unknown, it can be estimated by s^2 given in Equation 3.1. The null hypothesis H_0 is rejected at significance level α if

$$\frac{\sqrt{n}(\bar{x} - \mu_0 - \delta)}{s} < -t_{\alpha,n-1} \quad \text{and} \quad \frac{\sqrt{n}(\bar{x} - \mu_0 + \delta)}{s} > t_{\alpha,n-1}.$$

The power of this test can be estimated by

$$1 - T_{n-1}\left(t_{\alpha,n-1}\left|\frac{\sqrt{n}(\delta - \varepsilon)}{\sigma}\right|\right) - T_{n-1}\left(t_{\alpha,n-1}\left|\frac{\sqrt{n}(\delta + \varepsilon)}{\sigma}\right|\right).$$

Hence, the sample size needed to achieve power $1 - \beta$ can be obtained by setting the power to $1 - \beta$. Since the power is larger than

$$1 - 2T_{n-1}\left(t_{\alpha,n-1}\left|\frac{\sqrt{n}(\delta - |\varepsilon|)}{\sigma}\right|\right),$$

a conservative approximation to the sample size needed to achieve power $1 - \beta$ can be obtained by solving

$$T_{n-1}\left(t_{\alpha,n-1}\left|\frac{\sqrt{n}(\delta - |\varepsilon|)}{\sigma}\right|\right) = \frac{\beta}{2},$$

which can be done by using Table 3.1 with $\theta = (\delta - |\varepsilon|)/\sigma$. From approximation (3.5), formula (3.9) can be used to calculate the required sample size when n is sufficiently large.

3.1.4 An Example

To illustrate the use of sample size formulas derived above, we first consider an example concerning a study of osteoporosis in postmenopausal women. Osteoporosis and osteopenia (or decreased bone mass) most commonly develop in postmenopausal women. The consequences of osteoporosis are vertebral crush fractures and hip fractures. The diagnosis of osteoporosis is made when vertebral bone density is more than 10% below what is expected for sex, age, height, weight, and race. Usually, bone density is reported in terms of SD from mean values. The World Health Organization (WHO) defines osteopenia as bone density value greater than one SD below peak bone mass levels in young women and osteoporosis as a value of greater than 2.5 SD below the same measurement scale. In medical practice, most clinicians suggest therapeutic intervention should be begun in patients with osteopenia to prevent progression to osteoporosis.

3.1.4.1 Test for Equality

Suppose that the mean bone density before the treatment is 1.5 SD ($\mu_0 = 1.5$ SD) and after the treatment is expected to be 2.0 SD (i.e., $\mu_1 = 2.0$ SD). We have $\varepsilon = \mu_1 - \mu_0 = 2.0$ SD − 1.5

SD = 0.5 SD. By Equation 3.2, at $\alpha = 0.05$, the required sample size for having an 80% power (i.e., $\beta = 0.2$) for correctly detecting a difference of $\varepsilon = 0.5$ SD change from pretreatment to posttreatment can be obtained by normal approximation as

$$n = \frac{(z_{\alpha/2} + z_\beta)^2 \sigma^2}{\varepsilon^2} = \frac{(1.96 + 0.84)^2}{(0.5)^2} \approx 32.$$

On the other hand, the sample size can also be obtained by solving Equation 3.4. Note that

$$\theta = \frac{|\varepsilon|}{\sigma} = 0.5.$$

By referring to the column under $\alpha = 2.5\%$ (two-sided test) at the row with $\theta = 0.5$ in Table 3.1, it can be found that the sample size needed is 34.

3.1.4.2 Test for Noninferiority

For the prevention of progression from osteopenis to osteoporosis, we wish to show that the mean bone density posttreatment is no less than pretreatment by a clinically meaningful difference $\delta = -0.5$ SD. As a result, by Equation 3.6, at $\alpha = 0.05$, the required sample size for having an 80% power (i.e., $\beta = 0.20$) can be obtained by normal approximation as

$$n = \frac{(z_\alpha + z_\beta)^2 \sigma^2}{(\varepsilon - \delta)^2} = \frac{(1.64 + 0.84)^2}{(0.5 + 0.5)^2} \approx 7.$$

On the other hand, the sample size can also be obtained by using Table 3.1. Note that

$$\theta = \frac{\varepsilon - \delta}{\sigma} = 0.5 + 0.5 = 1.00.$$

By referring to the column under $\alpha = 5\%$ at the row with $\theta = 1.0$ in Table 3.1, it can be found that the sample size needed is 8.

3.1.4.3 Test for Equivalence

To illustrate the use of the sample size formula for testing equivalence, we consider another example concerning the effect of a test drug on body weight change in terms of body mass index (BMI) before and after the treatment. Suppose clinicians consider that a less than 5% change in BMI from baseline (pretreatment) to endpoint (posttreatment) is not a safety concern for the indication of the disease under study. Then, we consider $\delta = 5\%$ as the equivalence limit. The objective is then to demonstrate safety by testing equivalence in mean BMI between pretreatment and posttreatment of the test drug. Assume the true BMI difference is 0 ($\varepsilon = 0$) and the SD is 10% ($\sigma = 0.1$), by Equation 3.9 with $\alpha = 0.05$, the sample size required for achieving an 80% power can be obtained by normal approximation as

$$n = \frac{(z_\alpha + z_{\beta/2})^2 \sigma^2}{\delta^2} = \frac{(1.64 + 1.28)^2 0.10^2}{0.05^2} \approx 35.$$

On the other hand, the sample size calculation can also be performed by using Table 3.1. Note that

$$\theta = \frac{\delta}{\sigma} = \frac{0.05}{0.10} = 0.50.$$

By referring to the column under $\alpha = 5\%$ and $1 - \beta = 90\%$ at the row with $\theta = 0.50$ in Table 3.1, it can be found that the sample size needed is 36.

3.2 Two-Sample Parallel Design

Let x_{ij} be the response observed from the jth subject in the ith treatment group, $j = 1, \ldots, n_i$, $i = 1, 2$. It is assumed that $x_{ij}, j = 1, \ldots, n_i, i = 1, 2$, are independent normal random variables with mean μ_i and variance σ^2. Let

$$\bar{x}_{i\cdot} = \frac{1}{n_i} \sum_{j=1}^{n_i} x_{ij} \quad \text{and} \quad s^2 = \frac{1}{n_1 + n_2 - 2} \sum_{i=1}^{2} \sum_{j=1}^{n_i} (x_{ij} - \bar{x}_{i\cdot})^2 \tag{3.10}$$

be the sample means for the ith treatment group and the pooled sample variance, respectively. Also, let $\varepsilon = \mu_2 - \mu_1$ be the true mean difference between a test drug (μ_2) and a placebo-controlled or an active control agent (μ_1). Without loss of generality, consider $\varepsilon > 0$ ($\varepsilon < 0$) as an indication of *improvement* (*worsening*) of the test drug as compared to the placebo-controlled or active control agent. In practice, it may be desirable to have an unequal treatment allocation, that is, $n_1/n_2 = \kappa$ for some κ. Note that $\kappa = 2$ indicates a 2 to 1 test-control allocation, whereas $\kappa = 1/2$ indicates a 1 to 2 test-control allocation.

3.2.1 Test for Equality

The objective is to test whether there is a difference between the mean responses of the test drug and the placebo control or active control. Hence, the following hypotheses are considered:

$$H_0 : \varepsilon = 0 \quad \text{versus} \quad H_a : \varepsilon \neq 0.$$

When σ^2 is known, the null hypothesis H_0 is rejected at the significance level α if

$$\left| \frac{\bar{x}_{1\cdot} - \bar{x}_{2\cdot}}{\sigma \sqrt{\frac{1}{n_1} + \frac{1}{n_2}}} \right| > z_{\alpha/2}.$$

Under the alternative hypothesis that $\varepsilon \neq 0$, the power of the above test is

$$\Phi\left(\frac{\varepsilon}{\sigma\sqrt{\frac{1}{n_1}+\frac{1}{n_2}}} - z_{\alpha/2}\right) + \Phi\left(\frac{-\varepsilon}{\sigma\sqrt{\frac{1}{n_1}+\frac{1}{n_2}}} - z_{\alpha/2}\right)$$

$$\approx \Phi\left(\frac{|\varepsilon|}{\sigma\sqrt{\frac{1}{n_1}+\frac{1}{n_2}}} - z_{\alpha/2}\right),$$

after ignoring a small term of value $\leq \alpha/2$. As a result, the sample size needed to achieve power $1 - \beta$ can be obtained by solving the following equation:

$$\frac{|\varepsilon|}{\sigma\sqrt{\frac{1}{n_1}+\frac{1}{n_2}}} - z_{\alpha/2} = z_\beta.$$

This leads to

$$n_1 = \kappa n_2 \quad \text{and} \quad n_2 = \frac{(z_{\alpha/2}+z_\beta)^2\sigma^2(1+1/\kappa)}{\varepsilon^2}. \tag{3.11}$$

When σ^2 is unknown, it can be replaced by s^2 given in Equation 3.10. The null hypothesis H_0 is rejected if

$$\left|\frac{\bar{x}_{1\cdot}-\bar{x}_{2\cdot}}{s\sqrt{\frac{1}{n_1}+\frac{1}{n_2}}}\right| > t_{\alpha/2,n_1+n_2-2}.$$

Under the alternative hypothesis that $\varepsilon \neq 0$, the power of this test is

$$1 - T_{n_1+n_2-2}\left(t_{\alpha/2,n_1+n_2-2}\left|\frac{\varepsilon}{\sigma\sqrt{\frac{1}{n_1}+\frac{1}{n_2}}}\right.\right)$$

$$+ T_{n_1+n_2-2}\left(-t_{\alpha/2,n_1+n_2-2}\left|\frac{\varepsilon}{\sigma\sqrt{\frac{1}{n_1}+\frac{1}{n_2}}}\right.\right).$$

Thus, with $n_1 = \kappa n_2$, the sample size n_2 needed to achieve power $1 - \beta$ can be obtained by setting the power equal to $1 - \beta$.

After ignoring a small term of value $\leq \alpha/2$, the power is approximately

$$1 - T_{n_1+n_2-2}\left(t_{\alpha/2,n_1+n_2-2}\left|\frac{|\varepsilon|}{\sigma\sqrt{\frac{1}{n_1}+\frac{1}{n_2}}}\right.\right).$$

Hence, the required sample size n_2 can also be obtained by solving

$$T_{(1+\kappa)n_2-2}\left(t_{\alpha/2,(1+\kappa)n_2-2}\left|\frac{\sqrt{n_2}\,|\varepsilon|}{\sigma\sqrt{1+1/\kappa}}\right|\right)=\beta.$$

Table 3.2 can be used to obtain the solutions for $\kappa = 1$, 2, and some values of $\theta = |\varepsilon|/\sigma$, α, and β. When $\kappa = 1/2$, Table 3.2 can be used to find the required n_1 and $n_2 = 2n_1$.

From approximation (3.5), formula (3.2) can be used when both n_1 and n_2 are large.

TABLE 3.2

Smallest n with $T_{(1+\kappa)n-2}\left(t_{\alpha,(1+\kappa)n-2}\,|\,\sqrt{n}\theta/\sqrt{1+1/\kappa}\right)\le\beta$

	$\kappa = 1$				$\kappa = 2$			
	$\alpha = 2.5\%$		$\alpha = 5\%$		$\alpha = 2.5\%$		$\alpha = 5\%$	
	$1 - \beta =$		$1 - \beta =$		$1 - \beta =$		$1 - \beta =$	
θ	80%	90%	80%	90%	80%	90%	80%	90%
0.30	176	235	139	191	132	176	104	144
0.32	155	207	122	168	116	155	92	126
0.34	137	183	108	149	103	137	81	112
0.36	123	164	97	133	92	123	73	100
0.38	110	147	87	120	83	110	65	90
0.40	100	133	78	108	75	100	59	81
0.42	90	121	71	98	68	90	54	74
0.44	83	110	65	90	62	83	49	67
0.46	76	101	60	82	57	76	45	62
0.48	70	93	55	76	52	70	41	57
0.50	64	86	51	70	48	64	38	52
0.52	60	79	47	65	45	59	35	48
0.54	55	74	44	60	42	55	33	45
0.56	52	68	41	56	39	51	31	42
0.58	48	64	38	52	36	48	29	39
0.60	45	60	36	49	34	45	27	37
0.65	39	51	30	42	29	38	23	31
0.70	34	44	26	36	25	33	20	27
0.75	29	39	23	32	22	29	17	24
0.80	26	34	21	28	20	26	15	21
0.85	23	31	18	25	17	23	14	19
0.90	21	27	16	22	16	21	12	17
0.95	19	25	15	20	14	19	11	15
1.00	17	23	14	18	13	17	10	14
1.05	16	21	12	17	12	15	9	13
1.10	15	19	11	15	11	14	9	12
1.15	13	17	11	14	10	13	8	11
1.20	12	16	10	13	9	12	7	10
1.25	12	15	9	12	9	11	7	9
1.30	11	14	9	11	8	11	6	9
1.35	10	13	8	11	8	10	6	8
1.40	10	12	8	10	7	9	6	8
1.45	9	12	7	9	7	9	5	7
1.50	9	11	7	9	6	8	5	7

3.2.2 Test for Noninferiority/Superiority

The problem of testing noninferiority and superiority can be unified by the following hypotheses:

$$H_0 : \varepsilon \leq \delta \quad \text{versus} \quad H_a : \varepsilon > \delta,$$

where δ is the superiority or noninferiority margin. When $\delta > 0$, the rejection of the null hypothesis indicates the superiority of the test drug over the control. When $\delta < 0$, the rejection of the null hypothesis indicates the noninferiority of the test drug against the control.

When σ^2 is known, the null hypothesis H_0 is rejected at the α level of significance if

$$\frac{\bar{x}_1 - \bar{x}_2 - \delta}{\sigma\sqrt{\frac{1}{n_1} + \frac{1}{n_2}}} > z_\alpha.$$

Under the alternative hypothesis that $\varepsilon > \delta$, the power of the above test is given by

$$\Phi\left(\frac{\varepsilon - \delta}{\sigma\sqrt{\frac{1}{n_1} + \frac{1}{n_2}}} - z_\alpha\right).$$

The sample size needed to achieve power $1 - \beta$ can be obtained by solving

$$\frac{\varepsilon - \delta}{\sigma\sqrt{\frac{1}{n_1} + \frac{1}{n_2}}} - z_\alpha = z_\beta.$$

This leads to

$$n_1 = \kappa n_2 \quad \text{and} \quad n_2 = \frac{(z_\alpha + z_\beta)^2 \sigma^2 (1 + 1/\kappa)}{(\varepsilon - \delta)^2}. \tag{3.12}$$

When σ^2 is unknown, it can be replaced by s^2 given in Equation 3.10. The null hypothesis H_0 is rejected if

$$\frac{\bar{x}_1 - \bar{x}_2 - \delta}{s\sqrt{\frac{1}{n_1} + \frac{1}{n_2}}} > t_{\alpha, n_1 + n_2 - 2}.$$

Under the alternative hypothesis that $\varepsilon > \delta$, the power of this test is given by

$$1 - T_{n_1 + n_2 - 2}\left(t_{\alpha, n_1 + n_2 - 2} \,\middle|\, \frac{\varepsilon - \delta}{\sigma\sqrt{\frac{1}{n_1} + \frac{1}{n_2}}}\right).$$

The sample size needed to achieve power $1 - \beta$ can be obtained by solving the following equation:

$$T_{n_1+n_2-2}\left(t_{\alpha,n_1+n_2-2} \left| \frac{\varepsilon - \delta}{\sigma\sqrt{\frac{1}{n_1}+\frac{1}{n_2}}} \right| \right) = \beta.$$

By letting $\theta = (\varepsilon - \delta)/\sigma$, Table 3.2 can be used to find the required sample size.

From approximation (3.5), formula (3.12) can be used to calculate the required sample size when n_1 and n_2 are sufficiently large.

3.2.3 Test for Equivalence

The objective is to test the following hypotheses:

$$H_0 : |\varepsilon| \geq \delta \quad \text{versus} \quad H_a : |\varepsilon| < \delta.$$

The test drug is concluded to be equivalent to the control in average if the null hypothesis is rejected at significance level α.

When σ^2 is known, the null hypothesis H_0 is rejected at the α level of significance if

$$\frac{\bar{x}_1 - \bar{x}_2 - \delta}{\sigma\sqrt{\frac{1}{n_1}+\frac{1}{n_2}}} < -z_\alpha \quad \text{and} \quad \frac{\bar{x}_1 - \bar{x}_2 - \delta}{\sigma\sqrt{\frac{1}{n_1}+\frac{1}{n_2}}} > z_\alpha.$$

Under the alternative hypothesis that $|\varepsilon| < \delta$, the power of this test is

$$\Phi\left(\frac{\delta - \varepsilon}{\sigma\sqrt{\left(\frac{1}{n_1}\right)+\left(\frac{1}{n_2}\right)}} - z_\alpha\right) + \Phi\left(\frac{\delta + \varepsilon}{\sigma\sqrt{\left(\frac{1}{n_1}\right)+\left(\frac{1}{n_2}\right)}} - z_\alpha\right) - 1$$

$$\approx 2\Phi\left(\frac{\delta - |\varepsilon|}{\sigma\sqrt{\left(\frac{1}{n_1}\right)+\left(\frac{1}{n_2}\right)}} - z_\alpha\right) - 1.$$

As a result, the sample size needed to achieve power $1 - \beta$ can be obtained by solving the following equation:

$$\frac{\delta - |\varepsilon|}{\sigma\sqrt{\frac{1}{n_1}+\frac{1}{n_2}}} - z_\alpha = z_{\beta/2}.$$

This leads to

$$n_1 = \kappa n_2 \quad \text{and} \quad n_2 = \frac{(z_\alpha + z_{\beta/2})^2 \sigma^2 (1 + 1/\kappa)}{(\delta - |\varepsilon|)^2}. \tag{3.13}$$

When σ^2 is unknown, it can be replaced by s^2 given in Equation 3.10. The null hypothesis H_0 is rejected at the α level of significance if

$$\frac{\bar{x}_1 - \bar{x}_2 - \delta}{s\sqrt{\left(\frac{1}{n_1}\right) + \left(\frac{1}{n_2}\right)}} < -t_{\alpha, n_1 + n_2 - 2} \quad \text{and} \quad \frac{\bar{x}_1 - \bar{x}_2 - \delta}{s\sqrt{\left(\frac{1}{n_1}\right) + \left(\frac{1}{n_2}\right)}} > t_{\alpha, n_1 + n_2 - 2}.$$

Under the alternative hypothesis that $|\varepsilon| < \delta$, the power of this test is

$$1 - T_{n_1 + n_2 - 2}\left(t_{\alpha, n_1 + n_2 - 2} \left| \frac{\delta - \varepsilon}{\sigma\sqrt{\left(\frac{1}{n_1}\right) + \left(\frac{1}{n_2}\right)}} \right.\right)$$

$$-T_{n_1 + n_2 - 2}\left(t_{\alpha, n_1 + n_2 - 2} \left| \frac{\delta + \varepsilon}{\sigma\sqrt{\left(\frac{1}{n_1}\right) + \left(\frac{1}{n_2}\right)}} \right.\right).$$

Hence, with $n_1 = \kappa n_2$, the sample size n_2 needed to achieve power $1 - \beta$ can be obtained by setting the power to $1 - \beta$. Since the power is larger than

$$1 - 2T_{n_1 + n_2 - 2}\left(t_{\alpha, n_1 + n_2 - 2} \left| \frac{\delta - |\varepsilon|}{\sigma\sqrt{\left(\frac{1}{n_1}\right) + \left(\frac{1}{n_2}\right)}} \right.\right),$$

a conservative approximation to the sample size n_2 can be obtained by solving

$$T_{(1+\kappa)n_2 - 2}\left(t_{\alpha,(1+\kappa)n_2 - 2} \left| \frac{\sqrt{n_2}(\delta - |\varepsilon|)}{\sigma\sqrt{1 + 1/\kappa}} \right.\right) = \frac{\beta}{2}.$$

Table 3.2 can be used to calculate n_1 and n_2.

From approximation (3.5), formula (3.13) can be used to calculate the required sample size when n_1 and n_2 are sufficiently large.

3.2.4 An Example

Consider an example concerning a clinical trial for the evaluation of the effect of a test drug on cholesterol in patients with coronary heart disease (CHD). Cholesterol is the main lipid associated with arteriosclerotic vascular disease. The purpose of cholesterol testing is to identify patients at risk for arteriosclerotic heart disease. The liver metabolizes the cholesterol to its free form, which is transported in the bloodstream by lipoproteins. As indicated by Pagana and Pagana (1998), nearly 75% of the cholesterol is bound to low-density lipoproteins (LDLs) and 25% is bound to high-density lipoproteins (HDLs). Therefore, cholesterol is the main component of LDLs and only a minimal component of HDLs and very low-density lipoproteins. LDL is the most directly associated with increased risk of CHD.

A pharmaceutical company is interested in conducting a clinical trial to compare two cholesterol-lowering agents for the treatment of patients with CHD through a parallel design. The primary efficacy parameter is the LDL. In what follows, we will consider the situations where the intended trial is for (i) testing equality of mean responses in LDL, (ii) testing noninferiority or superiority of the test drug as compared to the active control agent, and (iii) testing for therapeutic equivalence. All sample size calculations in this section are performed for achieving an 80% power (i.e., $\beta = 0.20$) at the 5% level of significance (i.e., $\alpha = 0.05$).

3.2.4.1 Test for Equality

As discussed in Chapter 1, hypotheses for testing equality are point hypotheses. A typical approach for sample size calculation is to reject the null hypothesis of no treatment difference and conclude that there is a significant difference between treatment groups. Then, sample size can be chosen to achieve an 80% power for detecting a clinically meaningful difference (i.e., ε). In this example, suppose a difference of 5% (i.e., $\varepsilon = 5\%$) in percent change of LDL is considered to be of clinically meaningful difference. By Equation 3.11, assuming that the SD is 10% (i.e., $\sigma = 10\%$), the sample size by normal approximation can be determined by

$$n_1 = n_2 = \frac{2(z_{\alpha/2} + z_\beta)^2 \sigma^2}{\varepsilon^2} = \frac{2 \times (1.96 + 0.84)^2 \times 0.1^2}{0.05^2} \approx 63.$$

On the other hand, the sample size can also be obtained by using Table 3.2. Note that

$$\theta = \frac{\varepsilon}{\sigma} = \frac{0.05}{0.10} = 0.50.$$

By referring to the column under $\alpha = 2.5\%$ at the row with $\theta = 0.50$ in Table 3.2, it can be found that the sample size needed is 64.

3.2.4.2 Test for Noninferiority

Suppose that the pharmaceutical company is interested in establishing noninferiority of the test drug as compared to the active control agent. Similarly, we assume that the noninferiority margin is chosen to be 5% (i.e., $\delta = -0.05$). Also, suppose the true difference in mean LDL between treatment groups is 0% (i.e., $\varepsilon = \mu_2(\text{test}) - \mu_1(\text{control}) = 0.00$). Then, by Equation 3.12, the sample size by normal approximation can be determined by

$$n_1 = n_2 = \frac{2(z_\alpha + z_\beta)^2 \sigma^2}{(\varepsilon - \delta)^2} = \frac{2 \times (1.64 + 0.84)^2 \times 0.1^2}{(-0.00 - (-0.05))^2} \approx 50.$$

On the other hand, the sample size can also be obtained by using Table 3.2. Note that

$$\theta = \frac{|\varepsilon - \delta|}{\sigma} = \frac{0.05}{0.10} = 0.50.$$

By referring to the column under $\alpha = 5\%$ at the row with $\theta = 0.50$ in Table 3.2, it can be found that the sample size needed is 51.

3.2.4.3 Test for Equivalence

For establishment of equivalence, suppose the true mean difference is 1% (i.e., $\varepsilon = 0.01$) and the equivalence limit is 5% (i.e., $\delta = 0.05$). According to Equation 3.13, the sample size by normal approximation can be determined by

$$n_1 = n_2 = \frac{2(z_\alpha + z_{\beta/2})^2 \sigma^2}{(\delta - |\varepsilon|)^2} = \frac{2 \times (1.64 + 1.28)^2 \times 0.1^2}{(0.05 - 0.01)^2} \approx 107.$$

On the other hand, the sample size can also be obtained by using Table 3.2. Note that

$$\theta = \frac{\delta - |\varepsilon|}{\sigma} = \frac{0.04}{0.10} = 0.40.$$

By referring to the column under $\alpha = 5\%$, $\beta = 0.10$ at the row with $\theta = 0.40$ in Table 3.2, it can be found that the sample size needed is 108.

3.2.5 Remarks

The assumption that $\sigma_1^2 = \sigma_2^2$ may not hold. When $\sigma_1^2 \neq \sigma_2^2$, statistical procedures are necessarily modified. If σ_1^2 and σ_2^2 are unknown, this becomes the well-known Behrens–Fisher problem. Extensive literature have been devoted to this topic in the past several decades. Miller (1997) gave a comprehensive review of research work done in this area.

In practice, it is suggested that superiority be established by testing noninferiority first. Once the null hypothesis of inferiority is rejected, the test for superiority is performed. This test procedure controls the overall type I error rate at the nominal level α because it is a closed test procedure. More details regarding closed testing procedures can be found in Marcus, Peritz, and Gabriel (1976).

3.3 Two-Sample Crossover Design

In this section, we consider a $2 \times 2m$ replicated crossover design comparing mean responses of a test drug and a reference drug. Let y_{ijkl} be the lth replicate or response ($l = 1, ..., m$) observed from the jth subject ($j = 1, ..., n$) in the ith sequence ($i = 1, 2$) under the kth treatment ($k = 1, 2$). The following model is considered:

$$y_{ijkl} = \mu_k + \gamma_{ik} + s_{ijk} + e_{ijkl}, \tag{3.14}$$

where μ_k is the kth treatment effect, γ_{ik} is the fixed effect of the ith sequence under treatment k, and s_{ijk} is the random effect of the jth subject in the ith sequence under treatment k. (s_{ij1}, s_{ij2}), $i = 1, 2, j = 1, ..., n$, are assumed to be i.i.d. as bivariate normal random variables with mean 0 and covariance matrix

$$\Sigma = \begin{pmatrix} \sigma^2_{BT} & \rho\sigma_{BT}\sigma_{BR} \\ \rho\sigma_{BT}\sigma_{BR} & \sigma^2_{BR} \end{pmatrix}.$$

e_{ij1l} and e_{ij2l} are assumed to be independent normal random variables with mean 0 and variance σ^2_{WT} or σ^2_{WR} (depending on the treatment). Define

$$\sigma^2_D = \sigma^2_{BT} + \sigma^2_{BR} - 2\rho\sigma_{BT}\sigma_{BR}.$$

σ^2_D is the variability due to the effect of subject-by-treatment interaction, which reflects the heteroscedasticity of the subject random effect between the test drug and the reference drug.

Let $\varepsilon = \mu_2 - \mu_1$ (test−reference),

$$\bar{y}_{ijk.} = \frac{1}{m}(y_{ijk1} + \cdots + y_{ijkm}) \quad \text{and} \quad d_{ij} = \bar{y}_{ij1.} - \bar{y}_{ij2.}.$$

An unbiased estimate for ε is given by

$$\hat{\varepsilon} = \frac{1}{2n}\sum_{i=1}^{2}\sum_{j=1}^{n}d_{ij}.$$

Under model (3.14), $\hat{\varepsilon}$ follows a normal distribution with mean ε and variance $\sigma^2_m/(2n)$, where

$$\sigma^2_m = \sigma^2_D + \frac{1}{m}(\sigma^2_{WT} + \sigma^2_{WR}). \tag{3.15}$$

An unbiased estimate for σ^2_m can be obtained by

$$\hat{\sigma}^2_m = \frac{1}{2(n-1)}\sum_{i=1}^{2}\sum_{j=1}^{n}(d_{ij} - \bar{d}_{i.})^2,$$

where

$$\bar{d}_{i.} = \frac{1}{n}\sum_{j=1}^{n}d_{ij}.$$

Without loss of generality, consider $\varepsilon > 0$ ($\varepsilon < 0$) as an indication of *improvement* (*worsening*) of the test drug as compared to the reference drug. In practice, σ_m is usually unknown.

3.3.1 Test for Equality

The objective is to test the following hypotheses:

$$H_0 : \varepsilon = 0 \quad \text{versus} \quad H_a : \varepsilon \neq 0.$$

The null hypothesis H_0 is rejected at α level of significance if

$$\left| \frac{\hat{\varepsilon}}{\hat{\sigma}_m/\sqrt{2n}} \right| > t_{\alpha/2,2n-2}.$$

Under the alternative hypothesis that $\varepsilon \neq 0$, the power of this test is given by

$$1 - T_{2n-2}\left(t_{\alpha/2,2n-2} \left| \frac{\sqrt{2n}\varepsilon}{\sigma_m} \right| \right) + T_{2n-2}\left(-t_{\alpha/2,2n-2} \left| \frac{\sqrt{2n}\varepsilon}{\sigma_m} \right| \right).$$

As a result, the sample size needed to achieve power $1 - \beta$ can be obtained by setting the power to $1 - \beta$ or, after ignoring a small term $\leq \alpha/2$, by solving

$$T_{2n-2}\left(t_{\alpha/2,2n-2} \left| \frac{\sqrt{2n}\,|\varepsilon|}{\sigma_m} \right| \right) = \beta.$$

Table 3.2 with $\kappa = 1$ and $\theta = 2|\varepsilon|/\sigma_m$ can be used to obtain n. From approximation (3.5),

$$n = \frac{(z_{a/2} + z_\beta)^2 \sigma_m^2}{2\varepsilon^2} \tag{3.16}$$

for sufficiently large n.

3.3.2 Test for Noninferiority/Superiority

Similar to test for noninferiority/superiority under a parallel design, the problem can be unified by testing the following hypotheses:

$$H_0 : \varepsilon \leq \delta \quad \text{versus} \quad H_a : \varepsilon > \delta,$$

where δ is the noninferiority or superiority margin. When $\delta > 0$, the rejection of the null hypothesis indicates the superiority of the test drug against the control. When $\delta < 0$, the rejection of the null hypothesis indicates the noninferiority of the test drug over the control. The null hypothesis H_0 is rejected at the α level of significance if

$$\frac{\hat{\varepsilon} - \delta}{\hat{\sigma}_m/\sqrt{2n}} > t_{\alpha,2n-2}.$$

Under the alternative hypothesis that $\varepsilon > \delta$, the power of this test is given by

$$1 - T_{2n-2}\left(t_{\alpha,2n-2} \left| \frac{\sqrt{2n}\,(\varepsilon - \delta)}{\sigma_m} \right| \right).$$

As a result, the sample size needed to achieve power $1 - \beta$ can be obtained by solving

$$T_{2n-2}\left(t_{\alpha,2n-2} \left|\frac{\sqrt{2n}\,(\varepsilon-\delta)}{\sigma_m}\right|\right) = \beta,$$

which can be done by using Table 3.2 with $\kappa = 1$ and $\theta = 2(\varepsilon - \delta)/\sigma_m$. When n is sufficiently large, approximation (3.5) leads to

$$n = \frac{(z_\alpha + z_\beta)^2 \sigma_m^2}{2(\varepsilon - \delta)^2}. \tag{3.17}$$

3.3.3 Test for Equivalence

The objective is to test the following hypotheses:

$$H_{0:} |\varepsilon| \leq \delta \quad \text{versus} \quad H_a : |\varepsilon| < \delta.$$

The test drug is concluded to be equivalent to the control in average, that is, the null hypothesis H_0 is rejected at significance level α when

$$\frac{\sqrt{2n}\,(\hat{\varepsilon}-\delta)}{\hat{\sigma}_m} < -t_{\alpha,2n-2} \quad \text{and} \quad \frac{\sqrt{2n}\,(\hat{\varepsilon}+\delta)}{\hat{\sigma}_m} > t_{\alpha,2n-2}.$$

Under the alternative hypothesis that $|\varepsilon| < \delta$, the power of this test is

$$1 - T_{2n-2}\left(t_{\alpha,2n-2} \left|\frac{\sqrt{2n}(\delta-\varepsilon)}{\sigma_m}\right|\right)$$

$$-T_{2n-2}\left(t_{\alpha,2n-2} \left|\frac{\sqrt{2n}(\delta+\varepsilon)}{\sigma_m}\right|\right).$$

As a result, the sample size needed to achieve power $1 - \beta$ can be obtained by setting the power to $1 - \beta$. Since the power is larger than

$$1 - 2T_{2n-2}\left(t_{\alpha,2n-2} \left|\frac{\sqrt{2n}(\delta-|\varepsilon|)}{\sigma_m}\right|\right),$$

a conservative approximation of n can be obtained by solving

$$T_{2n-2}\left(t_{\alpha,2n-2} \left|\frac{\sqrt{2n}(\delta-|\varepsilon|)}{\sigma_m}\right|\right) = \frac{\beta}{2},$$

which can be done by using Table 3.2 with $\kappa = 1$ and $\theta = 2(\delta - |\varepsilon|))/\sigma_m$. When n is large, approximation (3.5) leads to

$$n = \frac{(z_\alpha + z_{\beta/2})^2 \sigma_m^2}{2(\delta - |\varepsilon|)^2}. \tag{3.18}$$

Note that an important application of testing for equivalence under crossover design is testing average bioequivalence (see Section 10.2). By applying a similar idea as introduced by Chow and Liu (2000), a different approximate sample size formula can be obtained as

$$n = \begin{cases} \dfrac{(z_\alpha + z_{\beta/2})^2 \sigma_m^2}{2\delta^2} & \text{if } \varepsilon = 0 \\ \dfrac{(z_\alpha + z_\beta)^2 \sigma_m^2}{2(\delta - |\varepsilon|)^2} & \text{if } \varepsilon \neq 0 \end{cases}$$

3.3.4 An Example

3.3.4.1 Therapeutic Equivalence

Consider a standard two-sequence, two-period crossover design ($m = 1$) for trials whose objective is to establish therapeutic equivalence between a test drug and a standard therapy. The sponsor is interested in having an 80% ($1 - \beta = 0.8$) power for establishing therapeutic equivalence. Based on the results from previous studies, it is estimated that the variance is 20% ($\sigma_m = 0.20$). Suppose the true mean difference is −10% (i.e., $\varepsilon = \mu_2(\text{test}) - \mu_1(\text{reference}) = -0.10$). Furthermore, we assume that the equivalence limit is 25% (i.e., $\delta = 0.25$). According to Equation 3.18,

$$n = \frac{(z_\alpha + z_{\beta/2})^2 \sigma_m^2}{2(\delta - |\varepsilon|)^2} = \frac{(1.64 + 1.28)^2 0.20^2}{2(0.25 - 0.10)^2} \approx 8.$$

On the other hand, the sample size calculation can also be performed by using Table 3.2. Note that

$$\theta = \frac{2(\delta - |\varepsilon|)}{\sigma_m} = \frac{2(0.25 - |-0.10|)}{0.20} = 1.50.$$

By referring to the column under $\alpha = 5\%$, $1 - \beta = 90\%$ at the row with $\theta = 1.50$ in Table 3.2, it can be found that the sample size needed is 9.

3.3.4.2 Noninferiority

Suppose that the sponsor is interested in showing noninferiority of the test drug against the reference with a noninferiority margin of −20% ($\delta = -20\%$). According to Equation 3.17, the sample size needed is given by

$$n = \frac{(z_\alpha + z_\beta)^2 \sigma_m^2}{2(\varepsilon - \delta)^2} = \frac{(1.64 + 0.84)^2 0.20^2}{2(-0.1 - (-0.2))^2} \approx 13.$$

On the other hand, the sample size calculation can also be performed by using Table 3.2. Note that

$$\theta = \frac{2(\varepsilon - \delta)}{\sigma_m} = \frac{2(-0.10 - (-0.20))}{0.20} = 1.00.$$

By referring to the column under $\alpha = 5\%$, $1 - \beta = 80\%$ at the row with $\theta = 1.00$ in Table 3.2, it can be found that the sample size needed is 14.

3.3.5 Remarks

Sample size calculation for the assessment of bioequivalence under higher-order crossover designs, including Balaam's design, two-sequence dual design, and four-period optimal design with or without log-transformation, can be found in Chen, Li, and Chow (1997). For the assessment of bioequivalence, the FDA requires that a log-transformation of the PK responses be performed before analysis.

In this section, we focus on $2 \times 2m$ replicated crossover designs. When $m = 1$, it reduces to the standard two-sequence, two-period crossover design. The standard 2×2 crossover design suffers the following disadvantages: (i) it does not allow independent estimates of the intrasubject variabilities because each subject only receives each treatment once and (ii) the effects of sequence, period, and carryover are confounded and cannot be separated under the study design. On the other hand, the $2 \times 2m$ ($m \geq 2$) replicated crossover design not only provides independent estimates of the intrasubject variabilities, but also allows separate tests of the sequence, period, and carryover effects under appropriate statistical assumption.

3.4 Multiple-Sample One-Way ANOVA

Let x_{ij} be the jth subject from the ith treatment group, $i = 1, ..., k, j = 1, ..., n$. Consider the following one-way ANOVA model:

$$x_{ij} = \mu_i + \varepsilon_{ij},$$

where μ_i is the fixed effect of the ith treatment and ε_{ij} is a random error in observing x_{ij}. It is assumed that ε_{ij} are i.i.d. normal random variables with mean 0 and variance σ^2. Let

$$\text{SSE} = \sum_{i=1}^{k} \sum_{j=1}^{n} (x_{ij} - \bar{x}_{i\cdot})^2$$

$$\text{SSA} = \sum_{i=1}^{k} (\bar{x}_{i\cdot} - \bar{x}_{\cdot\cdot})^2,$$

where

$$\bar{x}_{i\cdot} = \frac{1}{n} \sum_{j=1}^{n} x_{ij} \quad \text{and} \quad \bar{x}_{\cdot\cdot} = \frac{1}{k} \sum_{i=1}^{k} \bar{x}_{i\cdot}.$$

Then, σ^2 can be estimated by

$$\hat{\sigma}^2 = \frac{\text{SSE}}{k(n-1)}. \tag{3.19}$$

3.4.1 Pairwise Comparison

In practice, it is often of interest to compare means among treatments under study. Thus, the hypotheses of interest are

$$H_0 : \mu_i = \mu_j \quad \text{versus} \quad H_a : \mu_i \neq \mu_j$$

for some pairs (i, j). Under the above hypotheses, there are $k(k-1)/2$ possible comparisons. For example, if there are four treatments in the study, then we can also have a maximum of six pairwise comparisons. In practice, it is well recognized that multiple comparison will inflate the type I error. As a result, it is suggested that an adjustment be made for controlling the overall type I error rate at the desired significance level. Assume that there are τ comparisons of interest, where $\tau \leq k(k-1)/2$. We reject the null hypothesis H_0 at the α level of significance if

$$\left| \frac{\sqrt{n}(\bar{x}_{i.} - \bar{x}_{j.})}{\sqrt{2}\hat{\sigma}} \right| > t_{\alpha/(2\tau), k(n-1)}.$$

The power of this test is given by

$$1 - T_{k(n-1)}\left(t_{\alpha/(2\tau),k(n-1)} \left| \frac{\sqrt{n}\varepsilon_{ij}}{\sqrt{2}\sigma} \right. \right) + T_{k(n-1)}\left(-t_{\alpha/(2\tau),k(n-1)} \left| \frac{\sqrt{n}\varepsilon_{ij}}{\sqrt{2}\sigma} \right. \right)$$

$$\approx 1 - T_{k(n-1)}\left(t_{\alpha/(2\tau),k(n-1)} \left| \frac{\sqrt{n}\,|\varepsilon_{ij}|}{\sqrt{2}\sigma} \right. \right),$$

where $\varepsilon_{ij} = \mu_i - \mu_j$. As a result, the sample size needed to achieve power $1 - \beta$ for detecting a clinically meaningful difference between μ_i and μ_j is

$$n = \max\{n_{ij}, \text{for all interested comparison}\}, \tag{3.20}$$

where n_{ij} is obtained by solving

$$T_{k(n_{ij}-1)}\left(t_{\alpha/(2\tau),k(n_{ij}-1)} \left| \frac{\sqrt{n_{ij}}\,|\varepsilon_{ij}|}{\sqrt{2}\sigma} \right. \right) = \beta.$$

When the sample size is sufficiently large, approximately

$$n_{ij} = \frac{2(z_{\alpha/(2\tau)} + z_\beta)^2 \sigma^2}{\varepsilon_{ij}^2}.$$

3.4.2 Simultaneous Comparison

The hypotheses of interest is

$$H_0 : \mu_1 = \mu_2 = \cdots = \mu_k \quad \text{versus} \quad H_a : \mu_i \neq \mu_j \text{ for some } 1 \leq i < j \leq k.$$

The null hypothesis H_0 is rejected at the α level of significance if

$$F_A = \frac{n\,\mathrm{SSA}/(k-1)}{\mathrm{SSE}/[k(n-1)]} > F_{\alpha,k-1,k(n-1)},$$

where $F_{\alpha,k-1,k(n-1)}$ is the α upper quantile of the F-distribution with $k-1$ and $k(n-1)$ degrees of freedom. Under the alternative hypothesis, the power of this test is given by

$$P(F_A > F_{\alpha,k-1,k(n-1)}) \approx P(n\mathrm{SSA} > \sigma^2 \chi^2_{\alpha,k-1}),$$

where $x^2_{\alpha,k-1}$ is the αth upper quantile for a χ^2 distribution with $k-1$ degrees of freedom and the approximation follows from the fact that $\mathrm{SSE}/[k(n-1)]$ is approximately σ^2 and $x^2_{\alpha,k-1} \approx (k-1)F_{\alpha,k-1,k(n-1)}$ when $k(n-1)$ is large. Under the alternative hypothesis, $n\mathrm{SSA}/\sigma^2$ is distributed as a noncentral χ^2 distribution with degrees of freedom $k-1$ and noncentrality parameter $\lambda = n\Delta$, where

$$\Delta = \frac{1}{\sigma^2} \sum_{i=1}^{k} (\mu_i - \bar{\mu})^2, \quad \bar{\mu} = \frac{1}{k} \sum_{j=1}^{k} \mu_j.$$

Hence, the sample size needed to achieve power $1 - \beta$ can be obtained by solving

$$x^2_{k-1}(\chi^2_{\alpha,k-1} \mid \lambda) = \beta,$$

where $x^2_{k-1}(\cdot \mid \lambda)$ is the cumulative distribution function of the noncentral χ^2 distribution with degrees of freedom $k-1$ and noncentrality parameter λ. Some values of λ needed to achieve different power (80% and 90%) with different significance level (1% and 5%) for different number of treatment groups are listed in Table 3.3. Once an initial value Δ is given and λ is obtained from Table 3.3, the required sample size is $n = \lambda/\Delta$.

3.4.3 An Example

To illustrate the use of Table 3.3 for sample size determination when comparing more than two treatments, consider the following example. Suppose that we are interested in

TABLE 3.3

λ Values Satisfying $\chi^2_{k-1}(\chi^2_{\alpha,k-1}|\lambda) = \beta$

k	1 − β = 0.80		1 − β = 0.90	
	α = 0.01	α = 0.05	α = 0.01	α = 0.05
2	11.68	7.85	14.88	10.51
3	13.89	9.64	17.43	12.66
4	15.46	10.91	19.25	14.18
5	16.75	11.94	20.74	15.41
6	17.87	12.83	22.03	16.47
7	18.88	13.63	23.19	17.42
8	19.79	14.36	24.24	18.29
9	20.64	15.03	25.22	19.09
10	21.43	15.65	26.13	19.83
11	22.18	16.25	26.99	20.54
12	22.89	16.81	27.80	21.20
13	23.57	17.34	28.58	21.84
14	24.22	17.85	29.32	22.44
15	24.84	18.34	30.04	23.03
16	25.44	18.82	30.73	23.59
17	26.02	19.27	31.39	24.13
18	26.58	19.71	32.04	24.65
19	27.12	20.14	32.66	25.16
20	27.65	20.56	33.27	25.66

conducting a four-arm ($k = 4$) parallel group, double-blind, randomized clinical trial to compare four treatments. The comparison will be made with a significance level of $\alpha = 0.05$. Assume that the SD within each group is $\sigma = 3.5$ and that the true mean responses for the four treatment groups are given by

$$\mu_1 = 8.25, \quad \mu_2 = 11.75, \quad \mu_3 = 12.00, \quad \text{and} \quad \mu_4 = 13.00.$$

Then, $\Delta = 1.05$. From Table 3.3, for a four-group parallel design ($k = 4$), the noncentrality parameter λ needed to achieve a power of 80% ($\beta = 0.20$) at 5% level of significance is 10.91. As a result, the sample size per treatment group can be obtained as

$$n = \frac{10.91}{1.05} \approx 11.$$

3.4.4 Remarks

In practice, a question concerning when pairwise comparisons or a simultaneous comparison should be used often arises. To address this question, consider the following example. Suppose a sponsor is investigating a pharmaceutical compound for the treatment of patients with cancer. The investigator is not only interested in showing the efficacy of the test drug but also in establishing dose-response curve. To achieve this study objective, a four-group parallel trial is designed with four treatments: P (Placebo), A (10 mg), B (20 mg), and C (30 mg). Let μ_P, μ_A, μ_B, and μ_C represent the true mean of the clinical response under the four treatments, respectively. Since the primary objective of the trial is to demonstrate the efficacy of the test drug, the following hypotheses for pairwise comparison with the placebo are useful for the demonstration of the efficacy of the test drug.

$$H_0 : \mu_P = \mu_A \quad \text{versus} \quad H_a : \mu_P \neq \mu_A$$

$$H_0 : \mu_P = \mu_B \quad \text{versus} \quad H_a : \mu_P \neq \mu_B$$

$$H_0 : \mu_P = \mu_C \quad \text{versus} \quad H_a : \mu_P \neq \mu_C.$$

On the other hand, the following hypotheses for simultaneous comparison among doses is usually considered for studying dose-response:

$$H_0 : \mu_A = \mu_B = \mu_C \quad \text{versus} \quad H_a : \text{not } H_0.$$

Note that in practice, it is often of interest to test the null hypothesis of no treatment difference against an ordered alternative hypothesis, for example, $H_a: \mu_A < \mu_B < \mu_C$. In this case, some robust contrast-based trend tests can be used for sample size calculation.

3.5 Multiple-Sample Williams Design

In clinical research, crossover design is attractive because each subject serves as his/her control. In addition, it removes the intersubject variability from comparison under appropriate statistical assumption. For example, the FDA identifies crossover design as the design of choice for bioequivalence trials. As a result, a two-sequence, two-period crossover design comparing two treatments is often considered in clinical research. In practice, it is often of interest to compare more than two treatments under a crossover design. When there are more than two treatments, it is desirable to compare pairwise treatment effects with the same degrees of freedom. Hence, it is suggested that Williams design be considered. Under a Williams design, the following model is assumed:

$$y_{ijl} = P_{j'} + \gamma_i + \mu_l + e_{ijl}, \quad i, l = 1, ..., k, \ j = 1, ..., n,$$

where y_{ijl} is the response from the jth subject in the ith sequence under the lth treatment, $P_{j'}$ is the fixed effect for the j' period, j' is the number of the period for the ith sequence's lth treatment, $\sum_{j=1}^{a} P_j = 0$, γ_i is the fixed sequence effect, μ_l is the fixed treatment effect, and e_{ijl} is a normal random variable with mean 0 and variance σ_{il}^2. For fixed i and l, $e_{ijl}, j = 1, ..., n$ are independent and identically distributed. For fixed i and j, $e_{ijl}, l = 1, ..., a$ are usually correlated because they all come from the same subject.

In bioequivalence trials, Williams designs comparing three treatments (a 6×3 crossover design) or four treatments (a 4×4 crossover design) are commonly employed. The construction of a Williams design can be found in Jones and Kenward (1989) and Chow and Liu (1992, 2000). Note that if k is an odd integer, a Williams design results in a $2k \times k$

crossover design. On the other hand, if k is an even integer, a Williams design reduces to a $k \times k$ crossover design.

It should be noted that the sequence-by-period interaction is not included in the above model. This is because the responses from a given sequence's given treatment are all from the same period. Therefore, the fixed effect of the sequence-by-period interaction cannot be separated from the treatment effect without appropriate statistical assumption.

Without loss of generality, assume we want to compare treatments 1 and 2. Let

$$d_{ij} = y_{ij1} - y_{ij2}.$$

Then, the true mean difference between treatment 1 and 2 can be estimated by

$$\hat{\varepsilon} = \frac{1}{kn} \sum_{i=1}^{k} \sum_{j=1}^{n} d_{ij},$$

which is normally distributed with mean $\varepsilon = \mu_1 - \mu_2$ and variance $\sigma_d^2 / (kn)$, where σ_d^2 is defined to be the variance of d_{ij} and can be estimated by

$$\hat{\sigma}_d^2 = \frac{1}{k(n-1)} \sum_{i=1}^{k} \sum_{j=1}^{n} \left(d_{ij} - \frac{1}{n} \sum_{j'=1}^{n} d_{ij'} \right)^2.$$

3.5.1 Test for Equality

The objective is to test

$$H_0 : \varepsilon = 0 \quad \text{versus} \quad H_a : \varepsilon \neq 0.$$

The null hypothesis H_0 is rejected at α level of significance if

$$\left| \frac{\hat{\varepsilon}}{\hat{\sigma}_d / \sqrt{kn}} \right| > t_{\alpha/2, k(n-1)}.$$

Under the alternative hypothesis that $\varepsilon \neq 0$, the power of this test is approximately

$$1 - T_{k(n-1)} \left(t_{\alpha/2, k(n-1)} \left| \frac{\sqrt{kn}\varepsilon}{\sigma_d} \right| \right).$$

The sample size needed to achieve power $1 - \beta$ can be obtained by setting the power to $1 - \beta$. When n is sufficiently large, approximation (3.5) leads to

$$n = \frac{(z_{\alpha/2} + z_\beta)^2 \sigma_d^2}{k\varepsilon^2}. \tag{3.21}$$

3.5.2 Test for Noninferiority/Superiority

The problem of testing superiority and noninferiority can be unified by the following hypotheses:

$$H_0 : \varepsilon \leq \delta \quad \text{versus} \quad H_a : \varepsilon > \delta,$$

where δ is the superiority or noninferiority margin. When $\delta > 0$, the rejection of the null hypothesis indicates the superiority of the test drug over the control. When $\delta < 0$, the rejection of the null hypothesis indicates the noninferiority of the test drug against the control. The null hypothesis H_0 is rejected at α level of significance if

$$\frac{\hat{\varepsilon} - \delta}{\hat{\sigma}_d / \sqrt{kn}} > t_{\alpha, k(n-1)}.$$

Under the alternative hypothesis that $\varepsilon > \delta$, the power of this test is given by

$$1 - T_{k(n-1)} \left(t_{\alpha, k(n-1)} \left| \frac{\varepsilon - \delta}{\sigma_d / \sqrt{kn}} \right. \right).$$

As a result, the sample size needed to achieve power $1 - \beta$ can be obtained by setting the power to $1 - \beta$. When n is sufficiently large, approximation (3.5) leads to

$$n = \frac{(z_\alpha + z_\beta)^2 \sigma_d^2}{k(\varepsilon - \delta)^2}. \tag{3.22}$$

3.5.3 Test for Equivalence

The objective is to test the following hypotheses:

$$H_0 : |\varepsilon| \geq \delta \quad \text{versus} \quad H_a : |\varepsilon| < \delta.$$

The test drug is concluded equivalent to the control in average if the null hypothesis H_0 is rejected at significance level α, that is,

$$\frac{\sqrt{kn}(\hat{\varepsilon} - \delta)}{\hat{\sigma}_d} < -t_{\alpha, k(n-1)} \quad \text{and} \quad \frac{\sqrt{kn}(\hat{\varepsilon} - \delta)}{\hat{\sigma}_d} > t_{\alpha, k(n-1)}.$$

Under the alternative hypothesis that $|\varepsilon| < \delta$, the power of the above test is

$$1 - T_{k(n-1)} \left(t_{\alpha, k(n-1)} \left| \frac{\sqrt{kn}(\delta - \varepsilon)}{\sigma_d} \right. \right) - T_{k(n-1)} \left(t_{\alpha, k(n-1)} \left| \frac{\sqrt{kn}(\delta + \varepsilon)}{\sigma_d} \right. \right).$$

The sample size needed to achieve power $1 - \beta$ can be obtained by setting the power to $1 - \beta$. A conservative approximation to the required sample size can be obtained by solving

$$T_{k(n-1)}\left(t_{\alpha,k(n-1)} \left| \frac{\sqrt{kn}(\delta - |\varepsilon|)}{\sigma_d} \right| \right) = \frac{\beta}{2}.$$

When n is large, approximation (3.5) leads to

$$n = \frac{(z_\alpha + z_{\beta/2})^2 \sigma_d^2}{k(\delta - |\varepsilon|)^2}.$$

3.5.4 An Example

Consider a randomized, placebo-controlled, double-blind, three-way (three-sequence, three-period) crossover trial, which is known as a Williams 6×3 ($k = 3$) crossover trial comparing cardiovascular safety of three different treatments (A, B, and C). Based on the results from the pilot study, it is estimated that the SD is 0.10 (i.e., $\delta_d = 0.10$). Suppose the true mean for A, B, and C are given by 0.20, 0.15, and 0.25, respectively. At the 5% level of significance, the sample size needed for achieving a power of 80% to reject

$$H_0 : \mu_i = \mu_j \quad \text{versus} \quad H_a : \mu_i \neq \mu_j$$

can be obtained by

$$n_{AB} = \frac{(1.96 + 0.84)^2 0.10^2}{6(0.20 - 0.15)^2} \approx 6$$

$$n_{AC} = \frac{(1.96 + 0.84)^2 0.10^2}{6(0.20 - 0.25)^2} \approx 6$$

$$n_{BC} = \frac{(1.96 + 0.84)^2 0.10^2}{6(0.15 - 0.25)^2} \approx 2.$$

As a result, the sample size needed per sequence is given by

$$n = \max\{6, 6, 2\} = 6.$$

It should be noted that the sample size can also be obtained by using the noncentral t-distribution like before. However, since there are six sequences in this example, which alternates the degrees of freedom, both Tables 3.1 and 3.2 cannot be used.

3.6 Practical Issues

At the planning stage of a clinical trial, sample size calculation is necessarily performed based on an appropriate statistical test for the hypotheses that reflect the study objectives under a valid study design. In this section, some practical issues that are commonly encountered are discussed.

3.6.1 One-Sided versus Two-Sided Test

In this chapter, statistical tests used for sample size calculation under either a parallel design or a crossover design can be classified into either a one-sided test (i.e., test for non-inferiority and test for superiority) or a two-sided test (i.e., test for equality and test for equivalence). In clinical research, test for noninferiority or test for superiority are also known as one-sided equivalence test. As discussed in Chapter 1, it is very controversial to use a one-sided test or a two-sided test in clinical research. When switching from a two-sided test for therapeutic equivalence to a one-sided test for noninferiority under a parallel design with 1 to 1 allocation, sample size could be reduced substantially at a fixed α level of significance. Suppose that the true mean difference between two treatments is $\varepsilon = 0$. Based on Equations 3.12 and 3.13, the ratio of the sample sizes needed for noninferiority and therapeutic equivalence is given by

$$\frac{n_{\text{noninferiority}}}{n_{\text{equivalence}}} = \frac{(z_\alpha + z_\beta)^2}{(z_\alpha + z_{\beta/2})^2}.$$

Table 3.4 summarizes possible sample size reduction when switching from testing equivalence to testing noninferiority (one-sided equivalence). As it can be seen from Table 3.4, the sample size could be reduced by 27.8% when switching from testing equivalence to testing noninferiority at the $\alpha = 0.05$ level of significance but still maintain the same power of 80%.

3.6.2 Parallel Design versus Crossover Design

As indicated in the previous sections, sample size required for achieving a desired power under a crossover design may be less than that under a parallel design. Under a parallel design, treatment comparison is made based on both intersubject and intrasubject variabilities, whereas treatment comparison is made based on the intrasubject variability under a crossover design under appropriate statistical assumption. If both designs are equally *efficient* regardless of their relative merits and disadvantages, then the choice of the design should be based on an evaluation of the relative cost-effectiveness between the increase of an additional treatment period in a crossover design with respect to the increase of additional subjects in a parallel design.

TABLE 3.4

Sample Size Reduction from Testing Equivalence to Testing Noninferiority

α	β	Sample Size Reduction (%)
0.10	0.1	23.3
	0.2	31.4
0.05	0.1	20.9
	0.2	27.8
0.01	0.1	17.5
	0.2	22.9

TABLE 3.5

Sample Size Reduction from Parallel Design to Crossover Design

ρ	Sample Size Reduction (%)
0.0	0.00
0.1	0.05
0.2	0.10
0.3	0.15
0.4	0.20
0.5	0.25
0.6	0.30
0.7	0.35
0.8	0.40
0.9	0.45
1.0	0.50

Consider the sample size in testing equality or equivalence. The ratio of the sample size for a 2 × 2 crossover design ($m = 1$) over the sample size for a parallel design is given by

$$\frac{n_{\text{crossover}}}{n_{\text{parallel}}} = \frac{\sigma_{WT}^2 + \sigma_{WR}^2 + \sigma_D^2}{\sigma_{WR}^2 + \sigma_{WT}^2 + \sigma_{BR}^2 + \sigma_{BT}^2}.$$

Table 3.5 summarizes possible sample size reduction when switching from a parallel design to a crossover design under the assumption that $\sigma_{WT} = \sigma_{WR} = \sigma_{BR} = \sigma_{BR} = 1$. As it can be seen, the sample size could be reduced by 30% by switching a parallel design to a crossover design when $\rho = 0.6$.

3.6.3 Sensitivity Analysis

Sample size calculation is usually performed by using initial values of the difference in mean responses between treatment groups (i.e., ε), the SD (i.e., σ), and the clinically meaningful difference or a prespecified superiority/noninferiority margin or equivalence limit (i.e., δ). Any slight or moderate deviations from these initial values could result in a substantial change in the calculated sample sizes. Thus, it is suggested that a sensitivity analysis with respect to these initial values be performed. Sensitivity analysis provides useful information regarding what to expect if a deviation in any of the initial values shall occur. For example, consider a one-sample problem:

$$H_0 : \mu = \mu_0 \quad \text{versus} \quad H_a : \mu \neq \mu_0.$$

According to Equation 3.2, if the SD changes from σ to $c\sigma$ for some $c > 0$, the ratio of the sample sizes needed before and after the change is given by

$$\frac{n_{c\sigma}}{n_\sigma} = c^2,$$

TABLE 3.6

Sample Size Reduction When Variability Decreases

c	Sample Size Reduction (%)
1.0	0.00
0.9	0.19
0.8	0.36
0.7	0.51
0.6	0.64
0.5	0.75
0.4	0.84
0.3	0.91
0.2	0.96
0.1	0.99

which is independent of the choice of α and β. Table 3.6 summarizes possible sample size reduction when the SD changes from σ to $c\sigma$. People in practice may want to see how much the sample size would increase when the variability increases, which is equivalent to study how much sample size would be saved if the variability decreases. As a result, without loss of generality, we would assume $c < 1$.

From Table 3.6, when the SD decreases by 20% (i.e., $c = 0.8$), the sample size could be reduced by 36% when performing a test for equivalence at the $\alpha = 0.05$ level of significance but still maintain the same power of 80%.

4

Large Sample Tests for Proportions

In clinical research, primary clinical endpoints for the evaluation of the treatment effect of the compound under study could be discrete variables; for example, clinical response (e.g., complete response, partial response, and stable disease), survival in cancer trials, and the presence of adverse events in clinical trials. For the evaluation of the treatment effect based on discrete clinical endpoint, the proportions of events that have occurred between treatment groups are often compared. Under a given study design, statistical tests for specific hypotheses such as equality or equivalence/noninferiority can be carried out based on the large sample theory in a similar manner as continuous responses discussed in Chapter 3. In this chapter, our primary focus will be placed on comparing proportions between treatment groups with binary responses.

The remaining sections of this chapter are organized as follows. The next section discusses the general procedure of power analysis for sample size calculation for testing one-sample problem. Sections 4.2 and 4.3 summarize statistical procedures for sample size calculation for a two-sample problem under a parallel-group design and a crossover design, respectively. Sections 4.4 and 4.5 discuss statistical procedures for testing a multiple-sample problem under a parallel design and a crossover design (Williams design), respectively. Formulas for sample size calculation for comparing relative risks between treatment groups under a parallel design and a crossover design are given in Section 4.6 and 4.7, respectively. Section 4.8 provides some practical issues regarding sample size calculation for comparing proportions in clinical research.

4.1 One-Sample Design

Let x_i, $i = 1, \ldots, n$ be the binary response observed from the ith subject. In clinical research, x_i could be the indicator for the response of tumor in cancer trials, that is, $x_i = 1$ for responder (e.g., complete response plus partial response) or $x_i = 0$ for nonresponder. It is assumed that x_i's are i.i.d. with $P(x_i = 1) = p$, where p is the true response rate. Since p is unknown, it is usually estimated by

$$\hat{p} = \frac{1}{n}\sum_{i=1}^{n} x_i.$$

Also, let $\varepsilon = p - p_0$ be the difference between the true response rate of a test drug (p) and a reference value (p_0). Without loss of generality, consider $\varepsilon > 0$ ($\varepsilon < 0$) an indication of *improvement* (*worsening*) of the test drug as compared to the reference value. In practice, it is of interest to test for equality (i.e., $p = p_0$), noninferiority (i.e., $p - p_0$ is greater than or equal to a predetermined noninferiority margin), superiority (i.e., $p - p_0$ is greater than a predetermined superiority margin), and equivalence (i.e., the absolute difference between p and p_0 is within a difference of clinical importance). In what follows, formulas for sample size calculation for testing equality, noninferiority/superiority, and equivalence are derived. The formulas provide required sample sizes for achieving a desired power under the alternative hypothesis.

4.1.1 Test for Equality

To test whether there is a difference between the true response rate of the test drug and the reference value, the following hypotheses are usually considered:

$$H_0 : p = p_0 \quad \text{versus} \quad H_a : p \neq p_0.$$

or

$$H_0 : \varepsilon = 0 \quad \text{versus} \quad H_a : \varepsilon \neq 0.$$

Under the null hypothesis, the test statistic

$$\frac{\sqrt{n}\hat{\varepsilon}}{\sqrt{\hat{p}(1-\hat{p})}}, \tag{4.1}$$

where $\hat{\varepsilon} = \hat{p} - p_0$ approximately has a standard normal distribution for large n. Thus, we reject the null hypothesis at the α level of significance if

$$\left| \frac{\sqrt{n}\hat{\varepsilon}}{\sqrt{\hat{p}(1-\hat{p})}} \right| > z_{\alpha/2}.$$

Under the alternative hypothesis that $p = p_0 + \varepsilon$, where $\varepsilon \neq 0$, the power of the above test is approximately

$$\Phi\left(\frac{\sqrt{n}|\varepsilon|}{\sqrt{p(1-p)}} - z_{\alpha/2} \right).$$

As a result, the sample size needed for achieving a desired power of $1 - \beta$ can be obtained by solving the following equation:

$$\frac{\sqrt{n}|\varepsilon|}{\sqrt{p(1-p)}} - z_{\alpha/2} = z_{\beta}.$$

This leads to

$$n = \frac{(z_{\alpha/2} + z_{\beta})^2 p(1-p)}{\varepsilon^2}. \tag{4.2}$$

To use Equation 4.2, information regarding p is needed, which may be obtained through a pilot study or based on historical data. Note that $p(1-p)$ is a quadratic function symmetric about 0.5 on its domain (0, 1). Thus, using Equation 4.2 requires an upper bound on p and a lower bound on ε^2. For example, if we know that $p \leq \tilde{p}$, $1-\tilde{p}$, and $\varepsilon^2 \geq \tilde{\varepsilon}^2$, where $\tilde{p}$ is a known value between 0 and 0.5 and $\tilde{\varepsilon}^2$ is a known positive value, then $p(1-p) \leq \tilde{p}(1-\tilde{p})$ and a conservative n can be obtained by using Equation 4.2 with ε and p replaced by $\tilde{\varepsilon}$ and $\tilde{p}$, respectively.

4.1.2 Test for Noninferiority/Superiority

The problem of testing noninferiority and superiority can be unified by the following hypotheses:

$$H_0 : p - p_0 \leq \delta \quad \text{versus} \quad H_a : p - p_0 > \delta$$

or

$$H_0 : \varepsilon \leq \delta \quad \text{versus} \quad H_a : \varepsilon > \delta,$$

where δ is the noninferiority or superiority margin. When $\delta > 0$, the rejection of the null hypothesis indicates superiority over the reference value. When $\delta < 0$, the rejection of the null hypothesis implies noninferiority against the reference value.

When $p - p_0 = \delta$, the test statistic

$$\frac{\sqrt{n}(\hat{\varepsilon} - \delta)}{\sqrt{\hat{p}(1-\hat{p})}}$$

approximately has a standard normal distribution for large n. Thus, we reject the null hypothesis at the α level of significance if

$$\frac{\sqrt{n}(\hat{\varepsilon} - \delta)}{\sqrt{\hat{p}(1-\hat{p})}} > z_\alpha.$$

If $\varepsilon > \delta$, the power of the above test is given by

$$\Phi\left(\frac{\sqrt{n}(\varepsilon - \delta)}{\sqrt{p(1-p)}} - z_\alpha \right).$$

As a result, the sample size needed for achieving a power of $1 - \beta$ can be obtained by solving the following equation:

$$\frac{\sqrt{n}(\varepsilon - \delta)}{\sqrt{p(1-p)}} - z_\alpha = z_\beta.$$

This leads to

$$n = \frac{(z_\alpha + z_\beta)^2 p(1-p)}{(\varepsilon - \delta)^2}. \tag{4.3}$$

4.1.3 Test for Equivalence

To establish equivalence, the following hypotheses are usually considered:

$$H_0 : |p - p_0| > \delta \quad \text{versus} \quad H_a : |p - p_0| < \delta$$

or

$$H_0 : |\varepsilon| \geq \delta \quad \text{versus} \quad H_a : |\varepsilon| < \delta.$$

The proportion of the responses is concluded to be equivalent to the reference value of p_0 if the null hypothesis is rejected at a given significance level.

The above hypotheses can be tested using two one-sided test procedures as described in Chapter 3. The null hypothesis is rejected at approximately α level of significance if

$$\frac{\sqrt{n}(\hat{\varepsilon} - \delta)}{\sqrt{\hat{p}(1-\hat{p})}} < -z_\alpha \quad \text{and} \quad \frac{\sqrt{n}(\hat{\varepsilon} + \delta)}{\sqrt{\hat{p}(1-\hat{p})}} > z_\alpha.$$

When n is large, the power of this test is approximately

$$\Phi\left(\frac{\sqrt{n}(\delta - \varepsilon)}{\sqrt{p(1-p)}} - z_\alpha\right) + \Phi\left(\frac{\sqrt{n}(\delta + \varepsilon)}{\sqrt{p(1-p)}} - z_\alpha\right) - 1$$

$$\geq 2\Phi\left(\frac{\sqrt{n}(\delta - |\varepsilon|)}{\sqrt{p(1-p)}} - z_\alpha\right) - 1.$$

As a result, the sample size needed for achieving a power of $1 - \beta$ can be obtained by solving the following equations:

$$\frac{\sqrt{n}(\delta - |\varepsilon|)}{\sqrt{p(1-p)}} - z_\alpha = z_{\beta/2},$$

which leads to

$$n = \frac{(z_\alpha + z_{\beta/2})^2 p(1-p)}{(\delta - |\varepsilon|)^2}. \tag{4.4}$$

4.1.4 An Example

To illustrate the use of sample size formulas, consider the same example concerning a study of osteoporosis in postmenopausal women as described in Section 3.1.4. Suppose in addition to the study of the change in bone density posttreatment, it is also of interest to evaluate the treatment effect in terms of the response rate at the end of the study. Sample size calculation can then be carried out based on the response rate for achieving a desired power. The definition of a responder, however, should be given in the study protocol prospectively. For example, a subject may be defined as a responder if there is an improvement in bone density by more than one SD or 30% of the measurements of bone density.

4.1.4.1 Test for Equality

Suppose that the response rate of the patient population under study after treatment is expected to be around 50% (i.e., $p = 0.50$). By Equation 4.2, at $\alpha = 0.05$, the required sample size for having an 80% power (i.e., $\beta = 0.2$) for correctly detecting a difference between the posttreatment response rate and the reference value of 30% (i.e., $p_0 = 0.30$) is

$$n = \frac{(z_{\alpha/2} + z_\beta)^2 p(1-p)}{(p - p_0)^2} = \frac{(1.96 + 0.84)^2 0.5(1 - 0.5)}{(0.5 - 0.3)^2} = 49.$$

4.1.4.2 Test for Noninferiority

For the prevention of progression from osteopenia to osteoporosis, (we wish to show that the change in bone density after treatment of the majority of patients is at least as good as the reference value (30%) ($p_0 = 30\%$).) Also assume that a difference of 10% in the responder rate is considered to be of no clinical significance ($\delta = -10\%$). Assume the true response rate is 50% ($p = 50\%$). According to Equation 4.3, at $\alpha = 0.05$, the required sample size for having an 80% power (i.e., $\beta = 0.2$) is

$$n = \frac{(z_\alpha + z_\beta)^2 p(1-p)}{(p - p_0 - \delta)^2} = \frac{(1.64 + 0.84)^2 0.5(1 - 0.5)}{(0.5 - 0.3 + 0.1)^2} = 18.$$

4.1.4.3 Test for Equivalence

Assume that one brand name drug for osteoporosis on the market has a responder rate of 60% (i.e., $p_0 = 0.60$). It is believed that a 20% difference in the responder rate is of no clinical significance (i.e., $\delta = 0.2$). Hence, the investigator wants to show that the study drug is equivalent to the market drug in terms of the responder rate. By Equation 4.4, at $\alpha = 0.05$, assuming that the true response rate is 60% (i.e., $p = 0.60$), the sample size required for achieving an 80% power is

$$n = \frac{(z_\alpha + z_{\beta/2})^2 p(1-p)}{(\delta - |p - p_0|)^2} = \frac{(1.64 + 1.28)^2 \times 0.6(1 - 0.6)}{(0.2 - |0.6 - 0.6|)^2} \approx 52.$$

4.1.5 Remarks

For a one-sample test for equality, there exists another approach, which is very similar to Equation 4.1 but not exactly the same. This approach will reject the null hypothesis that $\varepsilon = 0$ if

$$\left| \frac{\sqrt{n}\hat{\varepsilon}}{\sqrt{p_0(1 - p_0)}} \right| > z_{\alpha/2}. \tag{4.5}$$

Since Equation 4.1 estimates the variance of $\sqrt{n}\hat{\varepsilon}$ without any constraints, we refer to Equation 4.1 as the unconditional method. On the other hand, since Equation 4.5 estimates the variance of $\sqrt{n}\hat{\varepsilon}$ conditional on the null hypothesis, we refer to Equation 4.5 as the conditional method. Note that both Equations 4.1 and 4.5 have asymptotic size α when n is sufficiently large. Then, which one should be used is always a dilemma because one is not necessarily more powerful than the other. For the purpose of completeness, the sample size calculation formula based on Equation 4.5 is given below. The same idea can be applied to the testing problems of noninferiority/ superiority.

Under the alternative hypothesis ($\varepsilon \neq 0$), the power of the test defined by Equation 4.5 is approximately

$$\Phi\left(\frac{\sqrt{n}\,|\varepsilon| - z_{\alpha/2}\sqrt{p_0(1-p_0)}}{\sqrt{p(1-p)}}\right).$$

As a result, the sample size needed for achieving a desired power of $1 - \beta$ can be obtained by solving the following equation:

$$\frac{\sqrt{n}\,|\varepsilon| - z_{\alpha/2}\sqrt{p_0(1-p_0)}}{\sqrt{p(1-p)}} = z_\beta.$$

This leads to

$$n = \frac{[z_{\alpha/2}\sqrt{p_0(1-p_0)} + z_\beta\sqrt{p(1-p)}]^2}{\varepsilon^2}.$$

4.2 Two-Sample Parallel Design

Let x_{ij} be a binary response from the jth subject in the ith treatment group, $j = 1, \ldots, n_i$, $i = 1, 2$. For a fixed i, it is assumed that x_{ij}'s are i.i.d. with $P(x_{ij} = 1) = p_i$. In practice, p_i is usually estimated by the observed proportion in the ith treatment group:

$$\hat{p}_i = \frac{1}{n_i}\sum_{j=1}^{n_i} x_{ij}.$$

Let $\varepsilon = p_1 - p_2$ be the difference between the true mean response rates of a test drug (p_1) and a control (p_2). Without loss of generality, consider $\varepsilon > 0$ ($\varepsilon < 0$) an indication of *improvement* (*worsening*) of the test drug as compared to the control value. In what follows, formulas for sample size calculation to achieve a desired power under the alternative hypothesis are derived for testing equality, noninferiority/superiority, and equivalence.

4.2.1 Test for Equality

To test whether there is a difference between the mean response rates of the test drug and the reference drug, the following hypotheses are usually considered:

$$H_0 : \varepsilon = 0 \quad \text{versus} \quad H_a : \varepsilon \neq 0.$$

We reject the null hypothesis at the α level of significance if

$$\left| \frac{\hat{p}_1 - \hat{p}_2}{\sqrt{\hat{p}_1(1 - \hat{p}_1)/n_1 + \hat{p}_2(1 - \hat{p}_2)/n_2}} \right| > z_{\alpha/2}. \tag{4.6}$$

Under the alternative hypothesis that $\varepsilon \neq 0$, the power of the above test is approximately

$$\Phi \left(\frac{|\varepsilon|}{\sqrt{p_1(1 - p_1)/n_1 + p_2(1 - p_2)/n_2}} - z_{\alpha/2} \right).$$

As a result, the sample size needed for achieving a power of $1 - \beta$ can be obtained by the following equation:

$$\frac{|\varepsilon|}{\sqrt{p_1(1 - p_1)/n_1 + p_2(1 - p_2)/n_2}} - z_{\alpha/2} = z_{\beta}.$$

This leads to

$$n_1 = \kappa n_2$$
$$n_2 = \frac{(z_{\alpha/2} + z_{\beta})^2}{\varepsilon^2} \left[\frac{p_1(1 - p_1)}{k} + p_2(1 - p_2) \right]. \tag{4.7}$$

4.2.2 Test for Noninferiority/Superiority

The problem of testing noninferiority and superiority can be unified by the following hypotheses:

$$H_0 : \varepsilon \leq \delta \quad \text{versus} \quad H_a : \varepsilon > \delta,$$

where δ is the superiority or noninferiority margin. When $\delta > 0$, the rejection of the null hypothesis indicates the superiority of the test drug over the control. When $\delta < 0$, the rejection of the null hypothesis indicates the noninferiority of the test drug against the control. We reject the null hypothesis at the α level of significance if

$$\left| \frac{\hat{p}_1 - \hat{p}_2 - \delta}{\sqrt{\hat{p}_1(1 - \hat{p}_1)/n_1 + \hat{p}_2(1 - \hat{p}_2)/n_2}} \right| > z_{\alpha}.$$

Under the alternative hypothesis that $\varepsilon > \delta$, the power of the above test is approximately

$$\Phi\left(\frac{\varepsilon - \delta}{\sqrt{p_1(1-p_1)/n_1 + p_2(1-p_2)/n_2}} - z_\alpha\right).$$

As a result, the sample size needed for achieving a power of $1 - \beta$ can be obtained by solving

$$\frac{\varepsilon - \delta}{\sqrt{p_1(1-p_1)/n_1 + p_2(1-p_2)/n_2}} - z_\alpha = z_\beta,$$

which leads to

$$n_1 = kn_2$$
$$n_2 = \frac{(z_\alpha + z_\beta)^2}{(\varepsilon - \delta)^2}\left[\frac{p_1(1-p_1)}{\kappa} + p_2(1-p_2)\right]. \qquad (4.8)$$

4.2.3 Test for Equivalence

The objective is to test the following hypotheses:

$$H_0: |\varepsilon| \geq \delta \quad \text{versus} \quad H_a : |\varepsilon| < \delta.$$

The null hypothesis is rejected and the test drug is concluded to be equivalent to the control if

$$\frac{\hat{p}_1 - \hat{p}_2 - \delta}{\sqrt{\hat{p}_1(1-\hat{p}_1)/n_1 + \hat{p}_2(1-\hat{p}_2)/n_2}} < -z_\alpha$$

and

$$\frac{\hat{p}_1 - \hat{p}_2 - \delta}{\sqrt{\hat{p}_1(1-\hat{p}_1)/n_1 + \hat{p}_2(1-\hat{p}_2)/n_2}} > z_\alpha.$$

Under the alternative hypothesis that $|\varepsilon| < \delta$, the power of the above test is approximately

$$2\Phi\left(\frac{\delta - |\varepsilon|}{\sqrt{p_1(1-p_1)/n_1 + p_2(1-p_2)/n_2}} - z_\alpha\right) - 1.$$

As a result, the sample size needed for achieving a power of $1 - \beta$ can be obtained by solving the following equation:

$$\frac{\delta - |\varepsilon|}{\sqrt{p_1(1-p_1)/n_1 + p_2(1-p_2)/n_2}} - z_\alpha = z_{\beta/2},$$

which leads to

$$n_1 = \kappa n_2$$

$$n_2 = \frac{(z_\alpha + z_{\beta/2})^2}{(\delta - |\varepsilon|)^2} \left[\frac{p_1(1-p_1)}{\kappa} + p_2(1-p_2) \right]. \tag{4.9}$$

4.2.4 An Example

Consider the following example concerning the evaluation of anti-infective agents in the treatment of patients with skin and skin structure infections. As it is well known, Gram-positive and Gram-negative pathogens are commonly associated with skin and skin structure infections such as streptococci, staphylococci, and various strains of enterobacteriaceae. For the evaluation of the effectiveness of a test antibiotic agent, clinical assessments and cultures are usually done at posttreatment visits (e.g., between four and eight days) after treatment has been completed but prior to treatment with another antimicrobial agent. If the culture is positive, the pathogen(s) is usually identified and susceptibility testing is performed. The effectiveness of therapy is usually assessed based on clinical and bacteriological responses at posttreatment visit. For example, clinical responses may include cure (e.g., no signs of skin infection at posttreatment visits), improvement (e.g., the skin infection has resolved to the extent that no further systemic antibiotic therapy is needed based on the best judgment of the investigator), and failure (e.g., lack of significant improvement in the signs and symptoms of the skin infection at or before posttreatment visits such that a change in antibiotic treatment is required). On the other hand, bacteriological responses may include cure (e.g., all pathogens eradicated at posttreatment day 4–8 or material suitable for culturing has diminished to a degree that proper cultures cannot be obtained), colonization (e.g., isolation of pathogen(s) from the original site of infection in the absence of local or systemic signs of infection at posttreatment visits), and failure (e.g., any pathogen(s) isolated at posttreatment visits coupled with the investigator's decision to prescribe alternate antibiotic therapy).

Suppose that a pharmaceutical company is interested in conducting a clinical trial to compare the efficacy, safety, and tolerability of two antimicrobial agents when administered orally once daily in the treatment of patients with skin and skin structure infections. In what follows, we will consider the situations where the intended trial is for (i) testing equality of mean cure rates, (ii) testing noninferiority or superiority of the test drug as compared to the active control agent, and (iii) testing for therapeutic equivalence. For this purpose, the following assumptions are made. First, sample size calculation will be performed for achieving an 80% power (i.e., $\beta = 0.2$) at the 5% level of significance (i.e., $\alpha = 0.05$).

4.2.4.1 Test for Equality

In this example, suppose that a difference of $\varepsilon = 20\%$ in the clinical response of cure is considered to be a clinically meaningful difference between the two antimicrobial agents. By Equation 4.7, assuming that the true cure rate for the active control agent is 65% ($p_1 = 0.80$ and $p_2 = p_1 + \varepsilon = 0.85$), respectively, the sample size with $\kappa = 1$ (equal allocation) can be determined by

$$n_1 = n_2 = \frac{(z_{\alpha/2} + z_\beta)^2 (p_1(1-p_1) + p_2(1-p_2))}{\varepsilon^2}$$

$$= \frac{(1.96 + 0.84)^2 (0.65(1-0.65) + 0.85(1-0.85))}{0.2^2}$$

$$\approx 70.$$

4.2.4.2 Test for Noninferiority

Now, suppose it is of interest to establish noninferiority of the test drug as compared to the active control agent. We consider the difference of less than 10% to be of no clinical importance. Thus, the noninferiority margin is chosen to be 10% (i.e., $\delta = -0.10$). Also, suppose the true mean cure rates of the treatment agents and the active control are 85% and 65%, respectively. Then, by Equation 4.8, the sample size with $\kappa = 1$ (equal allocation) can be determined by

$$n_1 = n_2 = \frac{(z_\alpha + z_\beta)^2 (p_1(1-p_1) + p_2(1-p_2))}{(\varepsilon - \delta)^2}$$

$$= \frac{(1.64 + 0.84)^2 (0.65(1-0.65) + 0.85(1-0.85))}{(0.20 + 0.10)^2}$$

$$\approx 25.$$

4.2.4.3 Test for Superiority

On the other hand, the pharmaceutical company may want to show the superiority of the test drug over the active control agent. Assume the superiority margin is 5% ($\delta = 0.05$). According to Equation 4.8, the sample size with $\kappa = 1$ (equal allocation) can be determined by

$$n_1 = n_2 = \frac{(z_\alpha + z_\beta)^2 (p_1(1-p_1) + p_2(1-p_2))}{(\varepsilon - \delta)^2}$$

$$= \frac{(1.64 + 0.84)^2 (0.65(1-0.65) + 0.85(1-0.85))}{(0.20 + 0.05)^2}$$

$$\approx 98.$$

As it can be seen, testing superiority usually requires larger sample size than testing noninferiority and equality.

4.2.4.4 Test for Equivalence

For the establishment of equivalence, suppose the true cure rate for the two agents are 75% ($p_1 = 0.75$) and 80% ($p_2 = 0.80$) and the equivalence limit is 20% (i.e., $\delta = 0.20$). According to Equation 4.9, the sample size with $\kappa = 1$ (equal allocation) can be determined by

$$
\begin{aligned}
n_1 = n_2 &= \frac{(z_\alpha + z_{\beta/2})^2(p_1(1-p_1) + p_2(1-p_2))}{(\delta - |\varepsilon|)^2} \\
&= \frac{(1.64 + 1.28)^2(0.75(1-0.75) + 0.80(1-0.80))}{(0.20 + 0.05)^2} \\
&\approx 132.
\end{aligned}
$$

4.2.5 Remarks

For a two-sample test for equality, there exists another approach, which is very similar to Equation 4.6 but not exactly the same. This approach will reject the null hypothesis that $\varepsilon = 0$ if

$$
\frac{\hat{p}_1 - \hat{p}_2}{\sqrt{(\frac{1}{n_1} + \frac{1}{n_2})\hat{p}(1-\hat{p})}}, \tag{4.10}
$$

where

$$
\hat{p} = \frac{n_1\hat{p}_1 + n_2\hat{p}_2}{n_1 + n_2}.
$$

Note that the difference between Equations 4.6 and 4.10 is the following. In Equation 4.6, the variance of $\hat{p}_1 - \hat{p}_2$ is estimated by maximum likelihood estimate (MLE) without any constraint, which is given by $\hat{p}_1(1-\hat{p}_1)/n_1 + \hat{p}_2(1-\hat{p}_2)/n_2$. On the other side, in Equation 4.10, the same quantity is estimated by MLE under the null hypothesis ($p_1 = p_2$), which gives $(1/n_1 + 1/n_2)\hat{p}(1-\hat{p})$. We will refer to Equation 4.6 as an unconditional approach and Equation 4.10 as a conditional approach. As to which test (conditional/unconditional) to be used is always a problem because one is not necessarily always more powerful than the other. However, a drawback of the conditional approach is that it is difficult to be generalized to other testing problems, for example, superiority and noninferiority/equivalence. Let

$$
p = \frac{n_1 p_1 + n_2 p_2}{n_1 + n_2}.
$$

When $n = n_1 = n_2$, which is a very important special case, it can be shown that

$$
\begin{aligned}
\left(\frac{1}{n_1} + \frac{1}{n_2}\right)\hat{p}(1-\hat{p}) &\approx \left(\frac{1}{n_1} + \frac{1}{n_2}\right)p(1-p) \\
&\geq \frac{p_1(1-p_1)}{n_1} + \frac{p_2(1-p_2)}{n_2} \\
&\approx \frac{\hat{p}_1(1-\hat{p}_1)}{n_1} + \frac{\hat{p}_2(1-\hat{p}_2)}{n_2},
\end{aligned}
$$

which implies that under the alternative hypothesis, the unconditional approach has more power than the conditional method. As a result, in this section and also the following section, we will focus on the unconditional method because it provides a unified approach for all the testing problems mentioned above.

Nevertheless, for the purpose of completeness, the conditional approach for a two-sample test of equality is also presented below. Under the alternative hypothesis that $\varepsilon \neq 0$ and $n_1 = \kappa n_2$, the power of Equation 4.10 is approximately

$$\Phi\left(\frac{|\varepsilon|}{\sqrt{(p_1(1-p_1)/n_1 + p_2(1-p_2)/n_2)}}\right.$$
$$\left. - z_{\alpha/2}\frac{\sqrt{(1/n_1 + 1/n_2)p(1-p)}}{\sqrt{(p_1(1-p_1)/n_1 + p_2(1-p_2)/n_2)}}\right),$$

where $p = (p_1 + \kappa p_2)/(1 + \kappa)$. As a result, the sample size needed for achieving a power of $1 - \beta$ can be obtained by solving the following equation:

$$\frac{|\varepsilon|}{\sqrt{(p_1(1-p_1)/n_1 + p_2(1-p_2)/n_2)}}$$
$$- z_{\alpha/2}\frac{\sqrt{(1/n_1 + 1/n_2)p(1-p)}}{\sqrt{(p_1(1-p_1)/n_1 + p_2(1-p_2)/n_2)}} = z_\beta.$$

This leads to

$$n_1 = \kappa n_2$$
$$n_2 = \frac{1}{\varepsilon^2}[z_{\alpha/2}\sqrt{(1+1/\kappa)p(1-p)} + z_\beta\sqrt{p_1(1-p_1)/\kappa + p_2(1-p_2)}]^2.$$

4.3 Two-Sample Crossover Design

In this section, we consider a $2 \times 2m$ replicated crossover design comparing mean response rates of a test drug and a reference drug. Let x_{ijkl} be the lth replicate of a binary response ($l = 1, ..., m$) observed from the jth subject ($j = 1, ..., n$) in the ith sequence ($i = 1, 2$) under the kth treatment ($k = 1, 2$). Assume that $(x_{ij11}, ..., x_{ij1m}, ..., x_{ijk1}, ..., x_{ijkm})$, $i = 1, 2, j = 1, ...,$ n are i.i.d. random vectors with each component's marginal distribution specified by $P(x_{ijkl} = 1) = p_k$. Note that the observations from the same subject can be correlated with each other. By specifying that $P(x_{ijkl} = 1) = p_k$, it implies that there are no sequence, period, and crossover effects. The statistical model incorporating those effects are more complicated for binary data compared with continuous data. Its detailed discussion is beyond the scope of this book.

Let $\varepsilon = p_2(\text{test}) - p_1(\text{reference})$,

$$\bar{x}_{ijk.} = \frac{1}{m}(x_{ijk1} + \cdots + x_{ijkm}) \quad \text{and} \quad d_{ij} = \bar{x}_{ij1.} - \bar{x}_{ij2.}.$$

An unbiased estimator of ε is given by

$$\hat{\varepsilon} = \frac{1}{2n}\sum_{i=1}^{a}\sum_{j=1}^{n}d_{ij}.$$

According to the central limit theorem, $\hat{\varepsilon}$ is asymptotically normally distributed as $N(0,\sigma_d^2)$, where $\sigma_d^2 = \mathrm{var}(d_{ij})$ and can be estimated by

$$\hat{\sigma}_d^2 = \frac{1}{2(n-1)}\sum_{i=1}^{a}\sum_{j=1}^{n}(d_{ij} - \bar{d}_{i.})^2,$$

where

$$\bar{d}_{i.} = \frac{1}{n}\sum_{j=1}^{n}d_{ij}.$$

Without loss of generality, consider $\varepsilon > 0$ ($\varepsilon < 0$) as an indication of *improvement* (*worsening*) of the test drug as compared to the reference drug.

4.3.1 Test for Equality

The objective is to test the following hypotheses:

$$H_0 : \varepsilon = 0 \quad \text{versus} \quad H_a : \varepsilon \neq 0.$$

Then, the null hypothesis will be rejected at α level of significance if

$$\left|\frac{\hat{\varepsilon}}{\hat{\sigma}_d/\sqrt{2n}}\right| > z_{a/2}.$$

Under the alternative hypothesis that $\varepsilon \neq 0$, the power of the above test is approximately

$$\Phi\left(\frac{\sqrt{2n}\varepsilon}{\sigma_d} - z_{\alpha/2}\right).$$

As a result, the sample size needed for achieving a power of $1 - \beta$ can be obtained by solving

$$\frac{\sqrt{2n}\,|\varepsilon|}{\sigma_d} - z_{\alpha/2} = t_{\beta}.$$

This leads to

$$n = \frac{(z_{\alpha/2} + z_\beta)^2 \sigma_d^2}{2\varepsilon^2}. \tag{4.11}$$

4.3.2 Test for Noninferiority/Superiority

Similar to the test for noninferiority/superiority under a parallel design, the problem can be unified by testing the following hypotheses:

$$H_0 : \varepsilon \leq \delta \quad \text{versus} \quad H_a : \varepsilon > \delta,$$

where δ is the noninferiority or superiority margin. When $\delta > 0$, the rejection of the null hypothesis indicates the superiority of the test drug against the control. When $\delta < 0$, the rejection of the null hypothesis indicates the noninferiority of the test drug over the control. The null hypothesis will be rejected at the α level of significance if

$$\frac{\hat{\varepsilon} - \delta}{\hat{\sigma}_d / \sqrt{2n}} > z_\alpha.$$

Under the alternative hypothesis that $\varepsilon > \delta$, the power of the above test is approximately

$$\Phi\left(\frac{\varepsilon - \delta}{\sigma_d / \sqrt{2n}} - z_{\alpha/2} \right).$$

As a result, the sample size needed for achieving a power of $1 - \beta$ can be obtained by solving

$$\frac{\varepsilon - \delta}{\sigma_d / \sqrt{2n}} - z_{\alpha/2} \geq z_\beta.$$

This leads to

$$n = \frac{(z_\alpha + z_\beta)^2 \sigma_d^2}{2(\varepsilon - \delta)^2}. \tag{4.12}$$

4.3.3 Test for Equivalence

The objective is to test the following hypotheses:

$$H_0 : |\varepsilon| \geq \delta \quad \text{versus} \quad H_a : |\varepsilon| < \delta.$$

The test drug will be concluded equivalent to the control on average if the null hypothesis is rejected at a given significance level. At the significance level of α, the null hypothesis will be rejected if

$$\frac{\sqrt{2n}(\hat{\varepsilon}-\delta)}{\hat{\sigma}_d} < -z_\alpha \quad \text{and} \quad \frac{\sqrt{2n}(\hat{\varepsilon}+\delta)}{\hat{\sigma}_d} > z_\alpha.$$

Under the alternative hypothesis that $|\varepsilon| < \delta$, the power of the above test is approximately

$$2\Phi\left(\frac{\sqrt{2n}(\delta-|\varepsilon|)}{\sigma_d} - z_\alpha\right) - 1.$$

As a result, the sample size needed for achieving a power of $1 - \beta$ can be obtained by solving

$$\frac{\sqrt{2n}(\delta-|\varepsilon|)}{\sigma_d} - z_\alpha \geq z_{\beta/2}.$$

This leads to

$$n \geq \frac{(z_\alpha + z_{\beta/2})^2 \sigma_d^2}{2(\delta-|\varepsilon|)^2}. \qquad (4.13)$$

4.3.4 An Example

Suppose a sponsor is interested in conducting an open label randomized crossover trial to compare an inhaled insulin formulation manufactured for commercial usage for patients with type 1 diabetes to the inhaled insulin formulation utilized in phase III clinical trials. Unlike subcutaneous injection, the efficiency and reproducibility of pulmonary insulin delivery is a concern. As a result, a replicated crossover consisting of two sequences of ABAB and BABA is recommended ($a = 2$, $m = 2$), where A is the inhaled insulin formulation for commercial usage and B is the inhaled insulin formulation utilized in phase III clinical trials. Qualified subjects are to be randomly assigned to receive one of the two sequences. In each sequence, subjects will receive single doses with a replicate of treatments A and B as specified in the sequence on days 1, 3, 5, and 7. In this trial, in addition to the comparison of pharmacokinetic parameters such as area under the blood concentration time curve and peak concentration (C_{max}), it is also of interest to compare the safety profiles between the two formulations in terms of the incidence rate of adverse events.

4.3.4.1 Test for Equality

Assuming $\sigma_d = 50\%$, according to Equation 4.11, the sample size needed to achieve 80% ($\beta = 0.2$) power in detecting 20% ($\varepsilon = 0.20$) difference in adverse events rate is given by

$$n = \frac{(z_{\alpha/2} + z_\beta)^2 \sigma_d^2}{2\varepsilon^2} = \frac{(1.96 + 0.84)^2 \times 0.5^2}{2 \times 0.2^2} = 24.5 \approx 25.$$

4.3.4.2 Test for Noninferiority

Assume $\sigma_d = 50\%$, no difference in the mean adverse event rates between the two treatments ($\varepsilon = 0$), and the noninferiority margin is $\delta = -20\%$. According to Equation 4.12, the sample size needed to achieve 80% ($\beta = 0.2$) power is given by

$$n = \frac{(z_\alpha + z_\beta)^2 \sigma_d^2}{2(\varepsilon - \delta)^2} = \frac{(1.64 + 0.84)^2 \times 0.5^2}{2 \times (0 - (-0.2))^2} = 19.2 \approx 20.$$

4.3.4.3 Test for Equivalence

Assume $\sigma_d = 50\%$, no difference in the mean adverse event rate between the two treatments ($\varepsilon = 0$), and the equivalence limit is 20% ($\delta = 0.2$). According to Equation 4.13, the sample size needed to achieve 80% ($\beta = 0.2$) is given by

$$n = \frac{(z_\alpha + z_{\beta/2})^2 \sigma_d^2}{2\delta^2} = \frac{(1.64 + 1.28)^2 \times 0.5^2}{0.2^2} = 26.6 \approx 27.$$

4.3.5 Remarks

For a crossover design, two ways exist to increase the power. One is to increase the number of subjects, that is, increase n. Another is to increase the number of replicates from each subject, that is, increase m. In practice, usually increasing m is more cost-effective compared to increasing n. The power of the test is mainly determined by the variability of $\hat{\varepsilon}$ under the alternative assumption. Heuristically, the variability of $\hat{\varepsilon}$ can be considered to consist of two parts, that is, intersubject and intrasubject variability components. From a statistical point of view, increasing n can decrease both intersubject and intrasubject components of $\hat{\varepsilon}$. As a result, as long as n is sufficiently large, the power can be arbitrarily close to 1. However, increasing the number of replicates (m) can only decrease the intrasubject variability component of $\hat{\varepsilon}$. When $m \to \infty$, the intrasubject variability will go to 0, but the intersubject variability still remains. Consequently, the power cannot be increased arbitrarily by increasing m.

In practice, if the intrasubject variability is relatively small compared with the intersubject variability, simply increasing the number of replicates may not provide sufficient power. In such a situation, the number of subjects should be sufficiently large to achieve the desired statistical power. On the other side, if the intrasubject variability is relatively large compared with the intersubject variability, it may be preferable to increase the number of replicates to achieve the desired power and retain a relatively low cost.

4.4 One-Way Analysis of Variance

Let x_{ij} be a binary response from the jth subject in the ith treatment group, $i = 1, \ldots, a, j = 1, \ldots, n$. Assume that $P(x_{ij} = 1) = p_i$. Define

$$\hat{p}_{i\cdot} = \frac{1}{n} \sum_{j=1}^{n} x_{ij}.$$

4.4.1 Pairwise Comparison

In practice, it is often of interest to compare proportions among treatments under study. Thus, the hypotheses of interest are

$$H_0 : \mu_i = \mu_j \quad \text{versus} \quad H_a : \mu_i \neq \mu_j, \quad \text{for some } i \neq j.$$

Under the above hypotheses, there are $a(a-1)/2$ possible comparisons. For example, if there are four treatments in the study, then we can have a maximum of six pairwise comparisons. In practice, it is well recognized that multiple comparison will inflate the type I error. As a result, it is suggested that an adjustment be made for controlling the overall type I error rate at the desired significance level. Assume that there are τ comparisons of interest, where $\tau \leq a(a-1)/2$. We reject the null hypothesis H_0 at the α level of significance if

$$\left| \frac{\sqrt{n}(\bar{p}_i - \bar{p}_j)}{\sqrt{\hat{p}_i(1-\hat{p}_i) + \hat{p}_j(1-\hat{p}_j)}} \right| > z_{\alpha/(2\tau)}.$$

The power of this test is approximately

$$\Phi \left(\frac{\sqrt{n}\,|\varepsilon_{ij}|}{\sqrt{p_i(1-p_i) + p_j(1-p_j)}} - z_{\alpha/(2\tau)} \right),$$

where $\varepsilon_{ij} = p_i - p_j$. As a result, the sample size needed for detecting a clinically meaningful difference between p_i and p_j can be obtained by solving

$$\frac{\sqrt{n}\,|\varepsilon_{ij}|}{\sqrt{p_i(1-p_i) + p_j(1-p_j)}} - z_{\alpha/(2\tau)} = z_\beta.$$

This leads to

$$n_{ij} = \frac{(z_{\alpha/(2\tau)} + z_\beta)^2 [p_1(1-p_1) + p_2(1-p_2)]}{\varepsilon_{ij}^2}. \tag{4.14}$$

The final sample size needed can be estimated by

$$n = \max\{n_{ij}, \text{all interested pairs } (i,\, j)\}. \tag{4.15}$$

4.4.2 An Example

Suppose an investigator is interested in conducting a parallel-group clinical trial comparing two active doses of a test compound against a standard therapy in patients with a specific carcinoma. Suppose the standard therapy, which is referred to as treatment 0, has a

20% response rate. For illustration purpose, the two active doses of the test compound are referred to as treatment 1 and treatment 2, respectively. Suppose the investigator would like to determine whether test treatments 1 and 2 will achieve the response rates of 40% and 50%, respectively. As a result, statistical comparisons of interest include the comparison between the standard therapy (treatment 0) versus. treatment 1 and between the standard therapy (treatment 0) versus treatment 2. In this case, $\tau = 2$. According to Equation 4.14, we have

$$n_{01} = \frac{(z_{0.05/(2\times2)} + z_{0.2})^2[0.2(1-0.2)+0.4(1-0.4)]}{(0.2-0.4)^2} \approx 95$$

and

$$n_{02} = \frac{(2.24+0.84)^2[0.2(1-0.2)+0.5(1-0.5)]}{0.09} \approx 44.$$

By Equation 4.15, the sample size needed to achieve an 80% power is given by $n = \max\{95, 44\} = 95$.

4.4.3 Remarks

It should be noted that the maximum approach described in this section is somewhat conservative in two aspects. First, the α adjustment based on the method of Bonferroni is conservative. Other less conservative methods for α adjustment may be used. Second, the formula is designed to detect statistically significant differences for all comparisons of interest. In practice, the comparisons of interest may not be equally important to the investigator. Hence, one of the comparisons is usually considered as the primary comparison and sample size calculation is performed based on the primary comparison. Once the sample size is determined, it can be justified under appropriate statistical assumption for other comparisons (secondary comparison) of interest.

4.5 Williams Design

We consider the Williams design described in Section 3.5. Let x_{ijl} be a binary response from the jth ($j = 1, ..., n$) subject in the ith ($i = 1, ..., a$) sequence under the lth ($l = 1, ..., b$) treatment. It is assumed that $(x_{ij1}, ..., x_{ijb})$, $i = 1, ..., a, j = 1, ..., n$ are i.i.d. random vectors with $P(x_{ijl} = 1) = p_l$. The observations from the same subject can be correlated with each other. By specifying that $P(x_{ijl} = 1) = p_l$, $l = 1, ..., m$, it implies that there is no sequence, period, or crossover effect. The statistical model incorporates those effects that are more complicated for binary data compared with continuous data. Its detailed discussion is beyond the scope of this book.

Without loss of generality, assume that we want to compare treatment 1 and treatment 2. Let

$$d_{ij} = y_{ij1} - y_{ij2}.$$

The true mean difference between treatment 1 and treatment 2 can be estimated by

$$\hat{\varepsilon} = \frac{1}{an}\sum_{i=1}^{a}\sum_{j=1}^{n}d_{ij},$$

which is asymptotically normally distributed with mean $\varepsilon = p_1 - p_2$ and variance σ_d^2/an, where σ_d^2 is defined to be the variance of d_{ij} and can be estimated by

$$\hat{\sigma}_d^2 = \frac{1}{a(n-1)}\sum_{i=1}^{a}\sum_{j=1}^{n}(d_{ij} - \frac{1}{n}\sum_{j'=1}^{n}d_{ij'})^2.$$

4.5.1 Test for Equality

Let $\varepsilon = \mu_1 - \mu_2$ be the true mean difference. The objective is to test the following hypotheses:

$$H_0 : \varepsilon = 0 \quad \text{versus} \quad H_a : \varepsilon \neq 0.$$

Then, the null hypothesis will be rejected at α level of significance if

$$\left| \frac{\hat{\varepsilon}}{\hat{\sigma}_d/\sqrt{an}} \right| > z_{\alpha/2}.$$

Under the alternative hypothesis that $\varepsilon \neq 0$, the power of this test is approximately

$$\Phi\left(\frac{\sqrt{an}\varepsilon}{\sigma_d} - z_{\alpha/2} \right).$$

As a result, the sample size needed for achieving a power of $1 - \beta$ can be obtained as

$$n = \frac{(z_{\alpha/2} + z_\beta)^2 \sigma_d^2}{a\varepsilon^2}. \tag{4.16}$$

4.5.2 Test for Noninferiority/Superiority

The problem of testing superiority and noninferiority can be unified by the following hypothesis:

$$H_0 : \varepsilon \leq \delta \quad \text{versus} \quad H_a : \varepsilon > \delta,$$

where δ is the superiority or noninferiority margin. When $\delta > 0$, the rejection of the null hypothesis indicates the superiority of the test drug over the control. When $\delta < 0$, the rejection of the null hypothesis indicates the noninferiority of the test drug against the control. The null hypothesis will be rejected at α level of significance if

$$\frac{\hat{\varepsilon} - \delta}{\hat{\sigma}_d / \sqrt{an}} > z_\alpha.$$

Under the alternative hypothesis that $\varepsilon > \delta$, the power of the above test is approximately

$$\Phi\left(\frac{\varepsilon - \delta}{\sigma_d / \sqrt{an}} - z_\alpha\right).$$

As a result, the sample size needed for achieving a power of $1 - \beta$ can be obtained by solving

$$\frac{\varepsilon - \delta}{\sigma_d / \sqrt{an}} - z_\alpha = z_\beta.$$

This leads to

$$n = \frac{(z_\alpha + z_\beta)^2 \sigma_d^2}{a(\varepsilon - \delta)^2}. \tag{4.17}$$

4.5.3 Test for Equivalence

The objective is to test the following hypotheses:

$$H_0 : |\varepsilon| \geq \delta \quad \text{versus} \quad H_a : |\varepsilon| < \delta.$$

The test drug will be concluded to be equivalent to the control on average if the null hypothesis is rejected at a given significance level. For example, at the significance level of α, the null hypothesis will be rejected if

$$\frac{\sqrt{an}(\hat{\varepsilon} - \delta)}{\hat{\sigma}_d} < -z_\alpha$$

and

$$\frac{\sqrt{an}(\hat{\varepsilon} + \delta)}{\hat{\sigma}_d} > z_\alpha.$$

Under the alternative hypothesis that $|\varepsilon| < \delta$, the power of the above test is approximately

$$2\Phi\left(\frac{\sqrt{an}(\delta - |\varepsilon|)}{\sigma_d} - z_\alpha\right) - 1.$$

As a result, the sample size needed for achieving a power of $1 - \beta$ can be obtained by solving

$$\frac{\sqrt{an}(\delta - |\varepsilon|)}{\sigma_d} - z_\alpha = z_{\beta/2}.$$

This leads to

$$n = \frac{(z_\alpha + z_{\beta/2})^2 \sigma_d^2}{a(\delta - |\varepsilon|)^2}. \tag{4.18}$$

4.5.4 An Example

Suppose that a sponsor is interested in conducting a 6×3 (Williams design) crossover experiment (i.e., $a = 6$) to compare two active doses (i.e., morning dose and evening dose) of a test compound against a placebo in patients with sleep disorder. We will refer to the placebo and the two active doses as treatment 0, treatment 1, and treatment 2, respectively. Qualified subjects will be randomly assigned to receive one of the six sequences of treatments. The trial consists of three visits. Each visit consists of two nights and three days with subjects in attendance at a designated Sleep Laboratory. On day two of each visit, the subject will receive one of the three treatments. Polysomnography will be applied to examine the treatment effect on sleep quality. Suppose the sponsor is interested in examining the existence of awakeness after the onset of sleep. As a result, sample size calculation is performed based on the proportion of subjects experiencing wakeness after the onset of sleep. Based on a pilot study, about 50%, 30%, and 35% of subjects receiving treatment 0, 1, and 2, respectively, experienced awakeness after the onset of sleep. As a result, for performing sample size calculation, we assume that the response rates for subjects receiving treatment 0, 1, and 2 are 50%, 30%, and 35%, respectively. Since the comparisons of interest include the comparison between treatment 1 and the placebo and between treatment 2 and the placebo, without loss of generality and for simplicity without adjusting type I error, we will focus on sample size calculation based on the comparison between treatment 1 and the placebo.

According to the information given above, it follows that the difference in proportion of subjects experiencing awakeness between treatment 1 and the placebo is given by 20% ($\varepsilon = 20\%$). From the pilot study, it is estimated that $\sigma_d = 75\%$. The significance level is fixed to be $\alpha = 5\%$.

4.5.4.1 Test for Equality

Since this is a 6×3 crossover design, the number of sequence is $a = 6$. According to Equation 4.16, the sample size needed to achieve 80% power ($\beta = 0.2$) is given by

$$n = \frac{(z_{\alpha/2} + z_\beta)^2 \sigma_d^2}{a\varepsilon^2} = \frac{(1.96 + 0.84)^2 0.75^2}{6 \times 0.2^2} \approx 19.$$

4.5.4.2 Test for Superiority

Assuming the superiority margin is 5% ($\delta = 0.05$), the sample size needed to achieve 80% power ($\beta = 0.2$) is given by

$$n = \frac{(z_\alpha + z_\beta)^2 \sigma_d^2}{a(\varepsilon - \delta)^2} = \frac{(1.64 + 0.84)^2 0.75^2}{6 \times (0.2 - 0.05)^2} \approx 27.$$

4.5.4.3 Test for Equivalence

Assuming the equivalence margin is 30% ($\delta = 0.30$), the sample size needed is given by

$$n = \frac{(z_\alpha + z_{\beta/2})^2 \sigma_d^2}{a(\delta - |\varepsilon|)^2} = \frac{(1.64 + 1.28)^2 0.75^2}{6 \times (0.3 - 0.2)^2} \approx 80.$$

4.6 Relative Risk—Parallel Design

In clinical trials, it is often of interest to investigate the relative effect (e.g., risk or benefit) of the treatments for the disease under study. Odds ratio has been frequently used to assess the association between a binary exposure variable and a binary disease outcome since it was introduced by Cornfield (1956). Let p_T be the probability of observing an outcome of interest for a patient treatment by a test treatment and p_C for a patient treated by a control. For a patient receiving the test treatment, the odds that he/she will have an outcome of interest over that he/she will not have an outcome are given by

$$O_T = \frac{p_T}{1 - p_T}.$$

Similarly, for a patient receiving the control, the odds are given by

$$O_C = \frac{p_C}{1 - p_C}.$$

As a result, the odds ratio between the test treatment and the control is defined as

$$OR = \frac{O_T}{O_C} = \frac{p_T(1 - p_C)}{p_C(1 - p_T)}.$$

The odds ratio is always positive and usually has a range from 0 to 4. $OR = 1$ (i.e., $pT = pC$) implies that there is no difference between the two treatments in terms of the outcome of interest. When $1 < OR < 4$, patients in the treatment group are more likely to have outcomes of interest than those in the control group. Note that $1 - OR$ is usually referred to as relative odds reduction in the literature. Intuitively, OR can be estimated by

$$\widehat{OR} = \frac{\hat{p}_T(1 - \hat{p}_C)}{\hat{p}_C(1 - \hat{p}_T)},$$

where $\hat{p}_T$ and $\hat{p}_C$ are the maximum likelihood estimators of p_T and p_C, respectively, given by

$$\hat{p}_T = \frac{x_T}{n_T} \quad \text{and} \quad \hat{p}_C = \frac{x_C}{n_C}, \tag{4.19}$$

and x_T and x_C are the observed numbers of patients in the respective treatment and control groups who have the outcome of interest. The asymptotic variance for $\log(\widehat{OR})$ can be obtained as

$$\text{var}[\log(\widehat{OR})] = \frac{1}{n_T p_T (1 - p_T)} + \frac{1}{n_C p_C (1 - p_C)},$$

which can be estimated by simply replacing p_T and p_C with their maximum likelihood estimator $\hat{p}_T$ and $\hat{p}_C$, respectively.

4.6.1 Test for Equality

The hypotheses of interest are given by

$$H_0 : OR = 1 \quad \text{versus} \quad H_a : OR \neq 1.$$

The test statistic is given by

$$T = \log(\widehat{OR}) \left[\frac{1}{n_T \hat{p}_T (1 - \hat{p}_T)} + \frac{1}{n_C \hat{p}_C (1 - \hat{p}_C)} \right]^{-1/2},$$

which approximately follows a standard normal distribution when n_T and n_C are sufficiently large. Thus, we reject the null hypothesis that $OR = 1$ if $|T| > z_{\alpha/2}$. Under the alternative hypothesis that $OR \neq 1$, the power of the above test can be approximated by

$$\Phi \left(|\log(OR)| \left[\frac{1}{n_T p_T (1 - p_T)} + \frac{1}{n_C p_C (1 - p_C)} \right]^{-1/2} - z_{\alpha/2} \right).$$

As a result, the sample size needed for achieving a desired power of $1 - \beta$ can be obtained by solving

$$|\log(OR)| \left[\frac{1}{n_T p_T (1 - p_T)} + \frac{1}{n_C p_C (1 - p_C)} \right]^{-1/2} - z_{\alpha/2} = z_\beta.$$

Under the assumption that $n_T/n_C = \kappa$ (a known ratio), we have

$$n_C = \frac{(z_{\alpha/2} + z_\beta)^2}{\log^2(OR)} \left(\frac{1}{\kappa p_T (1 - p_T)} + \frac{1}{p_C (1 - p_C)} \right). \tag{4.20}$$

4.6.2 Test for Noninferiority/Superiority

The problem of testing noninferiority and superiority can be unified by the following hypotheses:

$$H_0 : OR \le \delta' \quad \text{versus} \quad H_a : OR > \delta',$$

where δ' is the noninferiority or superiority margin in raw scale. The above hypotheses are the same as

$$H_0 : \log(OR) \le \delta \quad \text{versus} \quad H_a : \log(OR) > \delta,$$

where δ is the noninferiority or superiority margin in log scale. When $\delta > 0$, the rejection of the null hypothesis indicates superiority over the reference value. When $\delta < 0$, the rejection of the null hypothesis implies noninferiority against the reference value.
   Let

$$T = (\log(\widehat{OR}) - \delta) \left[\frac{1}{n_T \hat{p}_T (1 - \hat{p}_T)} + \frac{1}{n_C \hat{p}_C (1 - \hat{p}_C)} \right]^{-1/2}.$$

We reject the null hypothesis at the α level of significance if $T > z_\alpha$. Under the alternative hypothesis that $\log(OR) > \delta$, the power of the above test is approximately

$$\Phi \left((\log(OR) - \delta) \left[\frac{1}{n_T p_T (1 - p_T)} + \frac{1}{n_C p_C (1 - p_C)} \right]^{-1/2} - z_a \right).$$

As a result, the sample size needed for achieving a desired power of $1 - \beta$ can be obtained by solving

$$|\log(OR) = \delta| \left[\frac{1}{n_T p_T (1 - p_T)} + \frac{1}{n_C p_C (1 - p_C)} \right]^{-1/2} - z_{\alpha/2} = z_\beta.$$

Under the assumption that $n_T / n_C = \kappa$, we have

$$n_C = \frac{(z_\alpha + z_\beta)^2}{(\log(OR) - \delta)^2} \left(\frac{1}{\kappa p_T (1 - p_T)} + \frac{1}{p_C (1 - p_C)} \right). \tag{4.21}$$

4.6.3 Test for Equivalence

To establish equivalence, the following hypotheses are usually considered:

$$H_0 : |\log(OR)| \ge \delta \quad \text{versus} \quad H_a : |\log(OR)| < \delta.$$

The above hypotheses can be tested using the two one-sided test procedures as described in the previous sections. We reject the null hypothesis at α level of significance if

$$(\log(\widehat{OR}) - \delta) \left[\frac{1}{n_T \hat{p}_T (1 - \hat{p}_T)} + \frac{1}{n_C \hat{p}_C (1 - \hat{p}_C)} \right]^{-1/2} < -z_\alpha$$

and

$$(\log(\widehat{OR}) + \delta) \left[\frac{1}{n_T \hat{p}_T (1 - \hat{p}_T)} + \frac{1}{n_C \hat{p}_C (1 - \hat{p}_C)} \right]^{-1/2} > z_\alpha.$$

When $|\log(OR)| < \delta$, the power of this test is approximately

$$2\Phi \left((\delta - |\log(OR)|) \left[\frac{1}{n_T p_T (1 - p_T)} + \frac{1}{n_C p_C (1 - p_C)} \right]^{-1} - z_{\alpha/2} \right) - 1.$$

Under the assumption that $n_T / n_C = \kappa$, the sample size needed for achieving a desired power of $1 - \beta$ is given by

$$n_C = \frac{(z_\alpha + z_{\beta/2})^2}{(\delta - |\log(OR)|)^2} \left(\frac{1}{k p_T (1 - p_T)} + \frac{1}{p_C (1 - p_C)} \right). \tag{4.22}$$

4.6.4 An Example

Suppose that a sponsor is interested in conducting a clinical trial to study the relative risk between a test compound and a standard therapy for the prevention of relapse in subjects with schizophrenia and schizoaffective disorders. Based on the results from a previous study with 365 subjects (i.e., 177 subjects received the test compound and 188 received the standard therapy), about 25% (45/177) and 40% (75/188) of subjects receiving the test compound and the standard therapy experienced relapse after the treatment. Subjects who experienced first relapse may withdraw from the study or continue on. Among the subjects who experienced the first relapse and stayed on the study, about 26.7% (8/30) and 32.0% (16/50) experienced the second relapse. The sponsor is interested in studying the odds ratio of the test compound as compared to the standard therapy for the prevention of experiencing the first relapse. In addition, it is also of interest to examine the odds ratio for the prevention of experiencing the second relapse.

4.6.4.1 Test for Equality

Assume the responder rate in control group is 25% and the rate in test is 40%, which produces a relative risk

$$OR = \frac{0.40(1 - 0.25)}{(1 - 0.4)0.25} = 2.$$

According to Equation 4.20 and $n = n_T = n_C$ ($k = 1$), the sample size needed to achieve 80% ($\beta = 0.2$) at 0.05 ($\alpha = 0.05$) level of significance is given by

$$n = \frac{(1.96 + 0.84)^2}{\log^2(2)} \left[\frac{1}{0.4(1 - 0.4)} + \frac{1}{0.25(1 - 0.25)} \right] \approx 156.$$

4.6.4.2 Test for Superiority

Assume that 20% ($\delta = 0.2$) is considered as a clinically important superiority margin for log-scale relative risk. According to Equation 4.21, the sample size needed to achieve 80% power ($\beta = 0.2$) is given by

$$n = \frac{(1.64 + 0.84)^2}{(\log(2) - 0.2)^2} \left[\frac{1}{0.4(1 - 0.4)} + \frac{1}{0.25(1 - 0.25)} \right] \approx 241.$$

4.6.4.3 Test for Equivalence

Assume that the relapse rate of the study drug (25%) is approximately equal to a market drug ($\log(OR) = 0$) and that the equivalence margin in log-scale relative risk is 50% ($\delta = 0.50$). According to Equation 4.22, the sample size needed to achieve 80% ($\beta = 0.2$) power to establish equivalence is given by

$$n = \frac{(z_{0.05} + z_{0.2/2})^2}{0.5^2} \left[\frac{1}{0.25(1 - 0.25)} + \frac{1}{0.25(1 - 0.25)} \right] \approx 364.$$

4.7 Relative Risk—Crossover Design

Consider a 1×2 crossover design with no period effects. Without loss of generality, we assume that every subject will receive test first and then be crossed over to control. Let x_{ij} be a binary response from the jth subject in the ith period, $j = 1, ..., n$. The number of outcomes of interest under treatment is given by $x_T = \sum_{j=1}^{n} x_{1j}$. The number of outcomes of interest under control, x_C, is similarly defined. Then the true response rates under treatment and control can still be estimated according to Equation 4.19. According to Taylor's expansion, it can be shown that

$$\sqrt{n}(\log(\widehat{OR}) - \log(OR))$$

$$= \sqrt{n} \left[\frac{1}{p_T(1 - p_T)}(\hat{p}_T - p_T) - \frac{1}{p_C} p_C(1 - p_C)(\hat{p}_C - p_C) \right] + o_p(1)$$

$$= \frac{1}{\sqrt{n}} \sum_{j=1}^{n} \left[\frac{x_{1j} - p_T}{p_T(1 - p_T)} - \frac{x_{2j} - p_C}{p_C(1 - p_C)} \right] + o_p(1)$$

$$\to_d N(0, \sigma_d^2),$$

where

$$\sigma_d^2 = \text{var}\left(\frac{x_{1j} - p_T}{p_T(1-p_T)} - \frac{x_{2j} - p_C}{p_C(1-p_C)}\right).$$

Let

$$d_j = \left(\frac{x_{1j}}{\hat{p}_T(1-\hat{p}_T)} - \frac{x_{2j}}{\hat{p}_C(1-\hat{p}_C)}\right).$$

Then, σ_d^2 can be estimated by $\hat{\sigma}_d^2$, the sample variance based on $d_j, j = 1, ..., n$.

4.7.1 Test for Equality

The hypotheses of interest are given by

$$H_0 : \log(OR) = 0 \quad \text{versus} \quad H_a : \log(OR) \neq 0.$$

Under the null hypothesis, the test statistic

$$T = \frac{\sqrt{n}\log(\widehat{OR})}{\hat{\sigma}_d}$$

approximately follows a standard normal distribution when n_T and n_C are sufficiently large. Thus, we reject the null hypothesis that $OR = 1$ if $|T| > z_{\alpha/2}$. Under the alternative hypothesis that $OR \neq 1$, the power of the above test can be approximated by

$$\Phi\left(\frac{\sqrt{n}\,|\log(OR)|}{\sigma_d} - z_{\alpha/2}\right).$$

As a result, the sample size needed for achieving a desired power of $1 - \beta$ can be obtained by solving

$$\frac{\sqrt{n}\,|\log(OR)|}{\sigma_d} - z_{\alpha/2} = z_\beta.$$

This leads to

$$n = \frac{(z_{\alpha/2} + z_\beta)^2 \sigma_d^2}{\log^2(OR)}. \tag{4.23}$$

4.7.2 Test for Noninferiority/Superiority

The problem of testing noninferiority and superiority can be unified by the following hypotheses:

$$H_0 : \log(OR) \leq \delta \quad \text{versus} \quad H_a : \log(OR) > \delta,$$

where δ is the noninferiority or superiority margin in log scale. When $\delta > 0$, the rejection of the null hypothesis indicates superiority over the reference value. When $\delta < 0$, the rejection of the null hypothesis implies noninferiority against the reference value.

When $\log(OR) = \delta$, the test statistic

$$T = \frac{\sqrt{n}(\log(\widehat{OR}) - \delta)}{\hat{\sigma}_d}$$

approximately follows the standard normal distribution when n_T and n_C are sufficiently large. Thus, we reject the null hypothesis at the α level of significance if $T > z_\alpha$. Under the alternative hypothesis that $\log(OR) > \delta$, the power of the above test is approximately

$$\Phi\left(\frac{\log(OR) - \delta}{\sigma_d} - z_\alpha\right).$$

As a result, the sample size needed for achieving a desired power of $1 - \beta$ can be obtained by solving

$$\frac{\log(OR) - \delta}{\sigma_d} - z_\alpha = z_\beta.$$

It leads to

$$n = \frac{(z_\alpha + z_\beta)^2 \sigma_d^2}{[\log(OR) - \delta]^2}. \tag{4.24}$$

4.7.3 Test for Equivalence

To establish equivalence, the following hypotheses are usually considered:

$$H_0 : |\log(OR)| \geq \delta \quad \text{versus} \quad H_a : |\log(OR)| < \delta.$$

The above hypotheses can be tested using the two one-sided test procedures (see, e.g., Chow and Liu, 1998). We reject the null hypothesis at the α level of significance if

$$\frac{\sqrt{n}(\log(\widehat{OR}) - \delta)}{\hat{\sigma}_d} < -z_\alpha$$

and

$$\frac{\sqrt{n}(\log(\widehat{OR}) + \delta)}{\hat{\sigma}_d} > z_\alpha.$$

When $|\log(OR)| < \delta$, the power of the above test is approximately

$$2\Phi\left(\frac{(\delta-|\log(OR)|)}{\sigma_d} - z_\alpha\right) - 1.$$

Then, the sample size needed for achieving a desired power of $1 - \beta$ can be obtained by

$$n = \frac{(z_\alpha + z_{\beta/2})^2 \sigma_d^2}{(\delta-|\log(OR)|)^2}. \tag{4.25}$$

4.8 Practical Issues

4.8.1 Exact and Asymptotic Tests

It should be noted that all of the formulas for sample size calculation given in this chapter are derived based on asymptotic theory. In other words, the formulas are valid when the sample size is sufficiently large. However, "how large is considered sufficiently large" is always a question to researchers who are trying to determine the sample size at the planning stage of an intended study. Unfortunately, there is no simple rule that can be used to evaluate whether the sample size is sufficiently large. As an alternative, some exact tests may be useful when the expected sample size of the intended study is small (due to budget constraint and/or slow enrollment). Details of various commonly used exact tests, such as binomial test, Fisher's exact test, and multiple-stage optimal design will be discussed in Chapter 5.

4.8.2 Variance Estimates

For testing equality, noninferiority/superiority, and equivalence, the following test statistic is always considered:

$$Z = \frac{\hat{p}_1 - \hat{p}_2 + \varepsilon}{\hat{\sigma}},$$

where $\hat{p}_1$ and $\hat{p}_2$ are observed response rates from treatment 1 and treatment 2, respectively, and $\hat{\sigma}$ is an estimate of the standard error σ, which is given by

$$\sigma = \sqrt{\frac{p_1(1-p_1)}{n_1} + \frac{p_2(1-p_2)}{n_2}}.$$

Under the null hypothesis, Z is asymptotically normally distributed with mean 0 and standard deviation 1. As an example, for testing noninferiority between an active treatment (treatment 1) and an active control (treatment 2), large Z values (i.e., treatment is better than control) favor the alternative hypothesis. Blackwelder (1982) recommended σ^2 be estimated by the observed variance, which is given by

$$\hat{\sigma}^2 = \frac{\hat{p}_1(1-\hat{p}_1)}{n_1} + \frac{\hat{p}_2(1-\hat{p}_2)}{n_2}.$$

In practice, however, σ^2 can be estimated by different methods. For example, Dunnett and Gent (1977) proposed to estimate variance from fixed marginal totals. The idea is to estimate p_1 and p_2 under the null hypothesis restriction $p_1 - p_2 = \varepsilon$, subject to the marginal totals remaining equal to those observed. This approach leads to the estimates

$$\tilde{p}_1 = \left[\hat{p}_1 + \left(\frac{n_2}{n_1}\right)(\hat{p}_2 + \varepsilon)\right] \bigg/ \left(1 + \frac{n_2}{n_1}\right),$$

$$\tilde{p}_2 = \left[\hat{p}_1 + \left(\frac{n_2}{n_1}\right)(\hat{p}_2 - \varepsilon)\right] \bigg/ \left(1 + \frac{n_2}{n_1}\right).$$

As a result, an estimate of σ can then be obtained based on $\tilde{p}_1$ and $\tilde{p}_2$. Tu (1997) suggested σ^2 be estimated by the unbiased observed variance

$$\hat{\sigma}_u^2 = \frac{\hat{p}_1(1-\hat{p}_1)}{n_1 - 1} + \frac{\hat{p}_2(1-\hat{p}_2)}{n_2 - 1}.$$

In addition, Miettinen and Nurminen (1985) and Farrington and Manning (1990) considered estimating σ^2 using the constrained MLE as follows:

$$\hat{\sigma}_{MLE}^2 = \frac{\tilde{p}_1(1-\tilde{p}_1)}{n_1} + \frac{\tilde{p}_2(1-\tilde{p}_2)}{n_2},$$

where $\tilde{p}_1$ and $\tilde{p}_2$ are the constrained MLE of p_1 and p_2 under the null hypothesis. As indicated in Farrington and Manning (1990), $\tilde{p}_1$ can be obtained as the unique solution of the following maximum likelihood equation:

$$ax^3 + bx^2 + cx + d = 0,$$

where

$$a = 1 + \frac{n_2}{n_1},$$

$$b = -\left[1 + \frac{n_2}{n_1} + \hat{p}_1 + \left(\frac{n_2}{n_1}\right)\hat{p}_2 + \varepsilon\left(\frac{n_2}{n_1} + 2\right)\right],$$

$$c = \varepsilon^2 + \varepsilon\left(2\hat{p}_1 + \frac{n_2}{n_1} + 1\right)\hat{p}_1 + \left(\frac{n_2}{n_1}\right)\hat{p}_2,$$

$$d = -\hat{p}_1\varepsilon(1 + \varepsilon).$$

The solution is given by

$$\tilde{p}_1 = 2u \cos(w) - \left(\frac{b}{3a}\right) \quad \text{and} \quad \tilde{p}_2 = \tilde{p}_1 - \varepsilon,$$

where

$$w = \frac{1}{3}[\pi + \cos^{-1}(v/u^3)],$$
$$v = b^3/(3a)^3 - bc(6a^2) + d/(2a),$$
$$u = \text{sign}(v)\left[\frac{b^2}{(3a)^2} - \frac{c}{(3a)}\right]^{1/2}.$$

Biswas, Chan, and Ghosh (2000) showed that the method of the constrained MLE performs better than methods by Blackwelder (1982) and Dunnett and Gent (1977) in terms of controlling type I error rate, power, and confidence interval coverage through a simulation study. The power function (sample size calculation) is sensitive to the difference between true response rates. A small difference (i.e., $\varepsilon \neq 0$) will drop the power rapidly. Consequently, a large sample size is required for achieving a desired power.

4.8.3 Stratified Analysis

In clinical research, stratified randomization is often employed to isolate the possible confounding or interaction effects that may be caused by prognostic factors (e.g., age, weight, disease status, and medical history) and/or nonprognostic factors (e.g., study center). Responses in these strata are expected to be similar and yet they may be systematically different or subject to random fluctuation across strata. In the interest of a fair and reliable assessment of the treatment difference, it is suggested that the stratified analysis be performed. The purpose of the stratified analysis is to obtain an unbiased estimate of treatment difference with a desired precision.

Stratified analysis can be performed based on Blackwelder's approach or the method proposed by Miettinen and Nurminen (1985) and Farrington and Manning (1990) by adapting different weights in each strata. In practice, several weights are commonly considered. These weights include (i) equal weights, (ii) sample size, (iii) Cochran–Mantel–Haenszel, (iv) inverse of variance, and (v) minimum risk. Suppose there are K strata. Let n_{ik} be the sample size of the kth stratum in the ith treatment group and wk be the weight assigned to the kth stratum, where $k = 1, \ldots, K$. Basically, equal weights, that is, $w_k = w$ for all k imply that no weights are considered. Intuitively, one may consider using the weight based on sample size, that is,

$$w_k \propto (n_{1k} + n_{2k}).$$

In other words, larger strata will carry more weights as compared to smaller strata. Alternatively, we may consider the weight suggested by Cochran–Mantel–Haenszel as follows:

$$w_k \propto \frac{n_{1k} n_{2k}}{n_{1k} + n_{2k}}.$$

These weights, however, do not take into consideration the heterogeneity of variability across strata. To overcome this problem, the weight based on the inverse of variance for the kth stratum is useful, that is,

$$w_k \propto \sigma_k^{-1},$$

where σ_k^2 is the variance of the kth stratum. The weight of minimum risk is referred to as the weight that minimizes the mean squared error (Mehrotra and Railkar, 2000).

Biswas, Chan, and Ghosh (2000) conducted a simulation study to compare the relative performances of Blackwelder's approach and Miettinen and Nurminen's method with different weights for stratified analysis. The results indicate that Cochran–Mantel–Haenszel weight for Miettinen and Nurminen's method and minimum risk weight for Blackwelder's approach perform very well even in the case of extreme proportions and/or the presence of interactions. Inverse variance weight is biased, which leads to liberal confidence interval coverage probability.

4.8.4 Equivalence Test for More Than Two Proportions

In clinical trials, it may be of interest to demonstrate therapeutic equivalence among a group of drug products for the treatment of certain disease under study. In this case, a typical approach is to perform a pairwise equivalence testing with or without adjusting the α level for multiple comparisons. Suppose a clinical trial was conducted to establish therapeutic equivalence among three drug products (A, B, and C) for the treatment of women with advanced breast cancer. For a given equivalence limit, equivalence test can be performed for testing (i) drug A versus drug B, (ii) drug A versus drug C, and (iii) drug B versus drug C. It is very likely that we may conclude that drug A is equivalent to drug B and drug B is equivalent to drug C, but drug A is not equivalent to drug C based on pairwise comparison. In this case, equivalence among the three drug products cannot be established. As an alternative approach, Wiens, Heyse, and Matthews (1996) consider the following hypotheses for testing equivalence among a group of treatments:

$$H_0: \max_{1 \le i \le j \le K} |p_i - p_j| \ge \delta \quad \text{versus} \quad H_a: \max_{1 \le i \le j \le K} |p_i - p_j| < \delta.$$

Testing the above hypotheses is equivalent to testing the following hypotheses:

$$H_0: \max_{1 \le i \le K} p_i - \max_{1 \le j \le K} p_j \ge \delta \quad \text{versus} \quad H_a: \max_{1 \le i \le K} p_i - \max_{1 \le j \le K} p_j < \delta.$$

Under the above hypotheses, formulas for sample size calculation can be similarly derived.

5

Exact Tests for Proportions

In Chapter 4, formulas for sample size calculation for comparing proportions were derived based on asymptotic approximations. In practice, sample sizes for some clinical trials such as phase II cancer trials are usually small and, hence, the formulas given in Chapter 4 may not be useful. In this chapter, our primary focus is placed on procedures for sample size calculation based on exact tests for small samples. Unlike the tests based on asymptotic distribution, the power functions of the exact tests usually do not have explicit forms. Hence, exact formulas for sample size calculation cannot be obtained. However, the sample size can be obtained numerically by greedy search over the sample space.

In the next three sections, procedures for obtaining sample sizes based on exact tests for comparing proportions such as the binomial test, negative binomial regression, and Fisher's exact test are discussed. In Section 5.4, procedures for sample size calculation under various optimal multiple-stage designs such as an optimal two-stage design, an optimal three-stage design, and a flexible optimal design for single-arm phase II cancer trials are given. Section 5.5 provides procedures for sample size calculation under a flexible design for multiple armed clinical trials. Some practical issues are presented in the last section.

5.1 Binomial Test

In this section, we describe the binomial test, which is probably the most commonly used exact test for one-sample testing problem with binary response in clinical research, and the related sample size calculation formula.

5.1.1 The Procedure

The test for equality and noninferiority/superiority can all be unified by the following hypotheses:

$$H_0 : p = p_0^* + \delta \quad \text{versus} \quad H_a : p = p_1, \tag{5.1}$$

where p_0^* is a predefined reference value and $p_1 > p_0^* + \delta$ is an unknown proportion. When $\delta = 0$, Equation 5.1 becomes the (one-sided) test for equality. When $\delta < 0$ ($\delta > 0$), it becomes the test for noninferiority (superiority). Let n be the sample size of a single-arm clinical study and m be the number of observed outcome of interest. When $p = p_0 = p_0^* + \delta$, m is distributed as a binomial random variable with parameters (p_0, n). If the number of the observed responses is greater than or equal to m, then it is considered at least as favorable as the observed outcome of H_a. The probability of observing these responses is defined as the exact p-value for the observed outcome. In other words,

$$\text{Exact } p\text{-value} = \sum_{i=m}^{n} \frac{n!}{i!(n-i)!} p_0^i (1-p_0)^{n-i}.$$

For a given significance level α, there exists a nonnegative integer r (called the critical value) such that

$$\sum_{i=r}^{n} \frac{n!}{i!(n-i)!} p_0^i (1-p_0)^{n-i} \le \alpha$$

and

$$\sum_{i=r-1}^{n} \frac{n!}{i!(n-i)!} p_0^i (1-p_0)^{n-i} > \alpha.$$

We then reject the null hypothesis at the α level of significance if $m \ge r$. Under the alternative hypothesis that $p = p_1 > p_0$, the power of this test can be evaluated as

$$P(m \ge r \,|\, H_a) = \sum_{i=r}^{n} \frac{n!}{i!(n-i)!} p_1^i (1-p_1)^{n-i}.$$

For a given power, the sample size required for achieving a desired power of $1 - \beta$ can be obtained by solving $P(m \ge r | H_a) \ge 1 - \beta$.

Tables 5.1 and 5.2 provide sample sizes required for achieving a desired power (80% or 90%) for $p_1 - p_0 = 0.15$ and $p_1 - p_0 = 0.20$, respectively. As an example, a sample size of 40 subjects is required for the detection of a 15% difference (i.e., $p_1 - p_0 = 0.15$) with a 90% power assuming that $p_0 = 0.10$. Note that with the selected sample size, we would reject the null hypothesis that $p_0 = 0.10$ at the α level of significance if there are 7 (out of 40) responses.

5.1.2 Remarks

The exact p-value is well defined only if the sample distribution is completely specified under the null hypothesis. On the other hand, the test for equivalence usually involves interval hypothesis, which means that, under the null hypothesis, we only know the parameter of interest is located within a certain interval but are unaware of its exact value. As a result, the distribution under the null hypothesis cannot be completely specified and, hence, exact test is not well defined in such a situation.

5.1.3 An Example

Suppose the investigator is interested in conducting a trial to study the treatment effect of a test compound in curing patients with certain types of cancer. The responder is defined to be the subject who is completely cured by the study treatment. According to literature, the standard therapy available on the market can produce a cure rate of 10% ($p_0 = 10\%$). A pilot study of the test compound shows that the test compound may produce a cure rate of 30% ($p_1 = 30\%$). The objective of the planning trial is to confirm such a difference truly exists.

TABLE 5.1

Sample Size n and Critical Value r for Binomial Test ($p_1 - p_0 = 0.15$)

α	p_0	p_1	$1 - \beta = 80\%$		$1 - \beta = 90\%$	
			r	n	r	n
0.05	0.05	0.20	3	27	4	38
	0.10	0.25	7	40	9	55
	0.15	0.30	11	48	14	64
	0.20	0.35	16	56	21	77
	0.25	0.40	21	62	27	83
	0.30	0.45	26	67	35	93
	0.35	0.50	30	68	41	96
	0.40	0.55	35	71	45	94
	0.45	0.60	38	70	52	98
	0.50	0.65	41	69	54	93
	0.55	0.70	45	70	58	92
	0.60	0.75	43	62	58	85
	0.65	0.80	41	55	55	75
	0.70	0.85	39	49	54	69
	0.75	0.90	38	45	46	55
	0.80	0.95	27	30	39	44
0.10	0.05	0.20	2	21	3	32
	0.10	0.25	5	31	6	40
	0.15	0.30	8	37	11	53
	0.20	0.35	12	44	16	61
	0.25	0.40	15	46	20	64
	0.30	0.45	19	50	26	71
	0.35	0.50	21	49	30	72
	0.40	0.55	24	50	35	75
	0.45	0.60	28	53	39	75
	0.50	0.65	31	53	41	72
	0.55	0.70	31	49	44	71
	0.60	0.75	32	47	43	64
	0.65	0.80	33	45	44	61
	0.70	0.85	29	37	41	53
	0.75	0.90	25	30	33	40
	0.80	0.95	22	25	28	32

It is desirable to have a sample size, which can produce 80% power at 5% level of significance. According to Table 5.1, the total sample size needed is given by 25. The null hypothesis should be rejected if there are at least five subjects who are classified as responders.

5.2 Negative Binomial

In clinical trials, recurrent events such as exacerbations in chronic obstructive pulmonary disease, relapses in multiple sclerosis, and seizures in epileptics are commonly

TABLE 5.2

Sample Size n and Critical Value r for Binomial Test ($p_1 - p_0 = 0.20$)

			$1 - \beta = 80\%$		$1 - \beta = 90\%$	
α	p_0	p_1	r	N	r	N
0.05	0.05	0.25	2	16	3	25
	0.10	0.30	5	25	6	33
	0.15	0.35	7	28	9	38
	0.20	0.40	11	35	14	47
	0.25	0.45	13	36	17	49
	0.30	0.50	16	39	21	53
	0.35	0.55	19	41	24	53
	0.40	0.60	22	42	28	56
	0.45	0.65	24	42	30	54
	0.50	0.70	23	37	32	53
	0.55	0.75	25	37	33	50
	0.60	0.80	26	36	32	45
	0.65	0.85	24	31	32	42
	0.70	0.90	23	28	30	37
	0.75	0.95	20	23	25	29
	0.80	1.00	13	14	13	14
0.10	0.05	0.25	2	16	2	20
	0.10	0.30	3	18	4	25
	0.15	0.35	5	22	7	32
	0.20	0.40	7	24	10	36
	0.25	0.45	9	26	13	39
	0.30	0.50	12	30	15	39
	0.35	0.55	13	29	19	44
	0.40	0.60	15	30	20	41
	0.45	0.65	16	29	24	44
	0.50	0.70	17	28	23	39
	0.55	0.75	19	29	25	39
	0.60	0.80	17	24	25	36
	0.65	0.85	16	21	24	32
	0.70	0.90	17	21	20	25
	0.75	0.95	13	15	17	20
	0.80	1.00	10	11	10	11

encountered. The primary endpoint is generally the total number of events experienced by the subjects. However, the recurrent event often exhibit overdispersion in the sense that the variance exceeds the mean. In this case, negative binomial (NB) regression is often considered to analyze recurrent event data since NB regression is able to account for over-dispersion (Keene et al., 2007; Friede and Schmidli, 2010; Zhu and Lakkis, 2014; Zhu, 2017).

5.2.1 Negative Binomial Distribution

The NB distribution, denoted by $Y \sim NB(\mu, \tau)$, is the probability distribution of the number of failures Y before $k = 1/\tau$ successes in a series of independent Bernoulli trials with the same probability $p = 1/(1 + \tau\mu)$ of success

$$\Pr(Y = y) = \frac{\Gamma(y + 1/\tau)}{y!\,\Gamma(1/\tau)} p^{1/\tau}(1 - p)^y,$$

where μ is the mean and τ is the dispersion parameter. Note that, for count data, the NB distribution is also known as a gamma mixture of Poisson distribution. That is, if Y follows a Poisson distribution with $\varepsilon\mu$ and ε is gamma distributed with mean 1 and variance τ, the marginal distribution of Y is NB(μ, τ). The NB distribution is useful to fit the overdispersed count data better than the Poisson distribution and its mean is always less than its variance

$$E(Y) = E[E(Y|\varepsilon)] = \mu,$$

$$\mathrm{Var}(Y) = E[\mathrm{Var}(Y|\varepsilon)] + \mathrm{Var}[E(Y|\varepsilon)] = \mu + \tau\mu^2.$$

5.2.2 Sample Size Requirement

In clinical trials, suppose n subjects are randomized to the experimental group ($g = 1$) or control group ($g = 0$). Let t_{gj} be the follow-up time and y_{gj} be the number of observed events for subject $j = 1, \ldots, n_g$ in group g. Without loss of generality, we assume constant event rate over time and $y_{gj} \sim \mathrm{NB}(\mu_{gj}, \tau)$, where $\mu_{gj} = \lambda_g t_{gj}$ and $\lambda_g = \exp(\xi_g)$ is the event rate for group g. The treatment effect is often measured by the rate ratio $\exp(\beta) = \lambda_1/\lambda_0$ or equivalently by the relative rate reduction (RRR)

$$\mathrm{RRR} = 1 - \frac{\lambda_1}{\lambda_0} = 1 - \exp(\beta).$$

Under asymptotic normality, the $100 \times (1 - \alpha)\%$ confidence interval for β, denoted by (C_L, C_U) can be obtained.

In a superiority trial, the control treatment is placebo or an active treatment, and the objective is to demonstrate that the experimental treatment can lower the event rate. Thus, the superiority hypothesis can be expressed as

$$H_0 : \lambda_0 = \lambda_1 \text{ or } \beta = 0 \quad \text{versus} \quad H_a : \lambda_0 \neq \lambda_1 \text{ or } \beta \neq 0.$$

If $C_U < 0$ or equivalently $\exp(C_U) < 1$, we can claim that the experimental treatment can significantly reduce the event rate compared with the control treatment. Under the above hypotheses, the required total sample size is given by

$$n_{total} = [(d_0 p_0)^{-1} + (d_1 p_1)^{-1}]\frac{(z_{1-\alpha/2} + z_P)^2}{\beta^2},$$

where P is the desired power and

$$d_g = E\left[\frac{\lambda_g t_{gj}}{1 + \tau \lambda_g t_{gj}}\right].$$

Under equal allocation ($p_0 = p_1 = 1/2$), the total sample size becomes

$$n_{total} = 2(d_0^{-1} + d_1^{-1})\frac{(z_{1-\alpha/2} + z_P)^2}{\beta^2}.$$

Note that the above formulas for sample size requirement are valid if $\lambda_1 > \lambda_0$ (i.e., $\beta > 0$) and the objective is to demonstrate that the experimental treatment can increase the event rate.

Similarly, in a noninferiority trial, the objective is to show that the experimental treatment is not worse than the active control by a margin, say M_0, where $M_0 > 1$ is a prespecified noninferiority margin. The margin is generally chosen to be close to 1 to demonstrate that the new treatment is not inferior to the active comparator. In noninferiority trials, the following hypotheses are often considered:

$$H_0: \frac{\lambda_1}{\lambda_0} \geq M_0 \quad \text{versus} \quad H_a: \frac{\lambda_1}{\lambda_0} < M_0,$$

Under the above hypotheses, the required total sample size is given by

$$n_{total} = [(d_0 p_0)^{-1} + (d_1 p_1)^{-1}]\frac{(z_{1-\alpha/2} + z_P)^2}{\beta^{*2}},$$

where

$$\beta^* = \log\left(\frac{M_0 \lambda_0}{\lambda_1}\right).$$

Note that the sample size formulas given in this subsection are derived based on the Wald test from the NB regression, which generally work well for moderate to large samples.

5.3 Fisher's Exact Test

For comparing proportions between two treatment groups, the hypotheses of interest are given by

$$H_0: p_1 = p_2 \quad \text{versus} \quad H_a: p_1 \neq p_2,$$

where p_1 and p_2 are the true proportions of treatment 1 and treatment 2, respectively. Unlike the one-sample binomial test, under the null hypothesis that $p_1 = p_2$, the exact values of p_1 and p_2 are unknown. Hence, it is impossible to track the marginal distribution of the events observed from different treatment groups. In this case, a conditional test such as Fisher's exact test is usually considered. In this section, we describe Fisher's exact test and the related sample size calculation formula.

5.3.1 The Procedure

Let m_i be the number of responses observed in the ith treatment group. Then, the total number of observed responses is $m = m_1 + m_2$. Under the null hypothesis that $p_1 = p_2$ and conditional on m, it can be shown that m_1 follows a hypergeometric distribution with parameters (m, n_1, n_2), that is,

$$P(m_1 = i \mid m, n_1, n_2) = \frac{\binom{n_1}{i}\binom{n_2}{m-i}}{\binom{n_1+n_2}{m}}.$$

Any outcomes with the same m but larger than m_1 would be considered at least as favorable to H_a as the observed outcomes. Then, the probability of observing these outcomes, which is at least as observed, is defined as the exact p-value. In other words,

$$\text{Exact } p\text{-value} = \sum_{i=m_1}^{m} \frac{\binom{n_1}{i}\binom{n_2}{m-i}}{\binom{n_1+n_2}{m}}.$$

We reject the null hypothesis at the α level of significance when the exact p-value is less than α. Under the alternative hypothesis that $p_1 \neq p_2$ and for a fixed n, the power of Fisher's exact test can be obtained by summing the probabilities of all the outcomes such that the exact p-value is less than α. However, it should be noted that no closed form exists for the power of Fisher's exact test. As a result, the sample size required for achieving a desired power can only be obtained numerically such as by greedy search for all possible outcomes.

Table 5.3 provides sample sizes required for achieving the desired power (80% or 90%) under various parameters (i.e., $p_2 - p_1$ ranging from 0.10 to 0.35) when testing the null hypothesis that $p_1 = p_2$. As an example, a sample size of 34 subjects is required for the detection of a 25% difference in proportion between treatment groups (i.e., $p_2 - p_1 = 0.25$) with an 80% power assuming that $p_1 = 0.15$.

5.3.2 Remarks

For Fisher's exact test, the exact p-value is well defined only if the conditional sample distribution is completely specified under the null hypothesis. On the other hand, the test for noninferiority/superiority and equivalence usually involves interval hypothesis, which means under the null hypothesis, we only know the parameter of interest is located within certain interval but are unaware of its exact value. As a result, the distribution under the null hypothesis cannot be completely specified and, hence, Fisher's exact test is not well defined in such a situation.

5.3.3 An Example

Suppose the investigator is interested in conducting a two-arm trial to study the treatment effect of a test compound in preventing the relapse rate in EAE score. The active control

TABLE 5.3

Sample Size for Fisher's Exact Test

			$\alpha = 0.10$		$\alpha = 0.05$	
$p_2 - p_1$	p_1	p_2	$\beta = 0.20$	$\beta = 0.10$	$\beta = 0.20$	$\beta = 0.10$
0.25	0.05	0.30	25	33	34	42
	0.10	0.35	31	41	39	52
	0.15	0.40	34	48	46	60
	0.20	0.45	39	52	49	65
	0.25	0.50	40	56	54	71
	0.30	0.55	41	57	55	72
	0.35	0.60	41	57	56	77
	0.40	0.65	41	57	56	77
	0.45	0.70	41	57	55	72
	0.50	0.75	40	56	54	71
	0.55	0.80	39	52	49	65
	0.60	0.85	34	48	46	60
	0.65	0.90	31	41	39	52
	0.70	0.95	25	33	34	42
0.30	0.05	0.35	20	26	25	33
	0.10	0.40	23	32	30	39
	0.15	0.45	26	35	34	45
	0.20	0.50	28	39	36	47
	0.25	0.55	29	40	37	51
	0.30	0.60	29	40	41	53
	0.35	0.65	33	40	41	53
	0.40	0.70	29	40	41	53
	0.45	0.75	29	40	37	51
	0.50	0.80	28	39	36	47
	0.55	0.85	26	35	34	45
	0.60	0.90	23	32	30	39
0.35	0.05	0.40	16	21	20	25
	0.10	0.45	19	24	24	31
	0.15	0.50	20	28	26	34
	0.20	0.55	23	29	27	36
	0.25	0.60	24	29	30	36
	0.30	0.65	24	33	31	40
	0.35	0.70	24	33	31	40
	0.40	0.75	24	29	30	36
	0.45	0.80	23	29	27	36
	0.50	0.85	20	28	26	34
	0.55	0.90	19	24	24	31
	0.60	0.95	16	21	20	25

involved in the trial is a standard therapy already available in the market. It is assumed that the responder rates for the test compound and the control are given by 10% ($p_1 = 10\%$) and 35% ($p_2 = 35\%$), respectively. The objective of the planning trial is to confirm such a difference truly exists. It is desirable to have a sample size, which can produce 80% power at 5% level of significance. According to Table 5.3, the sample size needed per arm is given by 39.

5.4 Optimal Multiple-Stage Designs for Single-Arm Trials

In phase II cancer trials, it is undesirable to stop a study early when the test drug is promising. On the other hand, it is desirable to terminate the study as early as possible when the treatment is not effective. For this purpose, an optimal multiple-stage design is often employed to determine whether a study drug holds sufficient promise to warrant further testing. In what follows, procedures for sample size calculation under various optimal multiple-stage designs are introduced.

5.4.1 Optimal Two-Stage Designs

The concept of an optimal two-stage design is to permit early stopping when a moderately long sequence of initial failures occurs. Denote the number of subjects studied in the first and second stage by n_1 and n_2, respectively. Under a two-stage design, n_1 patients are treated at the first stage. If there are less than r_1 responses, then stop the trial. Otherwise, stage 2 is implemented by including the other n_2 patients. A decision regarding whether the test drug is a promising compound is then made based on the response rate of the $N = n_1 + n_2$ subjects. Let p_0 be the undesirable response rate and p_1 be the desirable response rate ($p_1 > p_0$). If the response rate of a test drug is at the undesirable level, one wishes to reject it as an ineffective compound with a high probability (or the false-positive rate is low), and if its response rate is at the desirable level, not to reject it as a promising compound with a high probability (or the false-negative rate is low). As a result, it is of interest to test the following hypotheses:

$$H_0: p \leq p_0 \quad \text{versus} \quad H_a: p \geq p_1.$$

Rejection of H_0 (or H_a) means that further (or no further) study of the test drug should be carried out. Note that under the above hypotheses, the usual type I error is the false-positive rate in accepting an ineffective drug and the type II error is the false-negative rate in rejecting a promising compound.

To select among possible two-stage designs with specific type I and type II errors, Simon (1989) proposed to use the optimal design that achieves the minimum expected sample size when the response rate is p_0. Let EN be the expected sample size. Then, EN can be obtained as

$$EN = n_1 + (1 - PET)n_2,$$

where PET is the probability of early termination after the first stage, which depends upon the true probability of response p. At the end of the first stage, we would terminate the trial early and reject the test drug if r_1 or fewer responses are observed. As a result, PET is given by

$$PET = B(r_1; p, n_1),$$

where $B(\cdot; p, n_1)$ denotes the cumulative binomial distribution with parameter (p, n_1). We would reject the test drug at the end of the second stage if r or fewer responses are observed. Hence, the probability of rejecting the test drug with success probability p is given by

$$B(r_1; p, n_1) + \sum_{x=r_1+1}^{\min(n_1, r)} b(x; p, n_1) B(r - x; p, n_2),$$

where $b(\cdot; p, n_1)$ denotes the binomial probability function with parameter (p, n_1). For specified values of p_0, p_1, α, and β, Simon's optimal two-stage design can be obtained as the two-stage design that satisfies the error constraints and minimizes the expected sample size when the response probability is p_0. As an alternative design, Simon (1989) also proposed to seek the minimum total sample size first and then achieve the minimum expected sample size for the fixed total sample size when the response rate is p_0. This design is referred to as the minimax design.

Tables 5.4 and 5.5 provide sample sizes for optimal two-stage designs and minimax designs for a variety of design parameters, respectively. The tabulated results include the optimal sample size n_1 for the first stage, the maximum sample size n, the critical value r_1 at the end of the first stage, and the critical value r at the end of the trial. For example,

TABLE 5.4

Sample Sizes and Critical Values for Two-Stage Designs ($p_1 - p_0 = 0.15$)

				Optimal Design		Minimax Design	
p_0	p_1	α	β	r_1/n_1	r/N	r_1/n_1	r/N
0.05	0.20	0.10	0.10	0/12	3/37	0/13	3/32
		0.05	0.20	0/10	3/29	0/13	3/27
		0.05	0.10	1/21	4/41	1/29	4/38
0.10	0.25	0.10	0.10	2/21	7/50	2/27	6/40
		0.05	0.20	2/18	7/43	2/22	7/40
		0.05	0.10	2/21	10/66	3/31	9/55
0.20	0.35	0.10	0.10	5/27	16/63	6/33	15/58
		0.05	0.20	5/22	19/72	6/31	15/53
		0.05	0.10	8/37	22/83	8/42	21/77
0.30	0.45	0.10	0.10	9/30	29/82	16/50	25/69
		0.05	0.20	9/27	30/81	16/46	25/65
		0.05	0.10	13/40	40/110	27/77	33/88
0.40	0.55	0.10	0.10	16/38	40/88	18/45	34/73
		0.05	0.20	11/26	40/84	28/59	34/70
		0.05	0.10	19/45	49/104	24/62	45/94
0.50	0.65	0.10	0.10	18/35	47/84	19/40	41/72
		0.05	0.20	15/28	48/83	39/66	40/68
		0.05	0.10	22/42	60/105	28/57	54/93
0.60	0.75	0.10	0.10	21/34	47/71	25/43	43/64
		0.05	0.20	17/27	46/67	18/30	43/62
		0.05	0.10	21/34	64/95	48/72	57/84
0.70	0.85	0.10	0.10	14/20	45/59	15/22	40/52
		0.05	0.20	14/19	46/59	16/23	39/49
		0.05	0.10	18/25	61/79	33/44	53/68
0.80	0.95	0.10	0.10	5/7	27/31	5/7	27/31
		0.05	0.20	7/9	26/29	7/9	26/29
		0.05	0.10	16/19	37/42	31/35	35/40

TABLE 5.5

Sample Sizes and Critical Values for Two-Stage Designs ($p_1 - p_0 = 0.20$)

				Optimal Design		Minimax Design	
p_0	p_1	α	β	r_1/n_1	r/n	r_1/n_1	r/n
0.05	0.25	0.10	0.10	0/9	2/24	0/13	2/20
		0.05	0.20	0/9	2/17	0/12	2/16
		0.05	0.10	0/9	3/30	0/15	3/25
0.10	0.30	0.10	0.10	1/12	5/35	1/16	4/25
		0.05	0.20	1/10	5/29	1/15	5/25
		0.05	0.10	2/18	6/35	2/22	6/33
0.20	0.40	0.10	0.10	3/17	10/37	3/19	10/36
		0.05	0.20	3/13	12/43	4/18	10/33
		0.05	0.10	4/19	15/54	5/24	13/45
0.30	0.50	0.10	0.10	7/22	17/46	7/28	15/39
		0.05	0.20	5/15	18/46	6/19	16/39
		0.05	0.10	8/24	24/63	7/24	21/53
0.40	0.60	0.10	0.10	7/18	22/46	11/28	20/41
		0.05	0.20	7/16	23/46	17/34	20/39
		0.05	0.10	11/25	32/66	12/29	27/54
0.50	0.70	0.10	0.10	11/21	26/45	11/23	23/39
		0.05	0.20	8/15	26/43	12/23	23/37
		0.05	0.10	13/24	35/61	14/27	32/53
0.60	0.80	0.10	0.10	6/11	26/38	18/27	14/35
		0.05	0.20	7/11	30/43	8/13	25/35
		0.05	0.10	12/19	37/53	15/26	32/45
0.70	0.90	0.10	0.10	6/9	22/28	11/16	20/25
		0.05	0.20	4/6	22/27	19/23	21/26
		0.05	0.10	11/15	29/36	13/18	26/32

the first line in Table 5.5 corresponds to a design with $p_0 = 0.20$ and $p_1 = 0.40$. The optimal two-stage design gives $(r_1/n_1, r/n) = (3/13, 12/43)$ for achieving an 80% power at the 5% level of significance, that is, $(\alpha, \beta) = (0.05, 0.20)$. In other words, at the first stage, 13 subjects are tested. If no more than 3 subjects respond, then terminate the trial. Otherwise, accrual continues to a total of 43 subjects. We would conclude that the test drug is effective if there are more than 12 (out of 43 subjects) responses.

5.4.2 Flexible Two-Stage Designs

Chen and Ng (1998) proposed optimal multiple-stage flexible designs for phase II trials by simply assuming that the sample sizes are uniformly distributed on a set of k consecutive possible values. As an example, the procedure for obtaining an optimal two-stage flexible design is outlined below.

Let r_i and n_i be the critical value and the sample size for the first stage and R_j and N_j be the critical value and sample size for the second stage. Thus, for a given combination of (n_i, N_j), the expected sample size is given by

$$EN = n_i + (1 - PET)(N_j - n_i),$$

where

$$PET = B(r_i; p, n_i) = \sum_{x \leq r_i} b(x; p, n_i).$$

The probability of rejecting the test drug is then given by

$$B(r_i; p, n_i) + \sum_{x=r_i+1}^{\min(n_i, R_j)} b(x; p, n_i) B(R_j - x; p, N_j - n_i).$$

The average probability of an early termination ($APET$) is the average of PET for all possible n_i. The average total probability of rejecting the test drug ($ATPRT$) is the average of the above probability for all possible combinations of (n_i, N_j). The average expected sample size (AEN) is the average of EN. Chen and Ng (1998) considered the following criteria for obtaining an optimal flexible design. If the true response rate is p_0, we reject the test drug with a very high probability (i.e., $ATPRT \geq 1 - \alpha$). If the true response rate is p_1, we reject the test drug with a very low probability (i.e., $ATPRT \leq \beta$). There are many solutions of (r_i, n_i, R_j, N_j)'s that satisfy the α and β requirements for the specific p_0 and p_1. The optimal design is the one that has minimum AEN when $p = p_0$. The minimax design is the one that has the minimum N_k and the minimum AEN within this fixed N_k when $p = p_0$.

Tables 5.6 through 5.9 provide sample sizes for flexible two-stage designs and minimax designs for a variety of design parameters, respectively. The tabulated results include the optimal sample size n_i and the critical value r_i for the first stage and the total sample size N_j and critical value R_j at the end of the second stage. For example, the second line in Table 5.6 corresponds to a design with $p_0 = 0.10$ and $p_1 = 0.30$. The flexible two-stage design gives 1/11–17, 2/18 for the first stage and 3/24, 4/25–28, 5/29–31 for the second stage for achieving 90% power at the 10% level of significance. The optimal flexible two-stage design allows the first-stage sample size to range from 11 (n_1) to 18 (n_8). The critical value r_i is 1 if n_i ranges from 11 to 17, and 2 if n_i is 18. If the observed responses are greater than r_i, we accrue $27 - n_i$ additional subjects at the second stage. The flexible optimal two-stage design allows the total sample size to range from 24 (N_1) to 31 (N_8). The rejection boundary R_j is 3 if N_j is 24, 4 if N_j ranges from 25 to 28, and 5 if N_j ranges from 29 to 31.

5.4.3 Optimal Three-Stage Designs

The disadvantage of a two-stage design is that it does not allow early termination if there is a long run of failures at the start. To overcome this disadvantage, Ensign et al. (1994) proposed an optimal three-stage design, which modifies the optimal two-stage design. The optimal three-stage design is implemented by testing the following similar hypotheses:

$$H_0: p \leq p_0 \quad \text{versus} \quad H_a: p \geq p_1.$$

Rejection of H_0 (or H_a) means that further (or not further) study of the test drug should be carried out. At stage 1, n_1 patients are treated. We would reject H_a (i.e.,., the test treatment is not responding) and stop the trial if there is no response. If there are one or more responses, then proceed to stage 2 by including additional n_2 patients. We would reject H_a and stop

TABLE 5.6

Sample Sizes and Critical Values for Optimal Flexible Two-Stage Designs
($p_1 - p_0 = 0.15$)

p_0	p_1	α	β	r_i/n_i	R_j/N_j
0.05	0.20	0.10	0.10	0/15–16,1/17–22	2/30–31,3/32–37
		0.05	0.20	0/10–12,1/13–17	3/27–34
		0.05	0.10	1/17–24	4/41–46,5/47–48
0.10	0.25	0.10	0.10	2/19–25,3/26	6/44–45,7/46–51
		0.05	0.20	1/13–15,2/16–20	6/40,7/41–45,8/46–47
		0.05	0.10	2/21–24,3/25–28	9/57–61,10/62–64
0.20	0.35	0.10	0.10	6/28–31,7/32–35	15/62,16/63–65,
					17/66–68,18/69
		0.05	0.20	4/18–21,5/22–24,6/25	17/62–64,18/65–69,
		0.05	0.20	6/31,7/32–34,8/35–38	22/82–85,23/86–89
0.30	0.45	0.10	0.10	9/31,10/32–33	27/75–77,28/78–80
				11/34–37,12/38	29/81–82
		0.05	0.20	7/23,8/24–25,	27/73,28/74–76,
				9/26–29,10/30	29/77–78,30/79–80
		0.05	0.20	11/35–36,12/37–39,	36/98–99,37/100–102,
				13/40–42	38/103–104,39/105
0.40	0.55	0.10	0.10	12/30–31,13/32–33,	37/80–81,38/82–84,
				14/34–35,15/36–37	39/85–86,40/87
		0.05	0.20	11/25–26,12/27–29,	37/78,38/79–80,
				13/30–31,14/32	39/81–82,40/83–85
		0.05	0.10	16/38–39,17/40–41,	49/104–105,50/106–107,
				18/42–44,19/45	51/108–109,52/110–111
0.50	0.65	0.10	0.10	15/30,16/31–32,17/33–34,	44/78–79,45/80–81,
				18/35–36,19/37	46/82–83,47/84,48/85
		0.05	0.20	12/23,13/24–25,14/26–27,	45/77–78,46/79–80,
				15/28–29,16/30	47/81–82,48/83,49/84
		0.05	0.10	21/40,22/41–42,23/43–44,	59/103–104,60/105–106,
				24/45–46,25/47	61/107,62/108–109,63/110
0.60	0.75	0.10	0.10	16/27,17/28,18/29–30,	44/67,45/68,46/69–70,
				19/31–32,20/33,21/34	47/71,48/72,49/73–74
		0.05	0.20	14/22–23,15/24,16/25	46/68,47/69,48/70–71
		0.05	0.10	20/32–33,21/34,22/36–36,	61/90–91,62/92,63/93–94,
				23/37/24/38–39	64/95,65/96–97
0.70	0.85	0.10	0.10	13/19,14/20,15/21,	40/53,41/54,42/55,43/56,
				16/22–23,17/24,18/25–26	44/57–58,45/59,46/60
		0.05	0.20	9/13,10/14,11/15,12/16–17,	44/56–57,45/58,46/59,
				13/18,14/19,15/20	47/60,48/61–62,49/63
		0.05	0.10	17/24,18/26,19/26,	57/73–74,58/75,59/76–77,
				20/27–28,21/29,22/30,	60/78,61/79,62/80
				23/31	
0.80	0.95	0.10	0.10	8/10,9/11,10/12–13,	24/28,25/29,26/30,27/31,
				11/14,12/15,13/16,14/17	28/32,29/33,30/34–35
		0.05	0.20	7/9,8/10,9/11,10/12,	25/28,26/29,27/30,
				11/13,12/14,13/15,14/16	28/31–32,29/33,30/34,
					31/35
		0.05	0.10	10/12,11/13–14,12/15,	35/40,36/41,37/42,38/43,
				13/16,14/17,15/18,16/19	39/44,40/45–46,41/47

TABLE 5.7

Sample Sizes and Critical Values for Optimal Flexible Two-Stage Designs
($p_1 - p_0 = 0.20$)

p_0	p_1	α	β	r_i/n_i	R_j/N_j
0.05	0.25	0.10	0.10	0/8–13,1/14–15	1/18,2/19–25
		0.05	0.20	0/5–10,1/11–12	2/17–22,3/23–24
		0.05	0.10	0/8–13,1/14–15	2/24,3/25–31
0.10	0.30	0.10	0.10	1/11–17,2/18	3/24,4/25–28,5/29–31
		0.05	0.20	1/8–12,2/13–15	4/26,5/27–32,6/33
		0.05	0.10	1/12–14,2/15–19	6/36–39,7/40–43
0.20	0.40	0.10	0.10	2/14,3/15–17,4/18–21	9/35–36,10/37–38,11/39–42
		0.05	0.20	2/10–12,3/13–15,4/16–17	10/33–35,11/36–40
		0.05	0.10	4/18–20,5/21–24,6/25	13/48,14/49–51,15/52–55
0.30	0.50	0.10	0.10	4/14–16,5/17–19,	15/40–41,16/42–44,
				6/20–21	17/45–46,18/47
		0.05	0.20	3/11,4/12–14,	16/40–41,16/42–44,
				5/15–16/6/17–18	18/45–46,18/47
		0.05	0.10	6/19–20,7/21–23,	21/55,22/56–58,
				8/24–26	23/59–60,24/61–62
0.40	0.60	0.10	0.10	6/15–16,7/17–19,	21/44–45,22/46–47,
				8/20,9/21–22	23/48–49,24/50–51
		0.05	0.20	5/12–13,6/14,	22/44–45,23/46–47,24/48–49,
				7/15–16,8/17–19	25/50,26/51
		0.05	0.10	8/20,9/21–22,10/23–24,	28/58,29/59–60,30/61–62,
				11/25–26,12/27	31/63,32/64–65
0.50	0.70	0.10	0.10	7/15,8/16–17,9/18,	24/41–42,25/43/44,
				10/19–20,11/21,12/22	26/45,27/46–47,28/48
		0.05	0.20	5/10,6/11–12,	25/42,26/43–44,27–45,
				7/13–14,8/15,9/16–17	28/46–47,29/48,30/49
		0.05	0.10	10/19–20,11/21,	33/55–56,34/57–58,
				12/22–23,13/24–25,14/26	35/59,36/60–61,37/62
0.60	0.80	0.10	0.10	7/12,8/13–14,9/15,	24/35–36,25/37,26/38
				10/16–17,11/18,12/19	27/39–40,28/41,29/42
		0.05	0.20	5/8–9,6/10,7/11,	25/35–36,26/37,27/38,
				8/12–13,9/14–15	28/39–40,29/41,30/42
		0.05	0.10	11/17–18,12/19,13/20–21,	34/48–49,35/50–51,
				14/22,15/23,16/24	36/52,37/53–54,38/55
0.70	0.90	0.10	0.10	6/9,7/10,8/11,9/12–13,	18/23,19/24,20/25–26,
				10/14,11/15–16	21/27,22/28,23/29,24/30
		0.05	0.20	4/6,5/7,6/8,7/9,	22/27,23/28–29,24/30,
				8/10–11,9/12,10/13	25/31,26/32–33,27/34
		0.05	0.10	7/10,8/11,9/12–13,	27/34/28/35,29/36,30/37–38,
				10/14,11/15,12/16,13/17	31/39,32/40,33/41

the trial if the total number of responses is less than or equal to a prespecified number of r_2; otherwise continue to stage 3. At stage 3, n_3 more patients are treated. We would reject H_a if the total responses for the three stages combined is less than or equal to r_3. In this case, we conclude the test drug is ineffective. On the other hand, if there are more than r_3 responses, we reject H_0 and conclude the test drug is effective. Based on the concept of the

TABLE 5.8

Sample Sizes and Critical Values for Minimax Flexible Two-Stage Designs ($p_1 - p_0 = 0.15$)

p_0	p_1	α	β	r_i/n_i	R_j/N_j
0.05	0.20	0.10	0.10	0/16–22,1/23	2/26–28,3/29–33
		0.05	0.20	0/10–17	2/23,2/24–30
		0.05	0.10	0/22–27,1/28–29	3/33–34,4/35–40
0.10	0.25	0.10	0.10	1/25–27,2/28–32	5/37,6/38–42,7/43–44
		0.05	0.20	1/22–24,2/25–29	6/33–37,7/38–40
		0.05	0.10	2/25–29,3/30–32	8/49–52,9/53–56
0.20	0.35	0.10	0.10	6/37–39,7/40–42,8/43–440	14/54–55,15/56–59,16/60–61
		0.05	0.20	6/28,6/29–31,7/32–35	14/50–51,15/52–54,16/55–57
		0.05	0.10	8/41–45,9/46–48	19/71–72,20/73–74,21/75–78
0.30	0.45	0.10	0.10	11/43,12/44–46, 13/47–48,14/49–50	23/64,24/65–67, 25/68–69,26/70–71
		0.05	0.20	10/36,11/37, 12/38–39,13/40–43	23/60,24/61–63, 25/64–65,26/66–67
		0.05	0.10	15/50–52,16/53–55, 17/56–57	32/85–86,33/87–89, 34/90–91,35/92
0.40	0.55	0.10	0.10	16/43–44,17/45–46, 18/47,19/48–49,20/50	32/69–70,33/71, 34/72–73,35/74–75,36/76
		0.05	0.20	13/34–35,14/36, 15/37–39,16/40–41	32/65–66,33/67–68, 34/69–70,35/71,36/72
		0.05	0.10	23/60–61,24/62–63, 25/64–65,26/66–67	43/91,44/92–93, 45/94,46/95–96,47/97–98
0.50	0.65	0.10	0.10	19/41,20/42–43, 21/44–45,22/57–57,23/48	38/67,39/68–69,40/70–71, 41/72,42/73–74
		0.05	0.20	16/33,17/34–35, 18/36–37,19/38–39,20/40	38/64–65,39/66, 40/67–68,41/69,42/70–71
		0.05	0.10	26/53,27/54–55,28/56, 29/57,30/58–59,31/60	52/89–90,53/91–92, 54/93–94,55/95,56/96
0.60	0.75	0.10	0.10	22/38–39,23/40,24/41–42, 25/43,26/44–45	40/60,41/61,42/62–63, 43/64,44/65–66,45/67
		0.05	0.20	18/31,19/32,20/33–34, 21/35,22/36–37,23/38	40/57–58,41/59, 42/60–61,43/62,44/63–64
		0.05	0.10	23/39,24/40–41,25/42–43, 26/44,27/45,28/46	54/80,55/81,56/82, 57/83–84,58/85,59/86–87
0.70	0.85	0.10	0.10	19/28–29,20/30,21/31, 22/32,23/33–34,24/35	36/46–47,37/48,38/49, 39/50–51,40/52,41/53
		0.05	0.20	18/25,19/26,20/27–28, 21/29,22/30,23/31,24/32	36/45,37/46–47,38/48, 39/49,40/50–51,41/52
		0.05	0.10	26/38,27/39,28/40,29/41, 30/42–43,31/44,32/45	48/62,49/63,50/64,51/65, 52/66,53/67,54/68–69
0.80	0.95	0.10	0.10	9/12,10/13,11/14,12/15, 13/16,14/17,15/18,16/19	23/26,24/27–28,25/29, 26/30,27/31,28/32,29/33
		0.05	0.20	6/8,7/9,8/10,9/11, 10/12–13,11/14,12/15	23/26,24/27,25/28,26/29, 27/30,28/31,29/32,30/33
		0.05	0.10	22/26,23/27,24/28,25/29, 26/30,27/31,28/32,29/33	31/35,32/36,33/37,34/38, 35/39–40,36/41,37/42

TABLE 5.9

Sample Sizes and Critical Values for Minimax Flexible Two-Stage Designs
($p_1 - p_0 = 0.20$)

p_0	p_1	α	β	r_i/n_i	R_j/N_j
0.05	0.25	0.10	0.10	0/8–15	1/17,2/18–24
		0.05	0.20	0/6–12,1/13	2/14–21
		0.05	0.10	0/10–16,1/17	2/21–22,3/23–28
0.10	0.30	0.10	0.10	0/11–13,1/14–18	3/22–23,4/24–26,5/27–29
		0.05	0.20	0/11–14,1/15–18	3/19,4/20–22,5/23–26
		0.05	0.10	1/17–20,2/21–23,3/24	5/28–30,6/31–35
0.20	0.40	0.10	0.10	3/22–23,4/24,	8/30–31,9/32/33,
				5/25–27,6/28–29	10/34–37
		0.05	0.20	2/14,3/15–18,4/19–21	9/28–31,10/32–34,11/35
		0.05	0.10	5/27–29,11/40,12/41–42,	
				6/30–32,7/33–34	13/43–45,14/46–47
0.30	0.50	0.10	0.10	6/24–25,7/26–29,	13/35,14/36–37,
				8/30,9/31	15/38–40,16/41–42
		0.05	0.20	5/18–19,6/20–22,14/33–35,	
				15/36–37,7/23–24,8/25	16/38–39,17/40
		0.05	0.10	7/27,8/28–29,19/47–49,	
				20/50–51,9/30–31,10/32–34	21/52–53,22/54
0.40	0.60	0.10	0.10	8/23–24,9/25–26,	18/37–38,19/39–40,
				10/27,11/28–29,12/30	20/41,21/42–43,22/44
		0.05	0.20	6/18,7/19–20,8/21–22,	18/35–36,19/37,
				9/23,10/24–25	20/38–39,21/40,22/41–42
		0.05	0.10	10/26–27,11/28,	25/50–51,26/52,
				12/29–31,13/32–33	27/53–54,28/55–56,29/57
0.50	0.70	0.10	0.10	8/18,9/19–20,10/21,	21/36,22/37–38,23/39,
				11/22–23,12/24,13/25	24/40–41,25/42–43
		0.05	0.20	7/15,8/16–17,9/18–19,	21/34,22/35–36,23/37–38,
				10/20,11/21,12/22	24/39,25/40,26/41
		0.05	0.10	14/30–31,15/32,16/33–34,	18/47,29/48,30/49–50,
				17/35,18/36,19/37	31/51,32/52–53,33/54
0.60	0.80	0.10	0.10	9/17–18,10/19,11/20,	21/30–31,22/32,23/33,
				12/21,13/22–23,14/24	24/34–35,25/36,26/37
		0.05	0.20	6/11,7/12–13,8/14,	21/29,22/30–31,23/32,
				9/15–16,10/17,11/18	24/33,25/34–35,26/36
		0.05	0.10	11/19,12/20–21,	29/41,30/42–43,31/44,
				13/22,14/23,15/24–26	32/45,33/46–47,34/48
0.70	0.90	0.10	0.10	5/8,6/9,7/10–11,8/12,	17/22,18/23,19/24,20/25,
				9/13,10/14–15	21/26,22/27–28,23/29
		0.05	0.20	5/8,6/9,7/10,8/11,	17/21,18/22,19/23,20/24,
				9/12–13,10/14,11/15	21/25,22/26–27,23/28
		0.05	0.10	8/12,9/13,10/14,11/15,	24/30,35/31,26/32,27/33,
				12/16–17,13/18,14/19	28/34–35,29/36,30/37

above three-stage design, Ensign et al. (1994) considered the following to determine the sample size. For each value of n_1 satisfying

$$(1 - p_1)^{n_1} < \beta,$$

where

$$\beta = P(\text{reject } H_a \mid p_1),$$

computing the values of r_2, n_2, r_3, and n_3 that minimize the null expected sample size $EN(p_0)$ subject to the error constraints α and β, where

$$EN(p) = n_1 + n_2\{1 - \beta_1(p)\} + n_3\{1 - \beta_1(p) - \beta_2(p)\},$$

and β_i are the probability of making type II error evaluated at stage i. Ensign et al. (1994) use the value of

$$\beta - (1 - p_1)^{n_1}$$

as the type II error rate in the optimization along with type I error

$$\alpha = P(\text{reject } H_0 \mid p_0)$$

to obtain r_2, n_2, r_3, and n_3. Repeating this, n_i can then be chosen to minimize the overall $EN(p_0)$.

Tables 5.10 and 5.11 provide sample sizes for optimal three-stage designs based on the method proposed by Ensign et al. (1994) for a variety of design parameters. The tabulated results include the optimal size n_i and the critical value r_i of the ith stage. For example, the result in Table 5.10 corresponding to a design with $p_0 = 0.25$ and $p_1 = 0.40$ gives 0/6, 7/26, 24/75 for achieving an 80% power at the 5% level of significance. In other words, at the first stage, 6 subjects are treated. If there is no response, then the trial is terminated. Otherwise, accrual continues to a total of 26 subjects at the second stage. If no more than 7 subjects respond, then stop the trial. Otherwise, proceed to the third stage by recruiting 49 additional subjects. We would conclude that the test drug is effective if there are more than 24 responses for the subjects in the three stages combined.

Note that the optimal three-stage designs proposed by Ensign et al. (1994) restrict the rejection region in the first stage to be zero response, and the sample size to at least 5. As an alternative, Chen (1997b) also extended Simon's two-stage to a three-stage design without these restrictions. As a result, sample sizes can be obtained by computing the values of r_1, n_1, r_2, n_2, r_3, and n_3 that minimize the expected sample size:

$$EN = n_1 + (1 - PET_1)n_2 + (1 - PET_{all})n_3$$

$$PET_1 = B(r_1; n_1, p) = \sum_{x \leq r_1} b(x; n_1, p)$$

$$PET_{all} = PET_1 + \sum_{x = r_1 + 1}^{\min(n_1, r_2)} b(d; n, p)B(r_2 - x; n_2, p).$$

TABLE 5.10

Sample Sizes n_i and Critical Values r_i for Optimal Three-Stage Designs— Ensign et al. (1994) ($p_1 - p_0 = 0.15$)

p_0	p_1	α	β	Stage 1 r_1/n_1	Stage 2 $r_2/(n_1 + n_2)$	Stage 3 $r_3/(n_1 + n_2 + n_3)$
0.05	0.20	0.10	0.10	0/12	1/25	3/38
		0.05	0.20	0/10	2/24	3/31
		0.05	0.10	0/14	2/29	4/43
0.10	0.25	0.10	0.10	0/11	3/29	7/50
		0.05	0.20	0/9	3/25	7/43
		0.05	0.10	0/13	3/27	10/66
0.15	0.30	0.10	0.10	0/12	4/28	11/55
		0.05	0.20	0/9	5/27	12/56
		0.05	0.10	0/13	6/35	16/77
0.20	0.35	0.10	0.10	0/11	7/34	16/63
		0.05	0.20	0/6	6/28	18/67
		0.05	0.10	0/9	10/44	23/88
0.25	0.40	0.10	0.10	0/8	8/32	23/76
		0.05	0.20	0/6	7/26	24/75
		0.05	0.10	0/9	11/41	30/95
0.30	0.45	0.10	0.10	0/7	13/41	28/79
		0.05	0.20	0/7	9/27	31/84
		0.05	0.10	0/9	14/43	38/104
0.35	0.50	0.10	0.10	0/9	12/34	33/81
		0.05	0.20	0/5	12/31	37/88
		0.05	0.10	0/8	17/45	45/108
0.40	0.55	0.10	0.10	0/11	16/38	40/88
		0.05	0.20	0/5	14/32	40/84
		0.05	0.10	0/10	19/45	49/104
0.45	0.60	0.10	0.10	0/6	15/34	40/78
		0.05	0.20	0/5	12/25	47/90
		0.05	0.10	0/6	20/42	59/114
0.50	0.65	0.10	0.10	0/5	16/32	46.84
		0.05	0.20	0/5	12/25	47/90
		0.05	0.10	0/6	20/42	59/114
0.55	0.70	0.10	0.10	0/7	19/34	46/75
		0.05	0.20	0/5	15/26	48/76
		0.05	0.10	0/5	23/40	64/96
0.60	0.75	0.10	0.10	0/5	21/34	47/71
		0.05	0.20	0/5	13/21	49/72
		0.05	0.10	0/5	14/23	90/98
0.65	0.80	0.10	0.10	0/5	17/26	47/66
		0.05	0.20	0/5	12/18	49/67
		0.05	0.20	0/5	8/13	74/78
0.70	0.85	0.10	0.10	0/5	14/20	45/59
		0.05	0.20	0/5	14/19	46/59
		0.05	0.10	0/5	12/17	68/72
0.75	0.90	0.10	0.10	0/5	16/21	36/44
		0.05	0.20	0/5	10/13	40/48
		0.05	0.10	0/5	8/11	55/57
0.80	0.95	0.10	0.10	0/5	5/7	27/31
		0.05	0.20	0/5	7/9	26/29
		0.05	0.10	0/5	8/10	44/45

TABLE 5.11

Sample Sizes n_i and Critical Values r_i for Optimal Three-Stage Designs— Ensign et al. (1994) ($p_1 - p_0 = 0.20$)

p_0	p_1	α	β	Stage 1 r_1/n_1	Stage 2 $r_2/(n_1 + n_2)$	Stage 3 $r_3/(n_1 + n_2 + n_3)$
0.05	0.25	0.10	0.10	0/9	1/19	2/25
		0.05	0.20	0/7	1/15	3/26
		0.05	0.10	0/9	1/22	3/30
0.10	0.30	0.10	0.10	0/10	2/19	4/26
		0.05	0.20	0/6	2/17	5/29
		0.05	0.10	0/9	3/22	7/45
0.15	0.35	0.10	0.10	0/9	2/16	7/33
		0.05	0.20	0/5	3/17	9/41
		0.05	0.10	0/9	4/23	10/44
0.20	0.40	0.10	0.10	0/8	3/16	11/42
		0.05	0.20	0/5	4/17	12/43
		0.05	0.10	0/9	4/23	15/54
0.25	0.45	0.10	0.10	0/6	6/23	14/44
		0.05	0.20	0/5	5/17	16/48
		0.05	0.10	0/7	6/22	20/61
0.30	0.50	0.10	0.10	0/6	6/20	17/46
		0.05	0.20	0/5	5/15	19/49
		0.05	0.10	0/8	8/24	24/63
0.35	0.55	0.10	0.10	0/6	7/20	20/47
		0.05	0.20	0/6	8/20	19/42
		0.05	0.10	0/5	10/26	29/67
0.40	0.60	0.10	0.10	0/5	8/20	22/46
		0.05	0.20	0/5	7/16	24/48
		0.05	0.10	0/5	9/22	30/61
0.45	0.65	0.10	0.10	0/5	10/21	26/50
		0.05	0.20	0/5	7/15	24/43
		0.05	0.10	0/5	15/30	32/59
0.50	0.70	0.10	0.10	0/5	11/21	26/45
		0.05	0.20	0/5	8/15	26/43
		0.05	0.10	0/5	12/23	34/57
0.55	0.75	0.10	0.10	0/5	10/18	26/41
		0.05	0.20	0/5	9/15	28/43
		0.05	0.10	0/5	10/18	35/54
0.60	0.80	0.10	0.10	0/5	6/11	26/38
		0.05	0.20	0/5	7/11	30/43
		0.05	0.10	0/5	12/19	37/53
0.65	0.85	0.10	0.10	0/5	10/15	25/34
		0.05	0.20	0/5	10/14	25/33
		0.05	0.20	0/5	10/15	33/44
0.70	0.90	0.10	0.10	0/5	6/9	22/28
		0.05	0.20	0/5	4/6	22/27
		0.05	0.10	0/5	11/15	29/36
0.75	0.95	0.10	0.10	0/5	6/8	16/19
		0.05	0.20	0/5	9/11	19/22
		0.05	0.10	0/5	7/9	24/28

Tables 5.12 through 5.15 provide sample sizes for optimal three-stage designs and mini-max designs based on the method proposed by Chen (1997b) for a variety of design parameters. For example, the result in Table 5.12 corresponding to a design with $p_0 = 0.25$ and $p_1 = 0.40$ gives 4/17, 12/42, 25/79 for achieving an 80% power at the 5% level of significance. In other words, at the first stage, 17 subjects are treated. If no more than four responses are obtained, then the trial is terminated. Otherwise, accrual continues to a total of 42 subjects at the second stage. If no more than 12 subjects respond, then stop the trial. Otherwise, proceed to the third stage by recruiting 37 additional subjects. We would conclude that the test drug is effective if there are more than 25 responses for the subjects in the three stages combined.

5.5 Flexible Designs for Multiple-Arm Trials

In the previous section, we introduced procedures for sample size calculation under (flexible) optimal multiple-stage designs for single-arm phase II cancer trials. Sargent and Goldberg (2001) proposed a flexible optimal design considering a phase II trial that allows clinical scientists to select the treatment to proceed for further testing for a phase III trial based on other factors when the difference in the observed response rates between two treatments falls into the interval $[-\delta, \delta]$, where δ is a prespecified quantity. The proposed rule is that if the observed difference in the response rates of the treatments is larger than δ, then the treatment with the highest observed response rate is selected. On the other hand, if the observed difference is less than or equal to δ, other factors may be considered in the selection. In this framework, it is not essential that the very best treatment is definitely selected, rather it is important that a substantially inferior treatment is not selected when a superior treatment exists.

To illustrate the concept proposed by Sargent and Goldberg (2001), for simplicity, consider a two-arm trial. Let p_1 and p_2 denote the true response rates for the poor treatment and the better treatment, respectively. Without loss of generality, assume that $p_2 > p_1$.

Let $\hat{p}_1$ and $\hat{p}_2$ denote the corresponding observed response rates for treatment 1 and treatment 2, respectively. Sargent and Goldberg (2001) considered the probability of correctly choosing the better treatment, that is,

$$P_{Corr} = P\{\hat{p}_2 > \hat{p}_1 + \delta \,|\, p_1, p_2\}$$

and the probability of the difference between the two observed response rates falling into the ambiguous range of $[-\delta, \delta]$, that is,

$$P_{Amb} = P\{-\delta \leq \hat{p}_2 - \hat{p}_1 \leq \delta \,|\, p_1, p_2\}.$$

Assuming that each treatment arm has the same number of subjects (i.e., $n_1 = n_2 = n$), the above two probabilities are given by

$$P_{Corr} = \sum_{x=0}^{n}\sum_{y=0}^{n} I_{\{(x-y)/n > \delta\}} \binom{n}{x}\binom{n}{y}$$
$$\times p_2^x (1-p_2)^{n-x} p_1^y (1-p_1)^{n-y}$$

TABLE 5.12

Sample Sizes n_i and Critical Values r_i for Optimal Three-Stage Designs—Chen (1997b) ($p_1 - p_0 = 0.15$)

p_0	p_1	α	β	Stage 1 r_1/n_1	Stage 2 $r_2/(n_1 + n_2)$	Stage 3 $r_3/(n_1 + n_2 + n_3)$
0.05	0.20	0.10	0.10	0/13	1/22	3/37
		0.05	0.20	0/10	1/19	3/30
		0.05	0.10	0/14	2/29	4/43
0.10	0.25	0.10	0.10	1/17	3/29	7/50
		0.05	0.20	1/13	3/24	8/53
		0.05	0.10	1/17	4/34	10/66
0.15	0.30	0.10	0.10	2/20	5/33	11/55
		0.05	0.20	2/15	6/33	13/62
		0.05	0.10	3/23	8/46	16/77
0.20	0.35	0.10	0.10	3/21	8/37	17/68
		0.05	0.20	3/17	9/37	18/68
		0.05	0.10	5/27	11/49	23/88
0.25	0.40	0.10	0.10	4/20	10/39	24/80
		0.05	0.20	4/17	12/42	25/79
		0.05	0.10	6/26	15/54	32/103
0.30	0.45	0.10	0.10	6/24	14/44	28/79
		0.05	0.20	5/18	14/41	31/84
		0.05	0.10	8/29	19/57	38/104
0.35	0.50	0.10	0.10	7/23	18/49	34/84
		0.05	0.20	6/19	17/43	34/80
		0.05	0.10	9/28	23/60	45/108
0.40	0.55	0.10	0.10	7/21	19/46	38/83
		0.05	0.20	7/19	19/43	39/82
		0.05	0.10	12/31	28/64	54/116
0.45	0.60	0.10	0.10	12/28	27/56	43/85
		0.05	0.20	8/19	21/42	45/86
		0.05	0.10	13/30	29/60	58/112
0.50	0.65	0.10	0.10	10/22	25/48	48/86
		0.05	0.20	8/17	21/39	49/85
		0.05	0.10	14/29	34/63	62/109
0.55	0.70	0.10	0.10	13/25	25/44	47/77
		0.05	0.20	7/14	23/39	49/78
		0.05	0.10	15/28	36/61	65/105
0.60	0.75	0.10	0.10	11/20	26/42	57/71
		0.05	0.20	8/14	23/36	52/77
		0.05	0.10	14/24	36/56	70/105
0.65	0.80	0.10	0.10	11/18	27/40	49/69
		0.05	0.20	8/13	27/38	52/72
		0.05	0.20	16/25	35/50	66/92
0.70	0.85	0.10	0.10	14/20	18/37	45/59
		0.05	0.20	4/7	11/16	44/56
		0.05	0.10	12/18	28/38	58/75
0.75	0.90	0.10	0.10	10/14	23/29	38/47
		0.05	0.20	9/12	21/26	39/47
		0.05	0.10	10/14	23/29	55/67
0.80	0.95	0.10	0.10	5/7	16/19	30/35
		0.05	0.20	2/3	16/19	35/40
		0.05	0.10	6/8	24/28	41/47

TABLE 5.13

Sample Sizes n_i and Critical Values r_i for Optimal Three-Stage
Designs—Chen (1997b) ($p_1 - p_0 = 0.20$)

p_0	p_1	α	β	Stage 1 r_1/n_1	Stage 2 $r_2/(n_1 + n_2)$	Stage 3 $r_3/(n_1 + n_2 + n_3)$
0.05	0.25	0.10	0.10	0/9	1/18	2/26
		0.05	0.20	0/8	1/13	2/19
		0.05	0.10	0/10	1/17	3/30
0.10	0.30	0.10	0.10	0/10	2/19	4/26
		0.05	0.20	0/6	2/17	5/29
		0.05	0.10	1/13	3/23	7/45
0.15	0.35	0.10	0.10	1/12	3/21	7/33
		0.05	0.20	1/9	4/21	8/35
		0.05	0.10	2/15	5/27	11/51
0.20	0.40	0.10	0.10	1/10	6/26	11/43
		0.05	0.20	1/8	5/22	11/38
		0.05	0.10	3/17	7/30	14/50
0.25	0.45	0.10	0.10	3/16	7/25	13/41
		0.05	0.20	2/10	6/20	16/48
		0.05	0.10	4/18	10/33	19/58
0.30	0.50	0.10	0.10	3/13	9/28	17/46
		0.05	0.20	3/11	7/21	18/46
		0.05	0.10	4/16	11/32	23/60
0.35	0.55	0.10	0.10	6/18	13/33	20/48
		0.05	0.20	3/10	9/23	21/47
		0.05	0.10	6/18	15/38	27/62
0.40	0.60	0.10	0.10	7/18	9/26	22/46
		0.05	0.20	3/9	10/23	23/46
		0.05	0.10	6/16	17/38	32/66
0.45	0.65	0.10	0.10	5/13	13/27	26/50
		0.05	0.20	3/8	10/20	29/54
		0.05	0.10	6/15	17/34	34/63
0.50	0.70	0.10	0.10	4/10	12/24	26/45
		0.05	0.20	4/9	13/23	29/49
		0.05	0.10	7/15	19/34	38/65
0.55	0.75	0.10	0.10	5/11	12/21	27/43
		0.05	0.20	6/11	14/23	28/43
		0.05	0.10	7/14	16/27	36/56
0.60	0.80	0.10	0.10	6/11	14/22	29/43
		0.05	0.20	5/9	12/48	28/40
		0.05	0.10	6/11	19/29	38/55
0.65	0.85	0.10	0.10	5/9	13/19	25/34
		0.05	0.20	5/8	13/18	27/36
		0.05	0.20	6/10	16/23	35/47
0.70	0.90	0.10	0.10	5/8	11/15	22/28
		0.05	0.20	3/5	10/13	25/31
		0.05	0.10	6/9	16/21	31/39
0.75	0.95	0.10	0.10	3/5	6/8	16/19
		0.05	0.20	1/2	9/11	19/22
		0.05	0.10	6/8	13/16	24/28

TABLE 5.14

Sample Sizes n_i and Critical Values r_i for Minimax Three-Stage Designs—Chen (1997b) ($p_1 - p_0 = 0.15$)

p_0	p_1	α	β	Stage 1 r_1/n_1	Stage 2 $r_2/(n_1 + n_2)$	Stage 3 $r_3/(n_1 + n_2 + n_3)$
0.05	0.20	0.10	0.10	0/18	1/26	3/32
		0.05	0.20	0/14	1/20	3/27
		0.05	0.10	0/23	1/30	4/38
0.10	0.25	0.10	0.10	1/23	3/33	6/40
		0.05	0.20	1/17	3/30	7/40
		0.05	0.10	1/21	4/39	9/55
0.15	0.30	0.10	0.10	2/23	5/36	11/53
		0.05	0.20	2/19	6/36	11/48
		0.05	0.10	4/35	8/51	14/64
0.20	0.35	0.10	0.10	5/30	9/45	15/58
		0.05	0.20	3/22	7/35	15/53
		0.05	0.10	16/65	19/72	20/74
0.25	0.40	0.10	0.10	6/31	11/46	20/64
		0.05	0.20	7/30	12/42	20/60
		0.05	0.10	9/47	17/67	27/83
0.30	0.45	0.10	0.10	7/29	16/51	25/69
		0.05	0.20	8/29	14/42	25/65
		0.05	0.10	12/46	25/73	33/88
0.35	0.50	0.10	0.10	12/39	20/57	30/72
		0.05	0.20	10/33	18/48	29/66
		0.05	0.10	11/36	22/60	40/94
0.40	0.55	0.10	0.10	10/30	19/48	34/73
		0.05	0.20	13/33	30/63	34/70
		0.05	0.10	20/55	32/77	45/94
0.45	0.60	0.10	0.10	18/41	35/69	38/74
		0.05	0.20	13/32	25/53	38/70
		0.05	0.10	26/58	47/90	50/95
0.50	0.65	0.10	0.10	19/40	24/64	41/72
		0.05	0.20	18/36	36/62	40/68
		0.05	0.10	19/43	34/67	54/93
0.55	0.70	0.10	0.10	23/43	36/60	42/68
		0.05	0.20	18/33	41/64	42/66
		0.05	0.10	23/43	42/84	45/89
0.60	0.75	0.10	0.10	19/35	30/50	43/64
		0.05	0.20	19/32	40/58	42/61
		0.05	0.10	18/46	50/75	57/84
0.65	0.80	0.10	0.10	22/33	26/41	43/60
		0.05	0.20	16/26	27/40	41/55
		0.05	0.20	25/41	37/56	55/75
0.70	0.85	0.10	0.10	15/22	18/37	40/52
		0.05	0.20	11/17	16/24	39/49
		0.05	0.10	13/20	31/42	43/68
0.75	0.90	0.10	0.10	11/17	22/29	33/40
		0.05	0.20	8/12	16/21	33/39
		0.05	0.10	12/17	23/30	45/54
0.80	0.95	0.10	0.10	1/3	17/20	26/30
		0.05	0.20	7/9	16/19	26/29
		0.05	0.10	16/20	31/35	35/40

TABLE 5.15

Sample Sizes n_i and Critical Values r_i for Minimax Three-Stage Designs—Chen (1997b) ($p_1 - p_0 = 0.20$)

p_0	p_1	α	β	Stage 1 r_1/n_1	Stage 2 $r_2/(n_1 + n_2)$	Stage 3 $r_3/(n_1 + n_2 + n_3)$
0.05	0.25	0.10	0.10	0/13	1/18	2/20
		0.05	0.20	0/12	1/15	2/16
		0.05	0.10	0/15	1/21	3/25
0.10	0.30	0.10	0.10	0/12	1/16	4/25
		0.05	0.20	0/11	2/19	5/25
		0.05	0.10	0/14	2/22	6/33
0.15	0.35	0.10	0.10	1/13	3/22	7/32
		0.05	0.20	1/12	3/19	7/28
		0.05	0.10	1/16	4/28	9/38
0.20	0.40	0.10	0.10	2/16	5/26	10/36
		0.05	0.20	2/13	5/22	10/33
		0.05	0.10	2/16	6/28	13/45
0.25	0.45	0.10	0.10	3/18	8/31	13/39
		0.05	0.20	3/15	6/23	13/36
		0.05	0.10	4/21	9/35	17/45
0.30	0.50	0.10	0.10	6/26	11/35	15/39
		0.05	0.20	3/13	8/24	16/39
		0.05	0.10	5/20	12/36	21/53
0.35	0.55	0.10	0.10	2/11	10/27	18/42
		0.05	0.20	4/14	9/24	18/39
		0.05	0.10	10/34	17/45	24/53
0.40	0.60	0.10	0.10	5/17	9/26	20/41
		0.05	0.20	4/12	11/25	21/41
		0.05	0.10	7/20	17/39	27/54
0.45	0.65	0.10	0.10	6/16	13/29	22/41
		0.05	0.20	6/15	12/24	22/39
		0.05	0.10	15/32	28/51	29/53
0.50	0.70	0.10	0.10	7/17	14/28	23/39
		0.05	0.20	7/16	13/25	23/37
		0.05	0.10	8/18	18/34	32/53
0.55	0.75	0.10	0.10	13/23	22/35	24/38
		0.05	0.20	8/15	14/23	24/36
		0.05	0.10	12/22	21/35	32/49
0.60	0.80	0.10	0.10	8/15	14/22	24/35
		0.05	0.20	9/15	23/32	24/34
		0.05	0.10	15/26	24/37	32/45
0.65	0.85	0.10	0.10	4/8	11/17	23/31
		0.05	0.20	6/10	13/18	23/30
		0.05	0.20	16/24	28/37	30/40
0.70	0.90	0.10	0.10	5/9	13/18	20/25
		0.05	0.20	4/7	19/23	20/25
		0.05	0.10	5/9	12/17	26/32
0.75	0.95	0.10	0.10	3/5	6/8	16/19
		0.05	0.20	6/8	14/16	17/20
		0.05	0.10	9/12	19/22	22/26

and

$$P_{Amb} = \sum_{x=0}^{n} \sum_{y=0}^{n} I_{\{-\delta \le (x-y)/n \le \delta\}} \binom{n}{x} \binom{n}{y}$$
$$\times p_2^x (1-p_2)^{n-x} p_1^y (1-p_1)^{n-y},$$

where I_A is the indicator function of event A, that is, $I_A = 1$ if event A occurs and $I_A = 0$ otherwise. Sargent and Goldberg (2001) suggested that n be selected such that $P_{corr} + \rho P_{Amb} > \lambda$, a prespecified threshold. Table 5.16 provides results for $\rho = 0$ and $\rho = 0.5$ for different sample sizes for $p_2 = 0.35$ and $\delta = 0.05$.

Liu (2001) indicated that by the central limit theorem, we have

$$P_{Corr} \approx 1 - \Phi\left(\frac{\delta - \varepsilon}{\sigma}\right)$$

and

$$P_{Amb} \approx \Phi\left(\frac{\delta - \varepsilon}{\sigma}\right) - \Phi\left(\frac{-\delta - \varepsilon}{\sigma}\right),$$

where Φ is the standard normal cumulative distribution function, $\varepsilon = p_2 - p_1$ and

$$\sigma^2 = \frac{p_1(1-p_1) + p_2(1-p_2)}{n}.$$

As indicated in Chapter 4, the power of the test for the following hypotheses

$$H_0: p_1 = p_2 \quad \text{versus} \quad H_a: p_1 \ne p_2$$

is given by

$$1 - \beta = 1 - \Phi\left(z_{\alpha/2} - \frac{\varepsilon}{\delta}\right) + \Phi\left(-z_{\alpha/2} - \frac{\varepsilon}{\delta}\right).$$

Let $\lambda = P_{Corr} + \rho P_{Amb}$.
Then

$$\lambda = 1 - \Phi\left(z_{\alpha/2} - \frac{\varepsilon}{\delta}\right) + \rho\beta.$$

As a result, sample size per arm required for a given λ can be obtained. Table 5.17 gives sample sizes per arm for $\delta = 0.05$ and $\lambda = 0.80$ or 0.90 assuming $\rho = 0$ or $\rho = 0.5$ based on exact binomial probabilities.

TABLE 5.16

Probability of Various Outcomes for Different Sample Sizes ($\delta = 0.05$)

n	p_1	p_2	P_{Corr}	P_{Amb}	$P_{Corr} + 0.5P_{Amb}$
50	0.25	0.35	0.71	0.24	0.83
50	0.20	0.35	0.87	0.12	0.93
75	0.25	0.35	0.76	0.21	0.87
75	0.20	0.35	0.92	0.07	0.96
100	0.25	0.35	0.76	0.23	0.87
100	0.20	0.35	0.94	0.06	0.97

TABLE 5.17

Sample Sizes per Arm for Various λ Assuming $\delta = 0.05$ and $\rho = 0$ or $\rho = 0.5$

			$\rho = 0$	$\rho = 0.5$
p_1	p_2	$\lambda = 0.90$	$\lambda = 0.80$	$\lambda = 0.90$
0.05	0.20	32	13	16
0.10	0.25	38	15	27
0.15	0.30	0.53	17	31
0.20	0.35	57	19	34
0.25	0.40	71	31	36
0.30	0.45	73	32	38
0.35	0.50	75	32	46
0.40	0.55	76	33	47

TABLE 5.18

Sample Size per Arm for Trials with Three or Four
Arms for $\varepsilon = 0.15$, $\delta = 0.05$, and $\lambda = 0.80$ or 0.90

	$n(\rho = 0)$		$n(\rho = 0.5)$	
ε	$r = 3$	$r = 4$	$r = 3$	$r = 4$
$\lambda = 0.80$				
0.2	18	31	13	16
0.3	38	54	26	32
0.4	54	73	31	39
0.5	58	78	34	50
$\lambda = 0.90$				
0.2	39	53	30	34
0.3	77	95	51	59
0.4	98	119	68	78
0.5	115	147	73	93

Note that the method proposed by Sargent and Goldberg (2001) can be extended to the case where there are three or more treatments. The selection of the best treatment, however, may be based on pairwise comparison or a global test. Table 5.18 provides sample sizes per arm with three or four arms assuming $\delta = 0.05$ and $\lambda = 0.80$ or 0.90.

5.6 Remarks

Chen and Ng (1998) indicated that the optimal two-stage design described above is similar to the Pocock sequence design for randomized controlled clinical trials where the probability of early termination is high, the total possible sample size is larger, and the expected size under the alternative hypothesis is smaller (see also Pocock, 1977). The minimax design, on the other hand, is similar to the O'Brien—Fleming design where the probability of early termination is low, the total possible size is smaller, but the expected size under the alternative hypothesis is larger (O'Brien and Fleming, 1979). The minimum design is useful when the patient source is limited, such as a rare cancer or a single-site study.

Recently, multiple-stage designs have been proposed to monitor response and toxicity variables simultaneously. See, for example, Bryant and Day (1995), Conaway and Petroni (1996), Thall, Simon, and Estey (1995, 1996). In these designs, the multivariate outcomes are modeled and family-wise errors are controlled. It is suggested that this form of design should be frequently used in cancer clinical trials since delayed toxicity could be a problem in phase II trials. Chen (1997b) also pointed out that one can use optimal three-stage design for toxicity monitoring (not simultaneous with the response). The role of *response* with that of *no toxicity* can be exchanged and the designs are similarly optimal and minimax.

In practice, the actual size at each stage of the multiple-stage design may deviate slightly from the exact design. Green and Dahlberg (1992) reviewed various phase II designs and modified each design to have variable sample sizes at each stage. They compared various flexible designs and concluded that flexible designs work well across a variety of $p_0's$, $p_1's$, and powers.

6

Tests for Goodness-of-Fit and Contingency Tables

In clinical research, the range of a categorical response variable often contains more than two values. Also, the dimension of a categorical variable can often be multivariate. The focus of this chapter is on categorical variables that are nonbinary and on the association among the components of a multivariate categorical variable. A contingency table is usually employed to summarize results from multivariate categorical responses. In practice, hypotheses testing for goodness-of-fit, independence (or association), and categorical shift are usually conducted for the evaluation of clinical efficacy and safety of a test compound under investigation. For example, a sponsor may be interested in determining whether the test treatment has any influence on the performance of some primary study endpoints, for example, the presence/absence of a certain event such as disease progression, adverse event, or response (complete/partial) of a cancer tumor. It is then of interest to test the null hypothesis of independence or no association between the test treatment (e.g., before and after treatment) and the change in the study endpoint. In this chapter, formulas for sample size calculation for testing goodness-of-fit and independence (or association) under an $r \times c$ contingency table are derived based on various chi-square type test statistics such as Pearson's chi-square and likelihood ratio test statistics. In addition, procedures for sample size calculation for testing categorical shift using McNemar's test and/or Stuart–Maxwell test are also derived.

In the next section, a sample size calculation formula for goodness-of-fit based on Pearson's test is derived. Sample size calculation formulas for testing independence (or association) with single stratum and multiple strata are introduced based on various chi-square test statistics, respectively, in Sections 6.2 and 6.3. Test statistics and the corresponding sample size calculation for categorical shift is discussed in Section 6.4. Section 6.5 considers testing for carryover effect in a 2×2 crossover design. Some practical issues are presented in Section 6.6.

6.1 Tests for Goodness-of-Fit

In practice, it is often interesting to study the distribution of the primary study endpoint under the study drug with some reference distribution, which may be obtained from historical (control) data or literature review. If the primary study endpoint is a categorical response that is nonbinary, Pearson's chi-square test is usually applied.

6.1.1 Pearson's Test

For the ith subject, let X_i be the response taking values from $\{x_1, ..., x_r\}$, $i = 1, ..., n$. Assume that X_i's are i.i.d. Let

$$p_k = P(X_i = x_k),$$

where $k = 1, \ldots, r$. p_k can be estimated by $\hat{p}_k = n_k/n$, where n_k is the frequency count of the subjects with response value k. For testing goodness-of-fit, the following hypotheses are usually considered:

$$H_0: p_k = p_{k,0} \text{ for all } k \quad \text{versus} \quad p_k \neq p_{k,0} \text{ for some } k,$$

where $p_{k,0}$ is a reference value (e.g., historical control), $k = 1, \ldots, r$. Pearson's chi-square statistic for testing goodness-of-fit is given by

$$T_G = \sum_{k=1}^{r} \frac{n(\hat{p}_k - p_{k,0})^2}{p_{k,0}}.$$

Under the null hypothesis H_0, T_G is asymptotically distributed as a central chi-square random variable with $r - 1$ degrees of freedom. Hence, we reject the null hypothesis with approximate α level of significance if

$$T_G > \chi^2_{\alpha, r-1},$$

where $\chi^2_{\alpha, r-1}$ denotes the αth upper quantile of a central chi-square random variable with $r - 1$ degrees of freedom. The power of Pearson's chi-square test can be evaluated under some local alternatives. (A local alternative typically means that the difference between treatment effects in terms of the parameters of interest decreases to 0 at the rate of $1/\sqrt{n}$ when the sample size n increases to infinity.) More specifically, if

$$\lim_{n \to \infty} \sum_{k=1}^{r} \frac{n(p_k - p_{k,0})^2}{p_{k,0}} = \delta,$$

then T_G is asymptotically distributed as a noncentral chi-square random variable with $r - 1$ degrees of freedom and the noncentrality parameter δ. For a given degrees of freedom $r - 1$ and a desired power $1 - \beta$, δ can be obtained by solving for

$$\chi^2_{r-1}\left(\chi^2_{\alpha, r-1} \mid \delta\right) = \beta, \tag{6.1}$$

where $\chi^2_{r-1}(\cdot \mid \delta)$ denotes the noncentral chi-square distribution with $r - 1$ degrees of freedom and the noncentrality parameter δ. Note that $\chi^2_{r-1}(t \mid \delta)$ is decreasing in δ for any fixed t. Hence, Equation 6.1 has a unique solution. Let $\delta_{\alpha, \beta}$ be the solution of Equation 6.1 for given α and β. The sample size needed to achieve the desired power of $1 - \beta$ is then given by

$$n = \delta_{\alpha, \beta} \left[\sum_{k=1}^{r} \frac{(p_k - p_{k,0})^2}{p_{k,0}} \right]^{-1},$$

where p_k should be replaced by an initial value.

6.1.2 An Example

Suppose a sponsor is interested in conducting a pilot study to evaluate the clinical efficacy of a test compound on subjects with hypertension. The objective of the intended pilot study is to compare the distribution of the proportions of subjects whose blood pressures are below, within, and above some prespecified reference (normal) range with that from historical control. Suppose that it is expected that the proportions of subjects after treatments are 20% (below the reference range), 60% (within the reference range), and 20% (above the reference range), respectively. Thus, we have $r = 3$ and

$$(p_1, p_2, p_3) = (0.20, 0.60, 0.20).$$

Furthermore, suppose that based on historical data or literature review, the proportions of subjects whose blood pressures are below, within, and above the reference range are given by 25%, 45%, and 30%, respectively. This is,

$$(p_{10}, p_{20}, p_{30}) = (0.25, 0.45, 0.30).$$

The sponsor would like to choose a sample size such that the trial will have an 80% ($\beta = 0.20$) power for detecting such a difference at the 5% ($\alpha = 0.05$) level of significance. First, we need to find δ under the given parameters according to Equation 6.1:

$$\chi_2^2\left(\chi_{0.05,3-1}^2 \mid \delta\right) = 0.2.$$

This leads to $\delta_{0.05,0.2} = 9.634$. As a result, the sample size needed to achieve an 80% power is given by

$$n = 9.634\left[\frac{(0.20-0.25)^2}{0.25} + \frac{(0.60-0.45)^2}{0.45} + \frac{(0.20-0.30)^2}{0.30}\right]^{-1} \approx 104.$$

6.2 Test for Independence: Single Stratum

An $r \times c$ (two-way) contingency table is defined as a two-way table representing the cross-tabulation of observed frequencies of two categorical response variables. Let x_i, $i = 1, \ldots, r$ and y_j, $j = 1, \ldots, c$ denote the categories (or levels) of variables X and Y, respectively. Also, let n_{ij} be the cell frequency of $X = x_i$ and $Y = y_j$. Then, we have the following $r \times c$ contingency table:

	y_1	y_2	$\cdots$	y_c	
x_1	n_{11}	n_{12}	$\cdots$	n_{1c}	$n_1.$
x_2	n_{21}	n_{22}	$\cdots$	n_{2c}	$n_2.$
$\cdots$	$\cdots$	$\cdots$	$\cdots$	$\cdots$	$\cdots$
x_r	n_{r1}	n_{r2}	$\cdots$	n_{rc}	$n_r.$
	$n_{.1}$	$n_{.2}$	$\cdots$	$n_{.c}$	

where

$$n_{.j} = \sum_{i=1}^{r} n_{ij} \qquad \text{(the } j\text{th column total),}$$
$$n_{i.} = \sum_{j=1}^{c} n_{ij} \qquad \text{(the } i\text{th row total),}$$
$$n = \sum_{i=1}^{r} \sum_{j=1}^{c} n_{ij} \qquad \text{(the overall total).}$$

In practice, the null hypothesis of interest is that there is no association between the row variable and the column variable, that is, X is independent of Y. When $r = c = 2$, Fisher's exact test introduced in Chapter 5 can be applied. In the following, we consider some popular tests for general cases.

6.2.1 Pearson's Test

The following Pearson's chi-square test statistic is probably the most commonly employed test statistic for independence:

$$T_I = \sum_{i=1}^{r} \sum_{j=1}^{c} \frac{(n_{ij} - m_{ij})^2}{m_{ij}},$$

where

$$m_{ij} = \frac{n_{i.}n_{.j}}{n}.$$

Define $\hat{p}_{ij} = n_{ij}/n$, $\hat{p}_{i\cdot} = n_{i.}/n$, and $\hat{p}_{\cdot j} = n_{.j}/n$. Then, T_I can also be written as

$$T_I = \sum_{i=1}^{r} \sum_{j=1}^{c} \frac{n(\hat{p}_{ij} - \hat{p}_{i\cdot}\hat{p}_{\cdot j})^2}{\hat{p}_{i\cdot}\hat{p}_{\cdot j}}.$$

Under the null hypothesis that X and Y are independent, T_I is asymptotically distributed as a central chi-square random variable with $(r - 1)(c - 1)$ degrees of freedom. Under the local alternative with

$$\lim_{n \to \infty} \sum_{i=1}^{r} \sum_{j=1}^{c} \frac{n(p_{ij} - p_{i\cdot}p_{\cdot j})^2}{p_{i\cdot}p_{\cdot j}} = \delta, \tag{6.2}$$

where $p_{ij} = P(X = x_i, Y = y_j)$, $p_{i\cdot} = P(X = x_i)$, and $p_{\cdot j} = P(Y = y_j)$, T_1 is asymptotically distributed as a noncentral chi-square random variable with $(r - 1)(c - 1)$ degrees of freedom and the noncentrality parameter δ.

For given α and a desired power $1 - \beta$, δ can be obtained by solving

$$\chi^2_{(r-1)(c-1)}(\chi^2_{\alpha,(r-1)(c-1)} \,|\, \delta) = \beta. \tag{6.3}$$

Let the solution be $\delta_{\alpha,\beta}$. The sample size needed to achieve power $1 - \beta$ is then given by

$$n = \delta_{\alpha,\beta} \left[\sum_{i=1}^{r} \sum_{j=1}^{c} \frac{(p_{ij} - p_{i \cdot} p_{\cdot j})^2}{p_{i \cdot} p_{\cdot j}} \right]^{-1}.$$

6.2.2 Likelihood Ratio Test

Another commonly used test for independence is the likelihood ratio test. More specifically, the likelihood function for a two-way contingency table is given by

$$L = \prod_{i=1}^{r} \prod_{j=1}^{c} p_{ij}^{n_{ij}}.$$

Without any constraint, the above likelihood function is maximized at $p_{ij} = n_{ij}/n$, which leads to

$$\max_{p_{ij}} \log L = \sum_{i=1}^{r} \sum_{j=1}^{c} n_{ij} \log \frac{n_{ij}}{n}.$$

Under the null hypothesis that $p_{ij} = p_{i \cdot} p_{\cdot j}$, the likelihood function can be rewritten as

$$L = \prod_{i=1}^{r} p_{i \cdot}^{n_{i \cdot}} \prod_{j=1}^{c} p_{\cdot j}^{n_{\cdot j}}.$$

It is maximized at $p_{i \cdot} = n_{i \cdot}/n$ and $p_{\cdot j} = n_{\cdot j}/n$, which leads to

$$\max_{p_{ij} = p_{i \cdot} p_{\cdot j}} \log L = \sum_{i=1}^{r} \sum_{j=1}^{c} n_{ij} \log \frac{n_{i \cdot} n_{\cdot j}}{n^2}.$$

Hence, the likelihood ratio test statistic can be obtained as

$$T_L = 2 \left(\max_{p_{ij}} \log L - \max_{p_{ij} = p_{i \cdot} p_{\cdot j}} \log L \right) = \sum_{i=1}^{r} \sum_{j=1}^{c} n_{ij} \log \frac{n_{ij}}{m_{ij}},$$

where $m_{ij} = n_{i \cdot} n_{\cdot j}/n$. Under the null hypothesis, T_L is asymptotically distributed as a central chi-square random variable with $(r-1)(c-1)$ degrees of freedom. Thus, we reject the null hypothesis at approximate α level of significance if

$$T_L > \chi^2_{\alpha,(r-1)(c-1)}.$$

Note that the likelihood ratio test is asymptotically equivalent to Pearson's test for independence. Under the local alternative (6.2), it can be shown that T_L is still asymptotically equivalent to Pearson's test for testing independence in terms of the power. Hence, the sample size formula derived based on Pearson's statistic can be used for obtaining sample size for the likelihood ratio test.

6.2.3 An Example

A small-scale pilot study was conducted to compare two treatments (treatment and control) in terms of the categorized hypotension. The results are summarized in the following 2×3 ($r = 2$ and $c = 3$) contingency table:

	Hypotension			
	Below	Normal	Above	
Treatment	2	7	1	10
Control	2	5	3	10
	4	12	4	20

It can be seen that the treatment is better than the control in terms of lowering blood pressure. To confirm that such a difference truly exists, the investigator is planning a larger trial to confirm the finding by applying Pearson's chi-square test for independence. It is of interest to select a sample size such that there is an 80% ($\beta = 0.2$) power for detecting such a difference observed in the pilot study at the 5% ($\alpha = 0.05$) level of significance.

We first identify δ under the given parameters according to Equation 6.3 by solving

$$\chi^2_{(2-1)(3-1)}\left(\chi^2_{0.05,(2-1)(3-1)} \mid \delta\right) = 0.2.$$

This leads to $\delta_{0.05,0.2} = 9.634$. As a result, the sample size required for achieving an 80% power is given by

$$n = \delta_{0.05,0.2}\left[\sum_{i=1}^{r}\sum_{j=1}^{c}\frac{(p_{ij} - p_{i\cdot}p_{\cdot j})^2}{p_{i\cdot}p_{\cdot j}}\right]^{-1} = \frac{9.634}{0.0667} \approx 145.$$

6.3 Test for Independence: Multiple Strata

In clinical trials, multiple study sites (or centers) are usually considered not only to make sure that clinical results are reproducible but also to expedite patient recruitment so that the intended trials can be done within the desired time frame. In a multicenter trial, it is a concern whether the sample size within each center (or stratum) is sufficient for providing an accurate and reliable assessment of the treatment effect (and consequently for achieving the desired power) when there is significant treatment-by-center interaction. In practice, a typical approach is to pool data across all centers for an overall assessment of the treatment effect. However, how to control for center effect has become another issue in data analysis. When the data are binary, the Cochran–Mantel–Haenszel (CMH) test is probably the most commonly used test procedure, and can adjust for differences among centers.

6.3.1 Cochran–Mantel–Haenszel Test

To introduce the CMH method, consider summarizing data from a multicenter trial in the following series of 2×2 contingency tables:

	Binary	Reponse	
Treatment	0	1	Total
Treatment 1	$n_{h,10}$	$n_{h,11}$	$n_{h,1.}$
Treatment 2	$n_{h,20}$	$n_{h,21}$	$n_{h,2.}$
Total	$n_{h,.0}$	$n_{h,.1}$	$n_{h,..}$

where $h = 1, ..., H$, $n_{h,ij}$ is the number of patients in the hth center (stratum) under the ith treatment with response j. Let $p_{h,ij}$ be the probability that a patient in the hth stratum under the ith treatment has response j. The hypotheses of interest are given by

$$H_0: p_{h1j} = p_{h,2j} \text{ for all } h, j \quad \text{versus} \quad H_a: p_{h,1j} \neq p_{h,2j} \text{ for some } h, j.$$

The CMH test for the above hypotheses is defined as

$$T_{CMH} = \frac{\left[\sum_{h=1}^{H} (n_{h,11} - m_{h,11}) \right]^2}{\sum_{h=1}^{H} v_h},$$

where

$$m_{h,11} = \frac{n_{h,1.} n_{h,.1}}{n_h} \quad \text{and} \quad v_h = \frac{n_{h,1.} n_{h,2.} n_{h,.0} y_{h,.1}}{n_h^2 (n_h - 1)}, h = 1,...,H.$$

Under the null hypothesis H_0, T_{CMH} is asymptotically distributed as a chi-square random variable with one degree of freedom. Hence, we reject the null hypothesis at approximate α level of significance if

$$T_{CMH} > \chi^2_{\alpha,1}.$$

To evaluate the power of this test under the alternative hypothesis, we assume that $n_h \to \infty$ and $n_h/n \to \pi_h$, where $n = \sum_{h=1}^{H} n_h$. Then, under the local alternative

$$\lim \left| \frac{\sum_{h=1}^{H} \pi_h (p_{h,11} - p_{h,1.} p_{h,.1})}{\sqrt{\sum_{h=1}^{H} \pi_h p_{h,1.} p_{h,2.} p_{h,.0} p_{h,.1}}} \right| = \delta, \tag{6.4}$$

where $p_{h,i.} = p_{h,i0} + p_{h,i1}$ and $p_{h,.j} = p_{h,1j} + p_{h,2j}$. T_{CMH} is asymptotically distributed as a chi-square random variable with one degree of freedom and the noncentrality parameter δ^2. In such a situation, it can be noted that $T_{CMH} > \chi_{\alpha,1}$ is equivalent to $N(\delta, 1) > z_{\alpha/2}$. Hence, the sample size required for achieving a desired power of $1 - \beta$ at the α level of significance is given by

$$n = \frac{(z_{\alpha/2} + z_\beta)^2}{\delta^2}.$$

6.3.2 An Example

Consider a multinational, multicenter clinical trial conducted in four different countries (the United States, the United Kingdom, France, and Japan) for the evaluation of clinical efficacy and safety of a test compound. Suppose the objective of this trial is to compare the test compound with a placebo in terms of the proportions of patients who experience certain types of adverse events. Let 0 and 1 denote the absence and presence of the adverse event, respectively. It is expected that the sample size will be approximately evenly distributed across the four centers (i.e., $\pi_h = 25\%$, $h = 1, 2, 3, 4$). Suppose that based on a pilot study, the values of $p_{h,ij}$'s within each country are estimated as follows:

Center	Treatment	Binary Response 0	Binary Response 1	Total
1	Study drug	0.35	0.15	0.50
	Placebo	0.25	0.25	0.50
	Total	0.60	0.40	1.00
2	Study drug	0.30	0.20	0.50
	Placebo	0.20	0.30	0.50
	Total	0.50	0.50	1.00
3	Study drug	0.40	0.10	0.50
	Placebo	0.20	0.30	0.50
	Total	0.60	0.40	1.00
4	Study drug	0.35	0.15	0.50
	Placebo	0.15	0.35	0.50
	Total	0.50	0.50	1.00

By Equation 6.4, δ is given by 0.3030. Hence the sample size needed to achieve an 80% ($\beta = 0.20$) power at the 5% ($\alpha = 0.05$) level of significance is given by

$$n = \frac{(z_{\alpha/2} + z_\beta)^2}{\delta^2} = \frac{(1.96 + 0.84)^2}{0.3030^2} \approx 86.$$

6.4 Test for Categorical Shift

In clinical trials, it is often of interest to examine any change in laboratory values before and after the application of the treatment. When the response variable is categorical, this type of change is called a categorical shift. In this section, we consider testing for categorical shift.

6.4.1 McNemar's Test

For a given laboratory test, test results are usually summarized as either normal (i.e., the test result is within the normal range of the test) or abnormal (i.e., the test result is outside the normal range of the test). Let x_{ij} be the binary response ($x_{ij} = 0$: normal and $xij = 1$: abnormal) from the ith ($i = 1, \ldots, n$) subject in the jth ($j = 1$: pretreatment and $j = 2$: posttreatment) treatment. The test results can be summarized in the following 2×2 table:

	Posttreatment		
Pretreatment	Normal	Abnormal	
Normal	n_{00}	n_{01}	$n_{0.}$
Abnormal	n_{10}	n_{11}	$n_{1.}$
	$n_{.0}$	$n_{.1}$	$n_{.}$

where

$$n_{00} = \sum_{i=1}^{n} (1 - x_{i1})(1 - x_{i2}),$$

$$n_{01} = \sum_{i=1}^{n} (1 - x_{i1})x_{i2},$$

$$n_{10} = \sum_{i=1}^{n} x_{i1}(1 - x_{i2}),$$

$$n_{11} = \sum_{i=1}^{n} x_{i1}x_{i2}.$$

Define

$$p_{00} = P(x_{i1} = 0, x_{i2} = 0),$$
$$p_{01} = P(x_{i1} = 0, x_{i2} = 1),$$
$$p_{10} = P(x_{i1} = 1, x_{i2} = 0),$$
$$p_{11} = P(x_{i1} = 1, x_{i2} = 1),$$
$$p_{1+} = P(x_{i1} = 1) = p_{10} + p_{11},$$
$$p_{+1} = P(x_{i2} = 1) = p_{01} + p_{11}.$$

It is then of interest to test whether there is a categorical shift after treatment. A categorical shift is defined as either a shift from 0 (normal) in pretreatment to 1 (abnormal) in posttreatment or a shift from 1 (abnormal) in pretreatment to 0 (normal) in posttreatment. Consider

$$H_0: p_{1+} = p_{+1} \quad \text{versus} \quad H_a: p_{1+} \neq p_{+1},$$

which is equivalent to

$$H_0: p_{10} = p_{01} \quad \text{versus} \quad H_a: p_{10} \neq p_{01}.$$

The most commonly used test procedure to serve the purpose is McNemar's test, whose test statistic is given by

$$T_{MN} = \frac{n_{10} - n_{01}}{\sqrt{n_{10} + n_{01}}}.$$

Under the null hypothesis H_0, T_{MN} is asymptotically distributed as a standard normal random variable. Hence, we reject the null hypothesis at approximate α level of significance if

$$|T_{MN}| > z_{\alpha/2}.$$

Under the alternative hypothesis that $p_{01} \neq p_{10}$, it follows that

$$\begin{aligned} T_{MN} &= \frac{n_{01} - n_{10}}{\sqrt{n_{01} + n_{10}}} \\ &= \sqrt{\frac{n}{n_{01} + n_{10}}} \sqrt{n} \left(\frac{n_{10}}{n} - \frac{n_{01}}{n} \right) \\ &= \frac{1}{\sqrt{p_{10} + p_{01}}} \frac{1}{\sqrt{n}} \sum_{i=1}^{n} d_i, \end{aligned}$$

where $d_i = x_{i1} - x_{i2}$. Note that d_i's are independent and identically distributed random variables with mean $(p_{01} - p_{10})$ and variance $p_{01} + p_{10} - (p_{01} - p_{10})^2$. As a result, by the central limit theorem, the power of McNemar's test can be approximated by

$$\Phi \left(\frac{\sqrt{n}(p_{01} - p_{10}) - z_{\alpha/2}\sqrt{p_{01} + p_{10}}}{\sqrt{p_{01} + p_{10} - (p_{01} - p_{10})^2}} \right).$$

To achieve a power of $1 - \beta$, the sample size needed can be obtained by solving

$$\frac{\sqrt{n}(p_{01} - p_{10}) - z_{\alpha/2}\sqrt{p_{01} + p_{10}}}{\sqrt{p_{01} + p_{10} - (p_{01} - p_{10})^2}} = z_\beta.$$

which leads to

$$n = \frac{\left[z_{\alpha/2}\sqrt{p_{01} + p_{10}} + z_\beta\sqrt{p_{10} + p_{01} - (p_{01} - p_{10})^2} \right]^2}{(p_{01} - p_{10})^2}. \tag{6.5}$$

Define $\psi = p_{01}/p_{10}$ and $\pi_{\text{Discordant}} = p_{01} + p_{10}$. Then

$$n = \frac{\left[z_{\alpha/2}(\psi + 1) + z_\beta\sqrt{(\psi + 1)^2 - (\psi - 1)^2 \pi_{\text{Discordant}}} \right]^2}{(\psi - 1)^2 \pi_{\text{Discordant}}}.$$

6.4.2 Stuart–Maxwell Test

McNemar's test can be applied only to the case in which there are two possible categories for the outcome. In practice, however, it is possible that the outcomes are classified into more than two (multiple) categories. For example, instead of classifying the laboratory values into normal and abnormal (two categories), it is often preferred to classify them into three categories (i.e., below, within, and above the normal range). Let $x_{ij} \in \{1, \ldots, r\}$ be the categorical observation from the ith subject under the jth treatment ($j = 1$: pretreatment and $j = 2$: posttreatment). In practice, the data are usually summarized by the following $r \times r$ contingency table:

	Pretreatment				
Pretreatment	1	2	...	r	
1	n_{11}	n_{12}	...	n_{1r}	$n_{1.}$
2	n_{21}	n_{22}	...	n_{2r}	$n_{2.}$
...	...	...	...	...	...
r	n_{r1}	n_{r2}	...	n_{rr}	$n_{r.}$
	$n_{.1}$	$n_{.2}$	...	$n_{.r}$	$n_{..}$

Let

$$p_{ij} = P(x_{k1} = i, x_{k2} = j),$$

which is the probability that the subject will shift from i pretreatment to j posttreatment. If there is no treatment effect, one may expect that $p_{ij} = p_{ji}$ for all $1 \leq i, j, \leq r$. Hence, it is of interest to test the hypotheses

$$H_0: p_{ij} = p_{ji} \text{ for all } i \neq j \quad \text{versus} \quad H_a: p_{ij} \neq p_{ji} \text{ for some } i \neq j.$$

To test the above hypotheses, the test statistic proposed by Stuart and Maxwell is useful (see, e.g., Stuart, 1955). We refer to the test statistic as Stuart–Maxwell test, which is given by

$$T_{SM} = \sum_{i<j} \frac{(n_{ij} - n_{ji})^2}{n_{ij} + n_{ji}}.$$

Under the null hypothesis H_0, T_{SM} follows a standard chi-square distribution with $r(r-1)/2$ degrees of freedom. Hence, for a given significance level of α, the null hypothesis should be rejected if $T_{SM} > \chi^2_{\alpha, r(r-1)/2}$.

Under the local alternative given by

$$\lim_{n \to \infty} n \sum_{i<j} \frac{(p_{ij} - p_{ji})^2}{p_{ij} + p_{ji}} = \delta,$$

T_{SM} is asymptotically distributed as a noncentral chi-square random variable with $r(r-1)/2$ degrees of freedom and the noncentrality parameter δ. For a given degrees of freedom ($r(r-1)/2$) and a desired power $(1 - \beta)$, δ can be obtained by solving

$$\chi^2_{r(r-1)/2.}(\chi^2_{\alpha, r(r-1)/2} | \delta) = \beta, \tag{6.6}$$

where $\chi_a^2(\cdot\,|\,\delta)$ is the cumulative distribution function of the noncentral chi-square distribution with degrees freedom a and noncentrality parameter δ. Let $\delta_{\alpha,\beta}$ be the solution. Then, the sample size needed to achieve a power of $1 - \beta$ is given by

$$n = \delta_{\alpha,\beta} \left[\sum_{i<j} \frac{(p_{ij} - p_{ji})^2}{p_{ij} + p_{ji}} \right]^{-1}. \tag{6.7}$$

6.4.3 Examples

6.4.3.1 *McNemar's Test*

Suppose that an investigator is planning to conduct a trial to study a test compound under investigation in terms of the proportions of the patients with nocturnal hypoglycemia, which is defined to be the patients with the overnight glucose value ≤ 3.5 mgL on two consecutive visits (15 minutes per visit). At the first visit (pretreatment), patients' overnight glucose levels will be measured every 15 minutes. Whether or not the patient experiences nocturnal hypoglycemia will be recorded. At the second visit, patients will receive the study compound and the overnight glucose levels will be obtained in a similar manner. Patients' experience on nocturnal hypoglycemia will also be recorded. According to some pilot studies, it is expected that about 50% ($p_{10} = 0.50$) of patients will shift from 1 (nocturnal hypoglycemia pretreatment) to 0 (normal posttreatment) and 20% ($p_{01} = 0.20$) of patients will shift from 0 (normal pretreatment) to 1 (nocturnal hypoglycemia posttreatment). The investigator would like to select a sample size such that there is an 80% ($\beta = 0.20$) power for detecting such a difference if it truly exists at the 5% ($\alpha = 0.05$) level of significance. According to Equation 6.5, the required sample size can be obtained as follows:

$$n = \frac{\left[z_{\alpha/2}\sqrt{p_{01} + p_{10}} + z_\beta\sqrt{p_{10} + p_{01} - (p_{01} - p_{10})^2} \right]^2}{(p_{10} - p_{01})^2}$$

$$= \frac{\left[1.96\sqrt{0.20 + 0.50} + 0.84\sqrt{0.20 + 0.50 - (0.20 - 0.50)^2} \right]^2}{(0.20 - 0.50)^2}$$

$$\approx 59.$$

6.4.3.2 *Stuart–Maxwell Test*

A pilot study was conducted to study the treatment effect of a test compound based on the number of monocytes in the blood. The primary study endpoint is the number of monocytes (i.e., below, within, and above normal range) in the blood (i.e., $r = 3$). The results were summarized in the following contingency table:

Pretreatment	Posttreatment			
	Below	**Normal**	**Above**	
Below	3	4	4	11
Normal	2	3	3	8
Above	1	2	3	6
	6	9	10	25

From this pilot study, the results suggest that the test compound can increase the number of monocytes in the blood because the upper off-diagonal elements are always larger than those in the lower off-diagonal.

To confirm whether such a trend truly exists, a larger trial is planned to have an 80% ($\beta = 0.20$) power at the 5% ($\alpha = 0.05$) level of significance. For this purpose, we need to first estimate δ under the given parameters according to Equation 6.6.

$$\chi^2_{3(3-1)/2}\left(\chi^2_{0.05,3(3-1)/2}\,|\,\delta\right) = 0.20,$$

which leads to $\delta_{\alpha,\beta} = 10.903$. As a result, the sample size needed for achieving an 80% power is given by

$$n = \delta_{\alpha,\beta}\left[\sum_{i<j}\frac{(p_{ij}-p_{ji})^2}{p_{ij}+p_{ji}}\right]^{-1} = \frac{10.903}{0.107} \approx 102.$$

6.5 Carryover Effect Test

As discussed earlier, a standard 2×2 crossover design is commonly used in clinical research for the evaluation of clinical efficacy and safety of a test compound of interest. When the response is binary, under the assumption of no period and treatment-by-period interaction (carryover effect), McNemar's test can be applied to test for the treatment effect. In some cases, the investigator may be interested in testing the treatment-by-period interaction. In this section, statistical procedure for testing the treatment-by-period interaction, based on the model by Becker and Balagtas (1993), is introduced. The corresponding procedure for sample size calculation is derived.

6.5.1 Test Procedure

Consider a standard two-sequence, two-period crossover design, that is, (AB, BA). Let x_{ijk} be the binary response from the kth ($k = 1, \ldots, n_i$) subject in the ith sequence at the jth dosing period. Let $p_{ij} = P(x_{ijk} = 1)$. To separate the treatment, period, and carryover effects, Becker and Balagtas (1993) considered the following logistic regression model:

$$\log\frac{p_{11}}{1-p_{11}} = \alpha + \tau_1 + \rho_1,$$

$$\log\frac{p_{12}}{1-p_{12}} = \alpha + \tau_2 + \rho_2 + \gamma_1,$$

$$\log\frac{p_{21}}{1-p_{21}} = \alpha + \tau_2 + \rho_1,$$

$$\log\frac{p_{22}}{1-p_{22}} = \alpha + \tau_1 + \rho_2 + \gamma_2,$$

where τ_i is the ith treatment effect, ρ_j is the jth period effect, and γ_k is the carryover effect from the first period in the kth sequence. It is assumed that

$$\tau_1 + \tau_2 = 0,$$
$$\rho_1 + \rho_2 = 0,$$
$$\gamma_1 + \gamma_2 = 0.$$

Let

$$\gamma = \gamma_1 - \gamma_2$$
$$= \log \frac{p_{11}}{1 - p_{11}} + \log \frac{p_{12}}{1 - p_{12}} - \log \frac{p_{21}}{1 - p_{21}} - \log \frac{p_{22}}{1 - p_{22}}.$$

The hypotheses of interest are given by

$$H_0 : \gamma = 0 \quad \text{versus} \quad H_a : \gamma \neq 0.$$

Let $\hat{p}_{ij} = n_i^{-1} \sum_k x_{ijk}$. Then, γ can be estimated by

$$\hat{\gamma} = \log \frac{\hat{p}_{11}}{1 - \hat{p}_{11}} + \log \frac{\hat{p}_{12}}{1 - \hat{p}_{12}} - \log \frac{\hat{p}_{21}}{1 - \hat{p}_{21}} - \log \frac{\hat{p}_{22}}{1 - \hat{p}_{22}}.$$

It can be shown that $\hat{\gamma}$ is asymptotically distributed as a normal random variable with mean γ and variance $\sigma_1^2 n_1^{-1} + \sigma_2^2 n_2^{-1}$, where

$$\sigma_i^2 = \text{var} \left(\frac{x_{i1k}}{p_{i1}(1 - p_{i1})} + \frac{x_{i2k}}{p_{i2}(1 - p_{i2})} \right),$$

which can be estimated by $\hat{\sigma}_i^2$, the sample variance of

$$\frac{x_{i1k}}{\hat{p}_{i1}(1 - \hat{p}_{i1})} + \frac{x_{i2k}}{\hat{p}_{i2}(1 - \hat{p}_{i2})}, \quad k = 1, \ldots, n_i.$$

Hence, the test statistic is given by

$$T = \hat{\gamma} \left(\frac{\hat{\sigma}_1^2}{n_1} + \frac{\hat{\sigma}_2^2}{n_2} \right)^{-1/2}.$$

Under the null hypothesis H_0, T is asymptotically distributed as a standard normal random variable. We reject the null hypothesis at approximate α level of significance if

$$|T| > z_{\alpha/2}.$$

Under the alternative hypothesis that $\gamma \neq 0$, the power of this test procedure can be approximated by

$$\Phi\left(\gamma\left[\frac{\sigma_1^2}{n_1} + \frac{\sigma_2^2}{n_2}\right]^{-1/2} - z_{\alpha/2}\right).$$

For a given power of $1 - \beta$ and assuming that $n = n_1 = n_2$, the sample size needed can be obtained by solving

$$\gamma\left[\frac{\sigma_1^2}{n_1} + \frac{\sigma_2^2}{n_2}\right]^{-1/2} - z_{\alpha/2} = z_\beta,$$

which leads to

$$n = \frac{(z_{\alpha/2} + z_\beta)^2 \left(\sigma_1^2 + \sigma_2^2\right)}{\gamma^2}.$$

6.5.2 An Example

Consider a single-center, open, randomized, active-controlled, two-sequence, two-period crossover design with the primary efficacy endpoint of nocturnal hypoglycemia. The objective of the study is to compare the study drug with a standard therapy on the market-place in terms of the proportion of patients who will experience nocturnal hypoglycemia. As a result, the investigator is interested in conducting a larger trial to confirm whether such an effect truly exists. However, based on the results of a small-scale pilot study, no evidence of statistical significance in the possible carryover effect was detected. According to the pilot study, the following parameters were estimated: $\gamma = 0.89$, $\sigma_1 = 2.3$, and $\sigma_2 = 2.4$. The sample size needed to achieve an 80% ($\beta = 0.2$) power at the 5% ($\alpha = 0.05$) level of significance is given by

$$n = \frac{(z_{\alpha/2} + z_\beta)^2 \left(\sigma_1^2 + \sigma_2^2\right)}{\gamma^2} = \frac{(1.96 + 0.84)^2 (2.3^2 + 2.4^2)}{0.89^2} \approx 110.$$

Hence, a total of 110 subjects are required to have the desired power for the study.

6.6 Practical Issues

6.6.1 Local Alternative versus Fixed Alternative

In this chapter, we introduced various chi-square type test statistics for contingency tables. When the chi-square test has only one degree of freedom, it is equivalent to a Z-test (i.e., a test based on standard normal distribution). Hence, the formula for sample size calculation can be derived under the ordinary fixed alternative. When the degrees of freedom of the chi-square test is larger than one (e.g., Pearson's test for goodness-of-fit and independence), it can be verified that the chi-square test statistic is distributed as a weighted noncentral chi-square distribution under the fixed alternative hypothesis. In order words, it has the same distribution as the random variable $\sum_{i=1}^{k} \lambda_i \chi^2(\delta_i)$ for some k, λ_i, and δ_i, where $\chi^2(\delta_i)$ denotes a chi-square random variable with one degree of freedom and the noncentrality parameter δ_i. The power function based on a weighted noncentral chi-square random variable could be very complicated and no standard table/software is available. As a result, all the sample size formulas for the chi-square tests with more than one degree of freedom are derived under the local alternative. Under the concept of local alternative, one assumes that the difference in the parameters of interest between the null hypothesis and the alternative hypothesis shrinks to 0 at a speed of $1/\sqrt{n}$. In practice, however, it is more appealing to consider the alternative as fixed. In other words, the alternative hypothesis does not change as the sample size changes. Further research in sample size estimation based on a fixed alternative is needed.

6.6.2 Random versus Fixed Marginal Total

In randomized, controlled parallel clinical trials, the numbers of subjects assigned to each treatment group are usually fixed. Consequently, when the data are reported by a two-way contingency table (treatment and response), one of the margins (treatment) is fixed. However, it is not uncommon (e.g., in an observational study) that the number of subjects assigned to each treatment is also random. In this situation, Pearson's test for independence between treatment and response is still valid. Thus, Pearson's test is commonly used in the situation where the marginal distribution of the numbers of the subjects for each treatment is unknown. When the marginal distribution of the numbers of the subjects for each treatment is known or can be approximated by some distribution such as Poisson, Pearson's test may not be efficient. Alternatively, the likelihood ratio test may be useful (see, e.g., Shao and Chow, 1990).

6.6.3 $r \times c$ versus $p \times r \times c$

In this chapter, we focus on procedures for sample size calculation for testing goodness-of-fit, independence (or association), and categorical shift under an $r \times c$ contingency table. In practice, we may encounter the situation that involves a $p \times r \times c$ contingency table with third variable (e.g., sex, race, or age). In practice, handling this type of three-way contingency table is always a challenge. One simple solution is combining the third variable with the treatment variable and applying the standard procedure designed for a two-way contingency table. Further research regarding handling three-way contingency tables is necessary.

7

Comparing Time-to-Event Data

In clinical research, in addition to continuous and discrete study endpoints described in the previous two chapters, the investigator may also be interested in the occurrence of certain events such as adverse drug reactions and/or side effects, disease progression, relapse, or death. In most clinical trials, the occurrence of such an event is usually undesirable. Hence, one of the primary objectives of the intended clinical trials may be to evaluate the effect of the test drug on the prevention or delay of such events. The time to the occurrence of an event is usually referred to as the time to event. In practice, time to event has become a natural measure of the extent to which the event occurrence is delayed. When the event is the death, the time to event of a patient is the patient's survival time (i.e., time to death). Hence, in many cases, the analysis of time to event is sometimes referred to as *survival analysis*.

In practice, statistical methods for analysis of time-to-event data are very different from those commonly used for continuous variables (e.g., analysis of variance or analysis of covariance) due to the following reasons. First, time to event is usually subject to censoring, for example, right (left) or interval censoring, at which its exact value is unknown but we know that it is larger or smaller than an observed censoring time or within an interval. Second, time-to-event data are usually highly skewed, which violates the normality assumption of standard statistical methods such as the analysis of variance. In this chapter, for simplicity, we focus on sample size calculation based only on the most typical censor type (e.g., right censoring) and the most commonly used testing procedures such as the exponential model, Cox's proportional hazards model, and the weighted log-rank test.

The remainder of this chapter is organized as follows. In the next section, basic concepts regarding survival and hazard functions in the analysis of time-to-event data are provided. In Section 7.2, formulas for sample size calculation for testing equality, noninferiority/superiority, and equivalence in two-sample problems are derived under the commonly used exponential model. In Section 7.3, formulas for sample size calculation under Cox's proportional hazards model is presented. In Section 7.4, formulas for sample size estimation based on the weighted log-rank test are derived. Some practical issues are discussed in Section 7.5.

7.1 Basic Concepts

In this section, we introduce some basic concepts regarding survival and hazard functions, which are commonly used in the analysis of time-to-event data. In practice, hypotheses of clinical interest are often involved in comparing median survival times, survival functions, and hazard rates. Under a given set of hypotheses, appropriate statistical tests are then constructed based on consistent estimators of these parameters.

7.1.1 Survival Function

In the analysis of time-to-event data, the *survival function* is usually used to characterize the distribution of the time-to-event data. Let X be the variable of time to event and $S(x)$ be the corresponding survival function. Then, $S(x)$ is defined as

$$S(x) = P(X > x).$$

Thus, $S(x)$ is the probability that the event will occur after time x. When the event is death, $S(x)$ is the probability of a patient who will survive until x. Theoretically, X could be a continuous variable, a discrete response, or a combination of both. In this chapter, however, we consider only the case where X is a continuous random variable with a density function $f(x)$. The relationship between $S(x)$ and $f(x)$ is given by

$$f(x) = \frac{dS(x)}{dx}.$$

A commonly used nonparametric estimator for $S(x)$ is the Kaplan–Meier estimator, which is given by

$$\hat{S}(t) = \prod_i \left(1 - \frac{d_i}{n_i}\right), \tag{7.1}$$

where the product is over all event times, d_i is the number of events observed at the ith event time, and n_i is the number of subjects at risk just prior the ith event time.

7.1.2 Median Survival Time

In a clinical study, it is of interest to compare the median survival time, which is defined to be the 50% quantile of the survival distribution. In other words, if $m_{1/2}$ is the median survival time, then it should satisfy

$$P(X > m_{1/2}) = 0.5.$$

A commonly used nonparametric estimator for $m_{1/2}$ is given by $\hat{m}_{1/2} = \hat{S}^{-1}(0.5)$, where $\hat{S}$ is the Kaplan–Meier estimator. When the time-to-event data are exponentially distributed with hazard rate λ, it can be shown that the median survival time is given by $\log 2/\lambda$.

7.1.3 Hazard Function

Another important concept in the analysis of time-to-event data is the so-called *hazard function*, which is defined as

$$h(x) = \lim_{\Delta x \to 0} \frac{P(x \le X < x + \Delta x \mid X \ge x)}{\Delta x}.$$

As it can be seen, $h(x)$ can also be written as

$$h(x) = \frac{f(x)}{S(x)},$$

which implies that

$$S(x) = \exp\left\{-\int_0^x h(t)da\right\}.$$

If we assume a constant hazard rate (i.e., $h(t) = \lambda$ for some λ), $S(x)$ becomes

$$S(x) = \exp\{-\lambda x\}.$$

In this case, time-to-event variable X is distributed as an exponential variable with hazard rate λ.

7.1.4 An Example

A clinical trial was conducted to study a test treatment on patients with small cell lung cancer. The trial lasted for one year with 10 patients entering in the study simultaneously. The data are given in Table 7.1 with "+" indicating censored observations.

The Kaplan–Meier estimator can be obtained based on Equation 7.1 with median 0.310. The obtained Kaplan–Meier estimator is plotted in Figure 7.1. It can be seen from the Kaplan–Meier plot that approximately 80% patients in the patient population will live beyond 0.2 years. If we assume constant hazard rate over time, the estimated hazard rate according to Equation 7.2 in Section 7.2 is 1.59.

TABLE 7.1

Survival Data

Subject Number	Survival Time
1	0.29
2	0.25
3	0.12
4	0.69
5	1.00+
6	0.33
7	0.19
8	1.00+
9	0.23
10	0.93

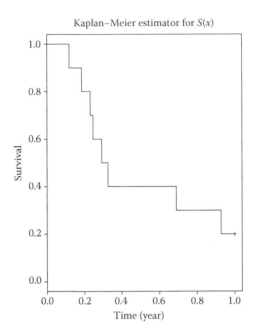

FIGURE 7.1
Kaplan–Meier estimator for $S(x)$.

7.2 Exponential Model

In what follows, formulas for sample size calculation based on hazard rates for median survival times and survival functions between treatment groups will be derived under an exponential model, which is the simplest parametric statistical model for time-to-event data. Under the exponential model, it is assumed that the time to event is exponentially distributed with a constant hazard rate. For survival analysis in clinical trials comparing two treatments, the hypothesis of interest could be either comparing the hazard rates or the median survival times. However, since the time to event is assumed to be exponentially distributed, the median survival time is determined by the hazard rate. As a result, comparing median survival times is equivalent to comparing hazard rates. Without loss of generality, we focus on comparing hazard rates between treatment groups. In this section, we will introduce the method by Lachin and Foulkes (1986).

Consider a two-arm parallel survival trial with accrual time period T_0 and the follow-up $T - T_0$. Let a_{ij} denote the entry time of the jth patient of the ith treatment group. It is assumed that a_{ij} follows a continuous distribution with the density function given by

$$g(z) = \frac{\gamma e^{-\gamma z}}{1 - e^{-\gamma T_0}}, \quad 0 \le z \le T_0.$$

When $\gamma > 0$, the entry distribution is convex, which implies fast patient entry at the beginning. When $\gamma < 0$, the entry distribution is concave, which implies lagging patient entry. For convenience, we define $g(z) = 1/T_0$ when $\gamma = 0$, which implies uniform patient entry. Let t_{ij}

be the time to event (i.e., the time from the patient entry to the time observing event) for the jth subject in the ith treatment group, $i = 1, 2, j = 1, \ldots, n_i$. It is assumed that t_{ij} follows an exponential distribution with hazard rate λ_i. The information observed from the sample is

$$(x_{ij}, \delta_{ij}) = (\min\{t_{ij}, T - a_{ij}\}, I\{t_{ij} \le T - a_{ij}\}).$$

For a fixed i, the joint likelihood for $x_{ij}, j = 1, \ldots, n_i$ can be written as

$$L(\lambda_i) = \frac{\gamma^n e^{-\gamma \sum_{j=1}^{n_i} a_{ij}}}{(1 - e^{-\gamma T_0})^n} \lambda_i^{\sum_{j=1}^{n_i} \delta_{ij}} e^{-\lambda_i \sum_{j=1}^{n_i} x_{ij}}.$$

It can be shown that the MLE for λ_i is given by

$$\hat{\lambda}_i = \frac{\sum_{j=1}^{n_i} \delta_{ij}}{\sum_{j=1}^{n_i} x_{ij}}. \tag{7.2}$$

According to the central limit theorem,

$$\sqrt{n_i}(\hat{\lambda}_i - \lambda_i) = \sqrt{n_i} \frac{\sum_{j=1}^{n_i}(\delta_{ij} - \lambda_i x_{ij})}{\sum_{j=1}^{n_i} x_{ij}}$$

$$= \frac{1}{E(x_{ij})} \frac{1}{\sqrt{n_i}} \sum_{j=1}^{n_i}(\delta_{ij} - \lambda_i x_{ij}) + o_p(1)$$

$$\to_d N(0, \sigma^2(\lambda_i)),$$

where

$$\sigma^2(\lambda_i) = \frac{1}{E^2(x_{ij})} \text{var}(\delta_{ij} - \lambda_i x_{ij})$$

and $\to_d$ denotes convergence in distribution. Note that

$$E(\delta_{ij}) = E(\delta_{ij}^2)$$

$$= 1 - \int_0^{T_0} g(a) e^{-\lambda_i(T-\alpha)} da$$

$$= 1 - \int_0^{T_0} \frac{\gamma e^{-\gamma a}}{1 - e^{-\gamma T_0}} e^{-\lambda_i(T-\alpha)} da$$

$$= 1 + \frac{\gamma e^{-\lambda_i T}(1 - e^{(\lambda_i - \gamma)T_0})}{(\lambda_i - \gamma)(1 - e^{-\gamma T_0})},$$

$$E(x_{ij}) = \int_0^{T_0} g(a)da \int_0^{T-a} \lambda_i x e^{-\lambda_i x} dx + (T-a)e^{-\lambda_i(T-a)}$$

$$= \int_0^{T_0} g(a) \frac{1-e^{-\lambda_i(T-a)}}{\lambda_i} da$$

$$= \frac{1}{\lambda_i} E(\delta_{ij}),$$

$$E(\delta_{ij} x_{ij}) = \int_0^{T_0} g(a)da \int_0^{T-a} \lambda_i x e^{-\lambda_i x} dx,$$

and

$$E\left(x_{ij}^2\right) = \int_0^{T_0} g(a)da \int_0^{T-a} \lambda_i x^2 e^{-\lambda_i x} dx + (T-a)^2 e^{-\lambda_i(T-a)}$$

$$= \int_0^{T_0} g(a) \int_0^{T-a} 2x e^{-\lambda_i x} dx$$

$$= \frac{2E(\delta_{ij} x_{ij})}{\lambda_i}.$$

Hence,

$$\mathrm{var}(\delta_{ij} - \lambda_i x_{ij}) = E\left(\delta_{ij}^2\right) - 2\lambda_i E(\delta_{ij} x_{ij}) + \lambda_i^2 E\left(x_{ij}^2\right)$$

$$= E\left(\delta_{ij}^2\right) - 2\lambda_i E(\delta_{ij} x_{ij}) + 2\lambda_i E(\delta_{ij} x_{ij})$$

$$= E\left(\delta_{ij}^2\right) - E(\delta_{ij}).$$

That is,

$$\sigma^2(\lambda_i) = \frac{\lambda_i^2}{E(\delta_{ij})} = \lambda_i^2 \left[1 + \frac{\gamma e^{-\lambda_i T}(1 - e^{(\lambda_i - \gamma)T_0})}{(\lambda_i - \gamma)(1 - e^{-\gamma T_0})} \right]^{-1}. \tag{7.3}$$

7.2.1 Test for Equality

Let $\varepsilon = \lambda_1 - \lambda_2$ be the difference between the hazard rates of a control and a test drug. To test whether there is a difference between the hazard rates of the test drug and the reference drug, the following hypotheses are usually considered:

$$H_0 : \varepsilon = 0 \quad \text{versus} \quad H_a : \varepsilon \neq 0.$$

Under the null hypothesis, test statistic

$$T = (\hat{\lambda}_1 - \hat{\lambda}_2) \left[\frac{\sigma^2(\hat{\lambda}_1)}{n_1} + \frac{\sigma^2(\hat{\lambda}_2)}{n_2} \right]^{-1/2}$$

approximately follows a standard normal distribution for large n_1 and n_2. We then reject the null hypothesis at approximate α level of significance if

$$\left| (\hat{\lambda}_1 - \hat{\lambda}_2) \left[\frac{\sigma^2(\hat{\lambda}_1)}{n_1} + \frac{\sigma^2(\hat{\lambda}_2)}{n_2} \right]^{-1/2} \right| > z_{\alpha/2}. \tag{7.4}$$

Under the alternative hypothesis that $\lambda_1 - \lambda_2 \neq 0$, the power of the above test is approximately

$$\Phi \left(|\lambda_1 - \lambda_2| \left[\frac{\sigma^2(\lambda_1)}{n_1} + \frac{\sigma^2(\lambda_2)}{n_2} \right]^{-1/2} - z_{\alpha/2} \right).$$

As a result, the sample size needed to achieve a desired power of $1 - \beta$ can be obtained by solving

$$|\lambda_1 - \lambda_2| \left[\frac{\sigma^2(\lambda_1)}{n_1} + \frac{\sigma^2(\lambda_2)}{n_2} \right]^{-1/2} - z_{\alpha/2} = z_{\beta}.$$

Under the assumption that $n_1 = kn_2$, we obtain that

$$n_2 = \frac{(z_{\alpha/2} + z_{\beta})^2}{(\lambda_1 - \lambda_2)^2} \left[\frac{\sigma^2(\lambda_1)}{k} + \sigma^2(\lambda_2) \right]. \tag{7.5}$$

7.2.2 Test for Noninferiority/Superiority

Since $\varepsilon = \lambda_1 - \lambda_2$, where λ_1 and λ_2 are the hazard rates of the control and test drug, respectively, in practice, a smaller hazard rate is considered a favor of the test drug. In other words, a negative value of ε implies a better performance of the test drug than the control. Hence, the problem of testing noninferiority and superiority can be unified by the following hypotheses:

$$H_0 : \varepsilon \leq \delta \quad \text{versus} \quad H_a : \varepsilon > \delta,$$

where δ is the superiority or noninferiority margin. When $\delta > 0$, the rejection of the null hypothesis indicates the superiority of the test drug over the control. When $\delta < 0$, the rejection of the null hypothesis is an indication of the noninferiority of the test drug to the control. Similarly, under the null hypothesis, test statistic

$$T = (\hat{\lambda}_1 - \hat{\lambda}_2 - \delta) \left[\frac{\sigma^2(\hat{\lambda}_1)}{n_1} + \frac{\sigma^2(\hat{\lambda}_2)}{n_2} \right]^{-1/2}$$

is asymptotically distributed as a standard normal random variable. Thus, we reject the null hypothesis at approximate α level of significance if

$$(\hat{\lambda}_1 - \hat{\lambda}_2 - \delta) \left[\frac{\sigma^2(\hat{\lambda}_1)}{n_1} + \frac{\sigma^2(\hat{\lambda}_2)}{n_2} \right]^{-1/2} > z_\alpha.$$

Under the alternative hypothesis that $\varepsilon > \delta$, the power of the above test is approximately

$$\Phi\left((\varepsilon - \delta) \left[\frac{\sigma^2(\lambda_1)}{n_1} + \frac{\sigma^2(\lambda_2)}{n_2} \right]^{-1/2} - z_\alpha \right).$$

As a result, the sample size needed to achieve a desired power of $1 - \beta$ can be obtained by solving

$$(\varepsilon - \delta) \left[\frac{\sigma^2(\lambda_1)}{n_1} + \frac{\sigma^2(\lambda_2)}{n_2} \right]^{-1/2} - z_\alpha = z_\beta.$$

Under the assumption that $n_1 = kn_2$, we have

$$n_2 = \frac{(z_\alpha + z_\beta)^2}{(\varepsilon - \delta)^2} \left[\frac{\sigma^2(\lambda_1)}{k} + \sigma^2(\lambda_2) \right]. \tag{7.6}$$

7.2.3 Test for Equivalence

The objective is to test the following hypotheses:

$$H_0 : |\varepsilon| \geq \delta \quad \text{versus} \quad H_a : |\varepsilon| < \delta.$$

The null hypothesis is rejected and the test drug is concluded to be equivalent to the control on average if

$$(\hat{\lambda}_1 - \hat{\lambda}_2 - \delta) \left[\frac{\sigma^2(\hat{\lambda}_1)}{n_1} + \frac{\sigma^2(\hat{\lambda}_2)}{n_2} \right]^{-1/2} < -z_\alpha$$

and

$$(\hat{\lambda}_1 - \hat{\lambda}_2 + \delta) \left[\frac{\sigma^2(\hat{\lambda}_1)}{n_1} + \frac{\sigma^2(\hat{\lambda}_2)}{n_2} \right]^{-1/2} > z_\alpha.$$

Under the alternative hypothesis ($|\varepsilon| < \delta$), the power of the above testing procedure is approximately

$$2\Phi\left((\delta - |\varepsilon|)\left[\frac{\sigma^2(\lambda_1)}{n_1} + \frac{\sigma^2(\lambda_2)}{n_2}\right]^{-1/2} - z_\alpha\right) - 1.$$

As a result, the sample size needed for achieving a power of $1 - \beta$ can be obtained by solving the following equation:

$$(\delta - |\varepsilon|)\left[\frac{\sigma^2(\lambda_1)}{n_1} + \frac{\sigma^2(\lambda_2)}{n_2}\right]^{-1/2} - z_\alpha = z_{\beta/2}.$$

As a result, the sample size needed for achieving a power of $1 - \beta$ is given by $n_1 = kn_2$ and

$$n_2 = \frac{(z_\alpha + z_{\beta/2})^2}{(\delta - |\varepsilon|)^2}\left(\frac{\sigma^2(\lambda_1)}{k} + \sigma^2(\lambda_2)\right). \tag{7.7}$$

7.2.4 An Example

Suppose that the sponsor is planning a trial among the patients with either Hodgkin's disease (HOD) or non-Hodgkin's lymphoma (NHL). The patients will be given either an allogeneic (allo) transplant from an HLA-matched sibling donor or an autologous (auto) transplant where their own marrow has been cleansed and returned to them after high dose of chemotherapy. The primary objective is to compare the patients with allo or auto transplant in terms of time to leukemia. The trial is planned to last for three ($T = 3$) years with one year accrual ($T_0 = 1$). Uniform patient entry for both allo and auto transplant groups is assumed ($\lambda = 0$). It is also assumed that the leukemia-free hazard rates for allo and auto transplant are given by 1 ($\lambda_1 = 1$) and 2 ($\lambda_2 = 2$), respectively. According to Equation 7.3, the variance function is given by

$$\sigma^2(\lambda_i) = \lambda_i^2\left[1 + \frac{e^{-\lambda_i T} - e^{-\lambda_i(T - T_0)}}{\lambda_i T_0}\right]^{-1}.$$

7.2.4.1 Test for Equality

Assume that $n_1 = n_2 = n$. According to Equation 7.5, the sample size needed to achieve an 80% ($\beta = 0.2$) power at 0.05 level of significance is

$$\begin{aligned} n &= \frac{(z_{\alpha/2} + z_\beta)^2}{(\lambda_2 - \lambda_1)^2}\left(\frac{\sigma^2(\lambda_1)}{k} + \sigma^2(\lambda_2)\right) \\ &= \frac{(1.96 + 0.84)^2}{(2 - 1)^2}(0.97 + 3.94) \\ &\approx 39. \end{aligned}$$

7.2.4.2 Test for Superiority

Assume that $n_1 = n_2 = n$ and the superiority margin $\delta = 0.2$. According to Equation 7.6, the sample size needed to achieve an 80% ($\beta = 0.2$) power at the 5% level of significance is

$$
\begin{aligned}
n &= \frac{(z_\alpha + z_\beta)^2}{(\lambda_2 - \lambda_1 - \delta)^2}\left[\frac{\sigma^2(\lambda_1)}{k} + \sigma^2(\lambda_2)\right] \\
&= \frac{(1.64 + 0.84)^2}{(2 - 1 - 0.2)^2}(0.97 + 3.94) \\
&\approx 48.
\end{aligned}
$$

7.2.4.3 Test for Equivalence

Assume that $n_1 = n_2 = n$, $\lambda_1 = \lambda_2 = 1$, and the equivalence margin is 0.5 ($\delta = 0.5$). According to Equation 7.7, the sample size needed to achieve an 80% ($\beta = 0.2$) power at the $\alpha = 0.05$ level of significance is

$$
\begin{aligned}
n &= \frac{(z_\alpha + z_{\beta/2})^2}{\delta^2}\left[\frac{\sigma^2(\lambda_1)}{k} + \sigma^2(\lambda_2)\right] \\
&= \frac{(1.64 + 1.28)^2}{(0.5 - 0)^2}(0.97 + 0.97) \\
&\approx 67.
\end{aligned}
$$

7.2.5 Remarks

7.2.5.1 Unconditional versus Conditional

According to Lachin (1981), for testing equality of hazard rates based on exponential model, there exists another way to construct the test statistic other than Equation 7.4. More specifically, the testing procedure can be modified to reject the null hypothesis if

$$
\left|(\hat{\lambda}_1 - \hat{\lambda}_2)\left[\sigma^2(\hat{\bar{\lambda}})\left(\frac{1}{n_1} + \frac{1}{n_2}\right)\right]^{-1/2}\right| > z_{\alpha/2}, \tag{7.8}
$$

where

$$
\hat{\bar{\lambda}} = \frac{n_1 \hat{\lambda}_1 + n_2 \hat{\lambda}_2}{n_1 + n_2}.
$$

As it can be seen, Equation 7.8 is very similar to Equation 7.4 except using a different estimate for the variance of $\hat{\lambda}_1 - \hat{\lambda}_2$. The difference is that the variance estimate used in Equation 7.4 is the MLE without constraint while the variance estimate used in Equation 7.8 is a pooled estimate of $\hat{\lambda}_1 - \hat{\lambda}_2$, which is a consistent conditional on H_0. We refer to Equation 7.4 as the unconditional method and Equation 7.8 as the conditional method. In practice, which method (unconditional/conditional) should be used to test equality is always a dilemma because one is not necessarily always more powerful than the other under the alternative hypothesis. However, it is difficult to generalize the conditional method to test noninferiority/superiority and equivalence because the MLE under H_0 is difficult to find. Although both unconditional and conditional methods have asymptotic size a under the null hypothesis, the powers under the alternative hypothesis are not equal to each other. Hence, the sample size formula for Equation 7.8 is different from the sample size formula for Equation 7.4. For the purpose of completeness, it is derived below.

Under the alternative hypothesis ($\varepsilon \neq 0$), the power of Equation 7.8 is approximately

$$\Phi\left(\left[|\varepsilon| - \left[\sigma^2(\bar{\lambda})\left(\frac{1}{n_1} + \frac{1}{n_2}\right)\right]z_{\alpha/2}\right]\left[\frac{\sigma^2(\lambda_1)}{n_1} + \frac{\sigma^2(\lambda_2)}{n_2}\right]^{-1/2}\right),$$

where

$$\bar{\lambda} = \frac{n_1\lambda_1 + n_2\lambda_2}{n_1 + n_2}.$$

Hence, the sample size needed to achieve a power of $1 - \beta$ can be achieved by solving

$$\left[|\varepsilon| - \left[\sigma^2(\bar{\lambda})\left(\frac{1}{n_1} + \frac{1}{n_2}\right)\right]z_{\alpha/2}\right]\left[\frac{\sigma^2(\lambda_1)}{n_1} + \frac{\sigma^2(\lambda_2)}{n_2}\right]^{-1/2} = z_\beta,$$

Under the assumption that $n_1 = kn_2$, it implies that

$$n_2 = \frac{1}{\varepsilon^2}\left|z_{\alpha/2}\sigma^2(\bar{\lambda})\left(\frac{1}{k} + 1\right) + z_\beta\left(\frac{\sigma(\lambda_1)}{k} + \sigma^2(\lambda_2)\right)^{1/2}\right|^2, \qquad (7.9)$$

where

$$\bar{\lambda} = \frac{k\lambda_1 + \lambda_2}{k + 1}.$$

7.2.5.2 Losses to Follow-Up, Dropout, and Noncompliance

If we further assume that the losses are exponentially distributed with loss hazard rate η_i in the ith treatment group, it has been shown by Lachin and Foulkes (1986) that variance of $\hat{\lambda}_1$ is given by

$$\sigma^2(\lambda_i, \eta_i, \gamma_i) = \lambda_i^2\left(\frac{\lambda_i}{\lambda_i + \eta_i} + \frac{\lambda_i\gamma_i e^{-(\lambda_i + \eta_i)T}[1 - e^{(\lambda_i + \eta_i - \gamma_i)T_0}]}{(1 - e^{-\gamma_i T_0})(\lambda_i + \eta_i)(\lambda_i + \eta_i - \gamma_i)}\right)^{-1}.$$

In such a situation, an appropriate test statistic can be constructed by replacing $\sigma(\hat{\lambda}_i)$ and $\sigma(\hat{\bar{\lambda}})$ by $\sigma(\hat{\lambda}_i, \hat{\eta}_i, \hat{\gamma}_i)$ and $\sigma(\hat{\bar{\lambda}}, \hat{\bar{\eta}}, \hat{\bar{\gamma}})$, respectively, where $\hat{\eta}_i$ and $\hat{\gamma}_i$ are the MLE of η_i and γ_i, respectively, and

$$\hat{\bar{\eta}}_i = \frac{n_1 \hat{\eta}_1 + n_2 \hat{\eta}_2}{n_1 + n_2} \quad \text{and} \quad \hat{\bar{\gamma}}_i = \frac{n_1 \hat{\gamma}_1 + n_2 \hat{\gamma}_2}{n_1 + n_2}.$$

Hence, appropriate sample size calculation formulas can be obtained by replacing $\sigma(\lambda_i)$ and $\sigma(\bar{\lambda})$ by $\sigma(\lambda_i, \eta_i, \gamma_i)$ and $\sigma(\bar{\lambda}, \bar{\eta}, \bar{\gamma})$, respectively, where

$$\bar{\eta}_i = \frac{n_1 \eta_1 + n_2 \eta_2}{n_1 + n_2} \quad \text{and} \quad \bar{\gamma}_i = \frac{n_1 \gamma_1 + n_2 \gamma_2}{n_1 + n_2}.$$

7.3 Cox's Proportional Hazards Model

The most commonly used regression model in survival analysis is Cox's proportional hazards model. Let t_i be the time to event for the ith subject and C_i be the corresponding censoring time. Besides t_i and C_i, each subject also provides a p-dimension column vector of covariates denoted by z_i. The most commonly encountered covariates include treatment indicator, demographical information, medical history, etc. Let $h(t|z)$ be the hazard rate at time t for an individual with covariate vector z. Cox's proportional hazard model assumes

$$h(t \,|\, z) = h(t \,|\, 0)e^{b'z},$$

where b, the coefficient vector with the same dimension as z, can be estimated by maximizing the following partial likelihood function:

$$L(b) = \prod_{i=1}^{d} \frac{e^{b'z_{(i)}}}{\sum_{j \in R_i} e^{b'z_j}},$$

The product is over all the observed deaths, $z_{(i)}$ is the covariate vector associated with the ith observed death, and R_i is the set of individuals at risk just prior to the ith observed death. Maximizing L is equivalent to solving $U(b) = 0$, where

$$U(b) = \sum_{i=1}^{d} z_{(i)} - \sum_{i=1}^{d} \frac{\sum_{j \in R_i} z_j e^{b'z_j}}{\sum_{j \in R_i} e^{b'z_j}}. \tag{7.10}$$

The corresponding information matrix is given by $I(b)$ with the (a, b)th element given by

$$I(b) = \sum_{i=1}^{d} \frac{\sum_{j \in R_i} z_j z_j' e^{b'z_j}}{\sum_{j \in R_i} e^{b'z_j}}$$

$$- \sum_{i=1}^{d} \left[\left(\frac{\sum_{j \in R_i} Z_j e^{b'z_j}}{\sum_{j \in R_i} e^{b'z_j}} \right) \left(\frac{\sum_{j \in R_i} Z_j e^{b'z_j}}{\sum_{j \in R_i} e^{b'z_j}} \right)' \right].$$

(7.11)

7.3.1 Test for Equality

In practice, it is of interest to test the following hypotheses:

$$H_0 : b = b_0 \quad \text{versus} \quad H_a : b \neq b_0.$$

To test $b = b_0$, the following score statistic proposed by Schoenfeld (1981) is used:

$$\chi^2_{SC} = U(b_0)' I^{-1}(b_0) U(b_0).$$

Under the null hypothesis of $b = b_0$, χ^2_{SC} is asymptotically distributed as a chi-square random variable with p degrees of freedom. The null hypothesis is rejected if $\chi^2_{SC} > \chi^2_{\alpha,p}$, where $\chi^2_{\alpha,p}$ is the αth upper quantile of a chi-square random variable with p degrees of freedom.

The most typical situation in practice is to compare two treatments without adjusting for other covariates. As a result, we consider the indicator as the only covariate ($z_i = 0$: treatment 1; $z_i = 1$: treatment 2). Then, according to Equations 7.12 and 7.13, it follows that

$$U(b) = d_1 - \sum_{i=1}^{d} \frac{Y_{2i} e^b}{Y_{1i} + Y_{2i} e^b},$$

(7.12)

and

$$I(b) = \sum_{i=1}^{d} \left[\frac{Y_i e^b}{Y_{1i} + Y_{2i} e^b} - \frac{Y_{2i}^2 e^{2b}}{(Y_{1i} + Y_{2i} e^b)^2} \right] = \sum_{i=1}^{d} \frac{Y_{1i} Y_{2i} e^b}{(Y_{1i} + Y_{2i} e^b)^2},$$

(7.13)

where Y_{ij} denotes the number of subjects at risk just prior to the ith observed event and $i = 1, 2$. To test for equality of two survival curves, the following hypotheses are usually considered:

$$H_0 : b = 0 \quad \text{versus} \quad H_a : b \neq 0.$$

Under the null hypothesis, we have

$$U(b) = d_1 - \sum_{i=1}^{d} \frac{Y_{2i}}{Y_{1i} + Y_{2i}},$$

and

$$I(b) = \sum_{i=1}^{d} \left[\frac{Y_{2i}}{Y_{1i} + Y_{2i}} - \frac{Y_{2i}^2}{(Y_{1i} + Y_{2i})^2} \right] = \sum_{i=1}^{d} \frac{Y_{2i} Y_{1i}}{(Y_{1i} + Y_{2i})^2}.$$

Note that the score test statistic $\chi_{sc}^2 = U(0)^2 / I(0)$ reduces to the following log-rank test statistic for two-sample problem:

$$L = \frac{\sum_{k=1}^{d} \left(I_k - \left(\frac{Y_{1i}}{Y_{1i} + Y_{2i}} \right) \right)}{\left[\sum_{k=1}^{d} \left(\frac{Y_{1i} + Y_{2i}}{(Y_{1i} + Y_{2i})^2} \right) \right]^{-\frac{1}{2}}},$$

where I_k is a binary variable indicating whether the kth event is from the first treatment group or not. Thus, we reject the null hypothesis at approximate α level of significance if $L > z_{\alpha/2}$. The formula for sample size calculation introduced below can be viewed as a special case of log-rank test under the assumption of proportional hazard.

Let p_i be the proportion of patients in the ith treatment group, $H_i(t)$ be the distribution function of censoring, and $\lambda_i(t), f_i(t)$, and $F_i(t)$ be the hazard, density, and distribution function of survival in group i, respectively. Define the functions

$$V(t) = p_1 f_0(t)(1 - H_1(t)) + p_2 f_1(t)(1 - H_2(t)),$$

and

$$\pi(t) = \frac{p_2(1 - F_1(t))(1 - H_2(t))}{p_1(1 - F_0(t))(1 - H_1(t)) + p_2(1 - F_1(t))(1 - H_2(t))}.$$

Then, L is asymptotically distributed as a normal random variable with variance 1 and mean given by

$$\frac{n^{1/2} \int_0^{\infty} \log(\lambda_2(t) / \lambda_1(t)) \pi(t)(1 - \pi(t)) V(t) dt}{\left[\int_0^{\infty} \pi(t)(1 - \pi(t)) V(t) dt \right]^{1/2}}. \tag{7.14}$$

Under the assumption of proportional hazard, $\log(\lambda_2(t)/\lambda_1(t)) = b$ is a constant. Assume that $H_2(t) = H_1(t)$. Let $d = \int_0^{\infty} V(t) dt$, which is the probability of observing an event. In practice, most commonly $F_1(t) \approx F_0(t)$. In such a situation, it can be noted that $\pi(t) \approx p_2$, then Equation 7.14 becomes

$$b(np_1 p_2 d)^{1/2}.$$

Therefore, the two-sided sample size formula with significance level α and power $1 - \beta$ is given by

$$n = \frac{(z_{\alpha/2} + z_{\beta})^2}{b^2 p_1 p_2 d}. \tag{7.15}$$

7.3.2 Test for Noninferiority/Superiority

We still assume that $z_i = 0$ for treatment 1 and $z_i = 1$ for treatment 2. The problem of testing noninferiority and superiority can be unified by the following hypotheses:

$$H_0 : b \le \delta \quad \text{versus} \quad H_a : b > \delta,$$

where δ is the superiority or noninferiority margin. When $\delta > 0$, the rejection of the null hypothesis indicates superiority over the reference value. When $\delta < 0$, the rejection of the null hypothesis implies noninferiority as compared to the reference value. When $b = \delta$, the test statistic

$$L = \frac{\sum_{k=1}^{d} \left[I_k - \left(\frac{Y_{1i} e^{\delta}}{Y_{1i} e^{\delta} + Y_{2i}} \right) \right]}{\left[\sum_{k=1}^{d} \left(\frac{Y_{1i} + Y_{2i} e^{\delta}}{Y_{1i} e^{\delta} + Y_{2i}} \right)^2 \right]^{\frac{1}{2}}}$$

follows a standard normal distribution when the sample size is sufficiently large. Hence, the null hypothesis should be rejected if $L > z_a$. Under the alternative hypothesis, L is asymptotically distributed as a normal random variable with variance 1 and mean given by

$$\frac{n^{1/2} \int_{0}^{\infty} \log(\lambda_2(t) / \lambda_1(t)) - \delta)\pi(t)(1 - \pi(t))V(t) dt}{\left[\int_{0}^{\infty} \pi(t)(1 - \pi(t))V(t) dt \right]^{\frac{1}{2}}}. \tag{7.16}$$

Under the assumption of proportional hazard, $\log(\lambda_2(t)/\lambda_1(t)) = b > \delta$ is a constant. Assume that $H_2(t) = H_1(t)$. Let $d = \int_{0}^{\infty} V(t) dt$, which is the probability of observing an event. In practice, most commonly $F_1(t) \approx F_0(t)$. In such a situation, it can be noted that $\pi(t) \approx p_2$, then Equation 7.16 becomes

$$(b - \delta)(np_1 p_2 d)^{1/2}.$$

Therefore, the sample size formula with significance level α and power $1 - \beta$ is given by

$$n = \frac{(z_{\alpha/2} + z_{\beta})^2}{(b - \delta)^2 p_1 p_2 d}. \tag{7.17}$$

7.3.3 Test for Equivalence

Assume that $z_i = 0$ for treatment 1 and $z_i = 1$ for treatment 2. To establish equivalence, the following hypotheses are usually considered:

$$H_0 : |b| > \delta \quad \text{versus} \quad H_a : |b| < \delta.$$

The above hypotheses can be tested using two one-sided test procedures. More specifically, the null hypothesis should be rejected if

$$\frac{\sum_{k=1}^{d} \left[I_k - \left(\dfrac{Y_{1i}e^{\delta}}{Y_{1i}e^{\delta} + Y_{2i}} \right) \right]}{\left[\sum_{k=1}^{d} \left(\dfrac{Y_{1i} + Y_{2i}e^{\delta}}{\left(Y_{1i}e^{\delta} + Y_{2i} \right)^2} \right) \right]^{-\frac{1}{2}}} < -z_\alpha$$

and

$$\frac{\sum_{k=1}^{d} \left[I_k - \left(\dfrac{Y_{1i}e^{-\delta}}{Y_{1i}e^{-\delta} + Y_{2i}} \right) \right]}{\left[\sum_{k=1}^{d} \left(\dfrac{Y_{1i}Y_{2i}e^{-\delta}}{\left(Y_{1i}e^{-\delta} + Y_{2i} \right)^2} \right) \right]^{-\frac{1}{2}}} > z_\alpha.$$

The power of the above procedure is approximately

$$2\Phi\left((\delta - |b|)\sqrt{np_1 p_2 d} - z_\alpha \right) - 1.$$

Hence, the sample size needed to achieve a power of $1 - \beta$ at α level of significance is given by

$$n = \frac{(z_\alpha + z_{\beta/2})^2}{(\delta - |b|)^2 p_1 p_2 d}. \tag{7.18}$$

7.3.4 An Example

Infection of a burn wound is a common complication resulting in extended hospital stays and in the death of severely burned patients. One of the important components of burn management is to prevent or delay the infection. Suppose an investigator is interested in conducting a trial to compare a new therapy with a routine bathing care method in terms of the time to infection. Assume that a hazard ratio of 2 (routine bathing care/test therapy) is considered of clinical importance ($b = \log (2)$). It is further assumed that about 80% of patients' infection may be observed during the trial period. Let $n = n_1 = n_2$ ($p_1 = p_2 = 0.5$).

7.3.4.1 Test for Equality

According to Equation 7.15, the sample size per treatment group needed to achieve a power of 80% ($\beta = 0.2$) at the 5% level of significance ($\alpha = 0.05$) is given by

$$n = \frac{(z_{\alpha/2} + z_\beta)^2}{b^2 p_1 p_2 d} = \frac{(1.96 + 0.84)^2}{\log^2(2) \times 0.5 \times 0.5 \times 0.8} \approx 82.$$

7.3.4.2 Test for Superiority

Assume that the superiority margin is 0.3 ($\delta = 0.3$). By Equation 7.17, the sample size per treatment group needed for achieving an 80% ($\beta = 0.2$) power at the 5% level of significance ($\alpha = 0.05$) is given by

$$n = \frac{(z_\alpha + z_\beta)^2}{(b - \delta)^2 p_1 p_2 d} = \frac{(1.64 + 0.84)^2}{(\log(2) - 0.3)^2 \times 0.5 \times 0.5 \times 0.8} \approx 200.$$

7.3.4.3 Test for Equivalence

Assume that the equivalence limit is 0.5 (i.e., $\delta = 0.5$) and $b = 0$. Then, by Equation 7.18, the sample size per treatment group required to achieve an 80% power ($\beta = 0.2$) at the 5% level of significance ($\alpha = 0.05$) is given by

$$n = \frac{(z_\alpha + z_{\beta/2})^2}{(\delta - |b|)^2 p_1 p_2 d} = \frac{(1.64 + 1.28)^2}{(0.5 - 0.0)^2 \times 0.5 \times 0.5 \times 0.8} \approx 171.$$

7.4 Weighted Log-Rank Test

When the time to event is not exponentially distributed and the assumption of proportional hazard does not hold, the treatment effects are usually evaluated by comparing survival curves ($S_i(t)$). To test whether there is a difference between the true survival curves, the following hypotheses are usually considered:

$$H_0 : S_1(t) = S_2(t) \quad \text{versus} \quad H_a : S_1(t) \neq S_2(t).$$

In such a situation, testing noninferiority/superiority or equivalence is usually difficult to be carried out, because $S_i(t)$ is an infinite-dimensional parameter and, hence, how to define noninferiority/superiority and equivalence is not clear. As a result, we provide sample size calculation formula for testing equality only in this section.

7.4.1 Tarone–Ware Test

To compare two survival curves, weighted log tests are usually considered. The test statistic of weighted log-rank test (the Tarone–Ware statistic) is given by

$$L = \frac{\sum_{k=1}^{d} w_i \left(I_i - \left(\frac{Y_{1i}}{Y_{1i} + Y_{2i}} \right) \right)}{\left[\sum_{i=1}^{d} w_i^2 \left(\frac{Y_{1i} + Y_{2i}}{(Y_{1i} + Y_{2i})^2} \right) \right]^{-\frac{1}{2}}},$$

where the sum is over all deaths, I_i is the indicator of the first group, w_i is the ith weight, and Y_{ij} is number of subjects at risk just before the jth death in the ith group. When $w_i = 1$, L becomes the commonly used log-rank statistic. Under the null hypothesis of $H_0 : S_1 = S_2$, L is asymptotically distributed as a standard normal random variable. Hence, we would reject the null hypothesis at approximate α level of significance if $|L| > z_{\alpha/2}$.

The sample size calculation formula we are going to introduce in this section was developed by Lakatos (1986, 1988). According to Lakatos' method, the trial period should be first partitioned into N equally spaced intervals. Let d_i denote the number of deaths within the ith interval. Define ϕ_{ik} to be the ratio of patients in the two treatment groups at risk just prior to the kth death in the ith interval. The expectation of L under a fixed local alternative can be approximated by

$$E = \frac{\sum_{i=1}^{N} \sum_{k=1}^{d_i} w_{ik} \left[\left(\frac{\phi_{ik}\theta_{ik}}{1+\phi_{ik}\theta_{ik}} \right) - \left(\frac{\phi_{ik}}{1+\phi_{ik}} \right) \right]}{\left[\sum_{i=1}^{N} \sum_{k=1}^{d_i} \left(\frac{w_{ik}^2 \phi_{ik}}{\left(1+\phi_{ik}\right)^2} \right) \right]^{\frac{1}{2}}}, \tag{7.19}$$

where the right summation of each double summation is over the d_i deaths in the ith interval, and the left summation is over the N intervals that partition the trial period. Treating this statistic as $N(E, 1)$, the sample size needed to achieve a power of $1 - \beta$ can be obtained by solving

$$E = z_{\alpha/2} + z_\beta.$$

When N is sufficiently large, we can assume that $\phi_{ik} = \phi_i$ and $w_{ik} = w_i$ for all k in the ith interval. Let $\rho_i = d_i/d$, where $d = \sum d_i$. Then, E can be written as

$$E = e(D)\sqrt{d},$$

where

$$e(D) = \frac{\sum_{i=1}^{N} w_i \rho_i \gamma_i}{\left(\sum_{i=1}^{N} w_i^2 \rho_i \eta_i \right)^{1/2}}, \tag{7.20}$$

$$\gamma_i = \frac{\phi_i \theta_i}{1+\phi_i \theta_i} - \frac{\phi_i}{1+\phi_i},$$

and

$$\eta_i = \frac{\phi_i}{\left(1+\phi_i\right)^2}. \tag{7.21}$$

It follows that

$$d = \frac{(z_{\alpha/2} + z_\beta)^2 \left(\sum_{i=1}^{N} w_i^2 \rho_i \eta_i \right)}{\left(\sum_{i=1}^{N} w_i \rho_i \gamma_i \right)^2}. \tag{7.22}$$

Let n_i denote the sample size in the ith treatment group. Under the assumption that $n = n_1 = n_2$, the sample size needed to achieve a power of $1 - \beta$ is given by

$$n = \frac{2d}{p_1 + p_2},$$

where p_i is the cumulative event rate for the ith treatment group.

7.4.2 An Example

To illustrate the method described above, the example given in Lakatos (1988) is considered. To carry out Lakatos' method, we need to first partition the trial period into N equal-length intervals. Then, parameters like γ_i, η_i, ρ_i, θ_i, and ϕ_i need to be specified. There are two ways we can specify them. One is to directly specify them for each interval or estimate them from a pilot study. Then the whole procedure becomes relatively easy. However, sometimes only yearly rates are given for the necessary parameters. Then we need to calculate all those parameters by ourselves.

For example, consider a two-year cardiovascular trial. It is assumed that the yearly hazard rates in the treatment group ($i = 1$) and the control group ($i = 2$) are given by 1 and 0.5, respectively. Hence, the yearly event rates in the two treatment groups are given by $1 - e^{-1} = 63.2\%$ and $1 - e^{-0.5} = 39.3\%$, respectively. It is also assumed that the yearly loss to follow-up and noncompliance rates are 3% and 4%, respectively. The rate at which patients assigned to control begin taking a medication with an efficacy similar to the experimental treatment is called "drop-in." In cardiovascular trials, drop-ins often occur when the private physician of a patient assigned to control detects the condition of interest, such as hypertension, and prescribes treatment. In this example, it is assumed that the yearly drop-in rate is 5%. Assume that the patient's status follows a Markov chain with four possible states, that is, lost to follow-up, event, active in treatment, and active in control, which are denoted by L, E, A_E, and A_C, respectively. Then, the yearly transition matrix of this Markov chain is given by

$$T = \begin{bmatrix} 1 & 0 & 0.03 & 0.03 \\ 0 & 1 & 0.3935 & 0.6321 \\ 0 & 0 & 1-\Sigma & 0.05 \\ 0 & 0 & 0.04 & 1-\Sigma \end{bmatrix}.$$

Entries denoted by $1 - \Sigma$ represent 1 minus the sum of the remainder of the column. Assume, however, we want to partition the two-year period into 20 equal-length intervals. Then we need the transition matrix within each interval. It can be obtained by replacing each off-diagonal entry x in T by $1 - (1 - x)^{1/K}$. The resulting transition matrix is given by

$$T_{1/20} = \begin{bmatrix} 1.0000 & 0.0000 & 0.0030 & 0.0030 \\ 0.0000 & 1.0000 & 0.0951 & 0.0488 \\ 0.0000 & 0.0000 & 0.8978 & 0.0051 \\ 0.0000 & 0.0000 & 0.0041 & 0.9431 \end{bmatrix}.$$

TABLE 7.2

Sample Size Calculation by Lakatos' Method

t_i	γ	η	ρ	θ	ϕ		L	E	A_C	A_E
0.1	0.167	0.250	0.098	2.000	1.000	C	0.003	0.095	0.897	0.005
						E	0.003	0.049	0.944	0.004
0.2	0.166	0.250	0.090	1.986	0.951	C	0.006	0.181	0.804	0.009
						E	0.006	0.095	0.891	0.007
0.3	0.166	0.249	0.083	1.972	0.905	C	0.008	0.258	0.721	0.013
						E	0.009	0.139	0.842	0.010
0.4	0.165	0.249	0.076	1.959	0.862	C	0.010	0.327	0.647	0.016
						E	0.011	0.181	0.795	0.013
0.5	0.164	0.248	0.070	1.945	0.821	C	0.013	0.389	0.580	0.018
						E	0.014	0.221	0.750	0.015
0.6	0.163	0.246	0.064	1.932	0.782	C	0.014	0.445	0.520	0.020
						E	0.016	0.259	0.708	0.016
0.7	0.162	0.245	0.059	1.920	0.746	C	0.016	0.496	0.466	0.022
						E	0.018	0.295	0.669	0.017
0.8	0.160	0.243	0.054	1.907	0.711	C	0.017	0.541	0.418	0.023
						E	0.020	0.330	0.632	0.018
0.9	0.158	0.241	0.050	1.894	0.679	C	0.019	0.582	0.375	0.024
						E	0.022	0.362	0.596	0.019
1.0	0.156	0.239	0.046	1.882	0.648	C	0.020	0.619	0.336	0.024
						E	0.024	0.393	0.563	0.019
1.1	0.154	0.236	0.043	1.870	0.619	C	0.021	0.652	0.302	0.025
						E	0.026	0.423	0.532	0.020
1.2	0.152	0.234	0.039	1.857	0.592	C	0.022	0.682	0.271	0.025
						E	0.028	0.450	0.502	0.020
1.3	0.149	0.231	0.036	1.845	0.566	C	0.023	0.709	0.243	0.025
						E	0.029	0.477	0.474	0.020
1.4	0.147	0.228	0.034	1.833	0.542	C	0.024	0.734	0.218	0.025
						E	0.031	0.502	0.448	0.020
1.5	0.144	0.225	0.031	1.820	0.519	C	0.025	0.755	0.195	0.025
						E	0.032	0.525	0.423	0.020
1.6	0.141	0.222	0.029	1.808	0.497	C	0.025	0.775	0.175	0.024
						E	0.033	0.548	0.399	0.019
1.7	0.138	0.219	0.027	1.796	0.477	C	0.026	0.793	0.157	0.024
						E	0.035	0.569	0.377	0.019
1.8	0.135	0.215	0.025	1.783	0.457	C	0.026	0.809	0.141	0.023
						E	0.036	0.589	0.356	0.018
1.9	0.132	0.212	0.023	1.771	0.439	C	0.027	0.824	0.127	0.023
						E	0.037	0.609	0.336	0.018
2.0	0.129	0.209	0.021	1.758	0.421	C	0.027	0.837	0.114	0.022
						E	0.038	0.627	0.318	0.018

C: control; *E:* experimental.

Then, the patient distribution at the end of the $i/10$th year is given by $T^i_{1/20}x$, where x is a four-dimension vector indicating the initial distribution of patients. So, for the treatment group, $x = (0, 0, 1, 0)'$ indicating that at the beginning of the trial, all patients are active in treatment. Similarly, for control, $x = (0, 0, 0, 1)$. For the purpose of illustration, consider at the time point 0.3 year, the patient distribution for the treatment group is given by

$$\begin{bmatrix} 1.0000 & 0.0000 & 0.0030 & 0.0030 \\ 0.0000 & 1.0000 & 0.0951 & 0.0488 \\ 0.0000 & 0.0000 & 0.8978 & 0.0051 \\ 0.0000 & 0.0000 & 0.0041 & 0.9431 \end{bmatrix}^3 \begin{pmatrix} 0 \\ 0 \\ 1 \\ 0 \end{pmatrix} = \begin{pmatrix} 0.0081 \\ 0.2577 \\ 0.7237 \\ 0.0104 \end{pmatrix}.$$

This indicates by the time of 0.3 year, we may expect that 0.81% of patients were lost to follow-up, 25.77% experienced events, 72.37% were still active in treatment, and 1.04% switched to some other medication with similar effects as the control (noncompliance). Hence, this vector becomes the third row and first four columns of Table 7.2. Similarly, we can produce all the rows for the first eight columns in Table 7.2.

Assuming equal allocation of patients across treatment groups, we have $\phi = 1$ when $t_i = 0.1$, which means just before time point $t_i = 0.1$, the ratio of patients at risk between treatment groups is 1. When $t_i = 0.2$, the patients at risk just prior to $t_i = 0.2$ in control groups are the patients still active ($A_C + A_E$) in the control group, which is given by $0.897 + 0.005 = 0.902$. Similarly, the patients at risk in the experimental group just prior to $t = 0.1$ is given by $0.944 + 0.004 = 0.948$. Hence, at $t_i = 0.2$, the value of ϕ can be determined by $\phi = 0.902/0.948 = 0.951$. Similarly, the values of ϕ at other time points can be calculated. Once ϕ is obtained, the value of η can be calculated according to formula (7.21).

In the next step, we need to calculate θ, which needs specification of hazard rates within each interval. First, we know before $t = 0.1$, all patients in the control group staying active in treatment (A_E) and all patients in the experimental group staying active in control (A_C). According to our assumption, the hazard rates for the two groups are given by 1 and 0.5, respectively. Hence, $\theta = 2$. When $t = 0.2$, we know in the control group the proportion of patients experiencing events is $0.181 - 0.095 = 0.086$. Hence, the hazard rate can be obtained by $\log (1 - 0.086)/0.1$. Similarly, the hazard rate in the experimental groups is given by $\log (1 - (0.095 - 0.049))/0.1 = \log (1 - 0.046)/0.1$. Hence, the value of θ is given by $\log (1 - 0.086)/\log (1 - 0.046) = 1.986$. The value of θ at other time points can be obtained similarly.

Finally, we need to calculate ρ_i. First, we can notice that the total events for the control and experimental groups are given by 0.837 and 0.627, respectively. The events experienced in the first interval ($t_i = 0.1$) for the two groups are given by 0.095 and 0.049, respectively. Hence, the value of ρ when $t_i = 0.1$ is given by $(0.095 + 0.049)/(0.837 + 0.627) = 0.098$. When $t_i = 0.2$, the event experienced in the control is given by $0.181 - 0.095 = 0.096$. The event experienced in experimental groups is given by $0.095 - 0.049 = 0.046$. Hence, the value of ρ can be obtained by $(0.086 + 0.046)/(0.837 + 0.627) = 0.090$. The value of ρ at other time points can be obtained similarly.

Owing to rounding error, readers may not be able to reproduce exactly the same number of the derived parameters ($\gamma, \eta, \rho, \theta, \phi$) as we do by performing appropriate operation on the first eight columns in Table 7.2. However, by keeping enough decimal digits and following our instructions given above, one should be able to reproduce exactly the same number as we do.

Once all the derived parameters are specified, we can calculate the desired number of events according to Equation 7.22, which gives $d = 101.684 \approx 102$. On the other hand, we can notice that the overall event rate for the control group is $P_C = 0.837$ and for experimental groups is $P_E = 0.627$. Hence, the total sample size needed is given by

$$n = \frac{2d}{P_E + P_C} = \frac{2 \times 102}{0.837 + 0.627} = 139.344 \approx 140.$$

7.5 Practical Issues

7.5.1 Binomial versus Time to Event

In clinical trials, it is not common to define the so-called responder based on the study endpoints. For example, in cancer trials, we may define a subject as a responder based on his/her time-to-disease progression. We can then perform an analysis based on the response rate to evaluate the treatment effect. This analysis reduces to a two-sample problem for comparing proportions as described in the previous chapters. However, it should be noted that the analysis using the response rate, which is defined based on the time-to-event data, is not as efficient as the analysis using the time-to-event data, especially when the underlying distribution for the time to event satisfies the exponential model or Cox's proportional hazards model.

7.5.2 Local Alternative versus Fixed Alternative

The sample size calculation formulas for both Cox's proportional hazards model and exponential model are all based on the so-called local alternatives (see Fleming and Harrington, 1991), which implies that the difference between treatment groups in terms of the parameters of interest (e.g., hazard function or survival function) decreases to 0 at the rate of $1/\sqrt{n}$, where n is the total sample size. In practice, this is a dilemma because the alternative hypothesis is always fixed, which does not change as the sample size changes. However, the sample size estimation for Cox's proportional hazard model and the weighted log-rank test are derived based on local alternatives. As a result, further research in sample size estimation based on a fixed alternative is an interesting but challenging topic for statisticians in the pharmaceutical industry.

7.5.3 One-Sample versus Historical Control

Historical control is often considered in survival analysis when the clinical trial involves only the test treatment. In practice, two approaches are commonly employed. First, it is to treat the parameters estimated from historical control (e.g., hazard rate, survival function, and median survival time) as fixed reference (true) values. Then, the objective of the study is to compare the corresponding parameters of the test treatment with the reference values. This is analogous to the one-sample problem discussed in the previous chapters. Under the assumption that the time to event is exponentially distributed, formulas can be similarly derived. Another approach is to utilize the whole sample from the historical study. Then, the standard testing procedure (e.g., log-rank test) will be used to assess the treatment effects. Some discussion on sample size determination for this approach can be found in Emrich (1989) and Dixon and Simon (1988).

8

Group Sequential Methods

Most clinical trials are longitudinal in nature. In practice, it is almost impossible to enroll and randomize all required subjects at the same time. Clinical data are accumulated sequentially over time. As a result, it is of interest to monitor the information for management of the study. In addition, it is of particular interest to obtain early evidence regarding efficacy, safety, and benefit/risk of the test drug under investigation for a possible early termination. Thus, it is not uncommon to employ a group sequential design with a number of planned interim analyses in a clinical trial. The rationale for interim analyses of accumulating data in clinical trials with group sequential designs have been well documented in the Greenberg Report (Heart Special Project Committee, 1988) more than three decades ago. Since then, the development of statistical methodology and decision processes for implementation of data monitoring and interim analyses for early termination has attracted considerable attention from academia, the pharmaceutical industry, and health authorities (see, e.g., Jennison and Turnbull, 2000).

Sections 8.1 through 8.4 introduce Pocock's test, O'Brien and Fleming's test, Wang and Tsiatis' test, and the inner wedge test for clinical trials with group sequential designs, respectively. Also included in these sections are the corresponding procedures for sample size calculation. The application of these tests to discrete study endpoints such as binary responses and time-to-event data is discussed in Sections 8.5 and 8.6, respectively. In Section 8.7, the concept of alpha-spending function in group sequential methods is outlined. Procedures for sample size reestimation at a given interim analysis without unblinding are examined in Section 8.8. In Section 8.9, conditional powers at interim analyses are derived for the cases when comparing means and proportions. Some practical issues are discussed in Section 8.10.

8.1 Pocock's Test

In clinical trials, a commonly employed statistical test for a group sequential design with a number of interim analyses is to analyze accumulating data at each interim analysis. This kind of test is referred to as a repeated significance test (Jennison and Turnbull, 2000). In this section, we introduce Pocock's test and the corresponding sample size calculation formula.

8.1.1 The Procedure

Pocock (1977) suggested performing a test at a constant nominal level to analyze accumulating data at each interim analysis over the course of the intended clinical trial. Suppose that the investigator is interested in conducting a clinical trial for comparing two treatment groups under a group sequential design with K planned interim analyses. Let x_{ij} be the observation from the jth subject in the ith treatment group, $i = 1, 2; j = 1, ..., n$. For a fixed i, it is assumed that x_{ij}'s are independent and identically distributed normal random variables with mean μ_i and variance σ_i^2. Denote by n_k the information (or the number of subjects) accumulated at the kth ($k = 1, ..., K$) interim analysis. For simplicity, we further assume that at each interim analysis, the numbers of subjects accumulated in each treatment group are the same. Note that in practice, this assumption may not hold. How to deal with unequal numbers of subjects accumulated in each treatment group is challenging to clinical scientists. One solution to this problem is using Lan and DeMets' alpha-spending function, which is discussed in Section 7.8. At each interim analysis, the following test statistic is usually calculated:

$$Z_k = \frac{1}{\sqrt{n_k \left(\sigma_1^2 + \sigma_2^2 \right)}} \left(\sum_{j=1}^{n_k} x_{1j} - \sum_{j=1}^{n_k} x_{2j} \right), k = 1, ..., K.$$

Note that σ_i^2 is usually unknown and it is usually estimated by the data available up to the point of the interim analysis. At each interim analysis, σ_i^2 is usually replaced by its estimates. Denote by $C_P(K, \alpha)$ the critical value for having an overall type I error rate of α. Pocock's test can be summarized as follows (see also Jennison and Turnbull, 2000):

1. After group $k = 1, ..., K - 1$
   a. If $|Z_k| > C_P(K, \alpha)$ then stop, reject H_0
   b. Otherwise continue to group $k + 1$
2. After group K
   a. If $|Z_K| > C_P(K, \alpha)$ then stop, reject H_0
   b. Otherwise stop, accept H_0

As an example, one Pocock type boundary for the standardized test statistic is plotted in Figure 8.1.

As it can be seen, the critical value $C_P(K, \alpha)$ only depends upon the type I error (α) and the total number of planned interim analysis (K), which is independent of the visit number (k). In other words, for each planned interim analysis, the same critical value is used for comparing treatment difference using the standard test statistic Z_k. The value of $C_P(K, \alpha)$ is choosing in such a way that the above test procedure has an overall type I error rate of a under the null hypothesis that $\mu_1 - \mu_2 = 0$. Since there exists no explicit formula for the calculation of $C_P(K, \alpha)$, a selection of various values of $C_P(K, \alpha)$ under different choices of parameters (K and α) is given in Table 8.1.

On the other hand, under the alternative hypothesis (i.e., $\theta \neq 0$), the power of the above test procedure can be determined by the total number of planned interim analysis (K), type I error rate (α), and type II error rate (β), and the proportion between σ^2 and δ^2 (i.e., σ^2/δ^2), where $\delta = |\mu_1 - \mu_2|$. As discussed in the previous chapters, if there are no interim analyses planned (i.e., $K = 1$), then the sample size is proportional to σ^2/δ^2. As a result, it is sufficient to specify the ratio $R_P(K, \alpha, \beta)$ of the maximum sample size of the group sequential test to the fixed sample size. The values of $R_P(K, \alpha, \beta)$ are given in Table 8.2. The maximum sample size needed for a group sequential trial with K interim analyses can be obtained by first calculating the fixed sample size without interim analyses, and then multiplying it by $R_P(K, \alpha, \beta)$.

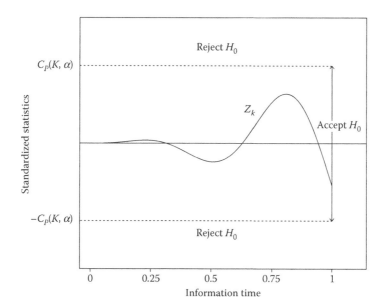

FIGURE 8.1
Pocock type stopping rule.

TABLE 8.1

$C_p(K, \alpha)$ for Two-Sided Tests with K Interim Analyses

K	$\alpha = 0.01$	$\alpha = 0.05$	$\alpha = 0.10$
1	2.576	1.960	1.645
2	2.772	2.178	1.875
3	2.873	2.289	1.992
4	2.939	2.361	2.067
5	2.986	2.413	2.122
6	3.023	2.453	2.164
7	3.053	2.485	2.197
8	3.078	2.512	2.225
9	3.099	2.535	2.249
10	3.117	2.555	2.270
11	3.133	2.572	2.288
12	3.147	2.588	2.304
15	3.182	2.626	2.344
20	3.225	2.672	2.392

8.1.2 An Example

Suppose that an investigator is interested in conducting a clinical trial with five interim analyses for comparing a test drug (T) with a placebo (P). Based on information obtained from a pilot study, data from the test drug and the placebo seem to have a common variance, that is, $\sigma^2 = \sigma_1^2 = \sigma_2^2 = 4$ with $\mu_T - \mu_P = 1$. Assuming these observed values are true, it is desirable to select a maximum sample size such that there is a 90% $(1 - \beta = 0.90)$ power for detecting such a difference between the test drug and the placebo at the 5% $(\alpha = 0.05)$ level of significance.

TABLE 8.2

$R_P(K, \alpha, \beta)$ for Two-Sided Tests with K Interim Analyses

	$1 - \beta = 0.8$			$1 - \beta = 0.9$		
K	$\alpha = 0.01$	$\alpha = 0.05$	$\alpha = 0.10$	$\alpha = 0.01$	$\alpha = 0.05$	$\alpha = 0.10$
1	1.000	1.000	1.000	1.000	1.000	1.000
2	1.092	1.110	1.121	1.084	1.100	1.110
3	1.137	1.166	1.184	1.125	1.151	1.166
4	1.166	1.202	1.224	1.152	1.183	1.202
5	1.187	1.229	1.254	1.170	1.207	1.228
6	1.203	1.249	1.277	1.185	1.225	1.249
7	1.216	1.265	1.296	1.197	1.239	1.266
8	1.226	1.279	1.311	1.206	1.252	1.280
9	1.236	1.291	1.325	1.215	1.262	1.292
10	1.243	1.301	1.337	1.222	1.271	1.302
11	1.250	1.310	1.348	1.228	1.279	1.312
12	1.257	1.318	1.357	1.234	1.287	1.320
15	1.272	1.338	1.381	1.248	1.305	1.341
20	1.291	1.363	1.411	1.264	1.327	1.367

By the formula for sample size calculation given in Chapter 3, the required fixed sample size when there are no planned interim analyses is

$$n_{\text{fixed}} = \frac{(z_{\alpha/2} + z_\beta)^2 (\sigma_1^2 + \sigma_2^2)}{(\mu_1 - \mu_2)^2} = \frac{(1.96 + 1.28)^2 (4 + 4)}{1^2} \approx 84.$$

From Table 8.2, we have

$$R_P(5, 0.05, 0.1) = 1.207.$$

Hence, the maximum sample size needed for the group sequential trial is given by

$$n_{\text{max}} = R_P(5, 0.05, 0.1) n_{\text{fixed}} = 1.207 \times 84 = 101.4.$$

Hence, it is necessary to have

$$n = \frac{n_{\text{max}}}{K} = \frac{101.4}{5} = 20.3 \approx 21,$$

subjects per group at each interim analysis.

8.2 O'Brien and Fleming's Test

Pocock's test is straightforward and simple. However, it is performed at a constant nominal level. As an alternative to Pocock's test, O'Brien and Fleming (1979) proposed a test, which is also based on the standardized statistics Z_k, by increasing the nominal significance level for rejecting H_0 at each analysis as the study progresses. As a result, it is difficult to reject the null hypothesis at early stages of the trial.

8.2.1 The Procedure

O'Brien and Fleming's test is carried out as follows (see, also Jennison and Turnbull, 2000):

1. After group $k = 1, ..., K - 1$
   a. If $|Z_k| > C_B(K, \alpha)\sqrt{K/k}$ then stop, reject H_0
   b. Otherwise continue to group $k + 1$
2. After group K
   a. If $|Z_K| > C_B(K, \alpha)$ then stop, reject H_0
   b. Otherwise stop, accept H_0

As an example, one O'Brien–Fleming type boundary is plotted in Figure 8.2. Note that the value of $C_B(K, \alpha)$ is chosen to ensure that the type I error rate is α. Like $C_P(K, \alpha)$, there exists no closed form for calculating $C_B(K, \alpha)$. For convenience, a selection of various values of $C_B(K, \alpha)$ under different choices of parameters are provided in Table 8.3.

Similar to the procedure for sample size calculation for Pocock's method, the maximum sample size needed to achieve a desired power at a given level of significance can be obtained by first calculating the sample size needed for a fixed sample size design, and then multiplying by a constant $R_B(K, \alpha, \beta)$. For various parameters, the values of $R_B(K, \alpha, \beta)$ are given in Table 8.4.

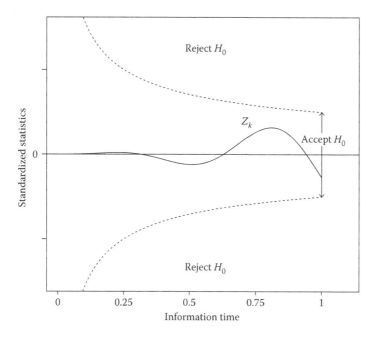

FIGURE 8.2
O'Brien–Fleming type stopping rule.

TABLE 8.3

$C_B(K, \alpha)$ for Two-Sided Tests with K Interim Analyses

K	$\alpha = 0.01$	$\alpha = 0.05$	$\alpha = 0.10$
1	2.576	1.960	1.645
2	2.580	1.977	1.678
3	2.595	2.004	1.710
4	2.609	2.024	1.733
5	2.621	2.040	1.751
6	2.631	2.053	1.765
7	2.640	2.063	1.776
8	2.648	2.072	1.786
9	2.654	2.080	1.794
10	1.660	2.087	1.801
11	2.665	2.092	1.807
12	2.670	2.098	1.813
15	2.681	2.110	1.826
20	2.695	2.126	1.842

8.2.2 An Example

Consider the example described in Section 8.1.2. Suppose that the investigator wishes to perform the same group sequential test using O'Brien and Fleming's test rather than Pocock's test. By Table 8.4,

$$R_B(5, 0.05, 0.1) = 1.026.$$

TABLE 8.4

$R_B(K, \alpha, \beta)$ for Two-Sided Tests with K Interim Analyses

	$1 - \beta = 0.8$			$1 - \beta = 0.9$		
K	$\alpha = 0.01$	$\alpha = 0.05$	$\alpha = 0.10$	$\alpha = 0.01$	$\alpha = 0.05$	$\alpha = 0.10$
1	1.000	1.000	1.000	1.000	1.000	1.000
2	1.001	1.008	1.016	1.001	1.007	1.014
3	1.007	1.017	1.027	1.006	1.016	1.025
4	1.011	1.024	1.035	1.010	1.022	1.032
5	1.015	1.028	1.040	1.014	1.026	1.037
6	1.017	1.032	1.044	1.016	1.030	1.041
7	1.019	1.035	1.047	1.018	1.032	1.044
8	1.021	1.037	1.049	1.020	1.034	1.046
9	1.022	1.038	1.051	1.021	1.036	1.048
10	1.024	1.040	1.053	1.022	1.037	1.049
11	1.025	1.041	1.054	1.023	1.039	1.051
12	1.026	1.042	1.055	1.024	1.040	1.052
15	1.028	1.045	1.058	1.026	1.042	1.054
20	1.030	1.047	1.061	1.029	1.045	1.057

Since the required fixed sample size is given by $n_{\text{fixed}} = 84$, the maximum sample size needed for each treatment group is given by

$$n_{\max} = R_B(5, 0.05, 0.1)n_{\text{fixed}} = 1.026 \times 84 = 86.2 \approx 87.$$

Therefore, $n = n_{\max}/K = 87/5 = 17.4 \approx 18$ subjects per treatment group at each interim analysis is required for achieving a 90% power at the 5% level of significance.

8.3 Wang and Tsiatis' Test

In addition to Pocock's test and O'Brien and Fleming's test, Wang and Tsiatis (1987) proposed a family of two-sided tests indexed by the parameter of Δ, which is also based on the standardized test statistic Z_k. Wang and Tsiatis' test include Pocock's and O'Brien–Fleming's boundaries as special cases.

8.3.1 The Procedure

Wang and Tsiatis' test can be summarized as follows (see also Jennison and Turnbull, 2000):

1. After group $k = 1, \ldots, K - 1$
   a. If $|Z_k| > C_{WT}(K, \alpha, \Delta)(k/K)^{\Delta - 1/2}$ then stop, reject H_0
   b. Otherwise continue to group $k + 1$
2. After group K
   a. If $|Z_K| > C_{WT}(K, \alpha, \Delta)$ then stop, reject H_0
   b. Otherwise stop, accept H_0

As an example, the Wang–Tsiatis type boundary when $\Delta = 0.25$ is given in Figure 8.3. As it can be seen, Wang and Tsiatis' test reduces to Pocock's test when $\Delta = 0.5$. When $\Delta = 0$, Wang and Tsiatis' test is the same as O'Brien and Fleming's test. As a result, values of C_{WT} (K, α, Δ) with $\Delta = 0$ and 0.5 can be obtained from Tables 8.1 and 8.3. Values of $C_{WT}(K, \alpha, \Delta)$ when $\Delta = 0.1, 0.25$, and 0.4 are given in Table 8.5.

Sample size calculation for Wang and Tsiatis' test can be performed in a similar manner as those for Pocock's test and O'Brien and Fleming's test. First, we need to calculate the sample size for a fixed sample size design with a given significance level and power. Then, we multiply this sample size by the constant of $R_{WT}(K, \alpha, \beta, \Delta)$ whose values are given in Table 8.6.

8.3.2 An Example

For illustration, consider the same example given in Section 8.1.2. Suppose that the investigator wishes to perform the same group sequential test using Wang and Tsiatis' test with $\Delta = 0.25$. By Table 8.6,

$$R_{WT}(5, 0.05, 0.1, 0.25) = 1.066.$$

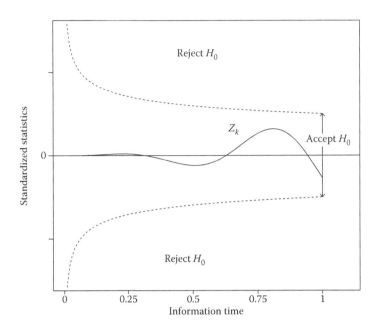

FIGURE 8.3
Wang–Tsiatis type stopping rule with $\Delta = 0.25$.

Since the fixed sample size is given by $n_{\text{fixed}} = 84$, the maximum sample size needed for each treatment group is given by

$$n_{\text{max}} = R_B(5, 0.05, 0.1)n_{\text{fixed}} = 1.066 \times 84 = 89.5 \approx 90.$$

Thus, at each interim analysis, sample size per treatment group required for achieving a 90% power at the 5% level of significance is given by

TABLE 8.5

$C_{WT}(K, \alpha, \Delta)$ for Two-Sided Tests with K Interim Analyses and $\alpha = 0.05$

K	$\Delta = 0.10$	$\Delta = 0.25$	$\Delta = 0.40$
1	1.960	1.960	1.960
2	1.994	2.038	2.111
3	2.026	2.083	2.186
4	2.050	2.113	2.233
5	2.068	2.136	2.267
6	2.083	2.154	2.292
7	2.094	2.168	2.313
8	2.104	2.180	2.329
9	2.113	2.190	2.343
10	2.120	2.199	2.355
11	2.126	2.206	2.366
12	2.132	2.213	2.375
15	2.146	2.229	2.397
20	2.162	2.248	2.423

TABLE 8.6

$R_{WT}(K, \alpha, \beta, \Delta)$ for Two-Sided Tests with K Interim Analyses and $\alpha = 0.05$

K	$1 - \beta = 0.8$			$1 - \beta = 0.9$		
	$\Delta = 0.01$	$\Delta = 0.05$	$\Delta = 0.10$	$\Delta = 0.01$	$\Delta = 0.05$	$\Delta = 0.10$
1	1.000	1.000	1.000	1.000	1.000	1.000
2	1.016	1.038	1.075	1.014	1.034	1.068
3	1.027	1.054	1.108	1.025	1.050	1.099
4	1.035	1.065	1.128	1.032	1.059	1.117
5	1.040	1.072	1.142	1.037	1.066	1.129
6	1.044	1.077	1.152	1.041	1.071	1.138
7	1.047	1.081	1.159	1.044	1.075	1.145
8	1.050	1.084	1.165	1.046	1.078	1.151
9	1.052	1.087	1.170	1.048	1.081	1.155
10	1.054	1.089	1.175	1.050	1.083	1.159
11	1.055	1.091	1.178	1.051	1.085	1.163
12	1.056	1.093	1.181	1.053	1.086	1.166
15	1.059	1.097	1.189	1.055	1.090	1.172
20	1.062	1.101	1.197	1.058	1.094	1.180

$$n = \frac{n_{\max}}{K} = \frac{90}{5} = 18.$$

8.4 Inner Wedge Test

As described above, the three commonly used group sequential methods allow early stop under the alternative hypothesis. In other words, the trial is terminated if there is substantial evidence of efficacy. In practice, however, if the trial demonstrates strong evidence that the test drug has no treatment effect, it is also of interest to stop the trial prematurely. For good medical practice, it may not be ethical to expose patients to a treatment with little or no efficacy but potential serious adverse effects. In addition, the investigator may want to put the resources on other promising drugs. To allow an early stop with either substantial evidence of efficacy or no efficacy, the most commonly used group sequential method is the so-called two-sided inner wedge test, which is also based on the standardized test statistics Z_k.

8.4.1 The Procedure

The inner wedge test can be carried out as follows (see also Jennison and Turnbull, 2000):

1. After group $k = 1, ..., K - 1$
   a. If $|Z_k| \geq b_k$ then stop and reject H_0
   b. If $|Z_k| < a_k$ then stop and accept H_0
   c. Otherwise continue to group $k + 1$

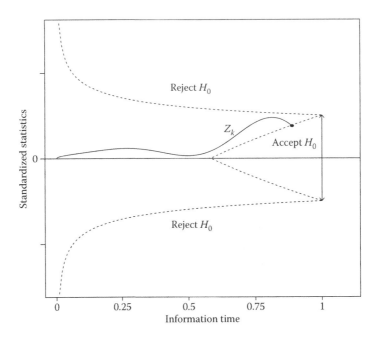

FIGURE 8.4
Inner wedge type stopping rule with $\Delta = 0.25$.

2. After group K,
    a. If $|Z_k| \geq b_K$ then stop and reject H_0
    b. If $|Z_k| < a_K$ then stop and accept H_0

The constants a_k and b_k are given by

$$a_k = [C_{W1}(K,\alpha,\beta,\Delta) + C_{W2}(K,\alpha,\beta,\Delta)] * \sqrt{k/K}$$
$$-C_{W2}(K,\alpha,\beta,\Delta)(k/K)^{\Delta-1/2},$$
$$b_k - C_{W1}(K,\alpha,\beta,\Delta)(k/K)^{\Delta-1/2}.$$

As an example, one inner wedge type boundary is given in Figure 8.4. For a given desired power $(1 - \beta)$, the sample size can be similarly determined. First, we calculate the sample size required for a fixed sample size design, denoted by n_{fixed}. Then, n_{fixed} is multiplied by $R_W(K, \alpha, \beta, \Delta)$. Values of $C_{W1}(K, \alpha, \beta, \Delta)$, $C_{W2}(K, \alpha, \beta, \Delta)$, and $R_W(K, \alpha, \beta, \Delta)$ are given in Tables 8.7 and 8.8.

8.4.2 An Example

To illustrate sample size calculation based on the inner wedge test, consider the following example. A group sequential trial with five $(K = 5)$ interim analyses is planned. The objective is to compare a test drug with a standard therapy through a parallel trial. An inner wedge test with $\Delta = 0.25$ is utilized. Based on a pilot study, the mean difference between the two treatments is 20% $(\mu_1 - \mu_2 = 0.2)$ and the standard deviation is 1.00 for both treatments $(\sigma_1 = \sigma_2 = 1)$. It is desirable to select a sample size to achieve an 80% $(1 - \beta = 0.80)$

TABLE 8.7

Constants $C_{W1}(K, \alpha, \beta, \Delta)$, $C_{W2}(K, \alpha, \beta, \Delta)$, and $R_W(K, \alpha, \beta, \Delta)$ with $\alpha = 0.05$ and $1 - \beta = 0.8$

Δ	K	C_{W1}	C_{W2}	R_W	Δ	K	C_{W1}	C_{W2}	R_W
−0.50	1	1.960	0.842	1.000	−0.25	1	1.960	0.842	1.000
	2	1.949	0.867	1.010		2	1.936	0.902	1.026
	3	1.933	0.901	1.023		3	1.932	0.925	1.040
	4	1.929	0.919	1.033		4	1.930	0.953	1.059
	5	1.927	0.932	1.041		5	1.934	0.958	1.066
	10	1.928	0.964	1.066		10	1.942	0.999	1.102
	15	1.931	0.979	1.078		15	1.948	1.017	1.120
	20	1.932	0.988	1.087		20	1.952	1.027	1.131
0.00	1	1.960	0.842	1.000	0.25	1	1.960	0.842	1.000
	2	1.935	0.948	1.058		2	1.982	1.000	1.133
	3	1.950	0.955	1.075		3	2.009	1.059	1.199
	4	1.953	0.995	1.107		4	2.034	1.059	1.219
	5	1.958	1.017	1.128		5	2.048	1.088	1.252
	10	1.980	1.057	1.175		10	2.088	1.156	1.341
	15	1.991	1.075	1.198		15	2.109	1.180	1.379
	20	1.998	1.087	1.212		20	2.122	1.195	1.40

power for detecting such a difference at the 5% ($\alpha = 0.05$) level of significance. The sample size needed for a fixed sample size design can be obtained as

$$n_{\text{fixed}} = \frac{(z_{0.975} + z_{0.80})^2 (\sigma_1^2 + \sigma_2^2)}{(\mu_1 - \mu_2)^2} = \frac{(1.96 + 0.84)^2 (1 + 1)}{0.2^2} = 392.$$

TABLE 8.8

Constants $C_{W1}(K, \alpha, \beta, \Delta)$, $C_{W2}(K, \alpha, \beta, \Delta)$, and $R_W(K, \alpha, \beta, \Delta)$ with $\alpha = 0.05$ and $1 - \beta = 0.9$

Δ	K	C_{W1}	C_{W2}	R_W	Δ	K	C_{W1}	C_{W2}	R_W
−0.50	1	1.960	1.282	1.000	−0.25	1	1.960	1.282	1.000
	2	1.960	1.282	1.000		2	1.957	1.294	1.006
	3	1.952	1.305	1.010		3	1.954	1.325	1.023
	4	1.952	1.316	1.016		4	1.958	1.337	1.033
	5	1.952	1.326	1.023		5	1.960	1.351	1.043
	10	1.958	1.351	1.042		10	1.975	1.379	1.071
	15	1.963	1.363	1.053		15	1.982	1.394	1.085
	20	1.967	1.370	1.060		20	1.988	1.403	1.094
0.00	1	1.960	1.282	1.000	0.25	1	1.960	1.282	1.000
	2	1.958	1.336	1.032		2	2.003	1.398	1.100
	3	1.971	1.353	1.051		3	2.037	1.422	1.139
	4	1.979	1.381	1.075		4	2.058	1.443	1.167
	5	1.990	1.385	1.084		5	2.073	1.477	1.199
	10	2.013	1.428	1.127		10	2.119	1.521	1.261
	15	2.026	1.447	1.148		15	2.140	1.551	1.297
	20	2.034	1.458	1.160		20	2.154	1.565	1.316

By Table 8.8,

$$n_{\max} = n_{\text{fixed}}R_W(5, 0.05, 0.2, 0.25) = 392 \times 1.199 = 470.$$

Hence, at each interim analysis, the sample size necessary per treatment group is given by

$$n = \frac{n_{\max}}{K} = \frac{470}{5} = 94.$$

8.5 Binary Variables

In this section we consider binary response variables.

8.5.1 The Procedure

Let x_{ij} be the binary response from the jth subject in the ith treatment group. Within each treatment group (i.e., for a fixed i), x_{ij}'s are assumed to be independent and identically distributed with mean p_i. Suppose that there are K planned interim analyses. Suppose also that at each interim analysis, equal number of subjects is accumulated in each treatment group. At each interim analysis, the following test statistic is usually considered:

$$Z_k = \frac{\sqrt{n_k}\,(\hat{p}_{1,k} - \hat{p}_{2,k})}{\sqrt{(\hat{p}_{1,k}(1 - \hat{p}_{1,k}) + (\hat{p}_{2,k}(1 - \hat{p}_{2,k})}},$$

where

$$\hat{p}_{1,k} = \frac{1}{n_{i,k}} \sum_{j=1}^{n_{i,k}} x_{ij}$$

and n_k is the number of subjects accumulated by the time of the kth interim analysis. Since Z_k, $k = 1, \ldots, K$, are asymptotically normally distributed with the same distribution as that of Z_k's for the continuous response, the repeated significance test procedures (e.g., Pocock, O'Brien–Fleming, and Wang–Tsiatis) can also be applied for binary responses. The resulting test procedure has an asymptotically type I error rate of α.

8.5.2 An Example

Suppose that an investigator is interested in conducting a group sequential trial comparing a test drug with a placebo. The primary efficacy study endpoint is a binary response. Based on information obtained in a pilot study, the response rates for the test drug and the placebo are given by 60% ($p_1 = 0.60$) and 50% ($p_2 = 0.50$), respectively. Suppose that a total of five ($K = 5$) interim analyses are planned. It is desirable to select a maximum sample size to have an 80% ($1 - \beta = 0.80$) power at the 5% ($a = 0.05$) level of significance. The sample size needed for a fixed sample size design is

$$n_{\text{fixed}} = \frac{(z_{\alpha/2} + z_{\beta})^2 (p_1(1-p_1) + p_2(1-p_2))}{(p_1 - p_2)^2}$$

$$= \frac{(1.96 + 0.84)^2 (0.6(1-0.6) + 0.5(1-0.5))}{(0.6 - 0.5)^2}$$

$$\approx 385.$$

If Pocock's test is used, then by Table 8.2, we have

$$n_{\max} = n_{\text{fixed}} R_P(5, 0.05, 0.20) = 385 \times 1.229 \approx 474.$$

Hence, at each interim analysis, the sample size per treatment group is $474/5 = 94.8 \approx 95$. On the other hand, if O'Brien and Fleming's method is employed, Table 8.4 gives

$$n_{\max} = n_{\text{fixed}} R_B(5, 0.05, 0.20) = 385 \times 1.028 \approx 396.$$

Thus, at each interim analysis, the sample size per treatment group is $396/5 = 79.2 \approx 80$. Alternatively, if Wang and Tsitis' test with $\Delta = 0.25$ is considered, Table 8.6 leads to

$$n_{\max} = n_{\text{fixed}} R_{WT}(5, 0.05, 0.20, 0.25) = 385 \times 1.072 \approx 413.$$

As a result, at each interim analysis, the sample size per treatment group is given by $413/5 = 82.6 \approx 83$.

8.6 Time-to-Event Data

To apply the repeated significance test procedures to time-to-event data, for simplicity, we only consider Cox's proportional hazard model.

8.6.1 The Procedure

As indicated in Chapter 7, under the assumption of proportional hazards, the log-rank test is usually used to compare the two treatment groups. More specifically, let $h(t)$ be the hazard function of treatment group A and $e^{\theta}h(t)$ be the hazard function of treatment group B. Let d_k denote the total number of uncensored failures observed when the kth interim analysis is conducted, $k = 1, \ldots, K$. For illustration purposes and without loss of generality, we assume that there are no ties. Let $T_{i,k}$ be the survival times of these subjects, $i = 1, \ldots, d_k$. Let $r_{iA,k}$ and $r_{iB,k}$ be the numbers of subjects who are still at risk at the kth interim analysis at time $T_{i,k}$ for treatments A and B, respectively. The log-rank test statistic at the kth interim analysis is then given by

$$S_k = \sum_{i=1}^{d_k} \left(\delta_{iB,k} - \frac{r_{iB,k}}{r_{iA,k} + r_{iB,k}} \right),$$

where $\delta_{iB,k} = 1$ if the failure at time $\tau_{i,k}$ is on treatment B and 0 otherwise. Jennison and Turnbull (2000) proposed to use $N(\theta I_k, I_k)$ to approximate the distribution of S_k, where I_k is the so-called observed information and is defined as

$$I_k = \sum_{i=1}^{d_k} \frac{r_{iA,k} r_{iB,k}}{(r_{iA,k} + r_{iB,k})^2}.$$

Then, the standardized test statistic can be calculated as

$$Z_k = \frac{S_k}{\sqrt{I_k}}.$$

As a result, Z_k can be used to compare with the commonly used group sequential boundaries (e.g., Pocock, O'Brien–Fleming, and Wang–Tsiatis). Under the alternative hypothesis, the sample size can be determined by first finding the information needed for a fixed sample size design with the same significance level and power. Then, calculate the maximum information needed for a group sequential trial by multiplying appropriate constants from Tables 8.2, 8.4, and 8.6.

8.6.2 An Example

Suppose that an investigator is interested in conducting a survival trial with five ($K = 5$) planned interim analyses at the 5% level of significance ($\alpha = 0.05$) with an 80% ($1 - \beta = 0.80$) power. Assume that $\theta = 0.405$. As indicated by Jennison and Turnbull (2000), the information needed for a fixed sample size design is given by

$$I_{\text{fixed}} = \frac{(z_{\alpha/2} + z_\beta)^2}{\theta^2} = \frac{(1.96 + 0.84)^2}{0.405^2} = 47.8.$$

If O'Brien and Fleming boundaries are utilized as a stopping rule, then the maximum information needed to achieve the desired power can be calculated as

$$I_{\text{max}} = I_{\text{fixed}} \times R_B(K, \alpha, \beta) = 47.8 \times 1.028 = 49.1,$$

where the value 1.028 of $R_B(5, 0.05, 0.2)$ is taken from Table 8.4. If θ is close to 0, which is true under the local alternative, it is expected that $r_{iA,k} \approx r_{iB,k}$ for each i. Hence, I_k can be approximated by $0.25 d_k$. It follows that the number of events needed is given by

$$n_d = \frac{I_{\text{max}}}{0.25} = 196.4 \approx 197.$$

Hence, a total number of 197 events are needed to achieve an 80% power for detecting a difference of $\theta = 0.405$ at the 5% level of significance. The corresponding sample size can be derived based on n_d by adjusting for some other factors, such as competing risk, censoring, and dropouts.

8.7 Alpha-Spending Function

One of the major disadvantages of the group sequential methods discussed in the previous sections is that they are designed for a fixed number of interim analyses with equally spaced information time. In practice, however, it is not uncommon that the interim analysis is actually planned based on calendar time. As a result, the information accumulated at each time point may not be equally spaced. The consequence is that the overall type I error may be far away from the target value.

As an alternative, Lan and DeMets (1983) proposed to distribute (or spend) the total probability of false-positive risk as a continuous function of the information time in group sequential procedures for interim analyses. If the total information scheduled to accumulate over the maximum duration T is known, the boundaries can be computed as a continuous function of the information time. This continuous function of the information time is referred to as the alpha-spending function, denoted by $\alpha(s)$. The alpha-spending function is an increasing function of information time. It is 0 when information time is 0; and is equal to the overall significance level when information time is 1. In other words, $\alpha(0) = 0$ and $\alpha(1) = a$. Let s_1 and s_2 be two information times, $0 < s_1 < s_2 \leq 1$. Also, denote $\alpha(s_1)$ and $\alpha(s_2)$ as their corresponding value of alpha-spending function at s_1 and s_2. Then, $0 < \alpha(s_1) < \alpha(s_2) \leq \alpha$ $\alpha(s_1)$ is the probability of type I error one wishes to spend at information time s_1. For a given alpha-spending function ($\alpha(s)$) and a series of standardized test statistic Z_k, $k = 1, ..., K$. The corresponding boundaries c_k, $k = 1, ..., K$ are chosen such that under the null hypothesis

$$P(|Z_1| < c_1, ..., |Z_{k-1}| < c_{k-1}, |Z_k| \geq c_k) = \alpha\left(\frac{k}{K}\right) - \alpha\left(\frac{k-1}{K}\right).$$

Some commonly used alpha-spending functions are summarized in Table 8.9 and Figure 8.5 is used to illustrate a true alpha-spending function.

We now introduce the procedure for sample size calculation based on Lan–DeMets' alpha-spending function, that is,

$$\alpha(s) = \alpha s^\rho, \rho > 0.$$

Although the alpha-spending function does not require a fixed maximum number and equally spaced interim analyses, it is necessary to make those assumptions to calculate the sample size under the alternative hypothesis. The sample size calculation can be performed in a similar manner. For a given significance level a and power $1 - \beta$, we can first calculate the sample size needed for a fixed sample size design and then multiply it by a constant $R_{LD}(K, \alpha, \beta, \rho)$. The values of $R_{LD}(K, \alpha, \beta, \rho)$ are tabulated in Table 8.10.

TABLE 8.9

Various Alpha-spending Functions

$\alpha_1(s) = 2\{1 - \Phi(z_{\alpha/2}/\sqrt{2})\}$	O'Brien–Fleming
$\alpha_2(s) = \alpha \log[1 + (e - 1)s]$	Pocock
$\alpha_3(s) = \alpha s^\rho, \rho > 0$	Lan–DeMets–Kim
$\alpha_4(s) = \alpha[(1 - e^{-\zeta s})/(1 - e^{-\zeta})], \zeta \neq 0$	Hwang–Shih

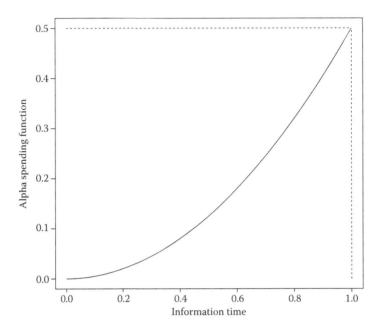

FIGURE 8.5
The alpha-spending function $\alpha(s)$.

TABLE 8.10

$R_{LD}(K, \alpha, \beta, \rho)$ for Two-Sided Tests with K Interim Analyses and $\alpha = 0.05$

	$1 - \beta = 0.8$			$1 - \beta = 0.9$		
K	$\rho = 0.01$	$\rho = 0.05$	$\rho = 0.10$	$\rho = 0.01$	$\rho = 0.05$	$\rho = 0.10$
1	1.000	1.000	1.000	1.000	1.000	1.000
2	1.082	1.028	1.010	1.075	1.025	1.009
3	1.117	1.045	1.020	1.107	1.041	1.018
4	1.137	1.056	1.027	1.124	1.051	1.025
5	1.150	1.063	1.032	1.136	1.058	1.030
6	1.159	1.069	1.036	1.144	1.063	1.033
7	1.165	1.073	1.039	1.150	1.067	1.036
8	1.170	1.076	1.041	1.155	1.070	1.039
9	1.174	1.079	1.043	1.159	1.073	1.040
10	1.178	1.081	1.045	1.162	1.075	1.042
11	1.180	1.083	1.046	1.164	1.077	1.043
12	1.183	1.085	1.048	1.166	1.078	1.044
15	1.188	1.088	1.050	1.171	1.082	1.047
20	1.193	1.092	1.054	1.176	1.085	1.050

Consider the same example as discussed in Section 8.1. To achieve a 90% power at the 5% level of significance, it is necessary to have $n_{\text{fixed}} = 84$ subjects per treatment group. Then, the maximum sample size needed for achieving the desired power with five interim analyses using the Lan–DeMets type alpha-spending function with $\rho = 2$ can be calculated as

$$n_{\max} = n_{\text{fixed}} \times R_{LD}(5, 0.05, 0.9, 2) = 84 \times 1.075 \approx 92.$$

Thus, a total of 92 subjects per treatment group is needed to have a 90% power at the 5% level of significance.

8.8 Sample Size Reestimation

In clinical trials with planned interim analyses, it is desirable to perform sample size reestimation at interim analyses. The objective is to determine whether the selected sample size is justifiable based on clinical data accumulated up to the time point of interim analysis. In practice, however, unblinding the treatment codes for sample size reestimation may introduce bias to remaining clinical trials. Shih (1993) and Shih and Zhao (1997) proposed some procedures without unblinding for sample size reestimation with interim data for double-blind clinical trials with binary outcomes.

8.8.1 The Procedure

Suppose that y_i, $i = 1, ..., n$ (treatment) and y_j, $j = n + 1, ..., N$ (control) are observations from a randomized, double-blind clinical trial. It is assumed that y_i and y_j are distributed as $B(1, p_1)$ and $B(1, p_2)$, respectively. Suppose that the hypotheses of interest are

$$H_0: p_1 = p_2 \quad \text{versus} \quad H_a: p_1 \neq p_2.$$

Further, suppose that $N = 2_n$ and a clinically meaningful difference is $\Delta = |p_1 - p_2|$. Then, as discussed in Chapter 4, the sample size required for achieving a desired power of $(1 - \beta)$ at α level of significance by an unconditional method is given by

$$n = \frac{(z_{\alpha/2} + z_\beta)^2 [p_1(1-p_1) + p_2(1-p_2)]}{\Delta^2}.$$

As discussed in Chapter 4, there are two methods available for comparing two proportions. They are, namely, conditional and unconditional methods. For illustration purposes, we only adopt the formula of the unconditional method. However, the procedure introduced below can be easily generalized to the conditional method. The estimated sample size can be obtained by simply replacing p_1 and p_2 with their estimates. As a result, sample size reestimation at an interim analysis without unblinding is to obtain estimates of p_1 and p_2 without revealing the treatment codes. For multicenter trials, Shih (1993) and Shih and Zhao (1997) suggested the following procedure for sample size reestimation without unblinding when 50% of the subjects as originally planned in the study protocol complete the trial.

First, within each center, each subject is randomly assigned to a dummy stratum, that is, either stratum A or stratum B. Note that this stratification is not based on any of the subjects' baseline characteristics. The use of dummy stratification is for sample size reestimation at the interim stage and statistical inference should not be affected at the end of the trial. Now, subjects in stratum A are randomly allocated to the treatment group with probability π and to the control group with probability $1 - \pi$, where $\pi \varepsilon (0, 0.5)$. Similarly, subjects in stratum B are randomly allocated to the treatment group with probability $1 - \pi$ and to the control group with probability π, where $\pi \in (0, 0.5)$. Based on the pooled events rates observed from each stratum, we then estimate p_1 and p_2 without unblinding the treatment codes as follows. We use the results from stratum A to estimate

$$\theta_1 = P(y_j = 1 \mid \text{subject } j \in \text{stratum A}) = \pi p_1 + (1-\pi)p_2$$

and that of stratum B to estimate

$$\theta_2 = P(y_j = 1 \mid \text{subject } j \in \text{stratum B}) = (1-\pi)p_1 + \pi p_2.$$

Based on the observed events rate θ_1 from stratum A and the observed event rate θ_2 from stratum B, p_1 and p_2 can be estimated by solving the following equations simultaneously:

$$\pi p_1 + (1-\pi)p_2 = \theta_1$$
$$(1-\pi)p_1 + \pi p_2 = \theta_2.$$

Thus, the estimates of p_1 and p_2 are given by

$$\hat{p}_1 = \frac{\pi \hat{\theta}_1 - (1-\pi)\hat{\theta}_2}{2\pi - 1}$$

and

$$\hat{p}_2 = \frac{\pi \hat{\theta}_2 - (1-\pi)\hat{\theta}_1}{2\pi - 1}.$$

Estimates $\hat{p}_1$ and $\hat{p}_2$ can then be used to update the sample size based on the formula for sample size calculation given above. Note that if the resultant sample size n^* is greater than the originally planned sample size n (i.e., $n^* > n$), it is suggested that an increase in sample size is necessary to achieve the desired power at the end of the trial. On the other hand, if $n^* < m$, a sample size reduction is justifiable. More details regarding sample size reestimation without unblinding the treatment codes can be found in Shih and Zhao (1997).

8.8.2 An Example

Consider a cancer trial comparing the response rates (i.e., complete response plus partial response) of patients between two treatments (i.e., test and control). The trial was conducted in two centers (A and B) with 18 patients each. At center A, each patient is assigned to the test treatment group with a probability of 0.6 and the control group with a probability of 0.4. At center B, each patient is assigned to the test treatment group with a probability of 0.4 and the control group with a probability of 0.6. It follows that $\pi = 0.4$. One interim analysis was planned when half of patients (i.e., nine patients per center) completed the trial. At the time of interim analysis, it is noted that the observed response rates for centers A and B are given by 0.6 ($\theta_1 = 0.6$) and 0.5 ($\theta_2 = 0.5$), respectively. It follows that

$$0.4p_1 + 0.6p_2 = 0.6$$
$$0.6p_1 + 0.4p_2 = 0.5.$$

This gives $p_1 = 0.3$ and $p_2 = 0.8$. Hence, the sample size needed to achieve a 90% ($\beta = 0.10$) at the 5% ($\alpha = 0.05$) level of significance is given by

$$
\begin{aligned}
n &= \frac{(z_{\alpha/2} + z_\beta)^2 (p_1(1-p_1) + p_2(1-p_2))}{(p_1 - p_2)^2} \\
&= \frac{(1.96 + 1.64)^2 (0.3(1-0.3) + 0.8(1-0.8))}{(0.3 - 0.8)^2} \\
&\approx 20.
\end{aligned}
$$

Thus, a total of 40 patients are needed to achieve the desired power. This sample size reestimation suggests that in addition to the planned 36 patients, four more patients are necessarily enrolled.

8.9 Conditional Power

Conditional power at a given interim analysis in group sequential trials is defined as the power of rejecting the null hypothesis at the end of the trial conditional on the observed data accumulated up to the time point of the planned interim analysis. For many repeated significance tests such as Pocock's test, O'Brien and Fleming's test, and Wang and Tsiatis' test, the trial can be terminated only under the alternative hypothesis. In practice, this is usually true if the test treatment demonstrates substantial evidence of efficacy. However, it should be noted that if the trial indicates a strong evidence of futility (lack of efficacy) during the interim analysis, it is unethical to continue the trial. Hence, the trial may also be terminated under the null hypothesis. However, except for the inner wedge test, most repeated significance tests are designed for early stop under the alternative hypothesis. In such a situation, the analysis of conditional power (or equivalently, futility analysis) can be used as a quantitative method for determining whether the trial should be terminated prematurely.

8.9.1 Comparing Means

Let x_{ij} be the observation from the jth subject ($j = 1, \ldots, n_i$) in the ith treatment group ($i = 1, 2$). $x_{ij}, j = 1, \ldots, n_i$, are assumed to be independent and identically distributed normal random variables with mean μ_i and variance σ_i^2. At the time of interim analysis, it is assumed that the first m_i of n_i subjects in the ith treatment group have already been observed. The investigator may want to evaluate the power for rejection of the null hypothesis based on the observed data and appropriate assumption under the alternative hypothesis. More specifically, define

$$
\bar{x}_{a,i} = \frac{1}{m_i} \sum_{j=1}^{m_i} x_{ij} \quad \text{and} \quad \bar{x}_{b,i} = \frac{1}{n_i - m_i} \sum_{j=m_i+1}^{n_i} x_{ij}.
$$

At the end of the trial, the following Z test statistic is calculated:

$$Z = \frac{\bar{x}_1 - \bar{x}_2}{\sqrt{s_1^2/n_1 + s_2^2/n_2}}$$

$$\approx \frac{\bar{x}_1 - \bar{x}_2}{\sqrt{\sigma_1^2/n_1 + \sigma_2^2/n_2}}$$

$$= \frac{(m_1\bar{x}_{a,1} + (n_1 - m_1)\bar{x}_{b,1})/n_1 - (m_2\bar{x}_{a,2} + (n_2 - m_2)\bar{x}_{a,2})/n_2}{\sqrt{\sigma_1^2/n_1 + \sigma_2^2/n_2}}.$$

Under the alternative hypothesis, we assume $\mu_1 > \mu_2$. Hence, the power for rejecting the null hypothesis can be approximated by

$$1 - \beta = P(Z > z_{\alpha/2})$$

$$= P\left(\frac{((n_1 - m_1)(\bar{x}_{b,1} - \mu_1)/n_1) - ((n_2 - m_2(\bar{x}_{b,2} - \mu_2)/n_2)}{\sqrt{((n_1 - m_1)\sigma_1^2/n_1^2) + ((n_2 - m_2)\sigma_2^2/n_2^2)}} > \tau\right)$$

$$= 1 - \Phi(\tau),$$

where

$$\tau = \left[z_{\alpha/2}\sqrt{\sigma_1^2/n_1 + \sigma_2^2/n_2} - (\mu_1 - \mu_2)\right.$$

$$\left. - \left(\frac{m_1}{n_1}(\bar{x}_{a,1} - \mu_1) - \frac{m_2}{n_2}(\bar{x}_{a,2} - \mu_2)\right)\right]$$

$$= \left[\frac{(n_1 - m_1)\sigma_1^2}{n_1^2} + \frac{(n_2 - m_2)\sigma_2^2}{n_2^2}\right]^{-1/2}.$$

As it can be seen from the above, the conditional power depends not only upon the assumed alternative hypothesis (μ_1, μ_2) but also upon the observed values ($\bar{x}_{a,1}, \bar{x}_{a,2}$) and the amount of information that has been accumulated (m_i/n_i) at the time of interim analysis.

8.9.2 Comparing Proportions

When the responses are binary, similar formulas can also be obtained. Let x_{ij} be the binary response observed from the jth subject ($j = 1, \ldots, n_i$) in the ith treatment group ($i = 1, 2$). Again, $x_{ij}, j = 1, \ldots, n_i$, are assumed to be independent and identically distributed binary variables with mean p_i. At the time of interim analysis, it is also assumed that the first m_i of n_i subjects in the ith treatment group have been observed. Define

$$\bar{x}_{a,i} = \frac{1}{m_i}\sum_{j=1}^{m_i} x_{ij} \quad \text{and} \quad \bar{x}_{b,i} = \frac{1}{n_i - m_i}\sum_{j=m_i+1}^{n_i} x_{ij}.$$

At the end of the trial, the following Z test statistic is calculated:

$$Z = \frac{\bar{x}_1 - \bar{x}_2}{\sqrt{\bar{x}_1(1-\bar{x}_1)/n_1 + \bar{x}_2(1-\bar{x}_2)/n_2}}$$

$$\approx \frac{\bar{x}_1 - \bar{x}_2}{\sqrt{p_1(1-p_1)/n_1 + p_2(1-p_2)/n_2}}$$

$$= \frac{(m_1\bar{x}_{a,1} + (n_1-m_1)\bar{x}_{b,1})/n_1 - (m_2\bar{x}_{a,2} + (n_2-m_2)\bar{x}_{a,2})/n_2}{\sqrt{p_1(1-p_1)/n_1 + p_2(1-p_2)/n_2}}.$$

Under the alternative hypothesis, we assume $p_1 > p_2$. Hence, the power for rejecting the null hypothesis can be approximated by

$$1 - \beta = P(Z > z_{\alpha/2})$$

$$= P\left[\frac{((n_1-m_1)(\bar{x}_{b,1}-\mu_1)/n_1) - ((n_2-m_2)(\bar{x}_{b,2}-\mu_2)/n_2)}{\sqrt{((n_1-m_1)p_1(1-p_1)/n_1^2) + ((n_2-m_2)p_2(1-p_2)/n_2^2)}} > \tau \right]$$

$$= 1 - \Phi(\tau),$$

where

$$\tau = \left[z_{\alpha/2}\sqrt{p_1(1-p_1)/n_1 + p_2(1-p_2)/n_2} - (\mu_1 - \mu_2) \right.$$

$$\left. - \left(\frac{m_1}{n_1}(\bar{x}_{a,1} - \mu_1) - \frac{m_2}{n_2}(\bar{x}_{a,2} - \mu_2) \right) \right]$$

$$= \left[\frac{(n_1-m_1)p_1(1-p_1)}{n_1^2} + \frac{(n_2-m_2)p_2(1-p_2)}{n_2^2} \right]^{-1/2}.$$

Similarly, the conditional power depends not only upon the assumed alternative hypothesis (p_1, p_2) but also upon the observed values ($\bar{x}_{a,1}$, $\bar{x}_{a,2}$) and the amount of information that has been accumulated (m_i/n_i) at the time of interim analysis.

8.10 Practical Issues

The group sequential procedures for interim analyses are basically in the context of hypothesis testing, which is aimed at pragmatic study objectives, that is, which treatment is better. However, most new treatments such as cancer drugs are very expensive or very toxic or both. As a result, if the degree of the benefit provided by the new treatment exceeds some minimum clinically significant requirement, only then will it be considered for the treatment of the intended medical conditions. Therefore, an adequate well-controlled trial should be able to provide not only the qualitative evidence, whether the experimental treatment is effective, but also the quantitative evidence from the unbiased estimation of the size of the effectiveness or safety over placebo given by the experimental therapy.

For a fixed sample design without interim analyses for early termination, it is possible to achieve both qualitative and quantitative goals with respect to the treatment effect. However, with group sequential procedure the size of benefit of the experimental treatment by the maximum likelihood method is usually overestimated because of the choice of stopping rule. Jennison and Turnbull (1990) pointed out that the sample mean might not even be contained in the final confidence interval. As a result, estimation of the size of treatment effect has received a lot of attention. Various estimation procedures have been proposed such as modified MLE, median unbiased estimator (MUE), and the midpoint of the equal-tailed 90% confidence interval. For more details, see Cox (1952), Tsiatis et al. (1984), Kim and DeMets (1987), Kim (1989), Chang and O'Brien (1986), Chang et al. (1989), Chang (1989), Hughes and Pocock (1988), and Pocock and Hughes (1989).

The estimation procedures proposed in the above literature require extensive computation. On the other hand, simulation results (Hughes and Pocock, 1988; Kim, 1989) showed that the alpha-spending function corresponding to the O'Brien–Fleming group sequential procedure is very concave and allocates only a very small amount of total nominal significance level to early stages of interim analyses, and hence, the bias, variance, and mean square error of the point estimator following O'Brien–Fleming procedure are also the smallest. Current research focuses mainly upon the estimation of the size of the treatment effect for the primary clinical endpoints on which the group sequential procedure is based. However, there are many other secondary efficacy and safety endpoints to be evaluated in the same trial. The impact of early termination of the trial based on the results from primary clinical endpoints on the statistical inference for these secondary clinical endpoints are unclear. In addition, group sequential methods and their followed estimation procedures so far are concentrated only on the population average. On the other hand, inference of variability is sometimes also of vital importance for certain classes of drug products and diseases. Research on estimation of variability following early termination is still lacking. Other areas of interest for interim analyses include clinical trials with more than two treatments and bioequivalence assessment. For group sequential procedures for the trials with multiple treatments, see Hughes (1993) and Proschan, Follmann, and Geller (1994). For group sequential bioequivalence testing procedure, see Gould (1995).

In practice, one of the most commonly used methods for sample size estimation in group sequential trials is to consider the most conservative scenario. In other words, we assume that the trial will not be stopped prematurely. Let C_K be the critical value for final analysis. Then, we reject the null hypothesis at the α level of significance if and only if $|Z_K| > C_K$. Under the null hypothesis, however, the type I error rate is no longer a. Instead it becomes

$$\alpha^* = P(|Z_K| > C_K).$$

Hence, the sample size for achieving a desired power can be estimated by adjusting α to α^*. This method has the merit of simplicity. Besides, it works well if $C_K \approx Z_{\alpha/2}$. However, if $C_K \gg Z_{\alpha/2}$, the resulting sample size could be very conservative.

9

Comparing Variabilities

In most clinical trials comparing a test drug and a control (e.g., a placebo control, standard therapy, or an active control), treatment effect is usually established by comparing mean response posttreatment change from the baseline of some primary study endpoints, assuming that their corresponding variabilities are comparable. In practice, however, variabilities associated with the test drug and the control could be very different. When the variability of the test drug is much larger than that of the reference drug, safety of the test drug could be a concern. Thus, in addition to comparing mean responses between treatments, it is also of interest to compare the variabilities associated with the responses between treatments.

In practice, the variabilities are usually classified into two categories, namely, the intrasubject (or within subject) variability and the intersubject (or between subject) variability. Intrasubject variability refers to the variability observed from repeated measurements from the same subject under the same experimental conditions. On the other hand, intersubject variability is the variability due to the heterogeneity among subjects. The total variability is simply the sum of the intrasubject and intersubject variabilities. In practice, it is of interest to test for equality, noninferiority/superiority, and similarity between treatments in terms of the intrasubject, intersubject, and/or total variabilities. The problem of comparing intrasubject variabilities is well studied by Chinchilli and Esinhart (1996) through an F statistic under a replicated crossover model. A similar idea can also be applied to compare total variabilities under a parallel design without replicates. However, how to compare intersubject and total variabilities under a crossover design is still challenging to biostatisticians in clinical research.

The remainder of this chapter is organized as follows. In Sections 9.1 and 9.2, formulas for sample size calculation for comparing intrasubject variabilities and intrasubject CVs are derived, respectively, under both replicated crossover designs and parallel designs with replicates. Sections 9.3 and 9.4 provide formulas for sample size calculation for comparing intersubject variabilities and total variabilities, respectively, under both crossover designs and parallel designs. Some practical issues are discussed in Section 9.5.

9.1 Comparing Intrasubject Variabilities

To assess intrasubject variability, replicates from the same subject are necessarily obtained. For this purpose, replicated crossover designs or parallel group designs with replicates are commonly employed. In what follows, statistical tests for comparing intrasubject variabilities under a parallel design with replicates and a replicated crossover design (e.g., a 2×4 replicated crossover design) are studied.

9.1.1 Parallel Design with Replicates

Let x_{ijk} be the observation of the kth replicate ($k = 1, \ldots, m$) of the jth subject ($j = 1, \ldots, n_i$) from the ith treatment ($i = T, R$). It is assumed that

$$x_{ijk} = \mu_i + S_{ij} + e_{ijk}, \tag{9.1}$$

where μ_i is the treatment effect, S_{ij} is the random effect due to the jth subject in the ith treatment group, and e_{ijk} is the intrasubject variability under the ith treatment. It is assumed that for a fixed i, S_{ij} are independent and identically distributed as normal random variables with mean 0 and variance σ_{Bi}^2, and e_{ijk}, $k = 1, \ldots, m$, are independent and identically distributed as a normal random variable with mean 0 and variance σ_{Wi}^2. Under this model, an unbiased estimator for σ_{Wi}^2 is given by

$$\hat{\sigma}_{Wi}^2 = \frac{1}{n_i(m-1)} \sum_{j=1}^{n_i} \sum_{k=1}^{m} (x_{ijk} - \bar{x}_{ij.})^2, \tag{9.2}$$

where

$$\bar{x}_{ij.} = \frac{1}{m} \sum_{k=1}^{m} x_{ijk}. \tag{9.3}$$

It can be seen that $n_i(m-1)\,\hat{\sigma}_{Wi}^2 / \sigma_{Wi}^2$ is distributed as a $\chi_{n_i(m-1)}^2$ random variable.

9.1.1.1 Test for Equality

In practice, it is often of interest to test whether two drug products have the same intrasubject variability. The following hypotheses are then of interest:

$$H_0 : \sigma_{WT}^2 = \sigma_{WR}^2 \quad \text{versus} \quad H_a : \sigma_{WT}^2 \neq \sigma_{WR}^2.$$

A commonly used test statistic for testing the above hypotheses is given by

$$T = \frac{\hat{\sigma}_{WT}^2}{\hat{\sigma}_{WR}^2}.$$

Under the null hypothesis, T is distributed as an F random variable with $n_T(m-1)$ and $n_R(m-1)$ degrees of freedom. Hence, we reject the null hypothesis at the α level of significance if

$$T > F_{\alpha/2, nT(m-1), nR(m-1)}$$

or

$$T < F_{1-\alpha/2, nT(m-1), nR(m-1)},$$

where $F_{\alpha/2, nT(m-1), nR(m-1)}$ is the upper $(\alpha/2)$th quantile of an F distribution with $n_T(m-1)$ and $n_R(m-1)$ degrees of freedom. Under the alternative hypothesis, without loss of generality, we assume that $\sigma_{WT}^2 < \sigma_{WR}^2$. The power of the above test is

$$
\begin{aligned}
\text{Power} &= P(T < F_{1-\alpha/2, nT(m-1), nR(m-1)}) \\
&= P(1/T > F_{\alpha/2, nR(m-1), nT(m-1)}) \\
&= P\left(\frac{\hat{\sigma}_{WR}^2/\sigma_{WR}^2}{\hat{\sigma}_{WT}^2/\sigma_{WT}^2} > \frac{\sigma_{WT}^2}{\sigma_{WR}^2} F_{\alpha/2, nR(m-1), nT(m-1)} \right) \\
&= P\left(F_{nR(m-1), nT(m-1)} > \frac{\sigma_{WT}^2}{\sigma_{WR}^2} F_{\alpha/2, nR(m-1), nT(m-1)} \right),
\end{aligned}
$$

where $F_{a,b}$ denotes an F random variable with a and b degrees of freedom. Under the assumption that $n = n_R = n_T$ and with a fixed σ_{WT}^2 and σ_{WR}^2, the sample size needed to achieve a desired power of $1 - \beta$ can be obtained by solving the following equation for n:

$$
\frac{\sigma_{WT}^2}{\sigma_{WR}^2} = \frac{F_{1-\beta, n(m-1), n(m-1)}}{F_{\alpha/2, n(m-1), n(m-1)}}.
$$

9.1.1.2 Test for Noninferiority/Superiority

The problem of testing noninferiority and superiority can be unified by the following hypotheses:

$$
H_0 : \frac{\sigma_{WT}}{\sigma_{WR}} \geq \delta \quad \text{versus} \quad H_a : \frac{\sigma_{WT}}{\sigma_{WR}} < \delta.
$$

When $\delta < 1$, the rejection of the null hypothesis indicates the superiority of the test product over the reference in terms of the intrasubject variability. When $\delta > 1$, the rejection of the null hypothesis indicates the noninferiority of the test product to the reference. The test statistic is given by

$$
T = \frac{\hat{\sigma}_{WT}^2}{\delta^2 \hat{\sigma}_{WR}^2}.
$$

Under the null hypothesis, T is distributed as an F random variable with $n_T(m-1)$ and $n_R(m-1)$ degrees of freedom. Hence, we reject the null hypothesis at the α level of significance if

$$
T < F_{1-\alpha, nT(m-1), nR(m-1)}.
$$

Under the alternative hypothesis that $\sigma_{WT}^2/\sigma_{WR}^2 < \delta$, the power of the above test is

$$
\begin{aligned}
\text{Power} &= P(T < F_{1-\alpha, nT(m-1), nR(m-1)}) \\
&= P(1/T > F_{\alpha, nR(m-1), nT(m-1)}) \\
&= P\left(\frac{\hat{\sigma}_{WR}^2/\sigma_{WR}^2}{\hat{\sigma}_{WT}^2/\sigma_{WT}^2} > \frac{\sigma_{WT}^2}{\delta^2 \sigma_{WR}^2} F_{\alpha/2, nR(m-1), nT(m-1)} \right) \\
&= P\left(F_{nR(m-1), nT(m-1)} > \frac{\sigma_{WT}^2}{\delta^2 \sigma_{WR}^2} F_{\alpha, nR(m-1), nT(m-1)} \right).
\end{aligned}
$$

Under the assumption that $n = n_T = n_R$, the sample size needed to achieve a desired power of $1 - \beta$ at the α level of significance can be obtained by solving the following equation for n:

$$\frac{\sigma_{WT}^2}{\delta^2 \sigma_{WR}^2} = \frac{F_{1-\beta, n(m-1), n(m-1)}}{F_{\alpha, n(m-1), n(m-1)}}.$$

9.1.1.3 Test for Similarity

For testing similarity, the following hypotheses are usually considered:

$$H_0 : \frac{\sigma_{WT}^2}{\sigma_{WR}^2} \geq \delta \quad \text{or} \quad \frac{\sigma_{WT}^2}{\sigma_{WR}^2} \leq \frac{1}{\delta} \quad \text{versus} \quad H_a : \frac{1}{\delta} < \frac{\sigma_{WT}^2}{\sigma_{WR}^2} < \delta,$$

where $\delta > 1$ is the similarity limit. The above hypotheses can be decomposed into the following two one-sided hypotheses:

$$H_{01} : \frac{\sigma_{WT}}{\sigma_{WR}} \geq \delta \quad \text{versus} \quad H_{a1} : \frac{\sigma_{WT}}{\sigma_{WR}} < \delta,$$

and

$$H_{02} : \frac{\sigma_{WT}}{\sigma_{WR}} \leq \frac{1}{\delta} \quad \text{versus} \quad H_{a2} : \frac{\sigma_{WT}}{\sigma_{WR}} > \frac{1}{\delta}.$$

These two one-sided hypotheses can be tested by the following two test statistics:

$$T_1 = \frac{\hat{\sigma}_{WT}^2}{\delta^2 \hat{\sigma}_{WR}^2} \quad \text{and} \quad T_2 = \frac{\delta^2 \hat{\sigma}_{WT}^2}{\hat{\sigma}_{WR}^2}.$$

We then reject the null hypothesis and conclude similarity at the α level of significance if

$$T_1 < F_{1-\alpha, nT(m-1), nR(m-1)} \quad \text{and} \quad T_2 > F_{\alpha, nT(m-1), nR(m-1)}.$$

Assuming that $n = n_T = n_R$ and $\sigma_{WT}^2 \leq \sigma_{WR}^2$, the power of the above test is

$$\begin{aligned}
\text{Power} &= P\left(\frac{F_{\alpha, n(m-1), n(m-1)}}{\delta^2} < \frac{\hat{\sigma}_{WT}^2}{\hat{\sigma}_{WR}^2} < \delta^2 F_{1-\alpha, n(m-1), n(m-1)} \right) \\
&= P\left(\frac{1}{F_{1-\alpha, n(m-1), n(m-1)} \delta^2} < \frac{\hat{\sigma}_{WT}^2}{\hat{\sigma}_{WR}^2} < \delta^2 F_{1-\alpha, n(m-1), n(m-1)} \right) \\
&= 1 - 2P\left(\frac{\hat{\sigma}_{WR}^2}{\hat{\sigma}_{WT}^2} > \delta^2 F_{1-\alpha, n(m-1), n(m-1)} \right) \\
&= 1 - 2P\left(F_{n(m-1), n(m-1)} > \frac{\delta^2 \sigma_{WT}^2}{\sigma_{WR}^2} F_{1-\alpha, n(m-1), n(m-1)} \right).
\end{aligned}$$

Thus, a conservative estimate for the sample size required for achieving a desired power of $1 - \beta$ can be obtained by solving the following equation for n:

$$\frac{\delta^2 \sigma_{WT}^2}{\sigma_{WR}^2} = \frac{F_{\beta/2, n(m-1), n(m-1)}}{F_{1-\alpha, n(m-1), n(m-1)}}.$$

9.1.1.4 An Example

Suppose that an investigator is interested in conducting a two-arm parallel trial with three ($m = 3$) replicates per subject to compare the variability of an inhaled formulation of a drug product (treatment) with a subcutaneous (SC) injected formulation (control) in terms of AUC. In practice, it is expected that the inhaled formulation has smaller intrasubject variability as compared to that of SC formulation. Based on PK data obtained from pilot studies, it is assumed the true standard deviation of treatment and control are given by 30% ($\sigma_{WT} = 0.30$) and 45% ($\sigma_{WR} = 0.45$), respectively. It is also believed that 10% ($\delta = 1.1$) of σ_{WR} is of no clinical importance. Hence, for testing noninferiority, the sample size per treatment needed to achieve an 80% power at the 5% level of significance can be obtained by solving the following equation:

$$\frac{0.30^2}{1.1^2 \times 0.45^2} = \frac{F_{0.80, 2n, 2n}}{F_{0.05, 2n, 2n}}.$$

The solution can be obtained by a standard numerical iteration technique, such as the simple grid search, which gives $n = 13$.

9.1.2 Replicated Crossover Design

Compared with the parallel design with replicates, the merit of a crossover design is the ability to make comparisons within subjects. In this section, without loss of generality, consider a $2 \times 2m$ replicated crossover design comparing two treatments. For convenience, we refer to the two treatments as a test formulation and a reference formulation. Under a $2 \times 2m$ replicated crossover design, in each sequence, each subject receives the test formulation m times and the reference formulation m times at different dosing periods. When $m = 1$, the $2 \times 2m$ replicated crossover design reduces to the standard two-sequence, two-period (2×2) crossover design. When $m = 2$, the $2 \times 2m$ replicated crossover design becomes the 2×4 crossover design recommended by the FDA for the assessment of population/individual bioequivalence (FDA, 2001).

Suppose that n_1 subjects are assigned to the first sequence and n_2 subjects are assigned to the second sequence. Let x_{ijkl} be the observation from the jth subject ($j = 1, \ldots, n_i$) in the ith sequence ($i = 1, 2$) under the lth replicate ($l = 1, \ldots, m$) of the kth treatment ($k = T, R$). As indicated in Chinchilli and Esinhart (1996), the following mixed effects model can best describe data observed from the $2 \times 2m$ replicated crossover design:

$$x_{ijkl} = \mu_k + \gamma_{ikl} + S_{ijk} + \varepsilon_{ijkl}, \tag{9.4}$$

where μ_k is the treatment effect for formulation k, γ_{ikl} is the fixed effect of the lth replicate on treatment k in the ith sequence with constraint

$$\sum_{i=1}^{2}\sum_{l=1}^{m}\gamma_{ikl}=0,$$

S_{ijT} and S_{ijR} are the random effects of the jth subject in the ith sequence, $(S_{ijT}, S_{ijR})'$'s are independent and identically distributed bivariate normal random vectors with mean $(0, 0)'$ and covariance matrix

$$\Sigma_B = \begin{pmatrix} \sigma_{BT}^2 & \rho\sigma_{BT}\sigma_{BR} \\ \rho\sigma_{BT}\sigma_{BR} & \sigma_{BR}^2 \end{pmatrix},$$

ε_{ijkl}'s are independent random variables from the normal distribution with mean 0 and variance σ_{WT}^2 or σ_{WR}^2, and the $(S_{ijT}, S_{ijR})'$ and ε_{ijkl} are independent. Note that σ_{BT}^2 and σ_{BR}^2 are the intersubject variances and σ_{WT}^2 and σ_{WR}^2 are intrasubject variances.

To obtain estimators of intrasubject variances, it is a common practice to use an orthogonal transformation, which is considered by Chinchilli and Esinhart (1996). A new random variable z_{ijkl} can be obtained by using the orthogonal transformation

$$\mathbf{z}_{ijk} = \mathbf{P}'\mathbf{x}_{ijk},\tag{9.5}$$

where

$$\mathbf{x}'_{ijk} = (x_{ijk1},\ x_{ijk2},...,x_{ijkm}),\quad \mathbf{z}'_{ijk} = (z_{ijk1},z_{ijk2},...,z_{ijkm})$$

and $\mathbf{P}$ is an $m \times m$ orthogonal matrix, that is, $\mathbf{P}'\mathbf{P}$ is an $m \times m$ diagonal matrix. The first column of $\mathbf{P}$ is usually defined by the vector $(1, 1,..., 1)'/\sqrt{m}$ to obtain $z_{ijk1} = \bar{x}_{ijk}$. The other columns can be defined to satisfy the orthogonality of $\mathbf{P}$ and $\text{var}(z_{ijkl}) = \sigma_{Wk}^2$ for $l = 2, \dots , m$. For example, in the 2×4 crossover design, the new random variable z_{ijkl} can be defined as

$$z_{ijk1} = \frac{x_{ijk1} + x_{ijk2}}{2} = \bar{x}_{ijk.} \quad \text{and} \quad z_{ijk2} = \frac{x_{ijk1} + x_{ijk2}}{\sqrt{2}}.$$

Now, the estimator of intrasubject variance can be defined as

$$\hat{\sigma}_{WT}^2 = \frac{1}{(n_1 + n_2 - 2)(m-1)}\sum_{i=1}^{2}\sum_{j=1}^{n_i}\sum_{l=2}^{m}(z_{ijTl} - \bar{z}_{i.Tl})^2,$$

$$\hat{\sigma}_{WR}^2 = \frac{1}{(n_1 + n_2 - 2)(m-1)}\sum_{i=1}^{2}\sum_{j=1}^{n_i}\sum_{l=2}^{m}(z_{ijRl} - \bar{z}_{i.Rl})^2,$$

where

$$\bar{z}_{i.kl} = \frac{1}{n_i} \sum_{j=1}^{n_i} z_{ijkl}.$$

It should be noted that $\hat{\sigma}_{WT}^2$ and $\hat{\sigma}_{WR}^2$ are independent.

9.1.2.1 Test for Equality

The following hypotheses are usually considered for testing equality in intrasubject variability:

$$H_0 : \sigma_{WT}^2 = \sigma_{WR}^2 \quad \text{versus} \quad H_a : \sigma_{WT}^2 \neq \sigma_{WR}^2.$$

Under the null hypothesis, the test statistic

$$T = \frac{\hat{\sigma}_{WT}^2}{\hat{\sigma}_{WR}^2}$$

is distributed as an F random variable with d and d degrees of freedom, where $d = (n_1 + n_2 - 2)(m - 1)$. Hence, we reject the null hypothesis at the α level of significance if

$$T > F_{\alpha/2,d,d}$$

or

$$T < F_{1-\alpha/2,d,d}.$$

Under the alternative hypothesis, without loss of generality, we assume that $\sigma_{WT}^2 < \sigma_{WR}^2$. The power of the above test is

$$
\begin{aligned}
\text{Power} &= P(T < F_{1-\alpha/2,d,d}) \\
&= P(1/T > F_{\alpha/2,d,d}) \\
&= P\left(\frac{\hat{\sigma}_{WR}^2/\sigma_{WR}^2}{\hat{\sigma}_{WT}^2/\sigma_{WT}^2} > \frac{\sigma_{WT}^2}{\sigma_{WR}^2} F_{\alpha/2,d,d} \right) \\
&= P\left(F_{(d,d)} > \frac{\sigma_{WT}^2}{\sigma_{WR}^2} F_{\alpha/2,d,d} \right).
\end{aligned}
$$

Under the assumption that $n = n_1 = n_2$ and with fixed σ_{WT}^2 and σ_{WR}^2, the sample size needed to achieve a desired power of $1 - \beta$ can be obtained by solving the following equation for n:

$$\frac{\sigma_{WT}^2}{\sigma_{WR}^2} = \frac{F_{1-\beta,(2n-2)(m-1),(2n-2)(m-1)}}{F_{\alpha/2,(2n-2)(m-1),(2n-2)(m-1)}}.$$

9.1.2.2 Test for Noninferiority/Superiority

The problem of testing noninferiority and superiority can be unified by the following hypotheses:

$$H_0 : \frac{\sigma_{WT}}{\sigma_{WR}} \geq \delta \quad \text{versus} \quad H_a : \frac{\sigma_{WT}}{\sigma_{WR}} < \delta.$$

When $\delta < 1$, the rejection of the null hypothesis indicates the superiority of the test product over the reference in terms of the intrasubject variability. When $\delta > 1$, the rejection of the null hypothesis indicates the noninferiority of the test product to the reference. Consider the following test statistic:

$$T = \frac{\hat{\sigma}_{WT}^2}{\delta^2 \hat{\sigma}_{WR}^2}.$$

Under the null hypothesis, T is distributed as an F random variable with d and d degrees of freedom. Hence, we reject the null hypothesis at the α level of significance if

$$T < F_{1-\alpha,d,d}.$$

Under the alternative hypothesis that $\sigma_{WT}^2 / \sigma_{WR}^2 < \delta$, the power of the above test is

$$
\begin{aligned}
\text{Power} &= P(T < F_{1-\alpha,d,d}) \\
&= P(1/T > F_{\alpha,d,d}) \\
&= P\left(\frac{\hat{\sigma}_{WR}^2 / \sigma_{WR}^2}{\hat{\sigma}_{WT}^2 / \sigma_{WT}^2} > \frac{\sigma_{WT}^2}{\delta^2 \sigma_{WR}^2} F_{\alpha,d,d} \right) \\
&= P\left(F_{(d,d)} > \frac{\sigma_{WT}^2}{\delta^2 \sigma_{WR}^2} F_{\alpha,d,d} \right).
\end{aligned}
$$

Thus, under the assumption that $n = n_1 = n_2$, the sample size needed to achieve a desired power of $1 - \beta$ at the α level of significance can be obtained by solving the following equation for n:

$$\frac{\sigma_{WT}^2}{\delta^2 \sigma_{WR}^2} = \frac{F_{1-\beta,(2n-2)(m-1),(2n-2)(m-1)}}{F_{\alpha,(2n-2)(m-1),(2n-2)(m-1)}}.$$

9.1.2.3 Test for Similarity

For testing similarity, the hypotheses of interest are given by

$$H_0 : \frac{\sigma_{WT}}{\sigma_{WR}} \geq \delta \text{ or } \frac{\sigma_{WT}}{\sigma_{WR}} \leq 1/\delta \quad \text{versus} \quad H_a : \frac{1}{\delta} < \frac{\sigma_{WT}}{\sigma_{WR}} < \delta,$$

where $\delta > 1$ is the equivalence limit. The above hypotheses can be decomposed into the following two one-sided hypotheses:

$$H_{01} : \frac{\sigma_{WT}}{\sigma_{WR}} \geq \delta \quad \text{versus} \quad H_{a1} : \frac{\sigma_{WT}}{\sigma_{WR}} < \delta,$$

and

$$H_{02} : \frac{\sigma_{WT}}{\sigma_{WR}} \leq \frac{1}{\delta} \quad \text{versus} \quad H_{a2} : \frac{\sigma_{WT}}{\sigma_{WR}} > \frac{1}{\delta}.$$

These two hypotheses can be tested by the following two test statistics:

$$T_1 = \frac{\hat{\sigma}_{WT}^2}{\delta^2 \hat{\sigma}_{WR}^2} \quad \text{and} \quad T_2 = \frac{\delta^2 \hat{\sigma}_{WT}^2}{\hat{\sigma}_{WR}^2}.$$

We then reject the null hypothesis and conclude similarity at the α level of significance if

$$T_1 < F_{1-\alpha,d,d} \quad \text{and} \quad T_2 > F_{\alpha,d,d}.$$

Assuming that $n = n_1 = n_2$, under the alternative hypothesis that $\sigma_{WT}^2 \leq \sigma_{WR}^2$, the power of the above test is

$$\begin{aligned}
\text{Power} &= P\left(\frac{F_{\alpha,d,d}}{\delta^2} < \frac{\hat{\sigma}_{WT}^2}{\hat{\sigma}_{WR}^2} < \delta^2 F_{1-\alpha,d,d} \right) \\[2mm]
&= P\left(\frac{1}{F_{1-\alpha,d,d}\delta^2} < \frac{\hat{\sigma}_{WT}^2}{\hat{\sigma}_{WR}^2} < \delta^2 F_{1-\alpha,d,d} \right) \\[2mm]
&\geq 1 - 2P\left(\frac{\hat{\sigma}_{WR}^2}{\hat{\sigma}_{WT}^2} > \delta^2 F_{1-\alpha,d,d} \right) \\[2mm]
&= 1 - 2P\left(F_{(d,d)} > \frac{\delta^2 \sigma_{WT}^2}{\sigma_{WR}^2} F_{1-\alpha,d,d} \right).
\end{aligned}$$

Hence, a conservative estimate for the sample size needed to achieve the power of $1 - \beta$ can be obtained by solving the following equation:

$$\frac{\delta^2 \sigma_{WT}^2}{\sigma_{WR}^2} = \frac{F_{\beta/2,(2n-2)(m-1),(2n-2)(m-1)}}{F_{1-\alpha,(2n-2)(m-1),(2n-2)(m-1)}}.$$

9.1.2.4 An Example

Consider the same example regarding comparison of intrasubject variabilities between two formulations (i.e., inhaled and SC) of a drug product as described in Section 9.1.2.3.

Suppose the intended study will be conducted under a 2×4 replicated crossover ($m = 2$) design rather than a parallel design with three replicates.

It is assumed that the true standard deviation of inhaled formulation and SC formulation are given by 30% ($\sigma_{WT} = 0.30$) and 45% ($\sigma_{WR} = 0.45$), respectively. It is also believed that 10% ($\delta = 1.10$) of σ_{WR} is of no clinical importance. Hence, the sample size needed per sequence to achieve an 80% power in establishing noninferiority at the 5% level of significance can be obtained by solving the following equation:

$$\frac{0.30^2}{1.1^2 \times 0.45^2} = \frac{F_{0.80, 2n-2, 2n-2}}{F_{0.05, 2n-2, 2n-2}}.$$

This gives $n = 14$.

9.2 Comparing Intrasubject CVs

In addition to comparing intrasubject variances, it is often of interest to study the intrasubject CV, which is a relative standard deviation adjusted for mean. In recent years, the use of intrasubject CV has become increasingly popular. For example, the FDA defines highly variable drug products based on their intrasubject CVs. A drug product is said to be a highly variable drug if its intrasubject CV is greater than 30%. The intrasubject CV is also used as a measure for reproducibility of blood levels (or blood concentration–time curves) of a given formulation when the formulation is repeatedly administered at different dosing periods. In addition, the information regarding the intrasubject CV of a reference product is usually used for performing power analysis for sample size calculation in bioavailability and bioequivalence studies. In practice, two methods are commonly used for comparing intrasubject CVs. One is proposed by Chow and Tse (1990), which is referred to as conditional random effects model. The other one is suggested by Quan and Shih (1996), which is a simple one-way random effects model. In this section, these two models are introduced and the corresponding formulas for sample size calculation are derived.

9.2.1 Simple Random Effects Model

Quan and Shih (1996) developed a method to estimate the intrasubject CV based on a simple one-way random mixed effects model. Comparing this model with model (9.6), it can be noted that the mixed effects model assumes that the intrasubject variability is a constant. An intuitive unbiased estimator for μ_i is given by

$$\hat{u}_i = \frac{1}{n_i m} \sum_{j=1}^{n_i} \sum_{k=1}^{m} x_{ijk}.$$

Hence, an estimator of the intrasubject CV can be obtained as

$$\widehat{CV}_i = \frac{\hat{\sigma}_{W_i}}{\hat{\mu}_i}.$$

By Taylor's expansion, it follows that

$$\widehat{CV}_i - CV_i = \frac{\hat{\sigma}_{Wi}}{\hat{\mu}_i} - \frac{\sigma_{Wi}}{\mu_i}$$

$$\approx \frac{1}{2\mu_i \sigma_{Wi}} \left(\hat{\sigma}_{Wi}^2 - \sigma_{Wi}^2 \right) - \frac{\sigma_{Wi}}{\mu_i^2} (\hat{\mu}_i - \mu_i).$$

Hence, by the central limit theorem, CV_i is asymptotically distributed as a normal random variable with mean CV_i and variance σ_i^{*2}/n_i, where

$$\sigma_i^{*2} = \frac{\sigma_{Wi}^2}{2m\mu_i^2} + \frac{\sigma_{Wi}^4}{\mu_i^4} = \frac{1}{2m} CV_i^2 + CV_i^4.$$

An intuitive estimator of σ_i^{*2} is given by

$$\hat{\sigma}_i^{*2} = \frac{1}{2m} \widehat{CV}_i^2 + \widehat{CV}_i^4.$$

9.2.1.1 Test for Equality

The following hypotheses are usually considered for testing equality in intrasubject CVs:

$$H_0 : CV_T = CV_R \quad \text{versus} \quad H_a : CV_T \neq CV_R.$$

Under the null hypothesis, the test statistic

$$T = \frac{\widehat{CV}_T - \widehat{CV}_R}{\sqrt{\hat{\sigma}_T^{*2}/n_T + \hat{\sigma}_R^{*2}/n_R}}$$

is asymptotically distributed as a standard normal random variable. Hence, we reject the null hypothesis at the α level of significance if $|T| > z_{\alpha/2}$. Under the alternative hypothesis, without loss of generality, it is assumed that $CV_T > CV_R$. The distribution of T can be approximated by a normal distribution with unit variance and mean

$$\frac{CV_T - CV_R}{\sqrt{\sigma_T^{*2}/n_T + \sigma_R^{*2}/n_R}}.$$

Thus, the power is approximately

$$P(|T| > z_{\alpha/2}) \approx P(T > z_{\alpha/2})$$

$$= 1 - \Phi \left(z_{\alpha/2} - \frac{CV_T - CV_R}{\sqrt{\sigma_T^{*2}/n_T + \sigma_R^{*2}/n_R}} \right).$$

Under the assumption that $n = n_1 = n_2$, the sample size needed to have a power of $1 - \beta$ can be obtained by solving the following equation:

$$z_{\alpha/2} - \frac{CV_T - CV_R}{\sqrt{\sigma_T^{*2}/n + \sigma_R^{*2}/n}} = -z_{\beta}.$$

This leads to

$$n = \frac{(\sigma_T^{*2} + \sigma_R^{*2})(z_{\alpha/2} + z_{\beta})^2}{(CV_T - CV_R)^2}.$$

9.2.1.2 Test for Noninferiority/Superiority

Similarly, the problem of testing noninferiority and superiority can be unified by the following hypotheses:

$$H_0 : CV_T - CV_R < \delta \quad \text{versus} \quad H_a : CV_T - CV_R \geq \delta,$$

where δ is the noninferiority/superiority margin. When $\delta > 0$, the rejection of the null hypothesis indicates the superiority of the test drug over the reference drug. When $\delta < 0$, the rejection of the null hypothesis indicates the noninferiority of the test drug to the reference drug.

Under the null hypothesis, the test statistic

$$T = \frac{\widehat{CV}_R - \widehat{CV}_T - \delta}{\sqrt{\hat{\sigma}_R^{*2}/n_R + \hat{\sigma}_T^{*2}/n_T}}$$

is asymptotically distributed as a standard normal random variable. Hence, we reject the null hypothesis at the α level of significance if $T > z_\alpha$. Under the alternative hypothesis, the distribution of T can be approximated by a normal distribution with unit variance and mean

$$\frac{CV_R - CV_T - \delta}{\sqrt{\sigma_R^{*2}/n_R + \sigma_T^{*2}/n_T}}.$$

Hence, the power is approximately

$$P(T > z_{\alpha/2}) = 1 - \Phi\left(z_{\alpha/2} - \frac{CV_R - CV_T - \delta}{\sqrt{\sigma_R^{*2}/n_R + \sigma_T^{*2}/n_T}} \right).$$

Under the assumption that $n = n_1 = n_2$, the sample size needed to have a power of $1 - \beta$ can be obtained by solving

$$z_{\alpha/2} - \frac{CV_R - CV_T - \delta}{\sqrt{\sigma_R^{*2}/n + \sigma_T^{*2}/n}} = -z_{\beta}.$$

This leads to

$$n = \frac{(\sigma_R^{*2} + \sigma_T^{*2})(z_{\alpha/2} + z_\beta)^2}{(CV_R - CV_T - \delta)^2}.$$

9.2.1.3 Test for Similarity

For testing similarity, the following hypotheses are usually considered:

$$H_0 : |CV_T - CV_R| \geq \delta \quad \text{versus} \quad H_a : |CV_T - CV_R| < \delta.$$

The two drug products are concluded to be similar to each other if the null hypothesis is rejected at a given significance level. The null hypothesis is rejected at the α level of significance if

$$\frac{\widehat{CV}_T - \widehat{CV}_R + \delta}{\sqrt{\sigma_T^{*2}/n_T + \sigma_R^{*2}/n_R}} > z_\alpha \quad \text{and} \quad \frac{\widehat{CV}_T - \widehat{CV}_R - \delta}{\sqrt{\sigma_T^{*2}/n_T + \sigma_R^{*2}/n_R}} < -z_\alpha.$$

Under the alternative hypothesis that $|CV_T - CV_R| < \delta$, the power of the above test procedure is approximately

$$2\Phi\left(\frac{\delta - |CV_T - CV_R|}{\sqrt{\sigma_T^{*2}/n_T + \sigma_R^{*2}/n_R}} - z_\alpha\right) - 1.$$

Hence, under the assumption that $n = n_1 = n_2$, the sample size needed to achieve $1 - \beta$ power at the α level of significance can be obtained by solving

$$\frac{\delta - |CV_T - CV_R|}{\sqrt{\sigma_T^{*2}/n_T + \sigma_R^{*2}/n_R}} - z_\alpha = z_{\beta/2}.$$

This gives

$$n = \frac{(z_\alpha + z_{\beta/2})^2(\sigma_T^{*2} + \sigma_R^{*2})}{(\delta - |CV_T - CV_R|)^2}.$$

9.2.1.4 An Example

Consider the same example as described in Section 9.2.1.3. Suppose the investigator is interested in conducting a parallel trial to compare intrasubject CVs between the inhaled formulation and SC formulation of the drug product under investigation rather than comparing intrasubject variabilities. Based on information obtained from a pilot study, it is assumed that the true CV of the treatment and control are given by 50% and 70%, respectively. Assume that 10% difference in CV is of no clinical importance.

The sample size needed per treatment group to establish noninferiority can be obtained as follows:

$$\sigma_1^2 = 0.25 \times 0.50^2 + 0.50^4 = 0.125,$$
$$\sigma_2^2 = 0.25 \times 0.70^2 + 0.70^4 = 0.363.$$

Hence, the sample size needed to achieve an 80% power for the establishment of noninferiority at the 5% level of significance is given by

$$n = \frac{(1.64 + 0.84)^2(0.125 + 0.363)}{(0.10 + 0.70 - 0.50)^2} \approx 34.$$

9.2.2 Conditional Random Effects Model

In practice, the variability of the observed response often increases as the mean increases. In many cases, the standard deviation of the intrasubject variability is approximately proportional to the mean value. To best describe this type of data, Chow and Tse (1990) proposed the following conditional random effects model:

$$x_{zjk} = A_{zj} + A_{ij}e_{ijk}, \tag{9.6}$$

where x_{ijk} is the observation from the kth replicate ($k = 1, \ldots, m$) of the jth subject ($j = 1, \ldots, n_i$) from the ith treatment ($i = T, R$), and A_{ij} is the random effect due to the jth subject in the ith treatment. It is assumed that A_{ij} is normally distributed as a normal random variable with mean μ_i and variance σ_{Bi}^2 and e_{ijk} is normally distributed as a normal random variable with mean 0 and variance σ_{Wi}^2. For a given subject with a fixed A_{ij}, x_{ijk} is normally distributed as a normal random variable with mean A_{ij} and variance $A_{ij}^2\sigma_{Wi}^2$. Hence, the CV for this subject is given by

$$CV_i = \frac{|A_{ij}\sigma_{Wi}|}{|A_{ij}|} = \sigma_{Wi}.$$

As it can be seen, the conditional random effects model assumes the CV is constant across subjects.

Define

$$\bar{x}_{i..} = \frac{1}{n_i m}\sum_{j=1}^{n_i}\sum_{k=1}^{m} x_{ijk},$$

$$M_{i1} = \frac{m}{n_i - 1}\sum_{j=1}^{n_i}(\bar{x}_{ij.} - \bar{x}_{i..})^2,$$

$$M_{i2} = \frac{1}{n_i(m-1)}\sum_{j=1}^{n_i}\sum_{k=1}^{m}(x_{ijk} - \bar{x}_{ij.})^2.$$

It can be verified that

$$E(\bar{x}_{i..}) = \mu_i,$$
$$E(M_{i1}) = (\mu_i^2 + \sigma_{Bi}^2)\sigma_{Wi}^2 + m\sigma_{iBi}^2 = \tau_{i1}^2,$$
$$E(M_{i2}) = (\mu_i^2 + \sigma_{Bi}^2)\sigma_{Wi}^2 = \tau_{i2}^2.$$

It follows that

$$CV_i = \sigma_{Wi} = \sqrt{\frac{E(M_{i2})}{E^2(\bar{x}_{i..}) + (M_{i1} - M_{i2})}}.$$

Hence, an estimator of CV_i is given by

$$\widehat{CV}_i = \sqrt{\frac{M_{i2}}{\bar{x}_{i..}^2 + (M_{i1} - M_{i2})m}}.$$

By Taylor's expansion,

$$\widehat{CV}_i - CV_i \approx \frac{1}{\tau_{i1}\tau_{i2}}(M_{i2} - \tau_{i2}^2) - \frac{\mu_i\tau_{i2}}{\tau_{i1}^3}(\bar{x}_{i..} - \mu_i)$$
$$- \frac{\tau_{i2}}{2m\tau_{i1}^3}(M_{i1} - M_{i2} - (\tau_{i1}^2 - \tau_{i2}^2))$$
$$= k_0(\bar{x}_{i..} - \mu_i) + k_1(M_{i1} - \tau_{i1}^2) + k_2(M_{i2} - \tau_{i2}^2),$$

where

$$k_0 = -\frac{\mu_i\tau_{i2}}{\tau_{i1}^3},$$
$$k_1 = -\frac{\tau_{i2}}{2m\tau_{i1}^3},$$
$$k_2 = \left(\frac{1}{\tau_{i1}\tau_{i2}} + \frac{\tau_{i2}}{2m\tau_{i1}^3}\right).$$

As a result, the distribution of $\widehat{CV}_i$ can be approximated by a normal random variable with mean CV_i and variance σ_i^{*2}/n_i, where

$$\sigma_i^{*2} = \text{var}\left[k_0\bar{x}_{ij.} + mk_1(\bar{x}_{ij.} - \bar{x}_{i..})^2 + \frac{k_2}{m-1}\sum_{k=1}^{m}(x_{ijk} - \bar{x}_{ij.})^2\right].$$

An intuitive estimator for σ_i^{*2} is the sample variance, denoted by $\hat{\sigma}_i^{*2}$, of

$$k_0\bar{x}_{ij.} + mk_1(\bar{x}_{ij.} - \bar{x}_{i..})^2 + \frac{k_2}{m-1}\sum_{k=1}^{m}(x_{ijk} - \bar{x}_{ij.})^2, \quad j = 1, ..., n_i.$$

9.2.2.1 Test for Equality

For testing equality, the following hypotheses are of interest:

$$H_0 : CV_T = CV_R \text{ versus } H_a : CV_T \neq CV_R.$$

Under the null hypothesis, test statistic

$$T = \frac{\widehat{CV}_T - \widehat{CV}_R}{\sqrt{\hat{\sigma}_T^{*2}/n_T + \hat{\sigma}_R^{*2}/n_R}}$$

is asymptotically distributed as a standard normal random variable. Hence, we reject the null hypothesis at the α level of significance if $|T| > z_{\alpha/2}$. Under the alternative hypothesis, without loss of generality, we assume that $CV_T > CV_R$. The distribution of T can be approximated by a normal distribution with unit variance and mean

$$\frac{CV_T - CV_R}{\sqrt{\sigma_T^{*2}/n_T + \sigma_R^{*2}/n_R}}.$$

Hence, the power is approximately

$$P(|T| > z_{\alpha/2}) \approx P(T > z_{\alpha/2})$$

$$= 1 - \Phi\left(z_{\alpha/2} - \frac{CV_T - CV_R}{\sqrt{\sigma_T^{*2}/n_T + \sigma_R^{*2}/n_R}}\right).$$

Under the assumption that $n = n_1 = n_2$, the sample size required for having a desired power of $1 - \beta$ can be obtained by solving

$$z_{\alpha/2} - \frac{CV_T - CV_R}{\sqrt{\sigma_T^{*2}/n + \sigma_R^{*2}/n}} = -z_\beta.$$

This leads to

$$n = \frac{(z_{\alpha/2} + z_\beta)^2(\sigma_T^{*2} + \sigma_R^{*2})}{(CV_T - CV_R)^2}.$$

9.2.2.2 Test for Noninferiority/Superiority

The problem of testing noninferiority and superiority can be unified by the following hypotheses:

$$H_0 : CV_R - CV_T < \delta \quad \text{versus} \quad H_a : CV_R - CV_T \geq \delta,$$

where δ is the noninferiority/superiority margin. When $\delta > 0$, the rejection of the null hypothesis indicates the superiority of the test drug over the reference drug. When $\delta < 0$, the rejection of the null hypothesis indicates noninferiority of the test drug to the reference drug. Under the null hypothesis, test statistic

$$T = \frac{\widehat{CV}_R - \widehat{CV}_T - \delta}{\sqrt{\hat{\sigma}_R^{*2}/n_R + \hat{\sigma}_T^{*2}/n_T}}$$

is asymptotically distributed as a standard normal random variable. Hence, we reject the null hypothesis at the α level of significance if $T > z_\alpha$. Under the alternative hypothesis, the distribution of T can be approximated by a normal distribution with unit variance and mean

$$\frac{CV_R - CV_T - \delta}{\sqrt{\sigma_R^{*2}/n_R + \sigma_T^{*2}/n_T}}.$$

Hence, the power is approximately

$$P(T > z_{\alpha/2}) = 1 - \Phi\left(z_{\alpha/2} - \frac{CV_R - CV_T - \delta}{\sqrt{\sigma_R^{*2}/n_R + \sigma_T^{*2}/n_T}}\right).$$

Under the assumption $n = n_1 = n_2$, the sample size needed to have the desired power $1 - \beta$ can be obtained by solving

$$z_{\alpha/2} - \frac{CV_R - CV_T - \delta}{\sqrt{\sigma_R^{*2}/n + \sigma_T^{*2}/n}} = -z_\beta.$$

This gives

$$n = \frac{(\sigma_R^{*2} + \sigma_T^{*2})(z_{\alpha/2} + z_\beta)^2}{(CV_R - CV_T - \delta)^2}.$$

9.2.2.3 Test for Similarity

For testing similarity, consider the following hypotheses:

$$H_0: \left|CV_T - CV_R\right| \geq \delta \quad \text{versus} \quad H_a: \left|CV_T - CV_R\right| < \delta.$$

The two drug products are concluded to be similar to each other if the null hypothesis is rejected at a given significance level. The null hypothesis is rejected at α level of significance if

$$\frac{\widehat{CV}_T - \widehat{CV}_R + \delta}{\sqrt{\hat{\sigma}_T^{*2}/n_T + \hat{\sigma}_R^{*2}/n_R}} > z_\alpha \quad \text{versus} \quad \frac{\widehat{CV}_T - \widehat{CV}_R - \delta}{\sqrt{\hat{\sigma}_T^{*2}/n_T + \hat{\sigma}_R^{*2}/n_R}} < -z_\alpha.$$

Under the alternative hypothesis that $|CV_T - CV_R| < \delta$, the power of the above test procedure is approximately

$$2\Phi\left(\frac{\delta - |CV_T - CV_R|}{\sqrt{\sigma_T^{*2}/n_T + \sigma_R^{*2}/n_R}} - z_\alpha\right) - 1.$$

Hence, under the assumption that $n = n_1 = n_2$, the sample size needed to achieve $1 - \beta$ power at the α level of significance can be obtained by solving

$$\frac{\delta - |CV_T - CV_R|}{\sqrt{\sigma_T^{*2}/n_T + \sigma_R^{*2}/n_R}} - z_\alpha = z_{\beta/2}.$$

This gives

$$n = \frac{(z_\alpha + z_{\beta/2})^2 (\sigma_T^{*2} + \sigma_R^{*2})}{(\delta - |CV_T - CV_R|)^2}.$$

9.2.2.4 An Example

Consider the same example as described in the previous subsection. Suppose it is found that the variability of the CV increases as the mean increases. In this case, the conditional random effects model is useful for comparing the two treatments. Again, we assume that CV of the test drug and the reference drug are given by 50% and 70%, respectively. Suppose it is also estimated from other studies that $\sigma_T^* = 0.30$ and $\sigma_R^* = 0.35$. Assume that 10% difference in CV is of no clinical importance. The sample size needed per treatment group to establish noninferiority can be obtained as follows:

$$n = \frac{(1.64 + 0.84)^2 (0.30^2 + 0.35^2)}{(0.10 + 0.70 - 0.50)^2} \approx 15.$$

As a result, 15 subjects per treatment group are needed to have an 80% power at the 5% level of significance.

9.2.2.5 Remarks

For comparing intrasubject variabilities and/or intrasubject CVs between treatment groups, replicates from the same subject are essential regardless of whether the study design is a parallel group design or a crossover design. In clinical research, data are often log-transformed before the analysis. It should be noted that the intrasubject standard deviation of log-transformed data is approximately equal to the intrasubject CV of the untransformed (raw) data. As a result, it is suggested that intrasubject variability be used when analyzing log-transformed data, while the intrasubject CV be considered when analyzing untransformed data.

9.3 Comparing Intersubject Variabilities

In addition to comparing intrasubject variabilities or intrasubject CVs, it is also of interest to compare intersubject variabilities. In practice, it is not uncommon that clinical results may not be reproducible from subject to subject within the target population or from subjects within the target population to subjects within a similar but slightly different population due to the intersubject variability. How to test a difference in intersubject and total variability between two treatments is a challenging problem to clinical scientists, especially biostatisticians, due to the following factors. First, unbiased estimators of the intersubject and total variabilities are usually not chi-square distributed under both parallel and crossover design with replicates. Second, the estimators for the intersubject and total variabilities under different treatments are usually not independent under a crossover design. As a result, unlike tests for comparing intrasubject variabilities, the standard F test is not applicable. Tests for comparing intersubject variabilities under a parallel design can be performed by using the method of a modified large sample (MLS) method. See Howe (1974), Graybill and Wang (1980), Ting et al. (1990), and Hyslop, et al. (2000). As indicated earlier, the MLS method is superior to many other approximation methods. Under crossover designs, however, the MLS method cannot be directly applied since estimators of variance components are not independent. Lee and Whitmore (2002) proposed an extension of the MLS method when estimators of variance components are not independent. In addition, tests for comparing intersubject and total variabilities under crossover designs are studied by Lee et al. (2002). Note that the MLS method by Hyslop et al. (2000) is recommended by the FDA (2001) as a statistical test for individual bioequivalence.

9.3.1 Parallel Design with Replicates

Under model (9.1), define

$$s_{Bi}^2 = \frac{1}{n_i - 1} \sum_{j=1}^{n_i} (\bar{x}_{ij.} - \bar{x}_{i..})^2, \tag{9.7}$$

where

$$\bar{x}_{i..} = \frac{1}{n_i} \sum_{j=1}^{n_i} \bar{x}_{ij.}$$

and $\bar{x}_{ij.}$ is given in Equation 9.3. Note that $E(s_{Bi}^2) = \sigma_{Bi}^2 + \sigma_{Wi}^2/m$. Therefore,

$$\hat{\sigma}_{Bi}^2 = s_{Bi}^2 - \frac{1}{m}\hat{\sigma}_{Wi}^2$$

are unbiased estimators for the intersubject variances, where $\hat{\sigma}_{Wi}^2$ is defined in Equation 9.2.

9.3.1.1 Test for Equality

For testing equality in intersubject variability, the following hypotheses are usually considered:

$$H_0 : \frac{\sigma_{BT}}{\sigma_{BR}} = 1 \quad \text{versus} \quad H_a : \frac{\sigma_{BT}}{\sigma_{BR}} \neq 1.$$

Testing the above hypotheses is equivalent to testing the following hypotheses:

$$H_0 : \sigma_{BT}^2 - \sigma_{BR}^2 = 0 \quad \text{versus} \quad H_0 : \sigma_{BT}^2 - \sigma_{BR}^2 \neq 0.$$

Let $\eta = \sigma_{BT}^2 - \sigma_{BR}^2$. An intuitive estimator of η is given by

$$\begin{aligned}
\hat{\eta} &= \hat{\sigma}_{BR}^2 - \hat{\sigma}_{BT}^2 \\
&= s_{BT}^2 - s_{BR}^2 - \hat{\sigma}_{WT}^2 / m + \hat{\sigma}_{WR}^2 / m.
\end{aligned}$$

A $(1 - \alpha) \times 100\%$ confidence interval for η is given by $(\hat{\eta}_L, \hat{\eta}_U)$, where

$$\hat{\eta}_L = \hat{\eta} - \sqrt{\Delta_L}, \quad \hat{\eta}_U = \hat{\eta} + \sqrt{\Delta_U}$$

and

$$\begin{aligned}
\Delta_L &= s_{BT}^4 \left(1 - \frac{n_T - 1}{\chi_{\alpha/2, n_T - 1}^2} \right)^2 + s_{BR}^4 \left(1 - \frac{n_R - 1}{\chi_{1-\alpha/2, n_R - 1}^2} \right)^2 \\
&\quad + \frac{\hat{\sigma}_{WT}^4}{m^2} \left(1 - \frac{n_T(m-1)}{\chi_{1-\alpha/2, n_T(m-1)}^2} \right)^2 + \frac{\hat{\sigma}_{WR}^4}{m^2} \left(1 - \frac{n_R(m-1)}{\chi_{\alpha/2, n_R(m-1)}^2} \right)^2, \\
\Delta_U &= s_{BT}^4 \left(1 - \frac{n_T - 1}{\chi_{1-\alpha/2, n_T - 1}^2} \right)^2 + s_{BR}^4 \left(1 - \frac{n_R - 1}{\chi_{\alpha/2, n_R - 1}^2} \right)^2 \\
&\quad + \frac{\hat{\sigma}_{WT}^4}{m^2} \left(1 - \frac{n_T(m-1)}{\chi_{\alpha/2, n_T(m-1)}^2} \right)^2 + \frac{\hat{\sigma}_{WR}^4}{m^2} \left(1 - \frac{n_R(m-1)}{\chi_{1-\alpha/2, n_R(m-1)}^2} \right)^2.
\end{aligned}$$

We reject the null hypothesis at the α level of significance if $0 \notin (\hat{n}_L, \hat{n}_U)$. Under the alternative hypothesis, without loss of generality, we assume that $\sigma_{BR}^2 > \sigma_{BT}^2$ and $n = n_T = n_R$. Thus, the power of the above test procedure can be approximated by

$$1 - \Phi \left(z_{\alpha/2} - \frac{\sqrt{n}(\sigma_{BT}^2 - \sigma_{BR}^2)}{\sigma^*} \right),$$

where

$$\sigma^{*2} = 2 \left[\left(\sigma_{BT}^2 + \frac{\sigma_{WT}^2}{m} \right)^2 + \left(\sigma_{BR}^2 + \frac{\sigma_{WR}^2}{m} \right)^2 \right.$$
$$\left. + \frac{\sigma_{WT}^4}{m^2(m-1)} + \frac{\sigma_{WR}^4}{m^2(m-1)} \right].$$

As a result, the sample size needed to achieve the desired power of $1 - \beta$ at the α level of significance can be obtained by solving

$$z_{\alpha/2} - \frac{\sqrt{n}(\sigma_{BT}^2 - \sigma_{BR}^2)}{\sigma^*} = -z_\beta.$$

This leads to

$$n = \frac{\sigma^{*2}(z_{\alpha/2} + z_\beta)^2}{(\sigma_{BT}^2 - \sigma_{BR}^2)^2}.$$

9.3.1.2 Test for Noninferiority/Superiority

Similar to testing intrasubject variabilities, the problem of testing noninferiority/superiority can be unified by the following hypotheses:

$$H_0 : \frac{\sigma_{BT}}{\sigma_{BR}} \geq \delta \quad \text{versus} \quad H_a : \frac{\sigma_{BT}}{\sigma_{BR}} < \delta.$$

Testing the above hypotheses is equivalent to testing the following hypotheses:

$$H_0 : \sigma_{BT}^2 - \delta^2 \sigma_{BR}^2 \geq 0 \quad \text{versus} \quad H_a : \sigma_{BT}^2 - \delta^2 \sigma_{BR}^2 < 0.$$

Define

$$\eta = \sigma_{BT}^2 - \delta^2 \sigma_{BR}^2.$$

For a given significance level α, similarly, the $(1 - \alpha) \times 100\%$th MLS upper confidence bound of η can be constructed as

$$\hat{\eta}_U = \hat{\eta} + \sqrt{\Delta_U},$$

where Δ_U is given by

$$\Delta_U = s_{BT}^4 \left(1 - \frac{n_T - 1}{\chi_{1-\alpha, n_T-1}^2} \right)^2 + \delta^4 s_{BR}^4 \left(1 - \frac{n_R - 1}{\chi_{\alpha, n_R-1}^2} \right)^2$$
$$+ \frac{\hat{\sigma}_{WT}^4}{m^2} \left(1 - \frac{n_T(m-1)}{\chi_{\alpha, n_T(m-1)}^2} \right)^2 + \frac{\delta^4 \hat{\sigma}_{WR}^4}{m^2} \left(1 - \frac{n_R(m-1)}{\chi_{1-\alpha, n_R(m-1)}^2} \right)^2.$$

We then reject the null hypothesis at the α level of significance if $\hat{\eta}_U < 0$. Under the assumption that $n = n_T = n_R$, using a similar argument to those in Section 9.2, the power of the above testing procedure can be approximated by

$$1 - \Phi\left(z_\alpha - \frac{\sqrt{n}(\sigma_{BT}^2 - \delta^2\sigma_{BR}^2)}{\sigma^*}\right),$$

where

$$\sigma^{*2} = 2\left[\left(\sigma_{BT}^2 + \frac{\sigma_{WT}^2}{m}\right)^2 + \delta^4\left(\sigma_{BR}^2 + \frac{\sigma_{WR}^2}{m}\right)^2 \right.$$
$$\left. + \frac{\sigma_{WT}^4}{m^2(m-1)} + \frac{\delta^4\sigma_{WR}^4}{m^2(m-1)}\right].$$

As a result, the sample size needed to achieve the power of $1 - \beta$ at the α level of significance can be obtained by solving

$$z_\alpha - \frac{\sqrt{n}(\sigma_{BT}^2 - \delta^2\sigma_{BR}^2)}{\sigma^*} = -z_\beta.$$

This gives

$$n = \frac{\sigma^{*2}(z_\alpha + z_\beta)^2}{(\sigma_{BT}^2 - \delta^2\sigma_{BR}^2)^2}.$$

9.3.1.3 An Example

For illustration purposes, consider the same example as described in Section 9.3.1.2 (i.e., a parallel design with three replicates). Suppose we are interested in testing difference in intersubject variabilities. In this case, we assume

$$\sigma_{BT} = 0.30, \quad \sigma_{BR} = 0.40,$$
$$\sigma_{WT} = 0.20, \quad \sigma_{WR} = 0.30.$$

Hence, the sample size needed to achieve an 80% power ($1 - \beta = 0.80$) in establishing noninferiority with a noninferiority margin 0.10 ($\delta = 1.10$) at the 5% level of significance ($\alpha = 0.05$) can be obtained as

$$\sigma^{*2} = 2\left[\left(0.30^2 + \frac{0.20^2}{3}\right)^2 + 1.1^4 \times \left(0.40^2 + \frac{0.30^2}{3}\right)^2\right.$$
$$\left. + \frac{0.20^4}{3^2(3-1)} + \frac{1.1^4 0.30^4}{3^2(3-1)}\right]$$
$$= 0.129.$$

Hence,

$$n = \frac{0.129(1.64 + 0.84)^2}{(0.30^2 - 1.1^2 \times 0.40)^2} \approx 74.$$

9.3.2 Replicated Crossover Design

Under model (9.4), estimators of intersubject variances can be defined by

$$s_{BT}^2 = \frac{1}{n_1 + n_2 - 2} \sum_{i=1}^{2} \sum_{j=1}^{n_i} (\bar{x}_{ijT.} - \bar{x}_{i.T.})^2,$$

$$s_{BR}^2 = \frac{1}{n_1 + n_2 - 2} \sum_{i=1}^{2} \sum_{j=1}^{n_i} (\bar{x}_{ijR.} - \bar{x}_{i.R.})^2,$$

where

$$\bar{x}_{i.k.} = \frac{1}{n_i} \sum_{j=1}^{n_i} \bar{x}_{ijk.}.$$

Note that $E(s_{BK}^2) = \sigma_{BK}^2 + \sigma_{BK}^2 / m$ for $k = T, R$. Therefore, unbiased estimators for the intersubject variance are given by

$$\hat{\sigma}_{BT}^2 = s_{BT}^2 - \frac{1}{m} \hat{\sigma}_{WT}^2, \tag{9.8}$$

$$\hat{\sigma}_{BR}^2 = s_{BR}^2 - \frac{1}{m} \hat{\sigma}_{WR}^2. \tag{9.9}$$

9.3.2.1 Test for Equality

For testing the equality in intersubject variability, the following hypotheses are considered:

$$H_0 : \frac{\sigma_{BT}}{\sigma_{BR}} = 1 \quad \text{versus} \quad H_a : \frac{\sigma_{BT}}{\sigma_{BR}} \neq 1$$

Testing the above hypotheses is equivalent to testing the following hypotheses:

$$H_0 : \sigma_{BT}^2 - \sigma_{BR}^2 = 0 \quad \text{versus} \quad H_a : \sigma_{BT}^2 - \sigma_{BR}^2 \neq 0.$$

Let $\eta = \sigma_{BK}^2 - \sigma_{BR}^2$. An intuitive estimator of η is given by

$$\hat{\eta} = \hat{\sigma}_{BT}^2 - \hat{\sigma}_{BR}^2$$
$$= s_{BT}^2 - s_{BR}^2 - \hat{\sigma}_{WT}^2 / m + \hat{\sigma}_{WR}^2 / m,$$

where $\hat{\sigma}_{BT}^2$ and $\hat{\sigma}_{BR}^2$ are given in Equations 9.8 and 9.9, respectively. Random vector $(\bar{x}_{ijT.}, \bar{x}_{ijR.})'$ for the jth subject in ith sequence has a bivariate normal distribution with covariance matrix given by

$$\Omega_B = \begin{pmatrix} \sigma_{BT}^2 + \sigma_{WT}^2/m & \rho\sigma_{BT}\sigma_{BR} \\ \rho\sigma_{BT}\sigma_{BR} & \sigma_{BR}^2 + \sigma_{WR}^2/m \end{pmatrix}. \tag{9.10}$$

An unbiased estimator of the covariance matrix Ω_B is

$$\hat{\Omega}_B = \begin{pmatrix} s_{BT}^2 & s_{BTR}^2 \\ s_{BTR}^2 & s_{BR}^2 \end{pmatrix}, \tag{9.11}$$

where

$$s_{BTR}^2 = \frac{1}{n_1 + n_2 - 2} \sum_{i=1}^{2} \sum_{j=1}^{n_i} (\bar{x}_{ijT.} - \bar{x}_{i.T.})(\bar{x}_{ijR.} - \bar{x}_{i.R.}) \tag{9.12}$$

is the sample covariance between $\bar{x}_{ijT.}$ and $\bar{x}_{ijR.}$. Let λ_i, $i = 1, 2$, be the two eigenvalues of the matrix $\Theta\Omega_B$, where

$$\Theta = \begin{pmatrix} 1 & 0 \\ 0 & -1 \end{pmatrix}. \tag{9.13}$$

Hence, λ_i can be estimated by

$$\hat{\lambda}_i = \frac{s_{BT}^2 - s_{BR}^2 \pm \sqrt{(s_{BT}^2 + s_{BR}^2)^2 - 4s_{BTR}^4}}{2} \quad \text{for} \quad i = 1, 2.$$

Without loss of generality, it can be assumed that $\hat{\lambda}_1 < 0 < \hat{\lambda}_2$. In Lee et al. (2002b), a $(1 - \alpha) \times 100\%$ confidence interval of η is given by $(\hat{n}_L, \hat{n}_U)$, where

$$\hat{\eta}_L = \hat{\eta} - \sqrt{\Delta_L}, \quad \hat{\eta}_U = \hat{\eta} + \sqrt{\Delta_U},$$

$$\Delta_L = \hat{\lambda}_1^2 \left(1 - \frac{n_S - 1}{\chi_{\alpha/2, n_S - 1}^2}\right)^2 + \hat{\lambda}_2^2 \left(1 - \frac{n_S - 1}{\chi_{1-\alpha/2, n_S - 1}^2}\right)^2$$
$$+ \frac{\hat{\sigma}_{WT}^4}{m^2} \left(1 - \frac{n_S(m-1)}{\chi_{\alpha/2, n_S(m-1)}^2}\right)^2 + \frac{\hat{\sigma}_{WR}^4}{m^2} \left(1 - \frac{n_S(m-1)}{\chi_{1-\alpha/2, n_S(m-1)}^2}\right)^2,$$

$$\Delta_U = \hat{\lambda}_1^2 \left(1 - \frac{n_S - 1}{\chi_{1-\alpha/2, n_S - 1}^2}\right)^2 + \hat{\lambda}_2^2 \left(1 - \frac{n_S - 1}{\chi_{\alpha/2, n_S - 1}^2}\right)^2$$
$$+ \frac{\hat{\sigma}_{WT}^4}{m^2} \left(1 - \frac{n_S(m-1)}{\chi_{1-\alpha/2, n_S(m-1)}^2}\right)^2 + \frac{\hat{\sigma}_{WR}^4}{m^2} \left(1 - \frac{n_S(m-1)}{\chi_{\alpha/2, n_S(m-1)}^2}\right)^2,$$

and $n_s = n_1 + n_2 - 2$. Then, we reject the null hypothesis at the α level of significance if $0 \notin (\hat{h}_L, \hat{h}_U)$.

Under the alternative hypothesis, the power of the above test can be approximated by

$$1 - \Phi\left(z_{\alpha/2} - \frac{\sqrt{n_s}(\sigma_{BT}^2 - \sigma_{BR}^2)}{\sigma^*}\right),$$

where

$$\sigma^{*2} = 2\left[\left(\sigma_{BT}^2 + \frac{\sigma_{WT}^2}{m}\right)^2 + \left(\sigma_{BR}^2 + \frac{\sigma_{WR}^2}{m}\right)^2 - 2\rho^2\sigma_{BT}^2\sigma_{BR}^2 \right. $$
$$\left. + \frac{\sigma_{WT}^4}{m^2(m-1)} + \frac{\sigma_{WR}^4}{m^2(m-1)}\right].$$

Thus, the sample size needed to achieve the power of $1 - \beta$ at the α level of significance can be obtained by solving

$$z_\alpha - \frac{\sqrt{n_s}(\sigma_{BT}^2 - \sigma_{BR}^2)}{\sigma^*} = -z_\beta.$$

This leads to

$$n_S = \frac{\sigma^{*2}(z_{\alpha/2} + z_\beta)^2}{(\sigma_{BT}^2 - \sigma_{BR}^2)^2}.$$

9.3.2.2 Test for Noninferiority/Superiority

Similar to testing intrasubject variabilities, the problem of testing noninferiority/superiority can be unified by the following hypotheses:

$$H_0 : \frac{\sigma_{BT}}{\sigma_{BR}} \geq \delta \quad \text{versus} \quad H_a : \frac{\sigma_{BT}}{\sigma_{BR}} < \delta.$$

Testing the above hypotheses is equivalent to testing the following hypotheses:

$$H_0 : \sigma_{BT}^2 - \delta^2\sigma_{BR}^2 \geq 0 \quad \text{versus} \quad H_a : \sigma_{BT}^2 - \delta^2\sigma_{BR}^2 < 0.$$

When $\delta < 1$, the rejection of the null hypothesis indicates the superiority of the test drug versus the reference drug. When $\delta > 1$, a rejection of the null hypothesis indicates the noninferiority of the test drug versus the reference drug. Let $\eta = \sigma_{BT}^2 - \delta^2\sigma_{BR}^2$. For a given significance level of α, similarly, the $(1 - \alpha)$th upper confidence bound of η can be constructed as

$$\hat{\eta}_U = \hat{\eta} + \sqrt{\Delta_U},$$

where Δ_U is given by

$$\Delta_U = \hat{\lambda}_1^2 \left(1 - \frac{n_S - 1}{\chi_{1-\alpha/2,n_S-1}^2}\right)^2 + \hat{\lambda}_2^2 \left(1 - \frac{n_S - 1}{\chi_{\alpha/2,n_S-1}^2}\right)^2$$

$$+ \frac{\hat{\sigma}_{WT}^4}{m^2}\left(1 - \frac{n_S(m-1)}{\chi_{1-\alpha/2,n_S(m-1)}^2}\right)^2 + \frac{\delta^4\hat{\sigma}_{WR}^4}{m^2}\left(1 - \frac{n_S(m-1)}{\chi_{\alpha/2,n_S(m-1)}^2}\right)^2,$$

$n_s = n_1 + n_2 - 2$, and

$$\hat{\lambda}_i = \frac{s_{BT}^2 - \delta^2 s_{BR}^2 \pm \sqrt{(s_{BT}^2 + \delta^2 s_{BR}^2)^2 - 4\delta^2 s_{BTR}^4}}{2}.$$

We then reject the null hypothesis at the α level of significance if $\hat{\eta}_U < 0$.

Using a similar argument to that in the previous section, the power of the above test procedure can be approximated by

$$1 - \Phi\left(z_\alpha - \frac{\sqrt{n_s}(\sigma_{BT}^2 - \delta^2\sigma_{BR}^2)}{\sigma^*}\right),$$

where

$$\sigma^{*2} = 2\left[\left(\sigma_{BT}^2 + \frac{\sigma_{WT}^2}{m}\right)^2 + \delta^4\left(\sigma_{BR}^2 + \frac{\sigma_{WR}^2}{m}\right)^2 - 2\delta^2\rho^2\sigma_{BT}^2\sigma_{BR}^2\right.$$

$$\left. + \frac{\sigma_{WT}^4}{m^2(m-1)} + \frac{\delta^4\sigma_{WR}^4}{m^2(m-1)}\right].$$

As a result, the sample size needed to achieve a power of $1 - \beta$ at the α level of significance can be obtained by solving

$$z_\alpha - \frac{\sqrt{n_s}(\sigma_{BT}^2 - \delta^2\sigma_{BR}^2)}{\sigma^*} = -z_\beta.$$

This leads to

$$n_S = \frac{\sigma^{*2}(z_{\alpha/2} + z_\beta)^2}{(\sigma_{BT}^2 - \delta^2\sigma_{BR}^2)^2}.$$

9.3.2.3 An Example

Suppose a 2×4 crossover design (ABAB, BABA) is used to compare two treatments (A and B) in terms of their intersubject variabilities. Information from pilot studies indicates that $\rho = 0.75$, $\sigma_{BT} = 0.3$, $\sigma_{BR} = 0.4$, $\sigma_{WT} = 0.2$, and $\sigma_{WR} = 0.3$. The objective is to establish noninferiority with a margin of 10% ($\delta = 1.10$). It follows that

$$
\begin{aligned}
\sigma^{*2} = 2\Bigg[&\left(0.30^2 + \frac{0.20^2}{2}\right)^2 + 1.1^4 \times \left(0.40^2 + \frac{0.30^2}{2}\right)^2 \\
&- 2 \times 1.1^2 (0.75 \times 0.3 \times 0.4)^2 + \frac{0.20^4}{2^2} + \frac{1.1^4 \times 0.30^4}{2^2}\Bigg] \\
= {}& 0.115.
\end{aligned}
$$

Hence, the sample size needed to achieve an 80% power ($1 - \beta = 0.80$) for the establishment of noninferiority at the 5% level of significance ($\alpha = 0.05$) is given by

$$
n_S = \frac{0.115(1.64 + 0.84)^2}{(0.30^2 - 1.1^2 \times 0.40^2)^2} \approx 66.
$$

Since $n_s = n_1 + n_2 - 2$, approximately 29 subjects per sequence are required for achieving an 80% power at the 5% level of significance.

9.4 Comparing Total Variabilities

In practice, it may also be of interest to compare total variabilities between drug products. For example, comparing total variability is required in assessing drug prescribability (FDA, 2001). Total variability can be estimated under a parallel group design with and without replicates and under various crossover designs (e.g., a 2×2 standard crossover design or a $2 \times 2m$ replicated crossover design). In this section, we focus on sample size calculation under a parallel group design with and without replicates, the standard 2×2 crossover design, and the $2 \times 2m$ replicated crossover design.

9.4.1 Parallel Designs without Replicates

For parallel design without replicates, the model in Equation 9.1 is reduced to

$$
x_{ij} = \mu_i + \varepsilon_{ij},
$$

where x_{ij} is the observation from the jth subject in the ith treatment group. Also, we assume that the random variable ε_{ij} has normal distribution with mean 0 and variance σ_{Ti}^2 for $i = T$, R. Hence, the total variability can be estimated by

$$
\hat{\sigma}_{Ti}^2 = \frac{1}{n_i - 1} \sum_{j=1}^{n_i} (x_{ij} - \bar{x}_{i.})^2,
$$

where

$$\bar{x}_{i.} = \frac{1}{n_i} \sum_{j=1}^{n_i} x_{ij}.$$

9.4.1.1 Test for Equality

For testing equality in total variability, the following hypotheses are considered:

$$H_0 : \sigma_{TT}^2 = \sigma_{TR}^2 \quad \text{versus} \quad H_a : \sigma_{TT}^2 \neq \sigma_{TR}^2.$$

Under the null hypothesis, the test statistic

$$T = \frac{\hat{\sigma}_{TT}^2}{\hat{\sigma}_{TR}^2}.$$

is distributed as an F random variable with $n_T - 1$ and $n_R - 1$ degrees of freedom. Hence, we reject the null hypothesis at the α level of significance if

$$T > F_{\alpha/2, nT-1, nR-1}$$

or

$$T < F_{1-\alpha/2, nT-1, nR-1}.$$

Under the alternative hypothesis (without loss of generality, we assume that $\sigma_{TT}^2 < \sigma_{TR}^2$), the power of the above test procedure is

$$\begin{aligned} \text{Power} &= P(T < F_{1-\alpha/2, n_T-1, n_R-1}) \\ &= P(1/T > F_{\alpha/2, n_R-1, n_T-1}) \\ &= P\left(\frac{\hat{\sigma}_{TR}^2 / \sigma_{TR}^2}{\hat{\sigma}_{TT}^2 / \sigma_{TT}^2} > \frac{\sigma_{TT}^2}{\sigma_{TR}^2} F_{\alpha/2, n_R-1, n_T-1} \right) \\ &= P\left(F_{n_R-1, n_T-1} > \frac{\sigma_{TT}^2}{\sigma_{TR}^2} F_{\alpha/2, n_R-1, n_T-1} \right). \end{aligned}$$

Under the assumption that $n = n_R = n_T$ and with a fixed σ_{TT}^2 and σ_{TR}^2, the sample size needed to achieve a desired power of $1 - \beta$ can be obtained by solving the following equation for n:

$$\frac{\sigma_{TT}^2}{\sigma_{TR}^2} = \frac{F_{1-\beta, n-1, n-1}}{F_{\alpha/2, n-1, n-1}}.$$

9.4.1.2 Test for Noninferiority/Superiority

The problem of testing noninferiority and superiority can be unified by the following hypotheses:

$$H_0 : \frac{\sigma_{TT}}{\sigma_{TR}} \geq \delta \quad \text{versus} \quad H_a : \frac{\sigma_{TT}}{\sigma_{TR}} < \delta.$$

When $\delta < 1$, the rejection of the null hypothesis indicates the superiority of the test product over the reference in terms of the total variability. When $\delta > 1$, the rejection of the null hypothesis indicates the noninferiority of the test product to the reference. The test statistic is given by

$$T = \frac{\hat{\sigma}_{TT}^2}{\delta^2 \hat{\sigma}_{TR}^2}.$$

Under the null hypothesis, T is distributed as an F random variable with n_T and n_R degrees of freedom. Hence, we reject the null hypothesis at the α level of significance if

$$T < F_{1-\alpha, n_T-1, n_R-1}.$$

Under the alternative hypothesis that $\sigma_{TT}^2 / \sigma_{TR}^2 < \delta^2$, the power of the above test procedure is

$$
\begin{aligned}
\text{Power} &= P(T < F_{1-\alpha, n_R-1, n_T-1}) \\
&= P(1/T > F_{\alpha, n_R-1, n_T-1}) \\
&= P\left(\frac{\hat{\sigma}_{TR}^2 / \sigma_{TR}^2}{\hat{\sigma}_{TT}^2 / \sigma_{TT}^2} > \frac{\sigma_{TT}^2}{\delta^2 \sigma_{TR}^2} F_{\alpha, n_R-1, n_T-1} \right) \\
&= P\left(F_{(n_R, n_T)} > \frac{\sigma_{TT}^2}{\delta^2 \sigma_{TR}^2} F_{\alpha, n_R-1, n_T-1} \right).
\end{aligned}
$$

Under the assumption that $n = n_T = n_R$, the sample size needed to achieve a desired power of $1 - \beta$ at the α level of significance can be obtained by solving

$$\frac{\sigma_{TT}^2}{\delta^2 \sigma_{TR}^2} = \frac{F_{1-\beta, n-1, n-1}}{F_{\alpha, n-1, n-1}}.$$

9.4.1.3 Test for Similarity

For testing similarity, the hypotheses of interest are given by

$$H_0 : \frac{\sigma_{TT}}{\sigma_{TR}} \geq \delta \quad \text{or} \quad \frac{\sigma_{TT}}{\sigma_{TR}} \leq 1/\delta \quad \text{versus} \quad H_a : \frac{1}{\delta} < \frac{\sigma_{TT}}{\sigma_{TR}} < \delta,$$

where $\delta > 1$ is the similarity limit. The above hypotheses can be decomposed into the following two one-sided hypotheses:

$$H_{01} : \frac{\sigma_{TT}}{\sigma_{TR}} \geq \delta \quad \text{versus} \quad H_{a1} : \frac{\sigma_{TT}}{\sigma_{TR}} < \delta,$$

and

$$H_{02} : \frac{\sigma_{TT}}{\sigma_{TR}} \leq \frac{1}{\delta} \quad \text{versus} \quad H_{a2} : \frac{\sigma_{TT}}{\sigma_{TR}} > \frac{1}{\delta}.$$

These two hypotheses can be tested by the following two test statistics:

$$T_1 : \frac{\hat{\sigma}_{TT}^2}{\delta^2 \hat{\sigma}_{TR}^2} \quad \text{and} \quad T_2 : \frac{\delta^2 \hat{\sigma}_{TT}^2}{\hat{\sigma}_{TR}^2}.$$

We reject the null hypothesis and conclude similarity at the α level of significance if

$$T_1 < F_{1-\alpha, n_T-1, n_R-1} \quad \text{and} \quad T_2 > F_{\alpha, n_T-1, n_R-1}.$$

Assuming that $n = n_T = n_R$, under the alternative hypothesis that $\sigma_{TT}^2 \leq \sigma_{TR}^2$, the power of the above test is

$$
\begin{aligned}
\text{Power} &= P\left(\frac{F_{\alpha, n-1, n-1}}{\delta^2} < \frac{\hat{\sigma}_{TT}^2}{\hat{\sigma}_{TR}^2} < \delta^2 F_{1-\alpha, n-1, n-1} \right) \\
&= P\left(\frac{1}{F_{1-\alpha, n-1, n-1} \delta^2} < \frac{\hat{\sigma}_{TT}^2}{\hat{\sigma}_{TR}^2} < \delta^2 F_{1-\alpha, n-1, n-1} \right) \\
&\geq 1 - 2P\left(\frac{\hat{\sigma}_{TR}^2/\sigma_{TR}^2}{\hat{\sigma}_{TT}^2/\sigma_{TT}^2} > \frac{\delta^2 \sigma_{TT}^2}{\sigma_{TR}^2} F_{1-\alpha, n-1, n-1} \right) \\
&= 1 - 2P\left(F_{n-1, n-1} > \frac{\delta^2 \sigma_{TT}^2}{\sigma_{TR}^2} F_{1-\alpha, n-1, n-1} \right).
\end{aligned}
$$

Hence, a conservative estimate for the sample size needed to achieve the desired power of $1 - \beta$ can be obtained by solving the following equation for n:

$$\frac{\delta^2 \sigma_{TT}^2}{\sigma_{TR}^2} = \frac{F_{\beta/2, n-1, n-1}}{F_{1-\alpha, n-1, n-1}}.$$

9.4.1.4 An Example

Consider the example discussed in the previous subsections. Suppose a parallel group design without replicates is to be conducted for comparing total variabilities between a test drug and a reference drug. It is assumed that $\sigma_{TT} = 0.55$ and $\sigma_{TR} = 0.75$. The sample size needed to achieve an 80% power ($1 - \beta = 0.80$) at the 5% level of significance ($\alpha = 0.05$) in establishing noninferiority with the noninferiority margin $\delta = 1.1$ can be obtained by solving

$$\frac{0.55^2}{1.10^2 \times 0.75^2} = \frac{F_{0.20, n-1, n-1}}{F_{0.05, n-1, n-1}}.$$

This gives $n = 40$.

9.4.2 Parallel Design with Replicates

In practice, parallel design with replicates can also be used to assess total variability. The merit of the parallel design with replicates is that it can serve more than just one purpose. For example, it can not only assess total variabilities, but also intersubject and intrasubject variabilities. Model (9.1) can be used to represent data here. Unbiased estimators for total variabilities are given by

$$\hat{\sigma}_{Ti}^2 = s_{Bi}^2 + \frac{m-1}{m}\hat{\sigma}_{Wi}^2,$$

where s_{Bi}^2 is defined in Section 9.3.1.

9.4.2.1 Test for Equality

Let $\eta = \sigma_{TT}^2 - \sigma_{TR}^2$; hence, a natural estimator for η is given by

$$\hat{\eta} = \hat{\sigma}_{TT}^2 - \hat{\sigma}_{TR}^2.$$

For testing equality in total variability, the following hypotheses are considered:

$$H_0 : \sigma_{TT}^2 = \sigma_{TR}^2 \quad \text{versus} \quad H_a : \sigma_{TT}^2 \neq \sigma_{TR}^2.$$

A $(1 - \alpha) \times 100\%$ confidence interval of η is given by $(\hat{\eta}_L, \hat{\eta}_U)$, where

$$\hat{\eta}_L = \hat{\eta} - \sqrt{\Delta_L}, \quad \hat{\eta}_U = \hat{\eta} + \sqrt{\Delta_U},$$

$$\Delta_L = s_{BT}^4 \left(1 - \frac{n_T - 1}{\chi_{\alpha/2,n_T-1}^2}\right)^2 + s_{BR}^4 \left(1 - \frac{n_R - 1}{\chi_{1-\alpha/2,n_R-1}^2}\right)^2$$
$$+ \frac{(m-1)^2 \hat{\sigma}_{WT}^4}{m^2} \left(1 - \frac{n_T(m-1)}{\chi_{1-\alpha/2,n_T(m-1)}^2}\right)^2$$
$$+ \frac{(m-1)^2 \hat{\sigma}_{WR}^4}{m^2} \left(1 - \frac{n_R(m-1)}{\chi_{\alpha/2,n_R(m-1)}^2}\right)^2,$$

and

$$\Delta_U = s_{BT}^4 \left(1 - \frac{n_T - 1}{\chi_{1-\alpha/2,n_T-1}^2}\right)^2 + s_{BR}^4 \left(1 - \frac{n_R - 1}{\chi_{\alpha/2,n_R-1}^2}\right)^2$$
$$+ \frac{(m-1)^2 \hat{\sigma}_{WT}^4}{m^2} \left(1 - \frac{n_T(m-1)}{\chi_{\alpha/2,n_T(m-1)}^2}\right)^2$$
$$+ \frac{(m-1)^2 \hat{\sigma}_{WR}^4}{m^2} \left(1 - \frac{n_R(m-1)}{\chi_{1-\alpha/2,n_R(m-1)}^2}\right)^2.$$

We reject the null hypothesis at the α level of significance if $0 \notin (\hat{\eta}_L, \hat{\eta}_U)$. Under the alternative hypothesis and assuming that $n = n_T = n_R$, the power of the above test procedure can be approximated by

$$\Phi\left(z_{\alpha/2} - \frac{\sqrt{n}(\sigma_{TT}^2 - \sigma_{TR}^2)}{\sigma^*}\right),$$

where

$$\sigma^{*2} = 2\left[\left(\sigma_{BT}^2 + \frac{\sigma_{WT}^2}{m}\right)^2 + \left(\sigma_{BR}^2 + \frac{\sigma_{WR}^2}{m}\right)^2 \right.$$
$$\left. + \frac{(m-1)\sigma_{WT}^4}{m^2} + \frac{(m-1)\sigma_{WR}^4}{m^2}\right].$$

As a result, the sample size needed to achieve $1 - \beta$ power at the α level of significance can be obtained by solving

$$z_{\alpha/2} - \frac{\sqrt{n}(\sigma_{TT}^2 - \sigma_{TR}^2)}{\sigma^*} = -z_\beta.$$

This gives

$$n = \frac{\sigma^{*2}(z_{\alpha/2} + z_\beta)^2}{(\sigma_{TT}^2 - \sigma_{TR}^2)^2}.$$

9.4.2.2 Test for Noninferiority/Superiority

The problem of testing noninferiority/superiority can be unified by the following hypotheses:

$$H_0 : \frac{\sigma_{TT}}{\sigma_{TR}} \geq \delta \quad \text{versus} \quad H_a : \frac{\sigma_{TT}}{\sigma_{TR}} < \delta.$$

Testing the above hypotheses is equivalent to testing the following hypotheses:

$$H_0 : \sigma_{TT}^2 - \delta^2 \sigma_{TR}^2 \geq 0 \quad \text{versus} \quad H_a : \sigma_{TT}^2 - \delta^2 \sigma_{TR}^2 < 0.$$

When $\delta < 1$, the rejection of the null hypothesis indicates the superiority of the test drug versus the reference drug. When $\delta > 1$, the rejection of the null hypothesis indicates the noninferiority of the test drug versus the reference drug. Let $\eta = \sigma_{TT}^2 - \delta^2 \sigma_{BR}^2$. For a given significance level of α, the $(1 - \alpha)$th upper confidence bound of η can be constructed as

$$\hat{\eta}_u = \hat{\eta} + \sqrt{\Delta_u}$$

where $\hat{\eta} = \hat{\sigma}_{TT}^2 - \delta^2 \hat{\sigma}_{TR}^2$ and Δ_u is given by

$$\Delta_u = s_{BT}^4 \left(1 - \frac{n_T - 1}{\chi_{1-\alpha/2, n_T-1}^2}\right)^2 + \delta^4 s_{BR}^4 \left(1 - \frac{n_R - 1}{\chi_{\alpha, n_R-1}^2}\right)^2$$

$$+ \frac{(m-1)^2 \hat{\sigma}_{WT}^4}{m^2} \left(1 - \frac{n_T(m-1)}{\chi_{\alpha, n_T(m-1)}^2}\right)^2$$

$$+ \frac{\delta^4 (m-1)^2 \hat{\sigma}_{WR}^4}{m^2} \left(1 - \frac{n_R(m-1)}{\chi_{1-\alpha, n_R(m-1)}^2}\right)^2.$$

We then reject the null hypothesis at the α level of significance if $\hat{\eta}_u < 0$. Using a similar argument to that in the previous section, the power of the above testing procedure can be approximated by

$$1 - \Phi \left(z_\alpha - \frac{\sqrt{n}(\sigma_{TT}^2 - \delta^2 \sigma_{TR}^2)}{\sigma^*} \right),$$

where

$$\sigma^{*2} = 2 \left[\left(\sigma_{BT}^2 + \frac{\sigma_{WT}^2}{m}\right)^2 + \delta^4 \left(\sigma_{BR}^2 + \delta^4 \frac{\sigma_{WR}^2}{m}\right)^2 \right.$$

$$\left. + \frac{(m-1)\sigma_{WT}^4}{m^2} + \delta^4 \frac{(m-1)\sigma_{WR}^4}{m^2} \right].$$

As a result, the sample size needed to achieve the desired power of $1 - \beta$ at the α level of significance can be obtained by solving

$$z_\alpha - \frac{\sqrt{n}(\sigma_{TT}^2 - \delta^2 \sigma_{TR}^2)}{\sigma^{*2}} = -z_\beta.$$

This gives

$$n = \frac{\sigma^{*2}(z_\alpha + z_\beta)^2}{(\sigma_{TT}^2 - \delta^2 \sigma_{TR}^2)^2}.$$

9.4.2.3 An Example

Consider the same example discussed in Section 9.4.2.2. Suppose a trial with a parallel design with three replicates ($m = 3$) is conducted to compare total variabilities between treatment groups. It is assumed that $\sigma_{BT} = 0.30$, $\sigma_{BR} = 0.40$, $\sigma_{WT} = 0.20$, and $\sigma_{WR} = 0.30$. The objective is to establish the noninferiority with $\delta = 1.1$. It follows that

$$\sigma^{*2} = 2\left[\left(0.30^2 + \frac{0.20^2}{3}\right)^2 + 1.1^4\left(0.4^2 + \frac{0.3^2}{3}\right)^2\right.$$

$$\left. + \frac{(3-1)0.20^4}{3^2} + 1.1^4\frac{(3-1)0.3^4}{3^2}\right]$$

$$= 0.133.$$

As a result, the sample size needed per treatment group to achieve an 80% ($1 - \beta = 0.80$) power at the 5% ($\alpha = 0.05$) level of significance is given by

$$n = \frac{0.133(1.64 + 0.84)^2}{(0.30^2 + 0.20^2 - 1.1^2(0.4^2 + 0.3^2))^2} \approx 28.$$

9.4.3 The Standard 2 × 2 Crossover Design

Under the standard 2 × 2 crossover design, model (9.4) is still useful. We omitted the subscript l since there are no replicates.

Intuitive estimators for the total variabilities are given by

$$\hat{\sigma}_{TT}^2 = \frac{1}{n_1 + n_2 - 2}\sum_{i=1}^{2}\sum_{j=1}^{n_i}(x_{ijT} - \bar{x}_{i\cdot T})^2$$

and

$$\hat{\sigma}_{TR}^2 = \frac{1}{n_1 + n_2 - 2}\sum_{i=1}^{2}\sum_{j=1}^{n_i}(x_{ijR} - \bar{x}_{i\cdot R})^2,$$

where

$$\bar{x}_{i\cdot T} = \frac{1}{n_i}\sum_{j=1}^{n_i}x_{ijT} \quad \text{and} \quad \bar{x}_{i\cdot R} = \frac{1}{n_i}\sum_{j=1}^{n_i}x_{ijR}.$$

9.4.3.1 Test for Equality

For testing the equality in total variability, again consider the following hypotheses:

$$H_0 : \frac{\sigma_{TT}^2}{\sigma_{TR}^2} = 1 \quad \text{versus} \quad H_a : \frac{\sigma_{TT}^2}{\sigma_{TR}^2} \neq 1.$$

Testing the above hypotheses is equivalent to testing the following hypotheses:

$$H_0 : \sigma_{TT}^2 - \sigma_{TR}^2 = 0 \quad \text{versus} \quad H_a : \sigma_{TT}^2 - \sigma_{TR}^2 \neq 0.$$

Let $\eta = \sigma_{TT}^2 - \sigma_{TR}^2$. An intuitive estimator of η is given by

$$\hat{\eta} = \hat{\sigma}_{TT}^2 - \hat{\sigma}_{TR}^2.$$

Let

$$\sigma_{BTR}^2 = \frac{1}{n_1 + n_2 - 2} \sum_{i=1}^{2} \sum_{j=1}^{n_i} (x_{ijT} - \bar{x}_{i \cdot T})(x_{ijR} - \bar{x}_{i \cdot R}).$$

Define $\hat{\lambda}_i$, $i = 1, 2$, by

$$\hat{\lambda}_i = \frac{\hat{\sigma}_{TT}^2 - \hat{\sigma}_{TR}^2 \pm \sqrt{(\hat{\sigma}_{TT}^2 + \hat{\sigma}_{TR}^2)^2 - 4\hat{\sigma}_{BTR}^4}}{2}.$$

Assume that $\hat{\lambda}_1 < 0 < \hat{\lambda}_2$. In Lee et al. (2002b), a $(1 - \alpha) \times 100\%$ confidence interval of η is given by $(\hat{\eta}_L, \hat{\eta}_U)$, where

$$\hat{\eta}_L = \hat{\eta} - \sqrt{\Delta_L}, \quad \hat{\eta}_U = \hat{\eta} + \sqrt{\Delta_U},$$

$$\Delta_L = \hat{\lambda}_1^2 \left(1 - \frac{n_1 + n_2 - 2}{\chi_{1-\alpha/2, n_1+n_2-2}^2} \right)^2 + \hat{\lambda}_2^2 \left(1 - \frac{n_1 + n_2 - 2}{\chi_{\alpha/2, n_1+n_2-2}^2} \right)^2$$

$$\Delta_U = \hat{\lambda}_1^2 \left(1 - \frac{n_1 + n_2 - 2}{\chi_{\alpha/2, n_1+n_2-2}^2} \right)^2 + \hat{\lambda}_2^2 \left(1 - \frac{n_1 + n_2 - 2}{\chi_{1-\alpha/2, n_1+n_2-2}^2} \right)^2.$$

We reject the null hypothesis at the α level of significance if $0 \notin (\hat{\eta}_L, \hat{\eta}_U)$.

Under the alternative hypothesis, without loss of generality, we assume $\sigma_{TR}^2 > \sigma_{TT}^2$. Let $n_s = n_1 + n_2 - 2$. The power of the above test can be approximated by

$$1 - \Phi \left(z_{\alpha/2} - \frac{\sqrt{n_s} (\sigma_{TT}^2 - \sigma_{TR}^2)}{\sigma^*} \right),$$

where

$$\sigma^{*2} = 2(\sigma_{TT}^2 + \sigma_{TR}^2 - 2\rho^2 \sigma_{BT}^2 \sigma_{BR}^2).$$

Hence, the sample size needed to achieve the power of $1 - \beta$ power at the α level of significance can be obtained by solving the following equation:

$$z_{\alpha/2} - \frac{\sqrt{n_s} (\sigma_{TT}^2 - \sigma_{TR}^2)}{\sigma^*} = -z_\beta,$$

which implies that

$$n_s = \frac{\sigma^{*2}(z_{\alpha/2} + z_\beta)^2}{(\sigma_{TT}^2 - \sigma_{TR}^2)^2}.$$

9.4.3.2 Test for Noninferiority/Superiority

The problem of testing noninferiority/superiority can be unified by the following hypotheses:

$$H_0 : \frac{\sigma_{TT}}{\sigma_{TR}} \geq \delta \quad \text{versus} \quad H_a : \frac{\sigma_{TT}}{\sigma_{TR}} < \delta.$$

Testing the above hypotheses is equivalent to testing the following hypotheses:

$$H_0 : \sigma_{TT}^2 - \delta^2 \sigma_{TR}^2 \geq 0 \quad \text{versus} \quad H_a : \sigma_{TT}^2 - \delta^2 \sigma_{TR}^2 < 0.$$

When $\delta < 1$, the rejection of the null hypothesis indicates the superiority of the test drug versus the reference drug. When $\delta > 1$, the rejection of the null hypothesis indicates the noninferiority of the test drug versus the reference drug. Let $\eta = \sigma_{TT}^2 - \delta^2 \sigma_{TR}^2$. For a given significance level of α, similarly, the $(1 - \alpha)$th upper confidence bound of η can be constructed as

$$\hat{\eta}_U = \hat{\eta} + \sqrt{\Delta_U},$$

where Δ_U is given by

$$\Delta_U = \hat{\lambda}_1^2 \left(\frac{n_1 + n_2 - 2}{\chi^2_{\alpha, n_1 + n_2 - 2}} - 1 \right)^2 + \hat{\lambda}_2^2 \left(\frac{n_1 + n_2 - 2}{\chi^2_{1-\alpha, n_1 + n_2 - 2}} - 1 \right)^2,$$

and $\hat{\lambda}_i$, $i = 1, 2$, are given by

$$\hat{\lambda}_i = \frac{\hat{\sigma}_{TT}^2 - \delta^4 \hat{\sigma}_{TR}^2 \pm \sqrt{(\hat{\sigma}_{TT}^2 + \delta^4 \hat{\sigma}_{TR}^2)^2 - 4\delta^2 \hat{\sigma}_{BTR}^4}}{2}.$$

We reject the null hypothesis at the α level of significance if $\hat{\eta}_U < 0$. Under the alternative hypothesis, the power of the above test procedure can be approximated by

$$1 - \Phi \left(z_\alpha - \frac{\sqrt{n_s}(\sigma_{BT}^2 - \delta^2 \sigma_{BR}^2)}{\sigma^*} \right),$$

where

$$\sigma^{*2} = 2(\sigma_{TT}^2 + \delta^4 \sigma_{TR}^2 - 2\delta^2 \rho^2 \sigma_{BT}^2 \sigma_{BR}^2).$$

Hence, the sample size needed to achieve $1 - \beta$ power at the α level of significance can be obtained by solving

$$z_\alpha - \frac{\sqrt{n_s}(\sigma_{TT}^2 - \delta^2 \sigma_{TR}^2)}{\sigma^*} = -z_\beta.$$

This gives

$$n_s = \frac{\sigma^{*2}(z_\alpha + z_\beta)^2}{(\sigma_{TT}^2 - \delta^2 \sigma_{TR}^2)^2}.$$

9.4.3.3 An Example

Consider the same example discussed in Section 9.4.3.2. Under the standard 2×2 crossover design, it is assumed that $\rho = 1$, $\sigma_{BT} = 0.30$, $\sigma_{BR} = 0.40$, $\sigma_{WT} = 0.20$, and $\sigma_{WR} = 0.30$. The objective is to establish noninferiority with a margin $\delta = 1.1$. It follows that

$$\sigma^{*2} = 2[(0.30^2 + 0.20^2)^2 + 1.1^4(0.4^2 + 0.3^2)^2$$
$$- 2 \times 1.1^2 \times 0.30^3 \times 0.4^2] = 0.147.$$

As a result, the sample size needed to achieve an 80% $(1 - \beta = 0.80)$ power at the 5% $(\alpha = 0.05)$ level of significance is given by

$$n_s = \frac{(0.153)(1.64 + 0.84)^2}{(0.30^2 + 0.20^2 - 1.1^2 \times (0.4^2 + 0.3^2))^2} \approx 31.$$

Since $n_s = n_1 + n_2 - 2$, approximately 17 subjects should be assigned to each sequence to achieve an 80% $(1 - \beta = 0.80)$ power.

9.4.4 Replicated $2 \times 2m$ Crossover Design

We can use a similar argument for test of intersubject variabilities under model (9.4) with the estimators

$$\hat{\sigma}_{Tk}^2 = s_{Bk}^2 + \frac{m-1}{m}\hat{\sigma}_{Wk}^2, \quad k = T, R.$$

9.4.4.1 Test for Equality

For testing the equality in total variability, consider the following hypotheses:

$$H_0 : \frac{\sigma_{TT}^2}{\sigma_{TR}^2} = 1 \quad \text{versus} \quad H_a : \frac{\sigma_{TT}^2}{\sigma_{TR}^2} \neq 1.$$

Testing the above hypotheses is equivalent to testing the following hypotheses:

$$H_0 : \sigma_{TT}^2 - \sigma_{TR}^2 = 0 \quad \text{versus} \quad H_a : \sigma_{TT}^2 - \sigma_{TR}^2 \neq 0.$$

Let $\hat{\eta} = \hat{\sigma}_{TT}^2 - \hat{\sigma}_{TR}^2$. In Lee et al. (2002b), a $(1 - \alpha) \times 100\%$ confidence interval of η is given by $(\hat{\eta}_L, \hat{\eta}_U)$, where

$$\hat{\eta}_L = \hat{\eta} - \sqrt{\Delta_L}, \quad \hat{\eta}_U = \hat{\eta} + \sqrt{\Delta_U},$$

$$\Delta_L = \hat{\lambda}_1^2 \left(1 - \frac{n_s - 1}{\chi_{1-\alpha/2, n_s - 1}^2}\right)^2 + \hat{\lambda}_2^2 \left(1 - \frac{n_s - 1}{\chi_{\alpha/2, n_s - 1}^2}\right)^2$$

$$+ \frac{(m-1)^2 \hat{\sigma}_{WT}^4}{m^2} \left(1 - \frac{n_s(m-1)}{\chi_{\alpha/2, n_s(m-1)}^2}\right)^2$$

$$+ \frac{(m-1)^2 \hat{\sigma}_{WR}^4}{m^2} \left(1 - \frac{n_s(m-1)}{\chi_{1-\alpha/2, n_s(m-1)}^2}\right)^2$$

$$\Delta_U = \hat{\lambda}_1^2 \left(1 - \frac{n_s - 1}{\chi_{\alpha/2, n_s - 1}^2}\right)^2 + \hat{\lambda}_2^2 \left(1 - \frac{n_s - 1}{\chi_{1-\alpha/2, n_s - 1}^2}\right)^2$$

$$+ \frac{(m-1)^2 \hat{\sigma}_{WT}^4}{m^2} \left(1 - \frac{n_s(m-1)}{\chi_{1-\alpha/2, n_s(m-1)}^2}\right)^2$$

$$+ \frac{(m-1)^2 \hat{\sigma}_{WR}^4}{m^2} \left(1 - \frac{n_s(m-1)}{\chi_{\alpha/2, n_s(m-1)}^2}\right)^2,$$

and $\hat{\lambda}_i$'s are the same as those used for the test of equality for intersubject variabilities. We reject the null hypothesis at the α level of significance if $0 \notin \hat{\eta}_L, \hat{\eta}_U$).

Under the alternative hypothesis, without loss of generality, we assume $\sigma_{TR}^2 > \sigma_{TT}^2$ and $n = n_T = n_R$. The power of the above test can be approximated by

$$1 - \Phi\left(z_{\alpha/2} - \frac{\sqrt{n}(\sigma_{TT}^2 - \sigma_{TR}^2)}{\sigma^*}\right),$$

where

$$\sigma^{*2} = 2\left[\left(\sigma_{BT}^2 + \frac{\sigma_{WT}^2}{m}\right)^2 + \left(\sigma_{BR}^2 + \frac{\sigma_{WR}^2}{m}\right)^2 - 2\rho^2 \sigma_{BT}^2 \sigma_{BR}^2 \right.$$

$$\left. + \frac{(m-1)\sigma_{WT}^4}{m^2} + \frac{(m-1)\sigma_{WR}^4}{m^2}\right].$$

Hence, the sample size needed to achieve the power of $1 - \beta$ at the α level of significance can be obtained by solving the following equation:

$$z_{\alpha/2} - \frac{\sqrt{n}(\sigma_{TT}^2 - \sigma_{TR}^2)}{\sigma^*} = z_\beta.$$

This leads to

$$n = \frac{\sigma^{*2}(z_{\alpha/2} + z_\beta)^2}{(\sigma_{TT}^2 - \sigma_{TR}^2)^2}.$$

9.4.4.2 Test for Noninferiority/Superiority

The problem of testing noninferiority/superiority can be unified by the following hypotheses:

$$H_0 : \frac{\sigma_{TT}}{\sigma_{TR}} \geq \delta \quad \text{versus} \quad H_a : \frac{\sigma_{TT}}{\sigma_{TR}} < \delta,$$

which is equivalent to

$$H_0 : \sigma_{TT}^2 - \delta^2 \sigma_{TR}^2 \geq 0 \quad \text{versus} \quad H_a : \sigma_{TT}^2 - \delta^2 \sigma_{TR}^2 < 0.$$

When $\delta < 1$, the rejection of the null hypothesis indicates the superiority of the test drug versus the reference drug. When $\delta > 1$, the rejection of the null hypothesis indicates the noninferiority of the test drug versus the reference drug. Let $\hat{\eta} = \hat{\sigma}_{TT}^2 - \delta^2 \hat{\sigma}_{TR}^2$. For a given significance level of α, similarly, the $(1 - \alpha)$th upper confidence bound of η can be constructed as

$$\hat{\eta}_U = \hat{\eta} + \sqrt{\Delta_U},$$

where $\hat{\eta} = \hat{\sigma}_{TT}^2 - \delta^2 \hat{\sigma}_{TR}^2$,

$$\begin{aligned}
\Delta_U = {}& \hat{\lambda}_1^2 \left(1 - \frac{n_s - 1}{\chi_{\alpha, n_s - 1}^2} \right)^2 + \hat{\lambda}_2^2 \left(1 - \frac{n_s - 1}{\chi_{1-\alpha, n_s - 1}^2} \right)^2 \\
& + \frac{(m-1)^2 \hat{\sigma}_{WT}^4}{m^2} \left(1 - \frac{n_s(m-1)}{\chi_{1-\alpha, n_s(m-1)}^2} \right)^2 \\
& + \frac{(m-1)^2 \hat{\sigma}_{WR}^4}{m^2} \left(1 - \frac{n_s(m-1)}{\chi_{\alpha, n_s(m-1)}^2} \right)^2,
\end{aligned}$$

and $\hat{\lambda}_i$'s are same as those used for the test of noninferiority for intersubject variabilities. We then reject the null hypothesis at the α level of significance if $\hat{\eta}_U < 0$. Using a similar argument to the previous section, the power of the above testing procedure can be approximated by

$$1 - \Phi \left(z_\alpha - \frac{\sqrt{n_s}(\sigma_{TT}^2 - \delta^2 \sigma_{TR}^2)}{\sigma^*} \right),$$

where

$$\sigma^{*2} = 2 \left[\left(\sigma_{BT}^2 + \frac{\sigma_{WT}^2}{m} \right)^2 + \delta^4 \left(\sigma_{BR}^2 + \frac{\sigma_{WR}^2}{m} \right)^2 - 2\delta^2 \rho^2 \sigma_{BT}^2 \sigma_{BR}^2 \right.$$
$$\left. + \frac{(m-1)\sigma_{WT}^4}{m^2} + \frac{\delta^4(m-1)\sigma_{WR}^4}{m^2} \right].$$

Hence, the sample size needed to achieve the power of $1 - \beta$ at the α level of significance can be obtained by solving the following equation:

$$z_{\alpha/2} - \frac{\sqrt{n_s}(\sigma_{TT}^2 - \delta^2 \sigma_{TR}^2)}{\sigma^*} = z_\beta.$$

This leads to

$$n_s = \frac{\sigma^{*2}(z_{\alpha/2} + z_\beta)^2}{(\sigma_{TT}^2 - \delta^2 \sigma_{TR}^2)^2}.$$

9.4.4.3 An Example

Suppose a 2×4 crossover design (ABAB, BABA) is used to compare two treatments (A and B) in terms of their total variabilities. Information from pilot studies indicates that $\rho = 0.75$, $\sigma_{BT}^2 = 0.3$, $\sigma_{BR}^2 = 0.4$, $\sigma_{WT}^2 = 0.2$, and $\sigma_{WR}^2 = 0.3$. The objective is to establish noninferiority with a margin $\delta = 1.1$. It follows that

$$\sigma^{*2} = 2 \left[\left(0.30^2 + \frac{0.20^2}{2} \right)^2 + 1.1^4 \left(0.40^2 + \frac{0.30^2}{2} \right)^2 \right.$$
$$- 2 \times 1.1^2 \times (0.75^3 \times 0.3 \times 0.4^2)$$
$$\left. + \frac{0.20^4}{2^2} + \frac{1.1^4 \times 0.30^4}{2^2} \right]$$
$$= 0.106.$$

Hence, the sample size needed to achieve an 80% power ($1 - \beta = 0.80$) at the 5% level of significance ($\alpha = 0.05$) is given by

$$n_s = \frac{(0.106)(1.64 + 0.84)^2}{(0.3^2 + 0.2^2 - 1.1^2 \times (0.4^2 + 0.3^2))^2} \approx 22.$$

Since $n_s = n_1 + n_2 - 2$, approximately 12 subjects per sequence are required for achieving an 80% power at the 5% level of significance.

9.5 Practical Issues

In recent years, the assessment of reproducibility in terms of intrasubject variability or intrasubject CV in clinical research has received much attention. Shao and Chow (2002) defined reproducibility of a study drug as a collective term that encompasses consistency, similarity, and stability (control) within the therapeutic index (or window) of a subject's clinical status (e.g., clinical response of some primary study endpoint, blood levels, or blood concentration–time curve) when the study drug is repeatedly administered at different dosing periods under the same experimental conditions. Reproducibility of clinical results observed from a clinical study can be quantitated through the evaluation of the so-called reproducibility probability, which will be briefly introduced in Chapter 12 (see also Shao and Chow, 2002).

For the assessment of intersubject variability and/or total variability, Chow and Tse (1991) indicated that the usual analysis of variance models could lead to negative estimates of the variance components, especially the intersubject variance component. In addition, the sum of the best estimates of the intrasubject variance and the intersubject variance may not lead to the best estimate for the total variance. Chow and Shao (1988) proposed an estimation procedure for variance components, which will not only avoid negative estimates but also provide a better estimate as compared to the maximum likelihood estimates. For estimation of total variance, Chow and Tse (1991) proposed a method as an alternative to the sum of estimates of individual variance components. These ideas could be applied to provide a better estimate of sample sizes for studies comparing variabilities between treatment groups.

10

Bioequivalence Testing

When a brand name drug is going off-patent, generic drug companies may file abbreviated new drug applications for generic drug approval. An approved generic drug can be used as a substitute for the brand name drug. In 1984, the FDA was authorized to approve generic drugs through *bioavailability* and *bioequivalence* studies under the Drug Price and Patent Term Restoration Act. Bioequivalence testing is usually considered as a surrogate for clinical evaluation of drug products based on the *Fundamental Bioequivalence Assumption* that when two formulations of the reference product (e.g., a brand name drug) and the test product (a generic copy) are equivalent in bioavailability, they will reach the same therapeutic effect. *In vivo* bioequivalence testing is commonly conducted with a crossover design on healthy volunteers to assess bioavailability through PK responses such as area under the blood or plasma concentration–time curve (AUC) and maximum concentration (C_{max}). For some locally acting drug products such as nasal aerosols (e.g., metered-dose inhalers) and nasal sprays (e.g., metered-dose spray pumps) that are not intended to be absorbed into the bloodstream, bioavailability may be assessed by measurements intended to reflect the rate and extent to which the active ingredient or active moiety becomes available at the site of action. Bioequivalence related to these products is called *in vitro* bioequivalence and is usually studied under a parallel design. Statistical procedures for some types of bioequivalence studies are described in the FDA guidance (FDA, 2000, 2001). Chow and Shao (2002) provided a review of statistical procedures for bioequivalence studies that are not provided by the FDA.

Most recently, for the assessment of biosimilar products, the FDA recommends that a stepwise approach be used for obtaining the totality of the evidence for the demonstration of biosimilarity between a proposed biosimilar product and an originator product (FDA, 2015). The stepwise approach starts with analytical studies for functional and structural characterization of critical quality attributes (CQAs) that are relevant to clinical outcomes at various stages of the manufacturing process followed by animal studies for the assessment of toxicity, clinical pharmacology PK or pharmacodynamics (PD) studies, and clinical studies for the assessment of immunogenicity, safety/tolerability, and efficacy (Christl, 2015; Tsong, 2015). For the analytical studies, FDA suggests that CQAs should be identified and classified into three tiers according to their criticality or risk ranking based on the mechanism of action (MOA) or PK using appropriate statistical models or methods. CQAs with CQAs that are most relevant to clinical outcomes will be classified to Tier 1, while CQAs that are less (mild-to-moderate) or least relevant to clinical outcomes will be classified to Tier 2 and Tier 3, respectively. Sample size requirement for analytical similarity assessment for CQAs in Tier 1 has received much attention because the similarity margin is not only data-driven but also depending upon the selection of the reference lots from the pool of the available lots.

In Section 10.1, we introduce various bioequivalence criteria. Section 10.2 introduces sample size calculation for the average bioequivalence. Sample size formulas for population bioequivalence and individual bioequivalence are provided in Sections 10.3 and 10.4, respectively. Section 10.5 focuses on sample size calculation for *in vitro* bioequivalence.

Sample size requirement for similarity assessment of CQAs in analytical studies for the demonstration of a proposed biosimilar product as compared to its originator product is discussed in the last section of this chapter for completeness.

10.1 Bioequivalence Criteria

In 1992, the FDA published its first guidance on statistical procedures for *in vivo* bioequivalence studies (FDA, 1992). The 1992 FDA guidance requires that the evidence of bioequivalence in average bioavailability in PK responses between the two drug products be provided. Let y_R and y_T denote PK responses (or log-PK responses if appropriate) of the reference and test formulations, respectively, and let $\delta = E(y_T) - E(y_R)$. Under the 1992 FDA guidance, two formulations are said to be bioequivalent if δ falls in the interval (δ_L, δ_U) with 95% assurance, where δ_L and δ_U are given limits specified in the FDA guidance. Since only the averages $E(y_T)$ and $E(y_R)$ are concerned in this method, this type of bioequivalence is usually referred to as *average bioequivalence* (ABE). In 2000, the FDA issued a guidance on general considerations of bioavailability and bioequivalence studies for orally administered drug products, which replaces the 1992 FDA guidance (FDA, 2000). Statistical design and analysis for the assessment of ABE as described in the 2000 FDA guidance are the same as those given in the 1992 FDA guidance.

The ABE approach for bioequivalence, however, has limitations for addressing drug interchangeability, since it focuses only on the comparison of population averages between the test and reference formulations (Chen, 1997a). Drug interchangeability can be classified as either drug prescribability or drug switchability. Drug prescribability is referred to as the physician's choice for prescribing an appropriate drug for his/her new patients among the drug products available, while drug switchability is related to the switch from a drug product to an alternative drug product within the same patient whose concentration of the drug product has been titrated to a steady, efficacious, and safe level. To assess drug prescribability and switchability, *population bioequivalence* (PBE) and *individual bioequivalence* (IBE) are proposed, respectively (see Anderson and Hauck, 1990; Sheiner, 1992; Schall and Luus, 1993; Esinhart and Chinchilli, 1994; Chow and Liu, 1995; Chen, 1997a). The concepts of PBE and IBE are described in the 1999 FDA draft guidance (FDA, 1999a) and the 2001 FDA guidance for industry (FDA, 2001). Let y_T be the PK response from the test formulation, y_R and y'_R be two identically distributed PK responses from the reference formulation, and

$$\theta = \frac{E(y_R - y_T)^2 - E(y_R - y'_R)^2}{\max\left\{\sigma_0^2, E(y_R - y'_R)^2/2\right\}},$$ (10.1)

where σ_0^2 is a given constant specified in the 2001 FDA guidance. If y_R, y'_R, and y_T are independent observations from different subjects, then the two formulations are PBE when $\theta < \theta_{PBE}$, where θ_{PBE} is an equivalence limit for the assessment of PBE as specified in the 2001 FDA guidance. If y_R, y'_R, and y_T are from the same subject ($E(y_R - y'_R)^2/2$ is then the within-subject variance), then the two formulations are IBE when $\theta < \theta_{IBE}$, where θ_{IBE} is an equivalence limit for IBE as specified in the 2001 FDA guidance. Note that θ in Equation 10.1 is a measure of the relative difference between the mean squared errors of $y_R - y_T$ and

$y_R - y'_R$. When y_R, y'_R, and y_T are from the same individual, it measures the drug switchability within the same individual. On the other hand, it measures drug prescribability when y_R, y'_R, and y_T are from different subjects. Thus, IBE addresses drug switchability, whereas PBE addresses drug prescribability. According to the 2001 FDA guidance, IBE or PBE can be claimed if a 95% upper confidence bound for θ is smaller than θ_{IBE} or θ_{PBE}, provided that the observed ratio of geometric means is within the limits of 80% and 125%.

For locally acting drug products such as nasal aerosols (e.g., metered-dose inhalers) and nasal sprays (e.g., metered-dose spray pumps) that are not intended to be absorbed into the bloodstream, the FDA indicates that bioequivalence may be assessed, with suitable justification, by *in vitro* bioequivalence studies alone (21 CFR 320.24). In the 1999 FDA guidance, *in vitro* bioequivalence can be established through six *in vitro* bioequivalence tests, which are for dose or spray content uniformity through container life, droplet or particle size distribution, spray pattern, plume geometry, priming and repriming, and tail off distribution. The FDA classifies statistical methods for the assessment of the six *in vitro* bioequivalence tests for nasal aerosols and sprays as either the nonprofile analysis or the profile analysis. For the nonprofile analysis, the FDA adopts the criterion and limit of the PBE. For the profile analysis, bioequivalence may be assessed by comparing the profile variation between test product and reference product bottles with the profile variation between reference product bottles.

10.2 Average Bioequivalence

It should be noted that testing ABE is a special case of testing equivalence. As a result, the formulas derived in Chapter 3 for testing equivalence under various designs are still valid for testing ABE. In practice, the most commonly used design for ABE is a standard two-sequence and two-period crossover design. Hence, for the sake of convenience, the sample size formula for ABE under such a design is presented here. Details regarding more general designs can be found in Chapter 3.

For the ABE, a standard two-sequence, two-period (2×2) crossover design is recommended by the FDA guidance. In a standard 2×2 crossover design, subjects are randomly assigned to one of the two sequences of formulations. In the first sequence, n_1 subjects receive treatments in the order of TR (T = test formulation, R = reference formulation) at two different dosing periods, whereas in the second sequence, n_2 subjects receive treatments in the order of RT at two different dosing periods. A sufficient length of washout between dosing periods is usually applied to wear off the possible residual effect that may be carried over from one dosing period to the next. Let y_{ijk} be the original or the log-transformation of the PK response of interest from the ith subject in the kth sequence at the jth dosing period. The following statistical model is considered:

$$y_{ijk} = \mu + F_l + P_j + Q_k + S_{ikl} + e_{ijk}, \tag{10.2}$$

where μ is the overall mean; P_j is the fixed effect of the jth period ($j = 1, 2$, and $P_1 + P_2 = 0$); Q_k is the fixed effect of the kth sequence ($k = 1, 2$, and $Q_1 + Q_2 = 0$); F_l is the fixed effect of the lth formulation (when $j = k$, $l = T$; when $j \neq k$, $l = R$; $F_T + F_R = 0$); S_{ikl} is the random effect of the ith subject in the kth sequence under formulation l and (S_{ikT}, S_{ikR}), $i = 1, \ldots, n_k$,

$k = 1, 2$, are independent and identically distributed bivariate normal random vectors with mean 0 and an unknown covariance matrix

$$
\begin{pmatrix}
\sigma_{BT}^2 & \rho\sigma_{BT}\sigma_{BR} \\
\rho\sigma_{BT}\sigma_{BR} & \sigma_{BR}^2
\end{pmatrix};
$$

e_{ijk}'s are independent random errors distributed as $N(0, \sigma_{Wl}^2)$; and S_{ikl}'s and e_{ijk}'s are mutually independent. Note that σ_{BT}^2 and σ_{BR}^2 are between-subject variances and σ_{WT}^2 and σ_{WR}^2 are within-subject variances, and that $\sigma_{TT}^2 = \sigma_{BT}^2 + \sigma_{WT}^2$ and $\sigma_{TR}^2 = \sigma_{BT}^2 + \sigma_{WR}^2$ are the total variances for the test and reference formulations, respectively. Under model (10.2), the ABE parameter δ defined in Section 10.1 is equal to $\delta = FT - FR$. According to the 2000 FDA guidance, ABE is claimed if the following null hypothesis H_0 is rejected at the 5% level of significance:

$$
H_0 : \delta \leq \delta_L \text{ or } \delta \geq \delta_U \quad \text{versus} \quad H_1 : \delta_L < \delta < \delta_U, \tag{10.3}
$$

where δ_L and δ_U are given bioequivalence limits. Under model (10.2),

$$
\hat{\delta} = \frac{\bar{y}_{11} - \bar{y}_{12} - \bar{y}_{21} + \bar{y}_{22}}{2} \sim N\left(\delta, \frac{\sigma_{1,1}^2}{4}\left(\frac{1}{n_1} + \frac{1}{n_2}\right)\right), \tag{10.4}
$$

where $\bar{y}_{jk}$ is the sample mean of the observations in the kth sequence at the jth period and $\sigma_{1,1}^2$ is

$$
\sigma_{a,b}^2 = \sigma_D^2 + a\sigma_{WT}^2 + b\sigma_{WR}^2 \tag{10.5}
$$

with $a = 1$ and $b = 1$. Let

$$
\hat{\sigma}_{1,1}^2 = \frac{1}{n_1 + n_2 - 2}\sum_{k=1}^{2}\sum_{i=1}^{n_k}(y_{i1k} - y_{i2k} - \bar{y}_{1k} - \bar{y}_{2k})^2. \tag{10.6}
$$

Then, $\hat{\sigma}_{1,1}^2$ is independent of $\hat{\delta}$ and

$$
(n_1 + n_2 - 2)\hat{\sigma}_{1,1}^2 \sim \sigma_{1,1}^2\chi_{n_1+n_2-2}^2,
$$

where χ_r^2 is the chi-square distribution with r degrees of freedom. Thus, the limits of a 90% confidence interval for δ are given by

$$
\hat{\delta}_{\pm} = \hat{\delta} \pm t_{0.05,n_1+n_2-2}\frac{\hat{\sigma}_{1,1}}{2}\sqrt{\frac{1}{n_1} + \frac{1}{n_2}},
$$

where $t_{0.05,r}$ is the upper 5th quantile of the t-distribution with r degrees of freedom. According to the 2000 FDA guidance, ABE can be claimed if and only if the 90% confidence interval falls within $(-\delta_L, \delta_U)$, that is, $\delta_L < \hat{\delta}_- < \hat{\delta}_+ < \delta_U$. Note that this is based on the two one-sided tests procedure proposed by Schuirmann (1987). The idea of Schuirmann's two one-sided tests is to decompose H_0 in Equation 10.3 into the following two one-sided hypotheses:

$$H_{01} : \delta \leq \delta_L \quad \text{and} \quad H_{02} : \delta \geq \delta_U$$

Apparently, both H_{01} and H_{02} are rejected at the 5% significance level if and only if $\delta_L < \hat{\delta}_- < \hat{\delta}_+ < \delta_U$. Schuirmann's two one-sided tests procedure is a test of size 5% (Berger and Hsu, 1996, Theorem 2).

Assume without loss of generality that $n_1 = n_2 = n$. Under the alternative hypothesis that $|\epsilon| < \delta$, the power of the above test is approximately

$$2\Phi\left(\frac{\sqrt{2n}(\delta - |\varepsilon|)}{\sigma_{1,1}} - t_{\alpha,2n-2}\right) - 1.$$

As a result, the sample size needed for achieving a power of $1 - \beta$ can be obtained by solving

$$\frac{\sqrt{2n}(\delta - |\varepsilon|)}{\sigma_{1,1}} - t_{\alpha,2n-2} = t_{\beta/2,2n-2}.$$

This leads to

$$n \geq \frac{(t_{\alpha,2n-2} + t_{\beta/2,2n-2})^2 \sigma_{1,1}^2}{2(\delta - |\varepsilon|)^2}. \tag{10.7}$$

Since the above equations do not have an explicit solution, for convenience, for 2×2 crossover design, the total sample size needed to achieve a power of 80% or 90% at 5% level of significance with various combinations of ε and δ is given in Table 10.1.

When sample size is sufficiently large, Equation 10.7 can be further simplified into

$$n = \frac{(z_\alpha + z_{\beta/2})^2 \sigma_{1,1}^2}{2(\delta - |\varepsilon|)^2}.$$

10.2.1 An Example

Suppose an investigator is interested in conducting a clinical trial with 2×2 crossover design to establish ABE between an inhaled formulation of a drug product (test) and a subcutaneous (SC) injected formulation (reference) in terms of log-transformed AUC. Based on PK data obtained from pilot studies, the mean difference of AUC can be assumed to be 5% ($\delta = 0.05$). Also, it is assumed that the standard deviation for intrasubject comparison is 0.40. By referring to Table 10.1, a total of 24 subjects per sequence is needed to achieve an

TABLE 10.1

Sample Size for Assessment of Equivalence under a 2×2 Crossover Design ($m = 1$)

		Power = 80%				Power = 90%			
	$\varepsilon =$	0%	5%	10%	15%	0%	5%	10%	15%
$\sigma_{1,1} =$	0.10	3	3	4	9	3	4	5	12
	0.12	3	4	6	13	3	4	7	16
	0.14	3	4	7	17	4	5	9	21
	0.16	4	5	9	22	4	6	11	27
	0.18	4	6	11	27	5	7	13	34
	0.20	5	7	13	33	6	9	16	42
	0.22	6	8	15	40	7	10	19	51
	0.24	6	10	18	48	8	12	22	60
	0.26	7	11	20	56	9	14	26	70
	0.28	8	13	24	64	10	16	29	81
	0.30	9	14	27	74	11	18	34	93
	0.32	10	16	30	84	13	20	38	105
	0.34	11	18	34	94	14	22	43	119
	0.36	13	20	38	105	16	25	48	133
	0.38	14	22	42	117	17	28	53	148
	0.40	15	24	47	130	19	30	59	164

Note: (1) The bioequivalence limit δ is 22.3%. (2) Sample size calculation was performed based on log-transformed data.

80% power at the 5% level of significance. On the other side, if we use normal approximation, the sample size needed can be obtained as

$$n = \frac{(z_{0.05} + z_{0.10})^2 \sigma_{1,1}^2}{2(\delta - |\varepsilon|)^2} = \frac{(1.96 + 0.84)^2 \times 0.40^2}{2(0.223 - 0.05)^2} \approx 21.$$

10.3 Population Bioequivalence

PBE can be assessed under the 2×2 crossover design described in Section 10.2. Under model (10.2), the parameter θ in Equation 10.1 for PBE is equal to

$$\theta = \frac{\delta^2 + \sigma_{TT}^2 - \sigma_{TR}^2}{\max\left\{\sigma_0^2, \sigma_{TR}^2\right\}}. \tag{10.8}$$

In view of Equation 10.8, PBE can be claimed if the null hypothesis in

$$H_0 : \lambda \geq 0 \quad \text{versus} \quad H_1 : \lambda < 0$$

is rejected at the 5% significance level provided that the observed ratio of geometric means is within the limits of 80% and 125%, where

$$\lambda = \delta^2 + \sigma_{TT}^2 - \sigma_{TR}^2 - \theta_{PBE} \max\left\{\sigma_0^2, \sigma_{TR}^2\right\},$$

and θ_{PBE} is a constant specified in FDA (2001).

Under model (10.2), an unbiased estimator of δ is $\hat{\delta}$ given in Equation 10.4. Commonly used unbiased estimators of σ_{TT}^2 and σ_{TR}^2 are, respectively,

$$\hat{\sigma}_{TT}^2 = \frac{1}{n_1 + n_2 - 2}\left[\sum_{i=1}^{n_1}(y_{i11} - \bar{y}_{11})^2 + \sum_{i=1}^{n_2}(y_{i22} - \bar{y}_{22})^2\right]$$
$$\sim \frac{\sigma_{TT}^2 \chi_{n_1+n_2-2}^2}{n_1 + n_2 - 2}$$

and

$$\hat{\sigma}_{TR}^2 = \frac{1}{n_1 + n_2 - 2}\left[\sum_{i=1}^{n_1}(y_{i21} - \bar{y}_{21})^2 + \sum_{i=1}^{n_2}(y_{i12} - \bar{y}_{12})^2\right]$$
$$\sim \frac{\sigma_{TR}^2 \chi_{n_1+n_2-2}^2}{n_1 + n_2 - 2}.$$

Applying linearization to the moment estimator

$$\hat{\lambda} = \hat{\delta}^2 + \hat{\sigma}_{TT}^2 - \hat{\sigma}_{TR}^2 - \theta_{PBE} \max\left\{\sigma_0^2, \hat{\sigma}_{TR}^2\right\},$$

Chow et al. (2003a) obtained the following approximate 95% upper confidence bound for λ. When $\sigma_{TR}^2 \geq \sigma_0^2$,

$$\hat{\lambda}_U = \hat{\delta}^2 + \hat{\sigma}_{TT}^2 - (1 + \theta_{PBE})\hat{\sigma}_{TR}^2 + t_{0.05, n_1+n_2-2}\sqrt{V}, \tag{10.9}$$

where V is an estimated variance of $\hat{\delta}^2 + \hat{\sigma}_{TT}^2 - (1 + \theta_{PBE})\hat{\sigma}_{TR}^2$ of the form

$$V = \left(2\hat{\delta}, 1, -(1 + \theta_{PBE})\right)C\left(2\hat{\delta}, 1, -(1 + \theta_{PBE})\right)',$$

and C is an estimated variance–covariance matrix of $\left(\hat{\delta}, +\hat{\sigma}_{TT}^2, \hat{\sigma}_{TR}^2\right)$. Since $\hat{\delta}$ and $\left(\hat{\sigma}_{TT}^2, \hat{\sigma}_{TR}^2\right)$ are independent,

$$C = \begin{pmatrix} \left(\hat{\sigma}_{1,1}^2/4\right)\left((1/n_1) + (1/n_2)\right) & (0,0) \\ (0,0)' & \frac{(n_1-1)C_1}{(n_1+n_2-2)^2} + \frac{(n_2-1)C_2}{(n_1+n_2-2)^2} \end{pmatrix},$$

where $\hat{\sigma}_{1,1}^2$ is defined by Equation 10.6, C_1 is the sample covariance matrix of $((y_{i11} - \bar{y}_{11})^2, (y_{i21} - \bar{y}_{21})^2), i = 1, \ldots, n_1$ and C_2 is the sample covariance matrix of $((y_{i22} - \bar{y}_{22})^2, (y_{i12} - \bar{y}_{12})^2), i = 1, \ldots, n_2$.

When $\sigma_{TR}^2 < \sigma_0^2$, the upper confidence bound for λ should be modified to

$$\hat{\lambda}_U = \hat{\delta}^2 + \hat{\sigma}_{TT}^2 - \hat{\sigma}_{TR}^2 - \theta_{PBE}\sigma_0^2 + t_{0.05,n_1+n_2-2}\sqrt{V_0}, \tag{10.10}$$

where

$$V_0 = (2\hat{\delta},1,-1)C(2\hat{\delta},1,-1)'.$$

The confidence bound $\hat{\lambda}_U$ in Equation 10.9 is referred to as the confidence bound under the reference-scaled criterion, whereas $\hat{\lambda}_U$ in Equation 10.10 is referred to as the confidence bound under the constant-scaled criterion. In practice, whether $\sigma_{TR}^2 \geq \sigma_0^2$ is usually unknown. Hyslop, Hsuan, and Holder (2000) recommend using the reference-scaled criterion or the constant-scaled criterion according to $\hat{\sigma}_{TR}^2 \geq \sigma_0^2$ or $\hat{\sigma}_{TR}^2 < \sigma_0^2$ which is referred to as the estimation method. Alternatively, we may test the hypothesis of $\sigma_{TR}^2 \geq \sigma_0^2$ versus $\sigma_{TR}^2 < \sigma_0^2$ to decide which confidence bound should be used; that is, if $\hat{\sigma}_{TR}^2(n_1 + n_2 - 2) \geq \sigma_0^2 \, \chi_{0.95,n_1+n_2-2}^2$, then $\hat{\lambda}_U$ in Equation 10.9 should be used; otherwise $\hat{\lambda}_U$ in Equation 10.10 should be used, where $\chi_{\alpha,r}^2$ denotes the αth upper quantile of the chi-square distribution with r degrees of freedom. This is referred to as the test method and is more conservative than the estimation method.

Based on an asymptotic analysis, Chow, Shao, and Wang (2003a) derived the following formula for sample size determination assuming that $n_1 = n_2 = n$:

$$n \geq \frac{\zeta(z_{0.05} + z_\beta)^2}{\lambda^2} \tag{10.11}$$

where

$$\zeta = 2\delta^2\sigma_{1,1}^2 + \sigma_{TT}^4 + (1+a)^2\sigma_{TR}^4 - 2(1+a)\rho^2\sigma_{BT}^2\sigma_{BR}^2,$$

δ, $\sigma_{1,1}^2$, σ_{TT}^2, σ_{TR}^2, σ_{BT}^2, σ_{BR}^2, and ρ are given initial values, z_t is the upper tth quantile of the standard normal distribution, $1 - \beta$ is the desired power, $a = \theta_{PBE}$ if $\sigma_{TR} \geq \sigma_0$, and $a = 0$ if $\sigma_{TR} < \sigma_0$.

Sample sizes n selected using Equation 10.11 with $1 - \beta = 80\%$ and the power P_n of the PBE test based on 10,000 simulations (assuming that the initial parameter values are the true parameter values) are listed in Table 10.2. It can be seen from Table 10.2 that the actual power P_n corresponding to each selected n is larger than the target value of 80%, although the sample size obtained by formula (10.11) is conservative since P_n is much larger than 80% in some cases.

10.3.1 An Example

Suppose an investigator is interested in conducting a clinical trial with 2×2 crossover design to establish PBE between an inhaled formulation of a drug product (test) and an SC injected formulation (reference) in terms of log-transformed AUC. Based on PK data obtained from pilot studies, the mean difference of AUC can be assumed to be 5% ($\delta = 0.00$). Also, it is assumed that the intersubject variability under the test and the reference are given by 0.40 and 0.40, respectively. The intersubject correlation coefficient (ρ) is

TABLE 10.2

Sample Size n Selected Using Equation 10.11 with $1 - \beta = 80\%$ and the Power P_n of the PBE Test Based on 10,000 Simulations

Parameter						$\rho = 0.75$		$\rho = 1$	
σ_{BT}	σ_{BR}	σ_{WT}	σ_{WR}	δ	λ	n	P_n	n	P_n
0.1	0.1	0.1	0.4	0.4726	−0.2233	37	0.8447	36	0.8337
				0.4227	−0.2679	24	0.8448	24	0.8567
				0.2989	−0.3573	12	0.8959	12	0.9035
				0.0000	−0.4466	7	0.9863	7	0.9868
0.1	0.4	0.1	0.1	0.4726	−0.2233	34	0.8492	32	0.8560
				0.3660	−0.3127	16	0.8985	15	0.8970
				0.2113	−0.4020	9	0.9560	8	0.9494
0.1	0.1	0.4	0.4	0.2983	−0.2076	44	0.8123	43	0.8069
				0.1722	−0.2670	23	0.8381	23	0.8337
				0.0000	−0.2966	17	0.8502	17	0.8531
0.1	0.4	0.4	0.4	0.5323	−0.4250	36	0.8305	35	0.8290
				0.4610	−0.4958	25	0.8462	24	0.8418
				0.2661	−0.6375	13	0.8826	13	0.8872
				0.0000	−0.7083	10	0.9318	10	0.9413
0.1	0.4	0.6	0.4	0.3189	−0.4066	39	0.8253	38	0.8131
				0.2255	−0.4575	29	0.8358	28	0.8273
				0.0000	−0.5083	22	0.8484	22	0.8562
0.1	0.4	0.6	0.6	0.6503	−0.6344	44	0.8186	44	0.8212
				0.4598	−0.8459	22	0.8424	22	0.8500
				0.3252	−0.9515	16	0.8615	16	0.8689
				0.0000	−1.057	12	0.8965	12	0.9000
0.4	0.4	0.1	0.1	0.3445	−0.1779	37	0.8447	22	0.8983
				0.2436	−0.2373	20	0.8801	12	0.9461
				0.1722	−0.2670	15	0.8951	9	0.9609
				0.0000	−0.2966	12	0.9252	7	0.9853
0.4	0.4	0.1	0.4	0.5915	−0.3542	44	0.8354	38	0.8481
				0.4610	−0.4958	21	0.8740	18	0.8851
				0.2661	−0.6375	12	0.9329	11	0.9306
				0.0000	−0.7083	9	0.9622	9	0.9698
0.4	0.4	0.6	0.4	0.0000	−0.3583	46	0.8171	43	0.8213
0.4	0.4	0.6	0.6	0.5217	−0.6351	41	0.8246	39	0.8252
				0.3012	−0.8166	22	0.8437	21	0.8509
				0.0000	−0.9073	17	0.8711	16	0.8755
0.4	0.6	0.4	0.4	0.6655	−0.6644	33	0.8374	30	0.8570
				0.5764	−0.7751	23	0.8499	21	0.8709
				0.3328	−0.9965	13	0.9062	12	0.9258
				0.0000	−1.107	10	0.9393	9	0.9488
0.4	0.6	0.4	0.6	0.9100	−0.8282	45	0.8403	42	0.8447
				0.7049	−1.159	21	0.8684	20	0.8874
				0.4070	−1.491	11	0.9081	11	0.9295
				0.0000	−1.656	9	0.9608	8	0.9577

(Continued)

TABLE 10.2 (*Continued*)

Sample Size n Selected Using Equation 10.11 with $1-\beta = 80\%$ and the Power P_n of the PBE Test Based on 10,000 Simulations

Parameter						$\rho = 0.75$		$\rho = 1$	
σ_{BT}	σ_{BR}	σ_{WT}	σ_{WR}	δ	λ	N	P_n	n	P_n
0.6	0.4	0.1	0.4	0.3905	−0.3558	41	0.8334	32	0.8494
				0.3189	−0.4066	30	0.8413	24	0.8649
				0.2255	−0.4575	23	0.8584	18	0.8822
				0.0000	−0.3583	17	0.8661	14	0.9009
0.6	0.4	0.4	0.4	0.0000	−0.3583	42	0.8297	35	0.8403
0.6	0.6	0.1	0.4	0.7271	−0.5286	47	0.8335	36	0.8584
				0.5632	−0.7401	23	0.8785	18	0.9046
				0.3252	−0.9515	13	0.9221	10	0.9474
				0.0000	−1.057	10	0.9476	8	0.9780
0.6	0.6	0.4	0.4	0.6024	−0.5444	47	0.8246	38	0.8455
				0.3012	−0.8166	19	0.8804	15	0.8879
				0.0000	−0.9073	14	0.8903	12	0.9147

assumed to be 0.75. It is further assumed that the intrasubject variability under the test and the reference are given by 0.10 and 0.10, respectively. The sample size needed to achieve an 80% power at the 5% level of significance is given by 12 subjects per sequence according to Table 10.2.

10.4 Individual Bioequivalence

For the IBE, the standard 2×2 crossover design is not useful because each subject receives each formulation only once and, hence, it is not possible to obtain unbiased estimators of within-subject variances. To obtain unbiased estimators of the within-subject variances, FDA (2001) suggested that the following 2×4 crossover design be used. In the first sequence, n_1 subjects receive treatments at four periods in the order of TRTR (or TRRT), while in the second sequence, n_2 subjects receive treatments at four periods in the order of RTRT (or RTTR). Let y_{ijk} be the observed response (or log-response) of the ith subject in the kth sequence at jth period, where $i = 1, \dots, n_k, j = 1, \dots, 4, k = 1, 2$. The following statistical model is assumed:

$$y_{ijk} = \mu + F_l + W_{ljk} + S_{ikl} + e_{ijk}, \tag{10.12}$$

where μ is the overall mean; F_l is the fixed effect of the lth formulation ($l = T, R$ and $F_T + F_R = 0$); W_{ljk}'s are fixed period, sequence, and interaction effects ($\Sigma_k \bar{W}_{lk} = 0$, where $\bar{W}_{lk}$ is the average of W_{ljk}'s with fixed (l, k), $l = T, R$); and S_{ikl}'s and e_{ijk}'s are similarly defined as those in Equation 10.2. Under model (10.12), θ in Equation 10.1 for IBE is equal to

$$\theta = \frac{\delta^2 + \sigma_D^2 + \sigma_{WT}^2 - \sigma_{WR}^2}{\max\left\{\sigma_0^2, \sigma_{WR}^2\right\}}, \tag{10.13}$$

where $\sigma_D^2 = \sigma_{BT}^2 + \sigma_{BR}^2 - 2\rho\sigma_{BT}\rho\sigma_{BR}$ is the variance of $S_{ikT} - S_{ikR}$, which is referred to as the variance due to the subject-by-formulation interaction. Then, IBE is claimed if the null hypothesis H_0: $\theta \geq \theta_{IBE}$ is rejected at the 5% level of significance provided that the observed ratio of geometric means is within the limits of 80% and 125%, where θ_{IBE} is the IBE limit specified in the 2001 FDA guidance. From Equation 10.13, we need a 5% level test for

$$H_0 : \gamma \geq 0 \quad \text{versus} \quad H_1 : \gamma < 0,$$

where

$$\gamma = \delta^2 + \sigma_D^2 + \sigma_{WT}^2 - \sigma_{WR}^2 - \theta_{IBE}\max\left\{\sigma_0^2, \sigma_{WR}^2\right\}.$$

Therefore, it suffices to find a 95% upper confidence bound $\hat{\gamma}_U$ for γ. IBE is concluded if $\hat{\gamma}_U < 0$.

The confidence bound $\hat{\gamma}_U$ recommended in FDA (2001) is proposed by Hyslop, Hsuan, and Holder (2000), which can be described as follows. For subject i in sequence k, let x_{ilk} and z_{ilk} be the average and the difference, respectively, of two observations from formulation l, and let $\bar{x}_{lk}$ and $\bar{z}_{lk}$ be, respectively, the sample mean based on x_{ilk}'s and z_{ilk}'s. Under model (10.12), an unbiased estimator of δ is

$$\hat{\delta} = \frac{\bar{x}_{T1} - \bar{x}_{R1} + \bar{x}_{T2} - \bar{x}_{R2}}{2} \sim N\left(\delta, \frac{\sigma_{0.5,0.5}^2}{4}\left(\frac{1}{n_1} + \frac{1}{n_2}\right)\right);$$

an unbiased estimator of $\sigma_{0.5,0.5}^2$ is

$$\sigma_{0.5,0.5}^2 = \frac{(n_1-1)s_{d1}^2 + (n_2-1)s_{d2}^2}{n_1 + n_2 - 2} \sim \frac{\sigma_{0.5,0.5}^2 \chi_{n_1+n_2-2}^2}{n_1 + n_2 - 2},$$

where s_{dk}^2 is the sample variance based on $x_{iTk} - x_{iRk}$, $i = 1, \ldots, n_k$; an unbiased estimator of σ_{WT}^2 is

$$\hat{\sigma}_{WT}^2 = \frac{(n_1-1)s_{T1}^2 + (n_2-1)s_{T2}^2}{2(n_1 + n_2 - 2)} \sim \frac{\sigma_{WT}^2 \chi_{n_1+n_2-2}^2}{n_1 + n_2 - 2},$$

where s_{Tk}^2 is the sample variance based on z_{iTk}, $i = 1, \ldots, n_k$; and an unbiased estimator of σ_{WR}^2 is

$$\hat{\sigma}_{WR}^2 = \frac{(n_1-1)s_{R1}^2 + (n_2-1)s_{R2}^2}{2(n_1 + n_2 - 2)} \sim \frac{\sigma_{WR}^2 \chi_{n_1+n_2-2}^2}{n_1 + n_2 - 2},$$

where s_{Rk}^2 is the sample variance based on z_{iRk}, $i = 1, \ldots, n_k$. Furthermore, estimators $\hat{\delta}, \hat{\sigma}_{0.5,0.5}^2, \hat{\sigma}_{WT}^2$, and $\hat{\sigma}_{WR}^2$ are independent. When $\sigma_{WR}^2 \geq \sigma_0^2$, an approximate 95% upper confidence bound for γ is

$$\hat{\gamma}_U = \hat{\delta}^2 + \hat{\sigma}_{0.5,0.5}^2 + 0.5\hat{\sigma}_{WT}^2 - (1.5 + \theta_{IBE})\hat{\sigma}_{WR}^2 \sqrt{U}, \tag{10.14}$$

where U is the sum of the following four quantities:

$$\left[\left(|\hat{\delta}| + t_{0.05,n_1+n_2-2}\,\frac{\hat{\sigma}_{0.5,0.5}}{2}\sqrt{\frac{1}{n_1}+\frac{1}{n_2}}\right)^2 - \hat{\delta}^2\right]^2,$$

$$\hat{\sigma}_{0.5,0.5}^4\left(\frac{n_1+n_2-2}{\chi_{0.95,n_1+n_2-2}^2}-1\right)^2,$$

$$0.5^2\hat{\sigma}_{WT}^4\left(\frac{n_1+n_2-2}{\chi_{0.95,n_1+n_2-2}^2}-1\right)^2,$$

and

$$(1.5+\theta_{IBE})^2\;\hat{\sigma}_{WR}^4\left(\frac{n_1+n_2-2}{\chi_{0.05,n_1+n_2-2}^2}-1\right)^2. \tag{10.15}$$

When $\sigma_0^2 > \sigma_{WR}^2$, an approximate 95% upper confidence bound for γ is

$$\hat{\gamma}_U = \hat{\delta}^2 + \hat{\sigma}_{0.5,0.5}^2 + 0.5\hat{\sigma}_{WT}^2 - 1.5\hat{\sigma}_{WR}^2 - \theta_{IBE}\sigma_0^2 + \sqrt{U_0}, \tag{10.16}$$

where U_0 is the same as U except that the quantity in Equation 10.15 should be replaced by

$$1.5^2\hat{\sigma}_{WR}^2\left(\frac{n_1+n_2-2}{\chi_{0.05,n_1+n_2-2}^2}-1\right)^2.$$

The estimation or test method for PBE described in Section 10.3 can be applied to decide whether the reference-scaled bound $\hat{\gamma}_U$ in Equation 10.14 or the constant-scaled bound $\hat{\gamma}_U$ in Equation 10.16 should be used.

Although the 2×2 crossover design and the 2×4 crossover design have the same number of subjects, the 2×4 crossover design yields four observations, instead of two, from each subject. This may increase the overall cost substantially. As an alternative to the 2×4 crossover design, Chow, Shao, and Wang (2002) recommended a 2×3 extra-reference design, in which n_1 subjects in sequence 1 receive treatments at three periods in the order of TRR, while n_2 subjects in sequence 2 receive treatments at three periods in the order of RTR. The statistical model under this design is still given by Equation 10.12. An unbiased estimator of δ is

$$\hat{\delta} = \frac{\bar{x}_{T1} - \bar{x}_{R1} + \bar{x}_{T2} - \bar{x}_{R2}}{2} \sim N\left(\delta, \frac{\sigma_{1,0.5}^2}{4}\left(\frac{1}{n_1}+\frac{1}{n_2}\right)\right),$$

where $\sigma_{a,b}^2$ is given by Equation 10.5; an unbiased estimator of $\sigma_{1,0.5}^2$ is

$$\hat{\sigma}_{1,0.5}^2 = \frac{(n_1 - 1)s_{d1}^2 + (n_2 - 1)s_{d2}^2}{n_1 + n_2 - 2} \sim \frac{\sigma_{1,0.5}^2 \chi_{n_1+n_2-2}^2}{n_1 + n_2 - 2};$$

an unbiased estimator of σ_{WR}^2 is

$$\hat{\sigma}_{WR}^2 = \frac{(n_1 - 1)s_{R1}^2 + (n_2 - 1)s_{R2}^2}{2(n_1 + n_2 - 2)} \sim \frac{\sigma_{WR}^2 \chi_{n_1+n_2-2}^2}{n_1 + n_2 - 2};$$

and estimators $\hat{\delta}, \hat{\sigma}_{1,0.5}^2$, and $\hat{\sigma}_{WR}^2$ are independent, since $x_{iT1} - x_{iR1}$, $x_{iT2} - x_{iR2}$, z_{iR1}, and z_{iR2} are independent. Chow et al. (2002) obtained the following approximate 95% upper confidence bound for γ. When $\hat{\sigma}_{WR}^2 \geq \sigma_0^2$,

$$\hat{\gamma}_U = \hat{\delta}^2 + \hat{\sigma}_{1,0.5}^2 (1.5 + \theta_{IBE}) \hat{\sigma}_{WR}^2 + \sqrt{U},$$

where U is the sum of the following three quantities:

$$\left[\left(|\hat{\delta}| + t_{0.05,n_1+n_2-2} \frac{\hat{\sigma}_{1,0.5}}{2} \sqrt{\frac{1}{n_1} + \frac{1}{n_2}} \right)^2 - \hat{\delta}^2 \right]^2,$$

$$\hat{\sigma}_{1,0.5}^4 \left(\frac{n_1 + n_2 - 2}{\chi_{0.95,n_1+n_2-2}^2} - 1 \right)^2,$$

and

$$(1.5 + \theta_{IBE})^2 \hat{\sigma}_{WR}^4 \left(\frac{n_1 + n_2 - 2}{\chi_{0.05,n_1+n_2-2}^2} - 1 \right)^2. \tag{10.17}$$

When $\sigma_{WR}^2 < \sigma_0^2$,

$$\hat{\gamma}_U = \hat{\delta}^2 + \hat{\sigma}_{1,0.5}^2 - 1.5\hat{\sigma}_{WR}^2 - \theta_{IBF}\sigma_0^2 + \sqrt{U_0},$$

where U_0 is the same as U except that the quantity in Equation 10.17 should be replaced by

$$1.5^2 \hat{\sigma}_{WR}^4 \left(\frac{n_1 + n_2 - 2}{\chi_{0.05,n_1+n_2-2}^2} - 1 \right)^2.$$

Again, the estimation or test method for PBE described in Section 10.3 can be applied to decide which bound should be used.

To determine sample sizes n_1 and n_2, we would choose $n_1 = n_2 = n$ so that the power of the IBE test reaches a given level $1 - \beta$ when the unknown parameters are set at some initial guessing values $\delta, \sigma_D^2, \sigma_{WT}^2,$ and σ_{WR}^2. For the IBE test based on the confidence bound $\hat{\gamma}_U$ its power is given by

$$P_n = P(\hat{\gamma}_U < \gamma)$$

when $\gamma < 0$. Consider first the case where $\sigma_{WR}^2 > \sigma_0^2$ Let U be given in the definition of the reference-scaled bound $\hat{\gamma}_U$, and U_β be the same as U but with 5% and 95% replaced by β and $1 - \beta$, respectively. Since

$$P\left(\hat{\gamma}_U < \gamma + \sqrt{U} + \sqrt{U_{1-\beta}}\right) \approx 1 - \beta,$$

the power P_n is approximately larger than β if

$$\gamma + \sqrt{U} + \sqrt{U_{1-\beta}} \leq 0.$$

Let $\tilde{\gamma}, \tilde{U},$ and $\tilde{U}_{1-\beta}$ be γ, U, and $U_{1-\beta}$, respectively, with parameter values and their estimators replaced by the initial values $\delta, \sigma_D^2, \sigma_{WT}^2,$ and σ_{WR}^2. Then, the required sample size n to have approximately power $1 - \beta$ is the smallest integer satisfying

$$\tilde{\gamma} + \sqrt{\tilde{U}} + \sqrt{\tilde{U}_{1-\beta}} \leq 0, \tag{10.18}$$

assuming that $n_1 = n_2 = n$ and the initial values are the true parameter values. When $\sigma_{WR}^2 < \sigma_0^2$, the previous procedure can be modified by replacing U by U_0 in the definition of constant-scaled bound $\hat{\gamma}_U$. If $\tilde{\sigma}_{WR}^2$ is equal or close to σ_0^2, then we recommend the use of U instead of U_0 to produce a more conservative sample size and the use of the test approach in the IBE test.

This procedure can be applied to either the 2×3 design or the 2×4 design.

Since the IBE tests are based on the asymptotic theory, n should be reasonably large to ensure the asymptotic convergence. Hence, we suggest that the solution greater than 10 from Equation 10.18 be used. In other words, a sample size of more than $n = 10$ per sequence that satisfies Equation 10.18 is recommended.

Sample sizes $n_1 = n_2 = n$ selected using Equation 10.18 with $1 - \beta = 80\%$ and the power P_n of the IBE test based on 10,000 simulations are listed in Table 10.3 for both the 2×3 extra-reference design and 2×4 crossover design. For each selected n that is smaller than 10, the power of the IBE test using $n^* = \max(n,10)$ as the sample size, which is denoted by P_{n^*}, is also included. It can be seen from Table 10.3 that the actual power P_n is larger than the target value of 80% in most cases and only in a few cases where n determined from Equation 10.18 is very small, the power P_n is lower than 75%. Using $n^* = \max(n, 10)$ as the sample size produces better results when selected by Equation 10.18 is very small, but in most cases it results in a power much larger than 80%.

10.4.1 An Example

Suppose an investigator is interested in conducting a clinical trial with 2×4 crossover design to establish IBE between an inhaled formulation of a drug product (test) and an

TABLE 10.3

Sample Size n Selected Using Equation 10.18 with $1-\beta = 80\%$ and the Power P_n of the IBE Test Based on 10,000 Simulations

Parameter				2 × 3 Extra-Reference				2 × 4 Crossover			
σ_D	σ_{WT}	σ_{WR}	δ	n	P_n	n^*	P_{n^*}	n	P_n	n^*	P_{n^*}
0	0.15	0.15	0	5	0.7226	10	0.9898	4	0.7007	10	0.9998
			0.1	6	0.7365	10	0.9572	5	0.7837	10	0.9948
			0.2	13	0.7718	13		9	0.7607	10	0.8104
0	0.2	0.15	0	9	0.7480	10	0.8085	7	0.7995	10	0.9570
			0.1	12	0.7697	12		8	0.7468	10	0.8677
			0.2	35	0.7750	35		23	0.7835	23	
0	0.15	0.2	0	9	0.8225	10	0.8723	8	0.8446	10	0.9314
			0.1	12	0.8523	12		10	0.8424	10	
			0.2	26	0.8389	26		23	0.8506	23	
0	0.2	0.2	0	15	0.8206	15		13	0.8591	13	
			0.1	20	0.8373	20		17	0.8532	17	
			0.2	52	0.8366	52		44	0.8458	44	
0	0.3	0.2	0	91	0.8232	91		71	0.8454	71	
0.2	0.15	0.15	0	20	0.7469	20		17	0.7683	17	
			0.1	31	0.7577	31		25	0.7609	25	
0.2	0.15	0.2	0	31	0.8238	31		28	0.8358	28	
			0.1	43	0.8246	43		39	0.8296	39	
0.2	0.2	0.2	0	59	0.8225	59		51	0.8322	51	
			0.2	91	0.8253	91		79	0.8322	79	
0	0.15	0.3	0	7	0.8546	10	0.9607	6	0.8288	10	0.9781
			0.1	7	0.8155	10	0.9401	7	0.8596	10	0.9566
			0.2	10	0.8397	10		9	0.8352	10	0.8697
			0.3	16	0.7973	16		15	0.8076	15	
			0.4	45	0.8043	45		43	0.8076	43	
0	0.3	0.3	0	15	0.7931	15		13	0.8162	13	
			0.1	17	0.7942	17		14	0.8057	14	
			0.2	25	0.8016	25		21	0.8079	21	
			0.3	52	0.7992	52		44	0.8009	44	
0	0.2	0.5	0	6	0.8285	10	0.9744	6	0.8497	10	0.9810
			0.1	6	0.8128	10	0.9708	6	0.8413	10	0.9759
			0.2	7	0.8410	10	0.9505	7	0.8600	10	0.9628
			0.3	8	0.8282	10	.9017	8	0.8548	10	0.9239
			0.4	10	0.8147	10		10	0.8338	10	
			0.5	14	0.8095	14		14	0.8248	14	
			0.6	24	0.8162	24		23	0.8149	23	
			0.7	51	0.8171	51		49	0.8170	49	
0	0.5	0.5	0	15	0.7890	15		13	0.8132	13	
			0.1	16	0.8000	16		13	0.7956	13	
			0.2	18	0.7980	18		15	0.8033	15	
			0.3	23	0.8002	23		19	0.8063	19	
			0.5	52	0.7944	52		44	0.8045	44	

(Continued)

TABLE 10.3 (*Continued*)

Sample Size n Selected Using Equation 10.18 with $1 - \beta = 80\%$ and the Power P_n of the IBE Test Based on 10,000 Simulations

Parameter				2 × 3 Extra-Reference				2 × 4 Crossover			
σ_D	σ_{WT}	σ_{WR}	δ	n	P_n	n^*	P_{n^*}	n	P_n	n^*	P_{n^*}
0.2	0.2	0.3	0	13	0.7870	13		12	0.7970	12	
			0.1	15	0.8007	15		14	0.8144	14	
			0.2	21	0.7862	21		20	0.8115	20	
			0.3	43	0.8037	43		40	0.8034	40	
0.2	0.3	0.3	0	26	0.7806	26		22	0.7877	22	
			0.1	30	0.7895	30		26	0.8039	26	
0.2	0.3	0.5	0	9	0.8038	10	0.8502	8	0.8050	10	0.8947
			0.1	9	0.7958	10	0.8392	9	0.8460	10	0.8799
			0.2	10	0.7966	10		9	0.7954	10	0.8393
			0.3	12	0.7929	12		11	0.8045	11	
			0.4	16	0.7987	16		15	0.8094	15	

$n^* = max(n, 10)$.

SC injected formulation (reference) in terms of log-transformed AUC. Based on PK data obtained from pilot studies, the mean difference of AUC can be assumed to be 0%. Also, it is assumed that the intrasubject standard deviation of test and reference are given by 60% and 40%, respectively. It is further assumed that the intersubject standard deviation of test and reference are given by 10% and 40%, respectively. The intersubject correlation coefficient (ρ) is assumed to be 0.75. According to Table 10.3, a total of 22 subjects per sequence is needed to achieve an 80% power at the 5% level of significance.

10.5 *In Vitro* Bioequivalence

Statistical methods for the assessment of *in vitro* bioequivalence testing for nasal aerosols and sprays can be classified as the nonprofile analysis and the profile analysis. In this section, we consider sample size calculation for nonprofile analysis.

The nonprofile analysis applies to tests for dose or spray content uniformity through container life, droplet size distribution, spray pattern, and priming and repriming. The FDA adopts the criterion and limit of the PBE for the assessment of *in vitro* bioequivalence in the nonprofile analysis. Let θ be defined in Equation 10.1 with independent *in vitro* bioavailabilities y_T, y_R, and y'_R , and let θ_{BE} be the bioequivalence limit. Then, the two formulations are *in vitro* bioequivalent if $\theta < \theta_{BE}$. Similar to the PBE, *in vitro* bioequivalence can be claimed if the hypothesis that $\theta \geq \theta_{BE}$ is rejected at the 5% level of significance provided that the observed ratio of geometric means is within the limits of 90% and 110%.

Suppose that m_T and m_R canisters (or bottles) from respectively the test and the reference products are randomly selected and one observation from each canister is obtained. The data can be described by the following model:

$$y_{jk} = \mu_k + \varepsilon_{jk}, \quad j = 1,\ldots,m_k,$$ (10.19)

where $k = T$ for the test product, $k = R$ for the reference product, μ_T and μ_R are fixed product effects, and ε_{jk}'s are independent random measurement errors distributed as $N(0, \sigma_k^2), k = T, R$. Under model (10.19), the parameter θ in Equation 10.1 is equal to

$$\theta = \frac{(\mu_T - \mu_R)^2 + \sigma_T^2 - \sigma_R^2}{\max\{\sigma_0^2, \sigma_R^2\}}. \tag{10.20}$$

and $\theta < \theta_{BE}$ if and only if $\zeta < 0$, where

$$\zeta = (\mu_T - \mu_R)^2 + \sigma_T^2 - \sigma_R^2 - \theta_{BE} \max\{\sigma_0^2, \sigma_R^2\}. \tag{10.21}$$

Under model (10.19), the best unbiased estimator of $\delta = \mu_T - \mu_R$ is

$$\hat{\delta} = \bar{y}_T - \bar{y}_R \sim N\left(\delta, \frac{\sigma_T^2}{m_T} + \frac{\sigma_R^2}{m_R}\right),$$

where $\bar{y}_k$ is the average of y_{jk} over j for a fixed k. The best unbiased estimator of σ_k^2 is

$$s_k^2 = \frac{1}{m_k - 1} \sum_{j=1}^{m_k} (\bar{y}_{jk} - \bar{y}_k)^2 \sim \frac{\sigma_k^2 \chi_{mk-1}^2}{m_k - 1}, \quad k = T, R.$$

Using the method for IBE testing (Section 10.4), an approximate 95% upper confidence bound for ζ in Equation 10.21 is

$$\tilde{\zeta}_U = \hat{\delta}^2 + s_T^2 - s_R^2 - \theta_{BE} \max\{\sigma_0^2, s_R^2\} + \sqrt{U_0},$$

where U_0 is the sum of the following three quantities:

$$\left[\left(|\hat{\delta}| + z_{0.05}\sqrt{\frac{s_T^2}{m_T} + \frac{s_R^2}{m_R}}\right)^2 - \hat{\delta}^2\right]^2,$$

$$s_T^4 \left(\frac{m_T - 1}{\chi_{0.95, m_T - 1}^2} - 1\right)^2,$$

and

$$(1 + c\theta_{BE})^2 s_R^4 \left(\frac{m_R - 1}{\chi_{0.05, m_R - 1}^2} - 1\right)^2,$$

and $c = 1$ if $s_R^2 \geq \sigma_0^2$ and $c = 0$ if $s_R^2 < \sigma_0^2$. Note that the estimation method for determining the use of the reference-scaled criterion or the constant-scaled criterion is applied here. *In vitro* bioequivalence can be claimed if $\tilde{\zeta}_U < 0$. This procedure is recommended by the FDA guidance.

To ensure that the previously described test has a significant level close to the nominal level 5% with a desired power, the FDA requires that at least 30 canisters of each of the test and reference products be tested. However, $m_k = 30$ may not be enough to achieve a desired power of the bioequivalence test in some situations (see Chow, Shao, and Wang, 2003b). Increasing m_k can certainly increase the power, but in some situations, obtaining replicates from each canister may be more practical, and/or cost-effective. With replicates from each canister, however, the previously described test procedure is necessarily modified to address the between-canister and within-canister variabilities.

Suppose that there are n_k replicates from each canister for product k. Let y_{ijk} be the ith replicate in the jth canister under product k, b_{jk} be the between-canister variation, and e_{ijk} be the within-canister measurement error. Then,

$$y_{ijk} = \mu_k + b_{jk} + e_{ijk}, \quad i = 1,\dots,n_k, j = 1,\dots,m_k \tag{10.22}$$

where $b_{jk} \sim N(0,\sigma_{Bk}^2), e_{ijk} \sim N(0,\sigma_{Wk}^2)$, and b_{jk}'s *and* e_{ijk}'s are independent. Under model (10.22), the total variances σ_T^2 and σ_R^2 in Equations 10.20 and 10.21 are equal to $\sigma_{BT}^2 + \sigma_{WT}^2$ and $\sigma_{BR}^2 + \sigma_{WR}^2$, respectively, that is, the sums of between-canister and within-canister variances. The parameter θ in Equation 10.1 is still given by Equation 10.20 and $\theta < \theta_{BE}$ if and only if $\zeta < 0$, where ζ is given in Equation 10.21.

Under model (10.22), the best unbiased estimator of $\delta = \mu_T - \mu_R$ is

$$\hat{\delta} = \bar{y}_T - \bar{y}_R \sim N\left(\delta, \frac{\sigma_{BT}^2}{m_T} + \frac{\sigma_{BR}^2}{m_R} + \frac{\sigma_{WT}^2}{m_T n_T} + \frac{\sigma_{WR}^2}{m_R n_R}\right),$$

where $\bar{y}_k$ is the average of y_{ijk} over i and j for a fixed k.

To construct a confidence bound for ζ in Equation 10.21 using the approach in IBE testing, we need to find independent, unbiased, and chi-square distributed estimators of σ_T^2 and σ_R^2. These estimators, however, are not available when $n_k > 1$. Note that

$$\sigma_k^2 = \sigma_{Bk}^2 + n_k^{-1}\sigma_{Wk}^2 + \left(1 - n_k^{-1}\right)\sigma_{Wk}^2, \quad k = T,R;$$

$\sigma_{BK}^2 + n_k^{-1}\sigma_{Wk}^2$ can be estimated by

$$s_{Bk}^2 = \frac{1}{m_k - 1}\sum_{j=1}^{m_k}\left(\bar{y}_{jk} - \bar{y}_k\right)^2 \sim \frac{\left(\sigma_{Bk}^2 + n_k^{-1}\sigma_{Wk}^2\right)\chi_{m_k-1}^2}{m_k - 1},$$

where $\bar{y}_{jk}$ is the average of y_{ijk} over i; σ_{Wk}^2 can be estimated by

$$s_{Wk}^2 = \frac{1}{m_k(n_k - 1)}\sum_{j=1}^{m_k}\sum_{i=1}^{n_k}\left(y_{ijk} - \bar{y}_{jk}\right)^2 \sim \frac{\sigma_{Wk}^2\chi_{m_k(n_k-1)}^2}{m_k(n_k - 1)};$$

and $\hat{\delta}, s_{Bk}^2, s_{Wk}^2, k = T,R$, are independent. Thus, an approximate 95% upper confidence bound for ζ in Equation 10.21 is

$$\tilde{\zeta}_U = \hat{\delta}^2 + s_{BT}^2 + \left(1 - n_T^{-1}\right)s_{WT}^2 - s_{BR}^2 - \left(1 - n_R^{-1}\right)s_{WR}^2$$

$$-\theta_{BE}\max\left\{\sigma_0^2, s_{BR}^2 + \left(1 - n_R^{-1}\right)s_{WR}^2\right\} + \sqrt{U},$$

where U is the sum of the following five quantities:

$$\left[\left(|\hat{\delta}| + z_{0.05}\sqrt{\frac{s_{BT}^2}{m_T} + \frac{s_{BR}^2}{m_R}}\right)^2 - \hat{\delta}^2\right]^2,$$

$$s_{BT}^2\left(\frac{m_T - 1}{\chi_{0.95, m_T - 1}^2} - 1\right)^2,$$

$$\left(1 - n_T^{-1}\right)^2 s_{WT}^4\left(\frac{m_T(n_T - 1)}{\chi_{0.95, m_T(n_T - 1)}^2} - 1\right)^2,$$

$$(1 + \theta_{BE})^2 s_{BR}^4\left|\frac{m_R - 1}{\chi_{0.95, m_R - 1}^2} - 1\right|^2,$$

and

$$(1 + \theta_{BE})^2\left(1 - n_R^{-1}\right)^2 s_{BR}^4\left|\frac{m_R(n_R - 1)}{\chi_{0.95, m_R(n_R - 1)}^2} - 1\right|^2,$$

and $c = 1$ if $s_{BR}^2 + (1 - n_R^{-1})s_{WR}^2 \geq \sigma_0^2$ and $c = 0$ if $s_{BR}^2 + (1 - n_R^{-1})s_{WR}^2 < \sigma_0^2$. *In vitro* bioequivalence can be claimed if $\tilde{\zeta}_U < 0$ provided that the observed ratio of geometric means is within the limits of 90% and 110%.

Note that the minimum sample sizes required by the FDA are $m_k = 30$ canisters and $n_k = 1$ observation from each canister. To achieve a desired power, Chow et al. (2003b) proposed the following procedure of sample size calculation. Assume that $m = m_T = m_R$ and $n = n_T = n_R$. Let $\psi = \left(\delta, \sigma_{BT}^2, \sigma_{BR}^2, \sigma_{WT}^2, \sigma_{WR}^2\right)$ be the vector of unknown parameters under model (10.22). Let U be given in the definition of $\hat{\zeta}_U$ and $U_{1-\beta}$ be the same as U but with 5% and 95% replaced by β and $1 - \beta$, respectively, where $1-\beta$ is a given power. Let $\tilde{U}$ and $\tilde{U}_{1-\beta}$ be U and $U_{1-\beta}$, respectively, with $\left(\hat{\delta}, s_{BT}^2, s_{BR}^2, s_{WT}^2, s_{WR}^2\right)$ replaced by $\tilde{\psi}$, an initial guessing value for which the value of ζ (denoted by $\tilde{\zeta}$) is negative. From the results in Chow et al. (2003b), it is advantageous to have a large m and a small n when mn, the total number of observations for one treatment, is fixed. Thus, the sample sizes m and n can be determined as follows.

In step 1, $m^* = 30$, $n^* = 1$.

Step 1. Set $m = 30$ and $n = 1$. If

$$\tilde{\zeta} + \sqrt{\tilde{U}} + \sqrt{\tilde{U}_{1-\beta}} \leq 0 \tag{10.23}$$

holds, stop and the required sample sizes are $m = 30$ and $n = 1$; otherwise, go to step 2.

Step 2. Let $n = 1$ and find a smallest integer m_* such that Equation 10.23 holds. If $m_* \leq m_+$ (the largest possible number of canisters in a given problem), stop and the required sample sizes are $m = m_*$ and $n = 1$; otherwise, go to step 3.

TABLE 10.4

Selected Sample Sizes m_* and n_* and the Actual Power p (10,000 Simulations)

σ_{BT}	σ_{BR}	σ_{WT}	σ_{WR}	δ	Step 1 p	Step 2 m_*, n_*	Step 2 p	Step 2' m_*, n_*	Step 2' p
0	0	0.25	0.25	0.0530	0.4893	55, 1	0.7658	30, 2	0.7886
				0	0.5389	47, 1	0.7546	30, 2	0.8358
		0.25	0.50	0.4108	0.6391	45, 1	0.7973	30, 2	0.8872
				0.2739	0.9138	–	–	–	–
		0.50	0.50	0.1061	0.4957	55, 1	0.7643	30, 2	0.7875
				0	0.5362	47, 1	0.7526	30, 2	0.8312
0.25	0.25	0.25	0.25	0.0750	0.4909	55, 1	0.7774	30, 3	0.7657
				0	0.5348	47, 1	0.7533	30, 2	0.7323
		0.25	0.50	0.4405	0.5434	57, 1	0.7895	30, 3	0.8489
				0.2937	0.8370	–	–	–	–
		0.50	0.50	0.1186	0.4893	55, 1	0.7683	30, 2	0.7515
				0	0.5332	47, 1	0.7535	30, 2	0.8091
0.50	0.25	0.25	0.50	0.1186	0.4903	55, 1	0.7660	30, 4	0.7586
				0	0.5337	47, 1	0.7482	30, 3	0.7778
0.25	0.50	0.25	.25	0.2937	.8357	–	–	–	–
		0.50	0.25	0.1186	0.5016	55, 1	0.7717	30, 4	0.7764
				0	0.5334	47, 1	0.7484	30, 3	0.7942
		0.25	0.50	0.5809	0.6416	45, 1	0.7882	30, 2	0.7884
				0.3873	0.9184	–	–	–	–
		0.50	0.50	0.3464	0.6766	38, 1	0.7741	30, 2	0.8661
				0.1732	0.8470	–	–	–	–
0.50	0.50	0.25	0.50	0.3464	0.6829	38, 1	0.7842	30, 2	0.8045
				0.1732	0.8450	–	–	–	–
		0.50	0.50	0.1500	0.4969	55, 1	0.7612	30, 3	0.7629
				0	0.5406	47, 1	0.7534	30, 2	0.7270

Step 3. Let $m = m_+$ and find a smallest integer n_* such that Equation 10.23 holds. The required sample sizes are $m = m_+$ and $n = n_*$.

If in practice it is much easier and inexpensive to obtain more replicates than to sample more canisters, then steps 2 and 3 in the previous procedure can be replaced by Step 2': Let $m = 30$ and find a smallest integer n_* such that Equation 10.23 holds. The required sample sizes are $m = 30$ and $n = n_*$.

Table 10.4 contains selected m_* and n_* according to Steps 1 through 3 or Steps 1 and 2' with $1 - \beta = 80\%$ and the simulated power p of the *in vitro* bioequivalence test using these sample sizes.

10.5.1 An Example

Suppose an investigator is interested in conducting a clinical trial with a parallel design with no replicates to establish *in vitro* bioequivalence between a generic drug product (test) and a brand name drug product (reference) in terms of *in vitro* bioavailability. Based on data obtained from pilot studies, the mean difference can be assumed to be 0% ($\delta = 0.00$).

Also, it is assumed that the intrasubject standard deviation of test and reference are given by 50% and 50%, respectively. It is further assumed that the intersubject standard deviation of the test and the reference are given by 50% and 50%, respectively. According to Table 10.4, 47 subjects per treatment group are needed to yield an 80% power at the 5% level of significance.

10.6 Sample Size Requirement for Analytical Similarity Assessment of Biosimilar Products

As indicated at the beginning of this chapter, the FDA recommends a stepwise approach for obtaining the totality of the evidence for demonstrating biosimilarity between a proposed biosimilar (test) product and an innovative (reference) biological product (FDA, 2015). The stepwise approach starts with analytical similarity assessment for CQA that are relevant to clinical outcomes at various stages of manufacturing process. CQAs that are most relevant to clinical outcomes will be classified to Tier 1, while CQAs that are less (mild to moderate) or least relevant to clinical outcomes will be classified to Tier 2 and Tier 3, respectively.

10.6.1 FDA's Tiered Approach

FDA proposes equivalence test for CQAs in Tier 1, quality range approach for CQAs in Tier 2, and raw data or graphical presentation for CQAs in Tier 3 to assist sponsors in analytical similarity assessment for obtaining totality of the evidence for demonstrating similarity between the proposed biosimilar product and the reference product. As indicated by the FDA, equivalence test for Tier 1 CQAs is more statistically rigorous than that of quality range approach for Tier 2 CQAs, which is in turn more rigorous than that of raw data or graphical presentation for Tier 3 CQAs. Although equivalence test is currently required by the FDA for the assessment of analytical similarity for CQAs in Tier 1 for regulatory submission, it is not clear (i) how many reference lots should be used for establishing equivalence acceptance criterion (EAC) or similarity margin and (ii) how many test lots should be considered for a valid assessment of similarity between the proposed biosimilar product and the innovative biologic product. To assist the sponsors, Dong, Tsong, and Wang (2017) proposed a rule for the selection of the number of reference lots for the establishment of EAC and consequently the number of test lots for equivalence test based on extensive simulation studies. This rule is easy to apply and yet lacks statistical justification. As a result, the purpose of this note is not only to provide statistical justification, but also to propose an alternative method to the FDA's proposal regarding sample size requirement for Tier 1 equivalence test.

10.6.2 Sample Size Requirement

For CQAs in Tier 1, FDA recommends that an equivalency test be performed to assess analytical similarity based on the following interval hypotheses:

$$H_0 : |\mu_T - \mu_R| > \delta \text{ vs. } H_a : |\mu_T - \mu_R| \leq \delta, \tag{10.24}$$

where $\delta > 0$ is the equivalence limit (or similarity margin), and μ_T and μ_R are the mean responses of the test (the proposed biosimilar) product and the reference product lots, respectively. Analytical equivalence (similarity) is concluded if the null hypothesis of nonequivalence (*dis*-similarity) is rejected. FDA further recommended that the EAC be selected as $1.5^*\sigma_R$, that is, $\delta = \text{EAC} = 1.5^*\sigma_R$, where σ_R is the variability of the reference product. Chow (2015) provided statistical justification for the selection of $c = 1.5$ in EAC following the idea of scaled average bioequivalence (SABE) criterion for highly variable drug products proposed by the FDA (Haidar et al., 2008).

For the establishment of EAC, FDA made the following assumptions. First, FDA assumes that the true difference in means is proportional to σ_R, that is, $\mu_T - \mu_R$ is proportional to σ_R. Second, σ_R is estimated by the sample standard deviation of test values from selected reference lots (one test value from each lot). Third, in the interest of achieving a desired power of the similarity test, FDA further recommends that an appropriate sample size be selected by evaluating the power under the alternative hypothesis at $\mu_T - \mu_R = (1/8)\,\sigma_R$. These assumptions, however, may not be true in practice. Thus, these debatable assumptions have generated tremendous discussion among FDA, biosimilar sponsors, and academia (see, e.g., Chow, Song, and Bai 2016).

The selection of n_R and $k = nT/n_R$: The selection of n_R for the establishment of EAC and n_T for having a desired power of passing the equivalence test in probably one the most important questions to the sponsors of biosimilar products. To assist the sponsors, Dong, Tsong, and Wang (2016) proposes to select n_R as follows:

$$n_R^* = \min\{1.5 * N_T, N_R\} \tag{10.25}$$

where N_R and N_T are the available reference lots and the available test lots, respectively. Dong, Tsong, and Wang (2017)'s proposal seems to suggest that n_T be selected as $n_T^* = n_R^*/1.5$ if $1.5\,N_T < N_R$ based on some extensive clinical trial simulation. Dong, Tsong, and Wang (2016)'s proposal does provide some guidance to the sponsors for the selection of n_R for the establishment of EAC. However, this proposal also raised the following questions. First, the selection of 1.5 (ratio between N_R and N_T) is not statistically justified. Second, how to select n_R^* from N_R if $1.5\,N_T < N_R$. Third, when $1.5\,N_T > N_R$, there is no rule for selecting n_T^*.

A Proposed Procedure: One of the most commonly asked questions for analytical similarity assessment is probably how many reference lots are required for establishing an acceptable EAC for achieving a desired power. For a given EAC, formulas for sample size calculation under different study designs are available in Chow, Shao, and Wang (2008). In general, under a parallel-group design and the hypotheses (10.24), sample size (the number of reference lots, n_R, and the number of test lots, n_T) required is a function of (i) overall type I error rate (α), (ii) type II error rate (β)or power ($1 - \beta$), (iii) clinically or scientifically meaningful difference (i.e., $\mu_T - \mu_R$), and (iv) the variability associated with the reference product (i.e., σ_R) assuming that $\sigma_T = \sigma_R$. Thus, we have

$$n_T = f(\alpha, \beta, \mu_T - \mu_R, k, \sigma_R)$$
$$= \frac{(z_\alpha + z_{\beta/2})^2 \sigma_R^2 (1 + 1/k)}{(\delta - |\mu_T - \mu_R|)^2}, \tag{10.26}$$

where $k = n_T/n_R$. In practice, for a given n_R (e.g., available reference lots), we select an appropriate k (and consequently n_T) for achieving a desired power of $1 - \beta$ for detecting a clinically meaningful difference of $\mu_T - \mu_R$ at a prespecified level of significance α assuming that the true variability is σ_R. FDA's recommendation attempts to control all parameters at the desired levels (e.g., $\alpha = 0.05$ and $1 - \beta = 0.8$) by *knowing* that $\mu_T - \mu_R$ and σ are varying. In practice, it is often difficult, if not impossible, to control (or find a balance point among) α (type I error rate), $1 - \beta$ (power), $\mu_T - \mu_R = \Delta$ (clinically meaningful difference), and σ_R (variability in observing the response) *at the same time*. For example, controlling α at a prespecified level of significance may be at the risk of decreasing power with a selected sample size.

If α, $\mu_T - \mu_R$, and σ_R are fixed, the above equation becomes $n_T = f(\beta, k)$. We can then select an appropriate k for achieving the desired power. Under the assumptions that (i) $\mu_T - \mu_R$ is proportional to σ_R, that is, $\mu_T - \mu_R = r\sigma_R$ and (ii) $\delta = EAC = 1.5*\sigma_R$, Equation 10.26 becomes

$$n_T = \frac{A^2\sigma_R^2(1+1/k)}{(1.5\sigma_R - r\sigma_R)^2} = \frac{A^2(1+1/k)}{(1.5-r)^2},$$
(10.27)

where $A = z_\alpha + z_{\beta/2}$. As indicated earlier, FDA allows a mean shift of $1/8\sigma_R = 0.125\sigma_R$. If we take this mean shift into consideration (i.e., $r = 0.125$ and assume that the desired power is 80% ($\beta = 0.2$) at the $\alpha = 5\%$ level of significance, that is, $A = z_{0.05} + z_{0.1} = 1.645 + 1.282 = 2.927$, then the above sample size requirement (10.27) becomes

$$n_T = \left(\frac{2.927}{1.5-0.125}\right)^2\left(1+\frac{1}{k}\right) = 2.762\left(1+\frac{1}{k}\right),$$

where $k = (n_T/n_R)$ If we choose $1/k = 1.5$, then $n_T = 6.095 \approx 7$. Thus, under the assumptions that (i) $\mu_T - \mu_R = r\sigma_R = (1/8)\sigma_R$ and (ii) $\delta = EAC = 1.5*\sigma_R$, seven test lots are required for achieving an 80% power at the 5% level of significance in an equivalence test for analytical similarity assessment for CQAs in Tier 1.

Based on the above discussion, we would like to propose the following strategy for the selection of n_T, k, and then n_R assuming that N_R (available reference lots) is much larger than n_R (required reference lots for establishing EAC in equivalence test):

Step 1: Selection of minimum number of test lots required, n_T. Under the assumptions that (i) $\mu_T - \mu_R = r\sigma_R = (1/8)\sigma_R$ and (ii) $\delta = EAC = 1.5*\sigma_R$ (FDA's recommendation), we can use Equation 10.26 to determine n_T for achieving a desired power (i.e., $1 - \beta$ at the α level of significance for various selections of k).

Step 2: Determination of k. Depending upon the availability of the reference lots (N_R) and test lots (N_T), carefully evaluate the tradeoff between controlling type I error rate and achieving desired power with various selection of different k's.

Step 3: Selection of n_R lots from N_R available reference lots. Once n_T and k have been determined, n_R can be obtained as $n_R = n_T/k$. The n_R lots, which will be randomly selected from the N_R available reference lots, will then be used for the establishment of EAC for equivalence test.

Note that the above strategy is developed under the assumptions that (i) $\mu_T - \mu_R = r\sigma_R = (1/8)\sigma_R$ and (ii) $\delta = EAC = 1.5*\sigma_R$ (a fixed margin). In practice, however, assumption that $\mu_T - \mu_R$ is proportional to σ_R may not be met. As a result, the above power calculation for sample size may be biased. For the second assumption, since it is a fixed approach, it is very likely that different sponsors may come up with different EACs using different available reference lots. It should also be noted that the proposed strategy may result in the situation where $n_T \neq n_R$.

11

Dose–Response Studies

As indicated in 21 CFR 312.21, the primary objectives of phase I clinical investigation are to (i) determine the metabolism and pharmacological activities of the drug, the side effects associated with increasing dose, and early evidence in effectiveness and (ii) obtain sufficient information regarding the drug's pharmacokinetics and pharmacological effects to permit the design of well-controlled and scientifically valid phase II clinical studies. Thus, phase I clinical investigation includes studies of drug metabolism, bioavailability, dose ranging, and multiple dose. The primary objectives of phase II studies are not only to initially evaluate the effectiveness of a drug based on clinical endpoints for a particular indication or indications in patients with disease or condition under study but also to determine the dosing ranges and doses for phase III studies and common short-term side effects and risks associated with the drug. In practice, the focus of phase I dose–response studies emphasize safety, while phase II dose–response studies emphasize the efficacy.

When studying the dose–response relationship of an investigational drug, a randomized, parallel-group trial involving a number of dose levels of the investigational drug and a control is usually conducted. Ruberg (1995a,b) indicated that some questions dictating design and analysis are necessarily addressed. These questions include the following: (i) Is there any evidence of the drug effect? (ii) How is the treatment response different from the control response? (iii) What is the nature of the dose–response? and (iv) What is the optimal dose? The first question is usually addressed by the analysis of variance. The second question can be addressed by the Williams' test for minimum effective dose (MED). The third question can be addressed by model-based approaches, either frequentist or Bayesian. The last question is a multidimensional issue involving efficacy as well as tolerability and safety such as the determination of maximum tolerable dose (MTD). In this chapter, we will limit our discussion to the sample size calculations for addressing the above questions.

In Sections 11.1 through 11.3, formulas for sample size calculation for continuous, binary response, and time-to-event study endpoints under a multiarm dose–response trial are derived, respectively. Section 11.4 provides the sample size formula for the determination of MED based on Williams' test. A sample size formula based on Cochran–Armitage's trend test for binary response is given in Section 11.5. In Section 11.6, sample size estimation and related operating characteristics of phase I dose escalation trials are discussed. A brief concluding remark is given in Section 11.7.

11.1 Continuous Response

To characterize the response curve, a multiarm design including a control group and K active dose groups are usually considered. This multiarm trial is informative for the drug candidates with a wide therapeutic window. The null hypothesis of interest is then given by

$$H_0 : \mu_0 = \mu_1 = \ldots = \mu_K, \tag{11.1}$$

where μ_0 is the mean response for the control group and μ_i is the mean response for the *i*th dose group. The rejection of hypothesis (11.1) indicated that there is a treatment effect. The dose–response relationship can then be examined under an appropriate alternative hypothesis. Under a specific alternative hypothesis, the required sample size per dose group can then be obtained. Spriet and Dupin-Spriet (1992) identified the following eight alternative hypotheses (H_a) for dose–responses:

1. $\mu_0 < \mu_1 < \ldots < \mu_{K-1} < \mu_K$;
2. $\mu_0 < \ldots < \mu_i = \ldots = \mu_j > \ldots > \mu_K$;
3. $\mu_0 < \ldots < \mu_i = \ldots = \mu_K$;
4. $\mu_0 = \ldots = \mu_i < \ldots < \mu_K$;
5. $\mu_0 < \mu_1 < \ldots = \mu_i = \ldots = \mu_K$;
6. $\mu_0 = \mu_1 = \ldots = \mu_i < \ldots < \mu_{K-1} < \mu_K$;
7. $\mu_0 = \mu_1 < \ldots < \mu_i = \ldots = \mu_K$;
8. $\mu_0 = \ldots = \mu_i < \ldots < \mu_{K-1} = \mu_K$.

In the subsequent sections, we will derive sample size formulas for various study end-points such as continuous response, binary response, and time-to-event data under a multiarm dose–response design, respectively.

11.1.1 Linear Contrast Test

Under a multiarm dose–response design, a linear contrast test is commonly employed. Consider the following one-sided hypotheses:

$$H_0 : L(\mu) = \sum_{i=0}^{K} c_i \mu_i \leq 0 \text{ versus } H_a : L(\mu) = \sum_{i=0}^{K} c_i \mu_i = \varepsilon > 0,$$

where μ_i could be mean, proportion, or ranking score in the *i*th arm, c_i is the contrast coefficient satisfying $\sum_{i=0}^{K} c_i = 0$, and ε is a constant. The test statistics under the null hypothesis and the alternative hypothesis can be expressed as

$$T(H) = \frac{L(\hat{\mu})}{\sqrt{\operatorname{var}(L(\hat{\mu}) \mid H_o)}}; \quad H \in H_o \cup H_a.$$

Under the alternative hypothesis, we have $\varepsilon = E(L(\hat{\mu}) \mid H_a)$. Denote $v_o^2 = \operatorname{var}(L(\hat{\mu}) \mid H_o)$ and $v_a^2 = \operatorname{var}(L(\hat{\mu}) \mid H_a)$. Then, under the null hypothesis, for large sample, we have

$$T(H_o) = \frac{L(\hat{\mu};\delta) \mid H_o}{v_0} \sim N(0,1).$$

Similarly, under the alternative hypothesis, it can be verified that

$$T(H_a) = \frac{L(\hat{\mu};\delta)}{v_0} \sim N\left(\frac{\varepsilon}{v_0}, \frac{v_a^2}{v_0^2}\right)$$

for large sample, where

$$v_0^2 = \text{var}\,(L(\hat{\mu};\delta) \mid H_o) = \sum_{i=0}^{k} c_i^2\, \text{var}\,(\hat{\mu}_i \mid H_o) = \sigma_0^2 \sum_{i=0}^{K} \frac{c_i^2}{n_i}$$

$$v_a^2 = \text{var}\,(L(\hat{\mu};\delta) \mid H_a) = \sum_{i=0}^{k} c_i^2\, \text{var}\,(\hat{\mu}_i \mid H_o) = \sum_{i=0}^{K} \frac{c_i^2 \sigma_i^2}{n_i}$$

That is,

$$\begin{cases} v_0^2 = \frac{\sigma_0^2}{n} \sum_{i=0}^{K} \frac{c_i^2}{f_i} \\ v_a^2 = \frac{1}{n} \sum_{i=0}^{K} \frac{c_i^2 \sigma_i^2}{f_i} \end{cases} \tag{11.2}$$

where the size fraction $f_i = n_i/n$ with $N = \sum_{i=0}^{K} n_i$. Note that σ_o and σ_a are the standard deviation of the response under H_0 and H_a, respectively. Let μ_i be the population mean for group i. The null hypothesis of no treatment effects can be written as follows:

$$H_o : L(\mu) = \sum_{i=0}^{K} c_i \mu_i = 0, \tag{11.3}$$

where $\sum_{i=0}^{K} c_i = 0$. Under the following alternative hypothesis

$$H_a : L(\mu) = \sum_{i=0}^{K} c_i \mu_i = \varepsilon, \tag{11.4}$$

and the assumption of homogeneous variances, the sample size can be obtained as

$$N = \left[\frac{(z_{1-\alpha} + z_{1-\beta})\sigma}{\varepsilon}\right]^2 \sum_{i=0}^{k} \frac{c_i^2}{f_i},$$

where f_i is the sample size fraction for the ith group and the population parameter σ. Note that, in practice, for the purpose of sample size calculation, one may use the pooled standard deviation if prior data are available.

An Example

Suppose that a pharmaceutical company is interested in conducting a dose–response study for a test drug developed for treating patients with asthma. A four-arm design consisting of a placebo control and three active dose levels (0, 20, 40, and 60 mg) of the test drug is proposed. The primary efficacy endpoint is percent change from baseline in FEV1. Based on data collected from pilot studies, it is expected that there are 5% improvement for the control group, 12%, 14%, and 16% improvements over baseline in the 20, 40, and 60 mg dose groups, respectively. Based on the data from the pilot studies, the homogeneous standard deviation for the FEV1 change from baseline is assumed to be $\sigma = 22\%$. Thus, we may consider the following contrasts for sample size calculation:

$$c_0 = -6, \quad c_1 = 1, \quad c_2 = 2, \quad c_3 = 3.$$

Note that $\sum c_i = 0$. Moreover, we have $\varepsilon = \sum_{i=0}^{3} c_i \mu_i = 58\%$. For simplicity, consider the balanced case (i.e., $f_i = 1/4$ for $i = 0, 1, \ldots, 3$) with one-sided at $\alpha = 0.05$, the sample size required for detecting the difference of $\varepsilon = 0.58$ with an 80% power is then given by

$$
\begin{aligned}
N &= \left[\frac{(z_{1-\alpha} + z_{1-\beta})\sigma}{\varepsilon} \right]^2 \sum_{i=0}^{K} \frac{c_i^2}{f_i} \\
&= \left[\frac{(1.645 + 0.842)0.22}{0.58} \right]^2 4((-6)^2 + 1^2 + 2^2 + 3^2) \\
&= 178.
\end{aligned}
$$

In other words, approximately 45 subjects per dose group are required for achieving an 80% power for detecting the specified clinical difference at the 5% level of significance.

Remark

Table 11.1 provides five different dose–response curves and the corresponding contrasts.
Sample sizes required for different dose–response curves and contrasts are given in Table 11.2. It can be seen from Table 11.2 that when the dose–response curve and the contrasts

TABLE 11.1

Response and Contrast Shapes

Shape	μ_0	μ_1	μ_2	μ_3	c_0	c_1	c_2	c_3
Linear	0.1	0.3	0.5	0.7	−3.00	−1.00	1.00	3.00
Step	0.1	0.4	0.4	0.7	−3.00	0.00	0.00	3.00
Umbrella	0.1	0.4	0.7	0.5	−3.25	−0.25	2.75	0.75
Convex	0.1	0.1	0.1	0.6	−1.25	−1.25	−1.25	3.75
Concave	0.1	0.6	0.6	0.6	−3.75	1.25	1.25	1.25

TABLE 11.2

Sample Size per Group for Various Contrasts

Response	Linear	Step	Contrast Umbrella	Convex	Concave
Linear	31	35	52	52	52
Step	39	35	81	52	52
Umbrella	55	74	33	825	44
Convex	55	50	825	33	297
Concave	55	50	44	297	33

Note: $\sigma = 1$, one-sided $\alpha = 0.05$.

have the same shape, a minimum sample size is required. If an inappropriate set of contrasts is used, the sample size could be 30 times larger than the optimal design.

11.2 Binary Response

Denote p_i the proportion of response in the ith group. Consider testing the following null hypothesis:

$$H_o : p_0 = p_1 = \ldots = p_k \qquad (11.5)$$

against the alternative hypothesis of

$$H_a : L(\mathbf{p}) = \sum_{i=0}^{k} c_i p_i = \varepsilon, \qquad (11.6)$$

where c_i are the contrasts satisfying $\sum_{i=1}^{k} c_i = 0$.

Similarly, by applying the linear contrast approach described above, the sample size required for achieving an 80% power for detecting a clinically significant difference of e at the 5% level of significance can be obtained as

$$N \geq \left[\frac{z_{1-\alpha} \sqrt{\sum_{i=0}^{k} \left(c_i^2 / f_i \right) \bar{p}(1 - \bar{p})} + z_{1-\beta} \sqrt{\sum_{i=0}^{k} \left(c_i^2 / f_i \right) p_i (1 - p_i)}}{\varepsilon} \right]^2, \qquad (11.7)$$

where $\bar{p}$ is the average of p_i.

Remark

Table 11.3 provides sample sizes required for different dose–response curves and contrasts. As it can be seen from Table 11.3, an appropriate selection of contrasts (i.e., it can

TABLE 11.3

Total Sample Size Comparisons for Binary Data

Response	Linear	Step	Contrast Umbrella	Convex	Concave
Linear	26	28	44	48	44
Step	28	28	68	48	40
Umbrella	48	68	28	792	36
Convex	28	36	476	24	176
Concave	36	44	38	288	28

Note: One-sided $\alpha = 0.05, \sigma_o^2 = \bar{p}(1-\bar{p}), \bar{p} = \sum_{i=0}^{k} f_i \hat{p}$

reflect the dose–response curve) yields a minimum sample size required for achieving the desired power.

11.3 Time-to-Event Endpoint

Under an exponential survival model, the relationship between hazard (λ), median (T_{median}), and mean (T_{mean}) survival time can be described as follows:

$$T_{Median} = \frac{\ln 2}{\lambda} = (\ln 2)T_{mean}. \tag{11.8}$$

Let λ_i be the population hazard rate for group i. The contrast test for multiple survival curves can be written as

$$H_o : L(\mu) = \sum_{i=0}^{k} c_i \lambda_i = 0 \text{ versus } L(\mu) = \sum_{i=0}^{k} c_i \lambda_i = \varepsilon > 0,$$

where contrasts satisfy the condition that $\sum_{i=0}^{k} c_i = 0$.

Similar to the continuous and binary endpoints, the sample size required for achieving the desired power of $1 - \beta$ is given by

$$N \geq \left[\frac{z_{1-\alpha}\sigma_0 \sqrt{\sum_{i=0}^{k} \frac{c_i^2}{f_i}} + z_{1-\beta} \sqrt{\sum_{i=0}^{k} \frac{c_i^2}{f_i} \sigma_i}}{\varepsilon} \right]^2. \tag{11.9}$$

where the variance σ_i^2 can be derived in several different ways. For simplicity, we may consider Lachin and Foulkes's maximum likelihood approach (Lachin and Foulkes, 1986).

Suppose we design a clinical trial with k groups. Let T_0 and T be the accrual time period and the total trial duration, respectively. We can then prove that the variance for uniform patient entry is given by

TABLE 11.4

Sample Sizes for Different Contrasts (Balanced Design)

Scenario	Contrast				Total n
Average dose effect	−3	1	1	1	666
Linear response trend	−6	1	2	3	603
Median time trend	−6	0	2	4	588
Hazard rate trend	10.65	−0.55	−3.75	−6.35	589

Note: Sample size ratios to the control group: 1, 1, 1, 1.

$$\sigma^2(\lambda_i) = \lambda_i^2 \left[1 + \frac{e^{-\lambda_i T}(1 - e^{\lambda_i T_0})}{T_0 \lambda_i}\right]^{-1}. \tag{11.10}$$

An Example

In a four-arm (the active control, lower dose of test drug, higher dose of test drug, and combined therapy) phase II oncology trial, the objective is to determine whether there is treatment effect with time to progression as the primary endpoint. Patient enrollment duration is estimated to be $T_0 = 9$ months and the total trial duration $T = 16$ months. The estimated median time for the four groups are 14, 20, 22, and 24 months (corresponding hazard rates of 0.0495, 0.0347, 0.0315, and 0.0289/month, respectively). For this phase II design, we use one-sided $a = 0.05$ and power = 80%. To achieve the most efficient design (i.e., minimum sample size), sample sizes from different contrasts and various designs (balanced or unbalanced) are compared. Table 11.4 provides the sample sizes for the balanced design. Table 11.5 gives sample sizes for unbalanced design with specific sample size ratios, that is, (Control: control, lower dose: Control, higher dose: control, and Combined: control)=(1, 2, 2, 2). This type of design is often seen in clinical trials where patients are assigned to the test group more than the control group due to the fact that the investigators are usually more interested in the response in the test groups. However, this unbalanced design is usually not an efficient design. An optimal design, that is, minimum variance design, where the number of patients assigned to each group is proportional to the variance of the group, is studied (Table 11.6). It can be seen from Table 11.5 that the optimal designs with sample size ratios (1, 0.711, 0.634, 0.574) are generally most powerful and requires fewer patients regardless of the shape of the contrasts. In all cases, the contrasts with a trend in median time or the hazard rate works well. The contrasts with linear trend also work well in most cases under the assumption of this particular trend of response

TABLE 11.5

Sample Sizes for Different Contrasts (Unbalanced Design)

Scenario	Contrast				Total n
Average dose effect	−3	1	1	1	1036
Linear dose–response	−6	1	2	3	924
Median time shape	−6	0	2	4	865
Hazard rate shape	10.65	−0.55	−3.75	−6.35	882

Note: Sample size ratios to the control group: 1, 2,2, 2.

TABLE 11.6

Sample Sizes for Different Contrasts (Minimum Variance Design)

Scenario	Contrast				Total n
Average dose effect	−3	1	1	1	548
Linear dose–response	−6	1	2	3	513
Median time shape	−6	0	2	4	525
Hazard rate shape	10.65	−0.55	−3.75	−6.35	516

Note: Sample size ratios (proportional to the variances): 1, 0.711, 0.634, 0.574.

(hazard rate). Therefore, the minimum variance design seems attractive with total sample sizes of 525 subjects, that is, 180, 128, 114, and 103 for the active control, lower dose, higher dose, and combined therapy groups, respectively. In practice, if more patients assigned to the control group is a concern and it is desirable to obtain more information on the test groups, a balanced design should be chosen with a total sample size of 588 subjects or 147 subjects per group.

11.4 Williams' Test for Minimum Effective Dose

Under the assumption of monotonicity in dose–response, Williams (1971, 1972) proposed a test to determine the lowest dose level at which there is evidence for a difference from control. Williams considered the following alternative hypothesis:

$$H_a : \mu_0 = \mu_1 = \ldots = \mu_{i-1} < \mu_i \leq \mu_{i+1} \leq \ldots \leq \mu_K$$

and proposed the following test statistic:

$$T_i = \frac{\hat{\mu}_i - \hat{Y}_o}{\hat{\sigma}\sqrt{1/n_i + 1/n_o}},$$

where $\hat{\sigma}^2$ is an unbiased estimate of σ^2, which is independent of $\hat{Y}_i$ and is distributed as $\sigma^2 \chi_v^2 / v$ and $\hat{\mu}_i$ is the maximum likelihood estimate of μ_i, which is given by

$$\hat{\mu}_i = \max_{1 \leq u \leq i} \min_{i \leq v \leq K} \left[\frac{\sum_{j=u}^{v} n_j \hat{Y}_j}{\sum_{j=1}^{v} n_j} \right].$$

When $n_i = n$ for $i = 0, 1, \ldots, K$, this test statistic can be simplified as

$$T_i = \frac{\hat{\mu}_i - \bar{Y}_0}{s\sqrt{2/n}},$$

which can be approximated by $(X_i - Z_0)/s$, where s^2 is an unbiased estimate of σ^2,

$$X_i = \max_{1 \le u \le i} \sum_{j=u}^{i} \frac{Z_j}{i - u + 1}$$

and Z_j follows a standard normal distribution. We then reject the null hypothesis of no treatment difference and conclude that the ith dose level is the minimum effective dose if

$$T_j > t_j(\alpha) \quad \text{for all} \quad j \ge i,$$

where $t_j(\alpha)$ is the upper ath percentile of the distribution of T_j. The critical values of $t_j(\alpha)$ are given in Tables 11.7 through 11.10.

Since the power function of the above test is rather complicated, as an alternative, consider the following approximation to obtain the required sample size per dose group:

TABLE 11.7

Upper 5 Percentile $t_k(\alpha)$ for T_k

	$k = $ Number of Dose Levels								
df/v	2	3	4	5	6	7	8	9	10
5	2.14	2.19	2.21	2.22	2.23	2.24	2.24	2.25	2.25
6	2.06	2.10	2.12	2.13	2.14	2.14	2.15	2.15	2.15
7	2.00	2.04	2.06	2.07	2.08	2.09	2.09	2.09	2.09
8	1.96	2.00	2.01	2.02	2.03	2.04	2.04	2.04	2.04
9	1.93	1.96	1.98	1.99	2.00	2.00	2.01	2.01	2.01
10	1.91	1.94	1.96	1.97	1.97	1.98	1.98	1.98	1.98
11	1.89	1.92	1.94	1.94	1.95	1.95	1.96	1.96	1.96
12	1.87	1.90	1.92	1.93	1.93	1.94	1.94	1.94	1.94
13	1.86	1.89	1.90	1.91	1.92	1.92	1.93	1.93	1.93
14	1.85	1.88	1.89	1.90	1.91	1.91	1.91	1.92	1.92
15	1.84	1.87	1.88	1.89	1.90	1.90	1.90	1.90	1.91
16	1.83	1.86	1.87	1.88	1.89	1.89	1.89	1.90	1.90
17	1.82	1.85	1.87	1.87	1.88	1.88	1.89	1.89	1.89
18	1.82	1.85	1.86	1.87	1.87	1.88	1.88	1.88	1.88
19	1.81	1.84	1.85	1.86	1.87	1.87	1.87	1.87	1.88
20	1.81	1.83	1.85	1.86	1.86	1.86	1.87	1.87	1.87
22	1.80	1.83	1.84	1.85	1.85	1.85	1.86	1.86	1.86
24	1.79	1.81	1.82	1.83	1.84	1.84	1.84	1.84	1.85
26	1.79	1.81	1.82	1.83	1.84	1.84	1.84	1.84	1.85
28	1.78	1.81	1.82	1.83	1.83	1.83	1.84	1.84	1.84
30	1.78	1.80	1.81	1.82	1.83	1.83	1.83	1.83	1.83
35	1.77	1.79	1.80	1.81	1.82	1.82	1.82	1.82	1.83
40	1.76	1.79	1.80	1.80	1.81	1.81	1.81	1.82	1.82
60	1.75	1.77	1.78	1.79	1.79	1.80	1.80	1.80	1.80
120	1.73	1.75	1.77	1.77	1.78	1.78	1.78	1.78	1.78
∞	1.739	1.750	1.756	1.760	1.763	1.765	1.767	1.768	1.768

TABLE 11.8

Upper 2.5 Percentile $t_k(\alpha)$ for T_k

	k = Number of Dose Levels						
df/v	2	3	4	5	6	8	10
5	2.699	2.743	2.766	2.779	2.788	2.799	2.806
6	2.559	2.597	2.617	2.628	2.635	2.645	2.650
7	2.466	2.501	2.518	2.528	2.535	2.543	2.548
8	2.400	2.432	2.448	2.457	2.463	2.470	2.475
9	2.351	2.381	2.395	2.404	2.410	2.416	2.421
10	2.313	2.341	2.355	2.363	2.368	2.375	2.379
11	2.283	2.310	2.323	2.330	2.335	2.342	2.345
12	2.258	2.284	2.297	2.304	2.309	2.315	2.318
13	2.238	2.263	2.275	2.282	2.285	2.292	2.295
14	2.220	2.245	2.256	2.263	2.268	2.273	2.276
15	2.205	2.229	2.241	2.247	2.252	2.257	2.260
16	2.193	2.216	2.227	2.234	2.238	2.243	2.246
17	2.181	2.204	2.215	2.222	2.226	2.231	2.234
18	2.171	2.194	2.205	2.211	2.215	2.220	2.223
19	2.163	2.185	2.195	2.202	2.205	2.210	2.213
20	2.155	2.177	2.187	2.193	2.197	2.202	2.205
22	2.141	2.163	2.173	2.179	2.183	2.187	2.190
24	2.130	2.151	2.161	2.167	2.171	2.175	2.178
26	2.121	2.142	2.151	2.157	2.161	2.165	2.168
28	2.113	2.133	2.143	2.149	2.152	2.156	2.159
30	2.106	2.126	2.136	2.141	2.145	2.149	2.151
35	2.093	2.112	2.122	2.127	2.130	2.134	2.137
40	2.083	2.102	2.111	2.116	2.119	2.123	2.126
60	2.060	2.078	2.087	2.092	2.095	2.099	2.101
120	2.037	2.055	2.063	2.068	2.071	2.074	2.076
∞	2.015	2.032	2.040	2.044	2.047	2.050	2.052

$$
\begin{aligned}
\text{Power} &= \Pr\left\{reject\right\}H_o|\,\mu_i \ge \mu_0 + \Delta \text{ for some } i\} \\
&= \{\Pr\left\{reject\right\}H_o|\,\mu_0 \ge \mu_1 = \ldots = \mu_K = \mu_0 + \Delta\} \\
&\ge \Pr\left\{\frac{\hat{Y}_K - \hat{Y}_0}{\sigma\sqrt{2/n}} > t_K(\alpha)\,|\,\mu_K = \mu_0 + \Delta\right\} \\
&= 1 - \Phi\left(t_K(\alpha) - \frac{\Delta}{\sigma\sqrt{2/n}}\right),
\end{aligned}
$$

where Δ is the clinically meaningful minimal difference. To have a power of $1 - \beta$, required sample size per group can be obtained by solving

$$
\beta = \Phi\left(t_K(\alpha) + \frac{\Delta}{\sigma\sqrt{2/n}} z_\beta\right).
$$

TABLE 11.9

Upper 1 Percentile $t_k(\alpha)$ for T_k

					k = Number of Dose Levels				
df/v	2	3	4	5	6	7	8	9	10
5	3.50	3.55	3.57	3.59	3.60	3.60	3.61	3.61	3.61
6	3.26	3.29	3.31	3.32	3.33	3.34	3.34	3.34	3.35
7	3.10	3.13	3.15	3.16	3.16	3.17	3.17	3.17	3.17
8	2.99	3.01	3.03	3.04	3.04	3.05	3.05	3.05	3.05
9	2.90	2.93	2.94	2.95	2.95	2.96	2.96	2.96	2.96
10	2.84	2.86	2.88	2.88	2.89	2.89	2.89	2.90	2.90
11	2.79	2.81	2.82	2.83	2.83	2.84	2.84	2.84	2.84
12	2.75	2.77	2.78	2.79	2.79	2.79	2.80	2.80	2.80
13	2.72	2.74	2.75	2.75	2.76	2.76	2.76	2.76	2.76
14	2.69	2.71	2.72	2.72	2.72	2.73	2.73	2.73	2.73
15	2.66	2.68	2.69	2.70	2.70	2.70	2.71	2.71	2.71
16	2.64	2.66	2.67	2.68	2.68	2.68	2.68	2.68	2.69
17	2.63	2.64	2.65	2.66	2.66	2.66	2.66	2.67	2.67
18	2.61	2.63	2.64	2.64	2.64	2.65	2.65	2.65	2.65
19	2.60	2.61	2.62	2.63	2.63	2.63	2.63	2.63	2.63
20	2.58	2.60	2.61	2.61	2.62	2.62	2.62	2.62	2.62
22	2.56	2.58	2.59	2.59	2.59	2.60	2.60	2.60	2.60
24	2.55	2.56	2.57	2.57	2.57	2.58	2.58	2.58	2.58
26	2.53	2.55	2.55	2.56	2.56	2.56	2.56	2.56	2.56
28	2.52	2.53	2.54	2.54	2.55	2.55	2.55	2.55	2.55
30	2.51	2.52	2.53	2.53	2.54	2.54	2.54	2.54	2.54
35	2.49	2.50	2.51	2.51	2.51	2.51	2.52	2.52	2.52
40	2.47	2.48	2.49	2.49	2.50	2.50	2.50	2.50	2.50
60	2.43	2.45	2.45	2.46	2.46	2.46	2.46	2.46	2.46
120	2.40	2.41	2.42	2.42	2.42	2.42	2.42	2.42	2.43
∞	2.366	2.377	2.382	2.385	2.386	2.387	2.388	2.389	2.389

Thus, we have

$$n = \frac{2\sigma^2 [t_k(\alpha) + z_\beta]^2}{\Delta^2},$$ (11.11)

where values of $t_K(\alpha)$ can be obtained from Tables 11.7 through 11.10. It should be noted that this approach is conservative.

An Example

We consider the previous example of an asthma trial with power = 80%, $\sigma = 0.22$, and one-sided $\alpha = 0.05$. (Note that there is no two-sided Williams' test.) Since the critical value $t_k(\alpha)$ is dependent on the degree of freedom v that is related to the sample size n, iterations are usually needed. However, for the current case, we know that $v > 120$ or ∞, which leads

TABLE 11.10

Upper 0.5 Percentile $t_k(a)$ for T_k

df/v	*k* = Number of Dose Levels						
	2	**3**	**4**	**5**	**6**	**8**	**10**
5	4.179	4.229	4.255	4.270	4.279	4.292	4.299
6	3.825	3.864	3.883	3.895	3.902	3.912	3.197
7	3.599	3.631	3.647	3.657	3.663	3.670	3.674
8	3.443	3.471	3.484	3.492	3.497	3.504	3.507
9	3.329	3.354	3.366	3.373	3.377	3.383	3.886
10	3.242	3.265	3.275	3.281	3.286	3.290	3.293
11	3.173	3.194	3.204	3.210	3.214	3.218	3.221
12	3.118	3.138	3.147	3.152	3.156	3.160	3.162
13	3.073	3.091	3.100	3.105	3.108	3.112	3.114
14	3.035	3.052	3.060	3.065	3.068	3.072	3.074
15	3.003	3.019	3.027	3.031	3.034	3.037	3.039
16	2.957	2.991	2.998	3.002	3.005	3.008	3.010
17	2.951	2.955	2.973	2.977	2.980	2.938	2.984
18	2.929	2.944	2.951	2.955	2.958	2.960	2.962
19	2.911	2.925	2.932	2.936	2.938	2.941	2.942
20	2.894	2.903	2.915	2.918	2.920	2.923	2.925
22	2.866	2.879	2.855	2.889	2.891	2.893	2.895
24	2.842	2.855	2.861	2.864	2.866	2.869	2.870
26	2.823	2.835	2.841	2.844	2.846	2.848	2.850
28	2.806	2.819	2.824	2.827	2.829	2.831	2.832
30	2.792	2.804	2.809	2.812	2.814	2.816	2.817
35	2.764	2.775	2.781	2.783	2.785	2.787	2.788
40	2.744	2.755	2.759	2.762	2.764	2.765	2.766
60	2.697	2.707	2.711	2.713	2.715	2.716	2.717
120	2.651	2.660	2.664	2.666	2.667	2.669	2.669
∞	2.607	2.615	2.618	2.620	2.621	2.623	2.623

to $t_3(0.05) = 1.75$. Thus, the sample size for 11% (9%–16%) treatment improvement over placebo in FEV1 is given by

$$n = \frac{2(0.22)^2(1.75 + 0.8415)}{0.11^2} = 53 \text{ per group.}$$

Note that this sample size formulation has a minimum difference from that based on the two-sample *t*-test with the maximum treatment difference as the treatment difference. For the current example, $n = 54$ from the two-sample *t*-test.

11.5 Cochran–Armitage's Test for Trend

Cochran–Armitage test (Cochran 1954; Armitage 1955) is a widely used test for monotonic trend with binary response since it is more powerful than the chi-square homogeneity test in identifying a monotonic trend (Nam 1987). This test requires preassigned fixed dose scores. Equally spaced scores are most powerful for linear response. When the true dose–response relationship is not linear, equally spaced scores may not be the best choice. Generally, single contrast-based test attains its greatest power when the dose–coefficient relationship has the same shape as the true dose–response relationship. Otherwise, it loses its power. Owing to limited information at the design stage of a dose–response trial, it is risky to use single contrast test. Note that the rejection of the null hypothesis does not mean the dose–response is linear or monotonic. It means that based on the data, it is unlikely that the dose–response is flat or all doses have the same response.

To test for monotonic trend with binary response, we consider the following hypotheses:

$$H_0 \ : \ p_0 = p_1 = ... = p_k$$
$$\text{versus} \quad H_\alpha \ : \ p_0 \le p_1 \le ... \le p_k \text{ with } p_0 < p_k. \tag{11.12}$$

Cochran (1954) and Armitage (1955) proposed the following test statistic:

$$T_{CA} = \sqrt{\frac{N}{(N-X)X}} \frac{\sum_{i=1}^{k}(x_i - (n_i X/N)c_i}{\sqrt{\sum_{i=0}^{k}\left(n_i c_i^2/N\right) - \left(\sum_{i=0}^{k}(n_i c/N)\right)^2}}, \tag{11.13}$$

where c_i is the predetermined scores ($c_0 < c_1 < ... < c_k$), x_i is the number of responses in group i ($i = 0$ for the control group), and p_i is the response rate in group i. n_i is the sample size for group i, where $X = \sum_{i=0}^{k} x_i$ and $N = \sum_{i=0}^{k} n_i$.

Note that the test (one-sided test) by Portier and Hoel (1984) was the modification (Neuhauser and Hothorn, 1999) from Armitages' (1955) original two-sided test, which is asymptotically distributed as a standard normal variable under the null hypothesis. Asymptotic power of test for linear trend and sample size calculation can be found in Nam (1998), which are briefly outlined below:

Let x_i be the $k+1$ mutually independent binomial variates representing the number of responses among n_i subjects at dose level d_i for $i = 0, 1, ... , k$. Define average response rate p_i

$$p_i = \frac{1}{N}\sum_i x_i, \bar{q} = 1 - \bar{p}, \quad \text{and} \quad \bar{d} = \frac{1}{N}\sum_{n_i} n_i d_i. \ U = \sum_i x_i(d_i - \bar{d}).$$

Assume that the probability of response follows a linear trend in logistic scale

$$p_i = \frac{e^{\gamma + \lambda d_i}}{1 + e^{\gamma + \lambda d_i}}.$$

An approximate test with continuity correction based on the asymptotically normal deviate is given by

$$z = \frac{(U - \frac{\Delta}{2})}{\sqrt{var(U | H_0 : \lambda = 0)}} = \frac{(U - \frac{\Delta}{2})}{\sqrt{\bar{p}\bar{q}\sum_i\left[\sum_i n_i(d_i - \bar{d})^2\right]}},$$

where $\Delta/2 = (d_i - d_{i-1})/2$ is the continuity correction for equally spaced doses. However, there is no constant Δ for unequally spaced doses.

The unconditional power is given by

$$\Pr\ (z \geq z_{1-\alpha}\ H_a) = 1 - \Phi\ (u),$$

where

$$u = E\left(U - \frac{\Delta}{2}\right) + z_{1-\alpha}\frac{\sqrt{\mathrm{var}(U|H_0)}}{\sqrt{\mathrm{var}(U|H_0)}}.$$

Thus, we have

$$E(U) - \frac{\Delta}{2} + z_{1-\alpha}\sqrt{\mathrm{var}(U|H_0)} + z_{1-\beta}\sqrt{\mathrm{var}(U|H_\alpha)}.$$

For $\Delta = 0$, that is, without continuity correction, the sample size is given by

$$n_0^* = \frac{1}{A^2}\left\{z_{1-\alpha}\sqrt{pq\left[\sum r_i(d_i - \bar{d})^2\right]} + z_{1-\beta}\sqrt{\left[\sum p_i q_i r_i(d_i - \bar{d})^2\right]}\right\}^2, \tag{11.14}$$

where $A = \sum r_i p_i(d_i - \bar{d})$, $p = 1/N\sum n_i p_i$, $d = 1-p$ and $r_i = n_i/n_0$ sample size ratio between the ith group and the control.

On the other hand, sample size with continuity correction is given by

$$n_0 = \frac{n_0^*}{4}\left[1 + \sqrt{1 + 2\frac{\Delta}{An_0^*}}\right]^2. \tag{11.15}$$

Note that the actual power of the test depends on the specified alternative. Thus, the sample size formula holds for any monotonic increasing alternative, that is, $p_i-1 < p_i$, $i = 1, \ldots, k$.

For balance design with equal size in each group, the formula for sample size per group is reduced to

$$n = \frac{n^*}{4}\left[1 + \sqrt{1 + \frac{2}{Dn^*}}\right]^2, \tag{11.16}$$

where

$$n^* = \left\{z_{1-\alpha}\sqrt{k(k^2 - 1)pq} + z_{1-\beta}\sqrt{\sum b_i^2 p_i q_i}\right\}^2 \tag{11.17}$$

TABLE 11.11

Sample Size from Nam Formula

Dose	1	2	3	4	Total n
	0.1	0.3	0.5	0.7	26
	0.1	0.4	0.4	0.7	32
Response	0.1	0.4	0.7	0.5	48
	0.1	0.1	0.1	0.6	37
	0.1	0.6	0.6	0.6	49

and $b_i = i - 0.5k$, and $D = \sum b_i p_i$.

Note that the above formula is based on one-sided test at the α level. For two-sided test, the Type I error rate is controlled at the 2α level. For equally spaced doses: 1, 2, 3, and 4, the sample sizes required for the five different sets of contracts are given in Table 11.11.

Neuhauser and Hothorn (1999) studied the power of Cochran–Armitage test under different true response shape through simulations (Table 11.12). These simulation results confirm that the most powerful test is achieved when contrast shape is consistent with the response shape.

Gastwirth (1985) and Podgor et al. (1996) proposed a single maximum efficiency robust test (MERT) statistic based on prior correlations between different contrasts, while Neuhauser and Hothorn (1999) proposed a maximum test among two or more contrasts and claim a gain in power.

11.6 Dose Escalation Trials

For non-life-threatening diseases, since the expected toxicity is mild and can be controlled without harm, phase I trials are usually conducted on healthy or normal volunteers. In life-threatening diseases such as cancer and AIDS, phase I studies are conducted with limited numbers of patients due to (i) the aggressiveness and possible harmfulness of treatments, (ii) possible systemic treatment effects, and (iii) the high interest in the new drug's efficacy in those patients directly.

TABLE 11.12

Power Comparisons with Cochran–Armitage Test

Shape	Equal Spaced Scores	Convex Scores	Concave Scores	MERT
No difference	0.03	0.03	0.03	0.03
Linear	0.91	0.82	0.89	0.92
Step	0.84	0.80	0.87	0.90
Umbrella	0.68	0.24	0.86	0.62
Convex	0.81	0.91	0.48	0.83
Concave	0.67	0.34	0.91	0.74

Source: Neuhauser, M. and Hothorn, L. 1999. *Computational Statistics and Data Analysis*, 30, 403–412.
Note: $n_i = 10$, $\alpha = 0.05$.

Drug toxicity is considered as tolerable if the toxicity is manageable and reversible. The standardization of the level of drug toxicity is the Common Toxicity Criteria (CTC) of the United States National Cancer Institute (NCI). Any adverse event (AE) related to treatment from the CTC category of Grade 3 and higher is often considered a dose-limiting toxicity (DLT). The MTD is defined as the maximum dose level with toxicity rates occurring no more than a predetermined value.

There are usually 5 to 10 predetermined dose levels in a dose escalation study. A commonly used dose sequence is the so-called modified Fibonacci sequence. Patients are treated with lowest dose first and then gradually escalated to higher doses if there is no major safety concern. The rules for dose escalation are predetermined. The commonly employed dose escalation rules are the traditional escalation rules (TER), also known as the "3 + 3" rule. The "3 + 3" rule is to enter three patients at a new dose level and enter another three patients when one toxicity is observed. The assessment of the six patients will be performed to determine whether the trial should be stopped at that level or to increase the dose. Basically, there are two types of the "3 + 3" rules, namely, TER and strict TER (or STER). TER does not allow dose de-escalation, but STER does when two of three patients have DLTs. The "3 + 3" STER can be generalized to the $A+B$ TER and STER escalation rules. To introduce the traditional $A + B$ escalation rule, let A, B, C, D, and E be integers. The notation A/B indicates that there are A toxicity incidences out of B subjects and $>A/B$ means that there are more than A toxicity incidences out of B subjects. We assume that there are n predefined doses with increasing levels and let p_i be the probability of observing a DLT at dose level i for $1 \leq i \leq n$. In what follows, general $A+B$ designs without and with dose de-escalation will be described. The closed forms of sample size calculation by Lin and Shih (2001) are briefly reviewed.

11.6.1 $A + B$ Escalation Design without Dose De-Escalation

The general $A + B$ designs without dose de-escalation can be described as follows. Suppose that there are A patients at dose level i. If less than C/A patients have DLTs, then the dose is escalated to the next dose level $i + 1$. If more than D/A (where $D \geq C$) patients have DLTs, then the previous dose $i - 1$ will be considered the MTD. If no less than C/A but no more than D/A patients have DLTs, B more patients are treated at this dose level i. If no more than E (where $E \geq D$) of the total of $A + B$ patients have DLTs, then the dose is escalated. If more than E of the total of $A + B$ patients have DLT, then the previous dose $i - 1$ will be considered the MTD. It can be seen that the traditional "3 + 3" design without dose de-escalation is a special case of the general $A + B$ design with $A = B = 3$ and $C = D = E = 1$.

Under the general $A + B$ design without dose escalation, the probability of concluding that MTD has reached at dose i is given by

$$P(MTD = dose\ i) = P\left(\begin{array}{l} \text{escalation at dose} \leq i \text{ and} \\ \text{stop escalation at dose } i+1 \end{array}\right)$$

$$= (1 - P_0^{i+1} - Q_0^{i+1})\left(\prod_{j=1}^{i}(P_0^j + Q_0^j)\right), 1 \leq i < n,$$

where

$$P_0^j = \sum_{k=0}^{C-1} \binom{A}{k} p_j^k (1-p_j)^{A-k},$$

and

$$Q_0^j = \sum_{k=C}^{D} \sum_{m=0}^{E-k} \binom{A}{k} p_j^k (1-p_j)^{A-k} \binom{B}{m} p_j^m (1-p_j)^{B-m},$$

in which

$$N_{ji} \begin{cases} \dfrac{AP_0^j + (A+B)Q_0^j}{P_0^j + Q_0^j} & \text{if } j < i+1 \\ \dfrac{A(1-P_0^j - P_1^j) + (A+B)(P_1^j - Q_0^j)}{1-P_0^j - Q_0^j} & \text{if } j = i+1 \\ 0 & \text{if } j > i+1 \end{cases}$$

An overshoot is defined as an attempt to escalate to a dose level at the highest level planned, while an undershoot is referred to as an attempt to de-escalate to a dose level at a lower dose than the starting dose level. Thus, the probability of undershoot is given by

$$P_1^* = P(MTD < dose\,1) = (1 - P_0^1 - Q_0^1), \tag{11.18}$$

and probability of overshot is given by

$$P_n^* = P(MTD \geq dose\,n) = \prod_{j=1}^{n} (P_0^j + Q_0^j). \tag{11.19}$$

The expected number of patients at dose level j is given by

$$N_j = \sum_{i=0}^{n-1} N_{ji} P_i^*. \tag{11.20}$$

Note that without consideration of undershoots and overshoots, the expected number of DLTs at dose i can be obtained as $N_i p_i$. As a result, the total expected number DLTs for the trial is given by $\sum_{i=1}^{n} N_i p_i$.

We can use Equation 11.20 to calculate the expected sample size at dose level for given toxicity rate at each dose level. We can also conduct a Monte Carlo study to simulate the trial and sample size required. Table 11.13 summarizes the simulation results. One can do two-stage design and Bayesian adaptive and other advanced design with the software.

11.6.2 *A* + *B* Escalation Design with Dose De-Escalation

Basically, the general $A + B$ design with dose de-escalation is similar to the design without dose de-escalation. However, it permits more patients to be treated at a lower dose (i.e., dose de-escalation) when excessive DLT incidences occur at the current dose level. The dose de-escalation occurs when more than D/A (where $D \geq C$) or more than $E/(A + B)$ patients have DLTs at dose level i. In this case, B more patients will be treated at dose level $i - 1$ provided that only A patients have been previously treated at this prior dose. If more than A patients have already been treated previously, then dose $i - 1$ is the MTD. The de-escalation may continue to the next dose level $i - 2$ and so on, if necessary. For this design, the MTD is the dose level at which no more than $E/(A + B)$ patients experience DLTs, and more than D/A or (no less than C/A and no more than D/A) if more than $E/(A + B)$ patients treated with the next higher dose have DLTs.

Similarly, under the general $A + B$ design with dose de-escalation, the probability of concluding that MTD has been reached at dose i is given by

$$P_i^* = P(MTD = dose\ i) = P\begin{pmatrix} \text{escalation at dose} \leq i \text{ and} \\ \text{stop escalation at dose } i+1 \end{pmatrix}$$

$$= \sum_{k=i+1}^{n} p_{ik},$$

where

$$p_{ik} = (Q_0^i + Q_0^i)(1 - P_0^k - Q_0^k)\left(\prod_{j=1}^{i-1}(P_0^j + Q_0^i)\right)\prod_{j=i+1}^{k-1} Q_2^j,$$

and

$$P_1^j = \sum_{i=C}^{D}\binom{A}{k}p_j^k(1-p_j)^{A-k},$$

$$Q_1^j = \sum_{k=0}^{C-1}\sum_{m=0}^{E-k}\binom{A}{k}p_j^k(1-p_j)^{A-k}\binom{B}{m}p_j^m(1-p_j)^{B-m},$$

TABLE 11.13

Simulation Results with 3+3 TER

Dose Level	1	2	3	4	5	6	7	Total
Dose	10	15	23	34	51	76	114	
DLT rate	0.01	0.014	0.025	0.056	0.177	0.594	0.963	
Expected n	3.1	3.2	3.2	3.4	3.9	2.8	0.2	19.7

Note: True MTD = 50, mean simulated MTD = 70, mean number of DLTs = 2.9.

$$Q_2^j = \sum_{k=0}^{C-1} \sum_{m=E+1-k}^{E-k} \binom{A}{k} p_j^k (1-p_j)^{A-k} \binom{B}{m} p_j^m (1-p_j)^{B-m},$$

$$N_{jn} = \frac{AP_0^j + (A+B)Q_0^j}{P_0^j + Q_0^j}.$$

Also, the probability of undershoot is given by

$$P_i^* = P(MTD < dose\ 1) = \sum_{k=1}^{n} \left\{ \left(\Pi_{j=1}^{k-1} Q_2^j \right)(1 - P_0^k - Q_0^k) \right\},$$

and the probability of overshooting is

$$P_n^* = P\ (MTD \geq dose\ n) = \prod_{j=1}^{n} (P_0^j + Q_0^j).$$

The expected number of patients at dose level j is given by

$$N_j = N_{jn}P_n^* + \sum_{i=0}^{n-1} \sum_{k=i+1}^{n} N_{jik}p_{ik},$$

where

$$N_{jik} = \begin{cases} \frac{AP_0^j + (A+B)Q_0^j}{P_0^j + Q_0^j} & \text{if } j < i \\ A+B & \text{if } i \leq j < k \\ \frac{A(1-P_1^j - P_1^j) + (A+B)(P_1^j - Q_0^j)}{1 - P_0^j - Q_0^j} & \text{if } j = k \\ 0 & \text{if } j > k \end{cases}.$$

Consequently, the total number of expected DLTs is given by $\sum_{i=1}^{n} N_i p_i$.

Table 11.14 is another example as in Table 11.13, but the simulation results are from STER rather than TER. In this example, we can see that the MTD is underestimated and the average sample size is 23 with STER, three patients more than that with TER. The excepted DLTs also increase with STER in this case. Note that the actual sample size varies from trial to trial. However, simulations will help in choosing the best escalation algorithm

TABLE 11.14

Simulation Results with 3 + 3 STER

Dose Level	1	2	3	4	5	6	7	Total
Dose	10	15	23	34	51	76	114	
DLT rate	0.01	0.014	0.025	0.056	0.177	0.594	0.963	
Expected n	3.1	3.2	3.5	4.6	5.5	3	0.2	23

Note: True MTD = 50, mean simulated MTD = 41. Mean number of DLTs = 3.3.

or optimal design based on the operating characteristics, such as accuracy and precision of the predicted MTD, expected DLTs and sample size, overshoots, undershoots, and the number of patients treated above MTD.

11.7 Concluding Remarks

In general, linear contrast tests are useful in detecting specific shapes of the dose–response curve. However, the selection of contrasts should be practically meaningful. It should be noted that the power of a linear contrast test is sensitive to the actual shape of the dose–response curve (Bretz and Hothorn, 2002). Alternatively, one may consider a slope approach to detect the shape of the dose–response curve (Cheng et al., 2006).

Williams' test is useful for identifying the minimum effective dose in the case of continuous response. Williams' test has a strong assumption of monotonic dose–response. The test may not be statistically valid if the assumption is violated. It should be noted that Williams' test is not a test for monotonicity. The sample size formula given in Equation 11.11 is rather conservative.

Nam's and Cochran–Armitage's methods are equivalent. Their methods are useful when the response is binary. Basically, both methods are regression-based methods for testing a monotonic trend. However, they are not rigorous tests for monotonicity. Testing to true monotonic response is practically difficult without extra assumptions (Chang and Chow, 2005).

The dose escalation trials are somewhat different because the sample size is not determined based on the error rates. Instead, it is determined by the escalation algorithm and dose–response (toxicity) relationship and predetermined dose levels. For the $A + B$ escalation rules, the sample size has a closed form as given in Section 11.6. For other designs, sample size will have to be estimated through computer simulations. It should be noted that the escalation algorithm and dose intervals not only have an impact on the sample size, but also affect other important operating characteristics such as the accuracy and precision of the estimation of the MTD and the number of DLTs.

12

Microarray Studies

One of the primary study objectives for microarray studies is to have a high probability of declaring genes to be differentially expressed if they are truly expressed, while keeping the probability of making false declarations of expression acceptably low (Lee and Whitmore, 2002). Traditional statistical testing approaches such as the two-sample t-test or Wilcoxon test are often used for evaluating the statistical significance of informative expressions but require adjustment for large-scale multiplicity. It is recognized that if a type I error rate of α is employed at each testing, then the probability to reject any hypothesis will exceed the overall α level. To overcome this problem, two approaches for controlling false discovery rate (FDR) and family-wise error rate (FWER) are commonly employed. In this chapter, formulas or procedures for sample size calculation for microarray studies derived under these two approaches are discussed.

In Section 1, a brief literature review is given. Section 12.2 gives a brief definition of FDR and introduces formulas and/or procedures for sample size calculation for the FDR approach given in Jung (2005). Also included in this section are some examples with and without constant effect sizes based on two-sided tests. Section 12.3 reviews multiple testing procedures and gives procedures for sample size calculation for microarray studies for the FWER approach (Jung, Bang, and Young, 2005). Also included in this section is an application to leukemia data given in Golub et al. (1999). A brief concluding remark is given in Section 12.4.

12.1 Literature Review

Microarray methods have been widely used for identifying differentially expressing genes in subjects with different types of disease. Sample size calculation plays an important role at the planning stage of a microarray study. Commonly considered standard microarray designs include a matched-pairs design, a completely randomized design, an isolated-effect design, and a replicated design. For a given microarray study, formulas and/or procedures for sample size calculation can be derived following the steps as described in the previous chapters. Several procedures for sample size calculation have been proposed in the literature in the microarray context (see, e.g., Simon, Radmacher, and Dobbin, 2002). Most of these procedures focused on exploratory and approximate relationships among statistical power, sample size (or the number of replicates), and effect size (often, in terms of fold-change), and used the most conservative Bonferroni adjustment for controlling FWER without taking into consideration the underlying correlation structure (see, e.g., Wolfinger et al., 2001; Black and Doerge, 2002; Pan, Lin, and Le, 2002; Cui and Churchill, 2003). Jung, Bang, and Young (2005) incorporated the correlation structure to derive a sample size formula, which is able to control the FWER efficiently.

As an alternative to the FWER approach, many researchers have proposed the use of the so-called FDR (see, e.g., Benjamini and Hochberg, 1995; Storey, 2002). It is believed that controlling *FDR* would relax the multiple testing criteria compared to controlling the *FWER*. Consequently, controlling *FDR* would increase the number of declared significant genes. Some operating and numerical characteristics of *FDR* are elucidated in recent publications (Genovese and Wasserman, 2002; Dudoit, Schaffer, and Boldrick, 2003). Lee and Whitmore (2002) considered multiple group cases, including the two-sample case, using ANOVA models and derived the relation between the effect sizes and the *FDR* based on a Bayesian perspective. Their power analysis approach, however, does not consider the issue of multiplicity. Müller et al. (2004) chose a pair of testing errors, including *FDR*, and minimized one while controlling the other at a specified level using a Bayesian decision rule. Müller et al. (2004) proposed using an algorithm to demonstrate the relationship between sample size and the chosen testing errors based on some asymptotic results for large samples. This approach, however, requires specification of complicated parametric models for prior and data distributions, and extensive computing for the Baysian simulations. Note that Lee and Whitmore (2002) and Gadbury et al. (2004) modeled a distribution of *p*-values from pilot studies to produce sample size estimates but did not provide an explicit sample size formula. Most of the existing methods for controlling *FDR* in microarray studies fail to show the explicit relationship between sample size and effect sizes due to various reasons. To overcome this problem, Jung (2005) proposed a sample size estimation procedure for controlling *FDR*, which will be introduced in the following section.

In this chapter, our emphasis will be on formulas or procedures for sample size calculation derived based on two approaches for controlling *FDR* and *FWER*.

12.2 *FDR* Control

Benjamini and Hochberg (1995) define the *FDR* as the expected value of the proportion of the nonprognostic genes among the discovered genes. It is then suggested that sample size should be selected to control the *FDR* at a prespecified level of significance.

12.2.1 Model and Assumptions

Suppose that we conduct m multiple tests, of which the null hypotheses are true for m_0 tests and the alternative hypotheses are true for $m_1 (= m - m_0)$ tests. The tests declare that, of the m_0 null hypotheses, A_0 hypotheses are null (true negative), and R_0 hypotheses are alternative (i.e., false rejection, false discovery, or false positive). Among the m_1 alternative hypotheses, A_1 are declared null (i.e., false negative), and R_0 are declared alternative (i.e., true rejection, true discovery, or true positive). Table 12.1 summarizes the outcome of m hypothesis tests.

TABLE 12.1

Outcomes of m Multiple Tests

	Accepted hypothesis		
True hypothesis	Null	Alternative	Total
Null	A_0	R_0	m_0
Alternative	A_1	R_1	m_1
Total	A	R	m

According to the definition by Benjamini and Hochberg (1995), the *FDR* is given by

$$FDR = E\left(\frac{R_0}{R}\right). \tag{12.1}$$

Note that this expression is undefined if $\Pr(R = 0) > 0$. To avoid this issue, Benjamini and Hochberg (1995) modified the definition of *FDR* as

$$FDR = \Pr(R > 0)E\left(\frac{R_0}{R} \mid R > 0\right). \tag{12.2}$$

These two definitions are identical if $\Pr(R = 0) = 0$, in which case we have $FDR = E(R_0/R \mid R > 0)$. Note that if $m = m_0$, then $FDR = 1$ for any critical value with $\Pr(R = 0) = 0$. As a result, Storey (2003) referred to the second term in the right-hand side of Equation 12.2 as *pFDR*, that is,

$$pFDR = E\left(\frac{R_0}{R} \mid R > 0\right)$$

and proposed controlling this quantity instead of *FDR*. Storey (2002) indicated that $\Pr(R > 0) \approx 1$ with a large m. In this case, *pFDR* is equivalent to *FDR*. Thus, throughout this chapter, we do not distinguish between *FDR* and *pFDR*. Hence, definitions (12.1) and (12.2) are considered to be equivalent. Benjamini and Hochberg (1995) proposed a multistep procedure to control the *FDR* at a specified level. Their methods, however, are conservative and the conservativeness increases as m_0 increases (Storey, Taylor, and Siegmund, 2004).

Suppose that, in the jth testing, we reject the null hypothesis H_j if the p-value p_j is smaller than or equal to $\alpha \in (0, 1)$. Assuming independence of the m p-values, we have

$$R_0 = \sum_{j=1}^{m} I(H_j \text{ true}, H_j \text{ rejected})$$

$$= \sum_{j=1}^{m} \Pr(H_j \text{ true})\Pr(H_j \text{ rejected} \mid H_j) + o_p(m),$$

which equals $m_0\alpha$, where $m^{-1}o_p(m) \to 0$ in probability as $m \to \infty$ (Storey, 2002). Ignoring the error term, we have

$$FDR(\alpha) = \frac{m_0\alpha}{R(\alpha)}, \tag{12.3}$$

where $R(\alpha) = \sum_{j=1}^{m} I(p_j \leq \alpha)$. Note that for a given α, the estimation of *FDR* by Equation 12.3 requires the estimation of m_0.

For the estimation of m_0, Storey (2002) considered that the histogram of m p-values is a mixture of (i) m_0 p-values that are corresponding to the true null hypotheses and following

$U(0,1)$ distribution and (ii) m_1 p-values that are corresponding to the alternative hypotheses and expected to be close to 0. Consequently, for a chosen constant λ away from 0, none (or few, if any) of the m_1 p-values will fall above λ, so that the number of p-values above λ, $\sum_{j=1}^{m} I(p_j > \lambda)$, can be approximated by the expected frequency among the m_0 p-values above λ from $U(0,1)$ distribution, that is, $m_0(1 - \lambda)$. Hence, for a given λ, m_0 can be estimated by

$$\hat{m}_0(\lambda) = \frac{\sum_{j=1}^{m} I(p_j > \lambda)}{1 - \lambda}.$$

By combining this m_0 estimator with Equation 12.3, Storey (2002) obtained the following estimator for $FDR(\alpha)$:

$$\widehat{FDR}(\alpha) = \frac{\alpha \times \hat{m}_0(\lambda)}{R(\alpha)} = \frac{\alpha \sum_{j=1}^{m} I(p_j > \lambda)}{(1 - \lambda)\sum_{j=1}^{m} I(p_j \le \alpha)}.$$

For an observed p-value p_j, Storey (2002) defined the minimum FDR level at which we reject H_j as q-value, which is given by

$$q_j = \inf_{\alpha \ge p_j} \widehat{FDR}(\alpha).$$

When $FDR(\alpha)$ is strictly increasing in α, the above formula can be reduced to

$$q_j = \widehat{FDR}(p_j).$$

It can be verified that this assumption holds if the power function of the individual tests is concave in α, which is the case when the test statistics follow a standard normal distribution under the null hypotheses. We would reject H_j (or, equivalently, discovered gene j) if q_j is smaller than or equal to the prespecified FDR level.

Note that the primary assumption of independence among m test statistics was relaxed to independence only among m_0 test statistics corresponding to the null hypotheses by Storey and Tibshirani (2001), and to weak independence among all m test statistics by Storey (2003) and Story, Taylor, and Siegmund (2004).

12.2.2 Sample Size Calculation

In this subsection, formulas and/or procedures for sample size calculation based on the approach for controlling FDR proposed by Jung (2005) will be introduced. Let M_0 and M_1 denote the set of genes for which the null and alternative hypotheses are true, respectively. Note that the cardinalities of M_0 and M_1 are m_0 and m_1, respectively. Since the estimated FDR is invariant to the order of the genes, we may rearrange the genes and set $M_1 = \{1, \ldots, m_1\}$ and $M_0 = \{m_1 + 1, \ldots, m\}$. By Storey (2002) and Storey and Tibshirani (2001), for large m and under independence (or weak dependence) among the test statistics, we have

$$R(\alpha) = E(R_0(\alpha)) + E(R_1(\alpha)) + o_p(m)$$
$$= m_0\alpha + \sum_{j \in M_1} \xi_j(\alpha) + o_p(m),$$

where $R_h(\alpha) = \sum_{j \in M_h} I(p_j \leq \alpha)$ for $h = 0, 1$, $\xi_j(\alpha) = P (p_j \leq \alpha)$ is the marginal power of the single α-test applied to gene $j \in M_1$. From Equation 12.3, we have

$$FDR(\alpha) = \frac{m_0 \alpha}{m_0 \alpha + \sum_{j \in M_1} \xi_j(\alpha)} \tag{12.4}$$

by omitting the error term.

Let X_{ij} (Y_{ij}) denote the expression level of gene j for subject i in group 1 (and group 2, respectively) with common variance σ_j^2. For simplicity, we consider two-sample t-tests,

$$T_j = \frac{\bar{X}_j - \bar{Y}_j}{\hat{\sigma}_j \sqrt{n_1^{-1} + n_2^{-1}}},$$

for hypothesis j (=1, ... , m), where n_k is the number of subjects in group k (=1, 2), $\bar{X}_j$ and $\bar{Y}_j$ are sample means of $\{X_{ij}, i = 1, ... , n_1\}$ and $\{Y_{ij}, i = 1, ... , n_2\}$, respectively, and $\hat{\sigma}_j^2$ is the pooled sample variance. We assume a large sample (i.e., $n_k \to \infty$), so that $T_j \sim N (0, 1)$ for $j \in M_0$. Let $n = n_1 + n_2$ denote the total sample size, and $a_k = n_k/n$ the allocation proportion for group k.

Let δ_j denote the effect size for gene j in the fraction of its standard error, that is,

$$\delta_j = \frac{E(X_j) - E(Y_j)}{\sigma_j}.$$

At the moment, we consider one-sided tests, $H_j : \delta_j = 0$ against $\bar{H}_j : \delta_j > 0$, by assuming $\delta_j > 0$ for $j \in M_1$ and $\delta_j = 0$ for $j \in M_0$. Note that, for large n,

$$T_j \sim N\left(\delta_j \sqrt{n a_1 a_2}, 1\right)$$

for $j \in M_1$. Thus, we have

$$\xi_j(\alpha) = \bar{\Phi}\left(z_\alpha - \delta_j \sqrt{n a_1 a_2}\right),$$

where $\bar{\Phi}(\cdot)$ denotes the survivor function and $z_\alpha = \bar{\Phi}^{-1}(\alpha)$ is the upper 100α-th percentile of $N (0, 1)$. Hence, Equation 12.2 is expressed as

$$FDR(\alpha) = \frac{m_0 \alpha}{m_0 \alpha + \sum_{j \in M_1} \bar{\Phi}\left(z_\alpha - \delta_j \sqrt{n a_1 a_2}\right)}. \tag{12.5}$$

From Equation 12.5, *FDR* is decreasing in δ_j, n, and $|a_1 - 1/2|$. Further, *FDR* is increasing in α. To verify this, it suffices to show that, for $j \in M_1$, $g(\alpha) = \xi_j(\alpha)/\alpha$ is decreasing in α, or $g'(\alpha) = \alpha^{-1}\{\xi_j'(\alpha) - \alpha^{-1}\xi_j(\alpha)\}$ is negative for all $\alpha \in (0, 1)$. Note that the latter condition holds if $\xi_j(\alpha)$ is concave in α. For this purpose, we assume that the test statistics follow the standard normal distribution under the null hypotheses. Let $\phi(t) = 1/\sqrt{2\pi} \exp(-z^2/2)$ and $\bar{\Phi} = \int_z^\infty dt$

denote the probability density function and the survivor function of the standard normal distribution, respectively. Noting that $\xi_j(\alpha) = \bar{\Phi}(z_\alpha - \delta_j\sqrt{na_1a_2})$ and $z_\alpha = \bar{\Phi}^{-1}(\alpha)$, we have

$$g'(\alpha) = \frac{\alpha\phi\left(\bar{\Phi}^{-1}(\alpha) - \delta_j\sqrt{na_1a_2}\right)/\phi\left(\bar{\Phi}^{-1}(\alpha)\right) - \bar{\Phi}\left(\bar{\Phi}^{-1}(\alpha) - \delta_j\sqrt{na_1a_2}\right)}{\alpha^2}$$

$$= \frac{\bar{\Phi}(z_\alpha)\phi\left(z_\alpha - \delta_j\sqrt{na_1a_2}\right)/\phi(z_\alpha) - \bar{\Phi}\left(z_\alpha - \delta_j\sqrt{na_1a_2}\right)}{\alpha^2}.$$

Showing $g'(\alpha) < 0$ is equivalent to showing

$$\frac{\phi\left(z_\alpha - \delta_j\sqrt{na_1a_2}\right)}{\bar{\Phi}\left(z_\alpha - \delta_j\sqrt{na_1a_2}\right)} < \frac{\phi(z_\alpha)}{\bar{\Phi}(z_\alpha)},$$

which holds since $\delta_j > 0$ and $\phi(z)/\bar{\Phi}(z)$ is an increasing function by the following lemma.

Lemma 12.2.1: $\phi(z)/\bar{\Phi}(z)$ *is an increasing function.*

Proof: Let us show that

$$\ell(z) \equiv \log\left\{\frac{\phi(z)}{\bar{\Phi}(z)}\right\} = -\frac{z^2}{2} - \log\int_z^\infty \exp\left(-\frac{t^2}{2}\right)dt$$

is an increasing function. Since

$$\ell'(z) = -z + \frac{\exp(-z^2/2)}{\int_z^\infty \exp(-t^2/2)dt},$$

$l'(z) > 0$ *for* $z \leq 0$. *For* $z > 0$, $l'(z) > 0$ *if and only if*

$$L(z) \equiv \frac{1}{2}\exp\left(\frac{-z^2}{2}\right) - \int_z^\infty \exp\left(\frac{-t^2}{2}\right)dt$$

is positive. We have

$$L'(z) = -\frac{1}{z^2}\exp\left(\frac{-z^2}{2}\right) - \exp\left(\frac{-z^2}{2}\right) + \exp\left(\frac{-z^2}{2}\right)$$

$$= -\frac{1}{z^2}\exp\left(\frac{-z^2}{2}\right) < 0.$$

Hence, for $z > 0$, L *is a decreasing function, and* $\lim_{z\to 0} L(z) = \infty$ *and* $\lim_{z\to\infty} L(z) = 0$, *so that* $L(z)$ *is positive. This completes the proof.* ∎

Note that if the effect sizes are equal among the prognostic genes, *FDR* is increasing in $\pi_0 = m_0/m$. It can be verified that *FDR* increases from 0 to m_0/m as α increases from 0 to 1.

At the design stage of a microarray study, m is usually determined by the microarray chips chosen for the experiment and m_1, $\{\delta_j, j \in M_1\}$, and α_1 are projected based on past experience or data from pilot studies if any. The only variables undecided in Equation 12.5 are α and n. With all other design parameters fixed, *FDR* is controlled at a certain level by the chosen α level. Thus, Jung (2005) proposed choosing the sample size n such that it will guarantee a certain number, say $r_1(\leq m_1)$, of true rejections with *FDR* controlled at a specified level f. Along this line, Jung (2005) derived a formula for sample size calculation as follows.

In Equation 12.5, the expected number of true rejections is

$$E\{R_1(\alpha)\} = \sum_{j \in M_1} \bar{\Phi}\left(z_a - \delta_j \sqrt{na_1a_2}\right).$$

$$(12.6)$$

In multiple testing controlling *FDR*, $E(R_1)/m_1$ plays the role of the power of a conventional testing; see Lee and Whitmore (2002) and van den Oord and Sullivan (2003). With $E(R_1)$ and the *FDR* level set at r_1 and f, respectively, Equation 12.5 is then expressed as

$$f = \frac{m_0\alpha}{m_0\alpha + r_1}.$$

By solving this equation with respect to α, we obtain

$$\alpha^* = \frac{r_1 f}{m_0(1 - f)}.$$

Given m_0, α^* is the marginal type I error level for r_1 true rejections with the *FDR* controlled at f. With α and $E(R_1)$ replaced by α^* and r_1, respectively, Equation 12.6 yields an equation $h(n) = 0$, where

$$h(n) = \sum_{j \in M_1} \bar{\Phi}\left(z_{\alpha^*} - \delta_j \sqrt{na_1a_2}\right) - r_1.$$

$$(12.7)$$

We can then obtain the sample size by solving this equation. Jung (2005) recommended solving the equation $h(n) = 0$ using the following bisection method:

1. Choose s_1 and s_2 such that $0 < s_1 < s_2$ and $h_1h_2 < 0$, where $h_k = h(s_k)$ for $k = 1, 2$. (If $h_1h_2 > 0$ and $h_1 > 0$, then choose a smaller s_1; if $h_1h_2 > 0$ and $h_2 < 0$, then choose a larger s_2.)
2. For $s_3 = (s_1 + s_2)/2$, calculate $h_3 = h(s_3)$.
3. If $h_1h_3 < 0$, then replace s_2 and h_2 with s_3 and h_3, respectively. Else, replace s_1 and h_1 with s_3 and h_3, respectively. Go to (2).
4. Repeat (2) and (3) until $|s_1 - s_3| < 1$ and $|h_3| < 1$, and obtain the required sample size $n = [s_3] + 1$, where $[s]$ is the largest integer smaller than s.

If we do not have prior information on the effect sizes, we may want to assume equal effect sizes $\delta_j = \delta(>0)$ for $j \in M_1$. In this case, Equation 12.7 is reduced to

$$h(n) = m_1\bar{\Phi}\left(z_{\alpha^*} - \delta\sqrt{na_1a_2}\right) - r_1$$

and, by solving $h(n) = 0$, we obtain the following formula:

$$n = \left\lceil \frac{(z_{\alpha^*} + z_{\beta^*})^2}{a_1 a_2 \delta^2} \right\rceil + 1, \qquad (12.8)$$

where $\alpha^* = r_1 f / \{m_0 (1 - f)\}$ and $\beta^* = 1 - r_1 / m_1$. Note that formula (12.8) is equivalent to the conventional sample size formula for detecting an effect size of δ with a desired power of $1 - \beta^*$ while controlling the type I error level at α^*.

As a result, the procedure for sample size calculation based on the approach of controlling *FDR* proposed by Jung (2005) can be summarized as follows:

- Step 1: Specify the input parameters:

  $f = FDR$ level

  $r_1 =$ number of true rejections

  $a_k =$ allocation proportion for group $k (=1, 2)$

  $m =$ total number of genes for testing

  $m_1 =$ number of prognostic genes $(m_0 = m - m_1)$

  $\{\delta_j, j \in M_1\} =$ effect sizes for prognostic genes

- Step 2: Obtain the required sample size:

  If the effect sizes are constant $\delta_j = \delta$ for $j \in M_1$,

$$n = \left\lceil \frac{(z_{\alpha^*} + z_{\beta^*})^2}{a_1 a_2 \delta^2} \right\rceil + 1,$$

where $\alpha^* = r_1 f / \{m_0 (1 - f)\}$ and $\beta^* = 1 - r_1 / m_1$.

Otherwise, solve $h(n) = 0$ using the bisection method, where

$$h(n) = \sum_{j \in M_1} \bar{\Phi}(z_{\alpha^*} - \delta_j \sqrt{n a_1 a_2}) - r_1$$

and $\alpha^* = r_1 f / \{m_0 (1 - f)\}$.

Remarks

Note that for given sample sizes n_1 and n_2, one may want to check how many true rejections are expected as if we want to check the power in a conventional testing. In this case, we may solve the equations for r_1. For example, when the effect sizes are constant, $\delta_j = \delta$ for $j \in M_1$, we solve the equation

$$z_{\alpha^*(r_1)} + z_{\beta^*(r_1)} = \delta \sqrt{n_1^{-1} + n_2^{-1}}$$

with respect to r_1, where $\alpha^*(r_1) = r_1 f / \{m_0 (1 - f)\}$ and $\beta^*(r_1) = 1 - r_1 / m_1$.

Examples

To illustrate the procedure for sample size calculation under the approach for controlling *FDR* proposed by Jung (2005), the examples based on one-sided tests with constant effect sizes and varied effect sizes described in Jung (2005) are used.

Example 12.1: One-Sided Tests and Constant Effect Sizes

Suppose that we want to design a microarray study on $m = 4000$ candidate genes, among which about $m_1 = 40$ genes are expected to be differentially expressing between two patient groups. Note that $m_0 = m - m_1 = 3960$. Constant effect sizes, $\delta_j = \delta = 1$, for the m_1 prognostic genes are projected. About equal number of patients are expected to enter the study from each group, that is, $a_1 = a_2 = 0.5$. We want to discover $r_1 = 24$ prognostic genes by one-sided tests with the *FDR* controlled at $f = 1\%$ level. Then,

$$\alpha^* = \frac{24 \times 0.01}{3960 \times (1 - 0.01)} = 0.612 \times 10^{-4}$$

and $\beta^* = 1 - 24/40 = 0.4$, so that $z_{\alpha^*} = 3.841$ and $z_{\beta^*} = 0.253$. Hence, from Equation 12.8, the required sample size is given as

$$n = \left[\frac{(3.841 + 0.253)^2}{0.5 \times 0.5 \times 1^2} \right] + 1 = 68,$$

or $n_1 = n_2 = 34$.

Example 12.2: One-Sided Tests and Varying Effect Sizes

We assume $(m, m_1, a_1, r_1, f) = (4000, 40, 0.5, 24, 0.01)$, $\delta_j = 1$ for $1 \leq j \leq 20$ and $\delta_j = 1/2$ for $21 \leq j \leq 40$. Then,

$$\alpha^* = \frac{24 \times 0.01}{3960 \times (1 - 0.01)} = 0.612 \times 10^{-4}$$

and $z_a{}^* = 3.841$, so that we have

$$h(n) = 20\bar{\Phi}\left(3.841 - \sqrt{n/4}\right) + 20\bar{\Phi}\left(3.841 - .5\sqrt{n/4}\right) - 24$$

Table 12.2 displays the bisection procedure with starting values $s_1 = 100$ and $s_2 = 200$. The procedure stops after seven iterations and gives $n = [147.7] + 1 = 148$.

Two-Sided Tests

Suppose one wants to test $H_j : \delta_j = 0$ against $\bar{H}_j : \delta_j \neq 0$. We reject H_j if $|T_j| > z_{\alpha/2}$ for a certain α level, and obtain the power function $\xi_j(\alpha) = \bar{\Phi}\left(z_{\alpha/2} - |\delta_j| \sqrt{na_1 a_2}\right)$. In this case, α^* is the same as that for the one-sided test case, that is,

$$\alpha^* = \frac{r_1 f}{m_0 (1 - f)},$$

TABLE 12.2

Bisection Procedure for Example 12.2

Step	s_1	s_2	s_3	h_1	h_2	h_3
1	100.0	200.0	150.0	−4.67	3.59	0.13
2	100.0	150.0	125.0	−4.67	0.13	−1.85
3	125.0	150.0	137.5	−1.85	0.13	−0.80
4	137.5	150.0	143.8	−0.80	0.13	−0.32
5	143.8	150.0	146.9	−0.32	0.13	−0.09
6	146.9	150.0	148.4	−0.09	0.13	0.02
7	146.9	148.4	147.7	−0.09	0.02	−0.04

Source: Jung, S.H. et al. 2005. *Biostatistics*, 6(1), 157–169.

but Equation 12.7 is changed to

$$h(n) = \sum_{j \in M_1} \bar{\Phi}\left(z_{\alpha^*/2} - |\delta_j| \sqrt{n a_1 a_2}\right) - r_1. \tag{12.9}$$

If the effect sizes are constant, that is, $\delta_j = \delta$ for $j \in M_1$, then we have a closed form formula

$$n = \left[\frac{(z_{\alpha^*/2} + z_{\beta^*})^2}{a_1 a_2 \delta^2}\right] + 1, \tag{12.10}$$

where $\alpha^* = r_1 f / \{m_0(1-f)\}$ and $\beta^* = 1 - r_1/m_1$.

Now we derive the relationship between the sample size for one-sided test case and that for two-sided test case. Suppose that the input parameters m, m_1, a_1, and $\{\delta_j, j \in M_1\}$ are fixed and we want r_1 true rejections in both cases. Without loss of generality, we assume that the effect sizes are nonnegative. The only difference between the two cases is the parts of α^* in Equation 12.7 and $\alpha^*/2$ in Equation 12.9. Let f_1 and f_2 denote the *FDR* levels for one- and two-sided testing cases, respectively. Then, the two formulas will give exactly the same sample size as far as these two parts are identical, that is,

$$\frac{r_1 f_1}{m_0(1 - f_1)} = \frac{r_1 f_2}{2 m_0(1 - f_2)},$$

which yields $f_1 = f_2/(2 - f_2)$. In other words, with all other parameters fixed, the sample size for two-sided tests to control the *FDR* at f can be obtained using the sample size formula for one-sided tests (12.7) by setting the target *FDR* level at $f/(2 - f)$. Note that this value is slightly larger than $f/2$. The same relationship holds when the effect sizes for prognostic genes are constant.

To illustrate the above procedure for sample size calculation under the approach for controlling *FDR* proposed by Jung (2005), the following example based on two-sided tests with constant effect sizes described in Jung (2005) is considered.

Example 12.3: Two-Sided Tests and Constant Effect Sizes

Jung (2005) considered $(m, m_1, \delta, a_1, r_1, f) = (4000, 40, 1, 0.5, 24, 0.01)$ as those given in Example 12.1, but use two-sided tests. Then,

$$\alpha^* = \frac{24 \times 0.01}{3960 \times (1 - 0.01)} = 0.612 \times 10^{-4}$$

and $\beta^* = 1 - 24/40 = 0.4$, so that $z_{\alpha^*/2} = 4.008$ and $z_{\beta^*} = 0.253$. Hence, from Equation 12.10, the required sample size is given as

$$n = \left\lceil \frac{(4.008 + 0.253)^2}{0.5 \times 0.5 \times 1^2} \right\rceil + 1 = 73.$$

By the above argument, we obtain exactly the same sample size using formula (12.8) and $f = 0.01/(2 - 0.01) = 0.005025$. Note that this sample size is slightly larger than $n = 68$, which was obtained for one-sided tests in Example 12.1.

Exact Formula Based on *t*-Distribution

Jung (2005) indicated that if the gene expression level, or its transformation, is a normal random variable and the available resources are so limited that only a small sample size can be considered, then one may want to use the exact formula based on t-distributions, rather than that based on normal approximation. In one-sided testing case, Jung (2005) suggested modifying Equation 12.5 as follows:

$$FDR(\alpha) = \frac{m_0 \alpha}{m_0 \alpha + \sum_{j \in M_1} T_{n-2, \delta_j \sqrt{na_1 a_2}}(t_{n-2,\alpha})},$$

where $T_{v,\eta}(t)$ is the survivor function for the noncentral t-distribution with v degrees of freedom and noncentrality parameter η, and $t_{v,\alpha} = T_{v,0}^{-1}(\alpha)$ is the upper 100α-th percentile of the central t-distribution with v degrees of freedom. The required sample size n for r_1 true rejections with the FDR controlled at f solves $h_T(n) = 0$, where

$$h_T(n) = \sum_{j \in M_1} T_{n-2, \delta_j \sqrt{na_1 a_2}}(t_{n-2,\alpha^*}) - r_1$$

and $\alpha^* = r_1 f / \{m_0 (1 - f)\}$. If the effect sizes are constant among the prognostic genes, then the equation reduces to

$$T_{n-2, \delta \sqrt{na_1 a_2}}(t_{n-2,\alpha^*}) = \frac{r_1}{m_1},$$

but, contrary to the normal approximation case, we do not have a closed form sample size formula since n is included in both the degrees of freedom and the noncentrality parameter of the t-distribution functions.

Similarly, the sample size for two-sided t-tests can be obtained by solving $h_T(n) = 0$, where

$$h_T(n) = \sum_{j \in M_1} T_{n-2, |\delta_j| \sqrt{na_1 a_2}}(t_{n-2,\alpha^*/2}) - r_1$$

and $\alpha^* = r_1 f/\{m_0(1-f)\}$. Note that the sample size for $FDR = f$ with two-sided testings is the same as that for $FDR = f/(2-f)$ with one-sided testings as in the testing based on normal approximation.

12.3 *FWER* Control

Microarray studies usually involve screening and monitoring of expression levels in cells for thousands of genes simultaneously for studying the association of the expression levels and an outcome or other risk factor of interest (Golub et al., 1999; Alizadeh and Staudt, 2000; Sander, 2000). A primary aim is often to reveal the association of the expression levels and an outcome or other risk factor of interest. Traditional statistical testing procedures, such as two-sample *t*-tests or Wilcoxon rank sum tests, are often used to determine statistical significance of the difference in gene expression patterns. These approaches, however, encounter a serious problem of multiplicity as a very large number—possibly 10,000 or more—of hypotheses are to be tested, while the number of studied experimental units is relatively small—tens to a few hundreds (West et al., 2001).

If we consider a per comparison type I error rate α in each test, the probability of rejecting any null hypothesis when all null hypotheses are true, which is called the *FWER*, will be greatly inflated. So as to avoid this pitfall, the Bonferroni test is used most commonly in this field despite its well-known conservativeness. Although Holm (1979) and Hochberg (1988) improved upon such conservativeness by devising multistep testing procedures, they did not exploit the dependency of the test statistics and consequently the resulting improvement is often minor. Westfall and Young (1989, 1993) proposed adjusting *p*-values in a state-of-the-art step-down manner using simulation or resampling method, by which dependency among test statistics is effectively incorporated. Westfall and Wolfinger (1997) derived exact adjusted *p*-values for a step-down method for discrete data. Recently, Westfall and Young's permutation-based test was introduced to microarray data analyses and was strongly advocated by Dudoit and her colleagues. Troendle, Korn, and McShane (2004) favor the permutation test over bootstrap resampling due to slow convergence in high-dimensional data. Various multiple testing procedures and error control methods applicable to microarray experiments are well documented in Dudoit, Schaffer, and Boldrick (2003). Which test to use among a bewildering variety of choices should be judged by relevance to research questions, validity (of underlying assumptions), type of control (strong or weak), and computability. Jung, Bang, and Young (2005) showed that the single-step test provides a simple and accurate method for sample size determination and that can also be used for multistep tests.

12.3.1 Multiple Testing Procedures

In this subsection, commonly employed single-step and multistep testing procedures are briefly described.

Single-Step versus Multistep

Suppose that there are n_1 subjects in group 1 and n_2 subjects in group 2. Gene expression data for m genes are measured from each subject. Furthermore, suppose that we would like to identify the informative genes, that is, those that are differentially expressed between the two groups. Let $(X_{1i1}, \ldots, X_{1im})$ and $(X_{2i1}, \ldots, X_{2im})$ denote the gene expression levels obtained from subject $i(=1, \ldots, n_1)$ in group 1 and subject $i(=1, \ldots, n_2)$ in group 2, respectively. Let $\mu_1 = (\mu_{11}, \ldots, \mu_{1m})$ and $\mu_2 = (\mu_{21}, \ldots, \mu_{2m})$ represent the respective mean vectors. To test whether gene j $(=1, \ldots, m)$ is not differentially expressed between the two conditions, that is, $H_j; \mu_{1j} - \mu_{2j} = 0$, the following t-test statistic is commonly considered:

$$T_j = \frac{\bar{X}_{1j} - \bar{X}_{2j}}{S_j \sqrt{n_1^{-1} + n_2^{-1}}},$$

where $\bar{X}_{kj}$ is the sample mean in group $k(=1, 2)$ and

$$S_j^2 = \frac{\left\{ \sum_{i=1}^{n_1}(X_{1ij} - \bar{X}_{1j})^2 + \sum_{i=1}^{n_2}(X_{2ij} - \bar{X}_{2j})^2 \right\}}{(n_1 + n_2 - 2)}$$

is the pooled sample variance for the jth gene.

Suppose that our interest is to identify any genes that are overexpressed in group 1. This can be formulated as multiple one-sided tests of H_j versus $\bar{H}_j : \mu_{1j} > \mu_{2j}$ for $j = 1, \ldots, m$. In this case, a single-step procedure, which adopts a common critical value c to reject H_j (in favor of $\bar{H}_j$, when $T_j > c$), is commonly employed. For this single-step procedure, the *FWER* fixed at α is given by

$$\alpha = P(T_1 > c \text{ or } T_2 > c, \ldots, \text{ or } T_m > c \mid H_0) = P\left(\max_{j=1,\ldots,m} T_j > c \mid H_0 \right), \tag{12.11}$$

Equation 12.11, where $H_0 : \mu_{1j} = \mu_{2j}$ for all $j = 1, \ldots, m$, or equivalently $H_0 = \cap_{j=1}^{m} H_j$, is the complete null hypothesis and the relevant alternative hypothesis is $H_a = \cap_{j=1}^{m} \bar{H}_j$. To control *FWER* at the nominal level α, the method of Bonferroni uses $c = c_\alpha = t_{n_1+n_2-2, \alpha/m}$, the upper α/m-quantile for the t-distribution with $n_1 + n_2 - 2$ degrees of freedom imposing normality for the expression data, or $c = z_{\alpha/m}$, the upper α/m-quantile for the standard normal distribution based on asymptotic normality. If gene expression levels are not normally distributed, the assumption of t-distribution may be violated. Furthermore, n_1 and n_2 usually may not be large enough to warrant a normal approximation. Note that the Bonferroni procedure is conservative for correlated data even when the assumed conditions are met. In practice, microarray data are collected from the same individuals and experience coregulation. Thus, they are expected to be correlated. To take into consideration the correlation structure under Equation 12.11, Jung, Bang, and Young (2005) derived the distribution of $W = \max_{j=1, \ldots, m} T_j$ under H_0 using the method permutation. Their method is briefly described below.

For a given total sample size n, there are $B = \binom{n}{n_1}$ different ways of partitioning the pooled sample of size $n = n_1 + n_2$ into two groups of sizes n_1 and n_2. The number of possible permutations B can be very large even with a small sample size. For the observed test statistic t_j of T_j from the original data, the unadjusted (or raw) p-values can be approximated by

$$p_j \approx B^{-1} \sum_{b=1}^{B} I\left(t_j^{(b)} \geq t_j \right)$$

where $I(A)$ is an indicator function of event A. For gene-specific inference, Jung, Bang, and Young (2005) define an adjusted p-value for gene j as the minimum *FWER* for which Hj will be rejected, that is,

$$\tilde{p}_j \approx P\left(\max_{j'=1,\dots,m} T_{j'} \geq t_j \mid H_0\right).$$

This probability can be estimated by the following algorithms for permutation distribution:

Algorithm 12.1: (Single-Step Procedure)
1. Compute the test statistics $t_1, \dots, t_m$ from the original data.
2. For the bth permutation of the original data $(b = 1, \dots, B)$, compute the test statistics $t_1^{(b)}, \dots, t_m^{(b)}$ and $w_b = \max_{j=1,\dots,m} t_j^{(b)}$.
3. Estimate the adjusted p-values by $\tilde{p}_j = \sum_{b=1}^{B} I(w_b \geq t_j)/B$ for $j = 1,\dots,m$.
4. Reject all hypotheses $H_j (j = 1,\dots,m)$ for that $\tilde{p}_j < \alpha$.

Alternatively, with steps (3) and (4) replaced, the cut-off value c_α can be determined:

Algorithm 12.1'
3'. Sort $w_1, \dots, w_B$ to obtain the order statistics $w_{(1)} \leq \cdots \leq w_{(B)}$ and compute the critical value $c_\alpha = w_{([B(1-\alpha)+1])}$, where $[\alpha]$ is the largest integer no greater than α. If there exist ties, $c_\alpha = w_{(k)}$, where k is the smallest integer such that $w_{(k)} \geq w_{([B(1-\alpha)+1])}$.
4'. Reject all hypotheses $H_j(j = 1, \dots, m)$ for which $t_j > c_\alpha$.

Below is a step-down analog suggested by Dudoit et al. (2002) and Dudoit, Schaffer, and Boldrick (2003), originally proposed by Westfall and Young (1989, 1993):

Algorithm 12.2: (Step-Down Procedure)
1. Compute the test statistics $t_1, \dots, t_m$ from the original data.

   i. Sort $t_1, \dots, t_m$ to obtain the ordered test statistics $t_{r_1} \geq \cdots \geq t_{r_m}$, where $H_{r_1},\dots,H_{r_m}$ are the corresponding hypotheses.
2. For the bth permutation of the original data $(b = 1, \dots, B)$, compute the test statistics $t_{r_1}^{(b)},\dots,t_{r_m}^{(b)}$ and $u_{b,j} = \max_{j'=j,\dots,m} t_{r_{j'}}^{(b)}$ for $j = 1, \dots, m$.
3. Estimate the adjusted p-values by $\tilde{p}_{r_j} = \sum_{b=1}^{B} I(u_{b,j} \geq t_{r_j})/B$ *for* $j = 1,\dots,m$.

   i. Enforce monotonicity by setting $\tilde{p}_{r_j} \leftarrow \max(\tilde{p}_{r_{j-1}}, \tilde{p}_{r_j})$ for $j = 2,\dots,m$.
4. Reject all hypotheses $H_{r_j} (j = 1, \dots, m)$ for which $\tilde{p}_{r_j} < \alpha$.

Remarks
Note that as indicated by Westfall and Young (1993), two-sided tests can be fulfilled by replacing t_j by $|t_j|$ in steps (B) and (C) in Algorithm 12.1. It can be shown that a single-step procedure, controlling the *FWER* weakly as in Equation 12.11, also controls the *FWER* strongly under the condition of subset pivotality.

12.3.2 Sample Size Calculation

Jung, Bang, and Young (2005) derived a procedure for sample size calculation using the single-step procedure. As indicated by Jung, Bang, and Young (2005), the calculated sample size is also applied to the step-down procedure since the two procedures have the same global power. In what follows, the procedure proposed by Jung, Bang, and Young (2005) using the single-step procedure is briefly described.

Algorithms for Sample Size Calculation

Suppose that one wishes to choose a sample size for achieving a desired global power of $1 - \beta$. Assuming that the gene expression data

$$\{(X_{ki1}, \ldots, X_{kim}), \text{ for } i = 1, \ldots, n_k, k = 1, 2\}$$

are random samples from an unknown distribution with $E(X_{kij}) = \mu_{kj}$, $\mathrm{Var}(X_{kij}) = \sigma_j^2$ and $\mathrm{Corr}(X_{kij}, X_{kij}) = \rho_{jj'}$. Let $R = (\rho_{jj'})_{j,j'=1,\ldots,m}$ be the $m \times m$ correlation matrix. Under H_a, the effect size is given by $\delta_j = (\mu_{1j} - \mu_{2j})/\sigma_j$. At the planning stage of a microarray study, we usually project the number of predictive genes D and set an equal effect size among them, that is,

$$
\begin{aligned}
\delta_j &= \delta \text{ for } j = 1, \ldots, D \\
&= 0 \text{ for } j = D+1, \ldots, m.
\end{aligned}
\tag{12.12}
$$

It can be verified that for large n_1 and n_2, $(T_1, \ldots, T_m)$ has approximately the same distribution as

$$(e_1, \ldots, e_m) \sim N(\mathbf{0}, R)$$

under H_0 and $(e_j + \delta_j \sqrt{npq}, R), j = 1, \ldots, m$ under H_a, where $p = n_1/n$ and $q = 1 - p$. Hence, at $FWER = \alpha$, the common critical value c_α is given as the upper α quantile of $\max_{j=1,\ldots,m} e_j$ from Equation 12.11. Similarly, the global power as a function of n is given by

$$h_a(n) = P\left\{ \max_{j=1,\ldots,m} \left(e_j + \delta_j \sqrt{npq} \right) > c_\alpha \right\}.$$

Thus, at a given $FWER = \alpha$, the sample size n required for detecting the specified effect sizes $(\delta_1, \ldots, \delta_m)$ with a global power $1 - \beta$ can be obtained as the solution to $h_a(n) = 1 - \beta$. Note that the analytic calculation of c_α and $h_a(n)$ is feasible only when the distributions of $\max_j e_j$ and $\max_j \left(e_j + S_j \sqrt{npq} \right)$ are available in simple forms. With a large m, however, it is almost impossible to derive the distributions. To avoid the difficulty, Jung, Bang, and Young suggested the following simulation be considered.

A simulation can be conducted to approximate c_α and $h_a(\cdot)$ by generating random vectors $(e_1, \ldots, e_m)$ from $N(\mathbf{0}, R)$. For simplicity, Jung, Bang, and Young (2005) suggested generating the random numbers assuming a simple, but realistic, correlation structure for the gene expression data such as block compound symmetry (BCS) or CS (i.e., with only one block). Suppose that m genes are partitioned into L blocks, and B_l denotes the set of genes belonging to block l ($l = 1, \ldots, L$). We may assume that $\rho_{jj'} = \rho$ if $j, j' \in B_l$ for some l, and $\rho_{jj'} = 0$ otherwise. Under the BCS structure, $(e_1, \ldots, e_m)$ can be generated as a function of *i.i.d.* standard normal random variates $u_1, \ldots, u_m, b_1, \ldots, b_L$:

$$e_j = u_j \sqrt{1 - \rho} + b_l \sqrt{\rho} \text{ for } j \in B_l. \tag{12.13}$$

As a result, the algorithm for sample size calculation can be summarized as follows:

1. Specify *FWER* (α), global power $(1 - \beta)$, effect sizes $(\delta_1, \ldots, \delta_m)$, and correlation structure ($\boldsymbol{R}$).

2. Generate K (say, 10,000) *i.i.d.* random vectors $\left\{I\left(e_1^{(k)}, \ldots, e_m^{(k)}\right), k = 1, \ldots, K\right\}$ from $N(0, \boldsymbol{R})$. Let $\bar{e}_k = \max_{j=1,\ldots,m} e_j^{(k)}$.

3. Approximate c_α by $\bar{e}_{[(1-\alpha)K+1]}$, the $[(1 - \alpha)K + 1]$-th order statistic of $\bar{e}_1, \ldots, \bar{e}_K$.

4. Calculate n by solving $\hat{h}_a(n) = 1 - \beta$ by the bisection method (Press et al., 1996), where $\hat{h}_a(n) = K^{-1} \sum_{k=1}^{K} I\left\{\max_{j=1,\ldots,m}\left(e_j^{(k)} + \delta_j \sqrt{npq}\right) > c_\alpha\right\}$.

Mathematically put, Step (4) is equivalent to finding $n^* = \min\left\{n: \hat{h}_a(n) \geq 1 - \beta\right\}$.

Note that the permutation procedure may alter the correlation structure among the test statistics under H_a. Suppose that there are m_1 genes in block 1, among which the first D are predictive. Then, under (12.13) and BCS, we have

$$Corr(T_j, T_{j'}) \approx \begin{cases} \dfrac{(\rho + pq\delta^2)}{(1 + pq\delta^2)} \equiv \rho_1 & \text{if } 1 \leq j < j' \leq D \\[2ex] \dfrac{\rho}{\sqrt{1 + pq\delta^2}} \equiv \rho_2 & \text{if } 1 \leq j \leq D < j' \leq m_1 \\[2ex] \rho & \text{if } D < j < j' \leq m_1 \\[1ex] \rho & \text{or } j, j' \in B_l, l \geq 2 \end{cases} \tag{12.14}$$

where the approximation is with respect to large n.

Let $\tilde{\boldsymbol{R}}$ denote the correlation matrix with these correlation coefficients. Note that $\tilde{\boldsymbol{R}} = \boldsymbol{R}$ under $H_0 : \delta = 0$, so that calculation of c_α is the same as in the naive method. However, $h_a(n)$ should be modified to

$$\tilde{h}_a(n) = P\left\{\max_{j=1,\ldots,m}\left(\tilde{e}_j + \delta_j \sqrt{npq}\right) > c_\alpha\right\}$$

where random samples of $(\tilde{e}_1, \ldots, \tilde{e}_m)$ can be generated using

$$\tilde{e}_j = \begin{cases} u_j\sqrt{1 - \rho_1} + b_1\sqrt{\rho_2} + b_{-1}\sqrt{\rho_1 - \rho_2} & \text{if } 1 \leq j \leq D \\[1ex] u_j\sqrt{1 - \rho} + b_1\sqrt{\rho_2} + b_0\sqrt{\rho - \rho_2} & \text{if } D < j \leq m_1 \\[1ex] u_j\sqrt{1 - \rho} + b_l\sqrt{\rho} & \text{if } j \in B_l \text{ for } l \geq 2 \end{cases}$$

with $u_1, \ldots, u_m, b_{-1}, b_0, b_1, \ldots, b_L$ independently from $N(0,1)$. Then, $\left\{\left(\tilde{e}_1^{(k)}, \ldots, \tilde{e}_m^{(k)}\right), k = 1, \ldots, K\right\}$ are *i.i.d.* random vectors from $N(0, \tilde{\boldsymbol{R}})$, and

$$\hat{\tilde{h}}_a(n) = K^{-1} \sum_{k=1}^{K} I\left\{\max_{j=1,\ldots,m}\left(\tilde{e}_j^{(k)} + \delta_j \sqrt{npq}\right) > c_\alpha\right\}.$$

The sample size can be obtained by solving $\hat{\tilde{h}}_a(n) = 1 - \beta$.

Remarks

Note that the methods discussed above are different from a pure simulation method in the sense that it does not require generating the raw data and then calculating test statistics. Thus, the computing time is not of an order of $n \times m$, but of m. Furthermore, we can share the random numbers $u_1, \ldots, u_m, b_{-1}, b_0, b_1, \ldots, b_L$ in the calculation of c_α and n.

We do not need to generate a new set of random numbers at each replication of the bisection procedures either. If the target n is not large, the large sample approximation may not perform well. In our simulation study, we examine how large n needs to be for an adequate approximation. If the target n is so small that the approximation is questionable, then we have to use a pure simulation method by generating raw data.

12.3.3 Leukemia Example

To illustrate the use of the procedure described above, the leukemia data from Golub et al. (1999) are reanalyzed. There are $n_{all} = 27$ patients with acute lymphoblastic leukemia (ALL) and $n_{aml} = 11$ patients with acute myeloid leukemia (AML) in the training set, and expression patterns in $m = 6810$ human genes are explored. Note that, in general, such expression measures are subject to preprocessing steps such as image analysis and normalization, and also to *a priori* quality control. Supplemental information and dataset can be found at the website http://www.genome.wi.mit.edu/MPR.

Gene-specific significance was ascertained for alternative hypotheses $\bar{H}_{1,j}: \mu_{ALL,j} \neq \mu_{AML,j}$, $\bar{H}_{2,j}: \mu_{ALL,j} < \mu_{AML,j}$, and $\bar{H}_{3,j}: \mu_{ALL,j} > \mu_{AML,j}$ by SDP and SSP. Jung, Bang, and Young (2005) implemented their algorithm as well as PROC MULTTEST in SAS with $B = 10,000$ permutations (Westfall, Zaykin, and Young, 2001). Owing to essentially identical results, we report the results from SAS. Table 12.3 lists 41 genes with two-sided adjusted p-values that are smaller than 0.05. Although adjusted p-values by SDP are slightly smaller than SSP, the results are extremely similar, confirming the findings from our simulation study. Note that Golub et al. (1999) and we identified 1100 and 1579 predictive genes without accounting for multiplicity, respectively. A Bonferroni adjustment declared 37 significant genes. This is not so surprising because relatively low correlations among genes were observed in these data. We do not show the results for $\bar{H}_{3,j}$; only four hypotheses are rejected. Note that the two-sided p-value is smaller than twice that of the smaller one-sided p-value as theory predicts and that the difference is not often negligible (Shaffer, 2002).

In Table 12.3, adjusted p-values from two-sided hypothesis less than 0.05 are listed in increasing order among total $m = 6810$ genes investigated. The total number of studied subjects n was 38 ($n_{all} = 27$ and $n_{aml} = 11$). $B = 10,000$ times of permutation were used. Note that C-myb gene has p-value of 0.015 against the hypothesis $\mu_{all} > \mu_{aml}$. Although some gene descriptions are identical, gene accession numbers are different.

Suppose that we would like to design a prospective study to identify predictive genes overexpressing in AML based on observed parameter values. Therefore, we assume $m = 6810$, $p = 0.3 (\approx 11/38)$, $D = 10$ or 100, $\delta = 0.5$ or 1, and BCS with block size 100 or CS with a common correlation coefficient of $\rho = 0.1$ or 0.4. We calculated the sample size using the modified formula under each parameter setting for *FWER* $\alpha = 0.05$ and a global power $1 - \beta = 0.8$ with $K = 5000$ replications. For $D = 10$ and $\delta = 1$, the minimal sample size required for BCS/CS are 59/59 and 74/63 for $\rho = 0.1$ and 0.4, respectively. If a larger number of genes, say $D = 100$, are anticipated to overexpress in AML with the same effect size, the respective sample sizes reduce to 34/34 and 49/41 to maintain the same power. With $\delta = 0.5$, the required sample size becomes nearly 3.5 to 4 times that for $\delta = 1$. Note that, with the same ρ, BCS tends to require a larger sample size than CS.

TABLE 12.3

Reanalysis of the Leukemia Data from Golub et al. (1999)

	Alternative Hypothesis			
	$\mu_{all} \neq \mu_{aml}$		$\mu_{all} < \mu_{aml}$	
Gene Index (Description)	SDP	SSP	SDP	SSP
1701 (FAH Fumarylacetoacetate)	0.0003	0.0003	0.0004	0.0004
3001 (Leukotriene C4 synthase)	0.0003	0.0003	0.0004	0.0004
4528 (Zyxin)	0.0003	0.0003	0.0004	0.0004
1426 (LYN V-yes-1 Yamaguchi)	0.0004	0.0004	0.0005	0.0005
4720 (LEPR Leptin receptor)	0.0004	0.0004	0.0005	0.0005
1515 (CD33 CD33 antigen)	0.0006	0.0006	0.0006	0.0006
402 (Liver mRNA for IGIF)	0.0010	0.0010	0.0009	0.0009
3877 (PRG1 Proteoglycan 1)	0.0012	0.0012	0.0010	0.0010
1969 (DF D component of complement)	0.0013	0.0013	0.0011	0.0011
3528 (GB DEF)	0.0013	0.0013	0.0010	0.0010
930 (Induced Myeloid Leukemia Cell)	0.0016	0.0016	0.0013	0.0013
5882 (IL8 Precursor)	0.0016	0.0016	0.0013	0.0013
1923 (Peptidyl-Prolyl Cis-Trans Isomerase)	0.0017	0.0017	0.0014	0.0014
2939 (Phosphotyrosine independent ligand p62)	0.0018	0.0018	0.0014	0.0014
1563 (CST3 Cystatin C)	0.0026	0.0026	0.0021	0.0021
1792 (ATP6C Vacuolar H+ ATPase proton channel subunit)	0.0027	0.0027	0.0023	0.0023
1802 (CTSD Cathepsin D)	0.0038	0.0038	0.0032	0.0032
5881 (Interleukin 8)	0.0041	0.0041	0.0036	0.0036
6054 (ITGAX Integrin)	0.0056	0.0055	0.0042	0.0041
6220 (Epb72 gene exon 1)	0.0075	0.0075	0.0062	0.0062
1724 (LGALS3 Lectin)	0.0088	0.0088	0.0071	0.0071
2440 (Thrombospondin-p50)	0.0091	0.0091	0.0073	0.0073
6484 (LYZ Lysozyme)	0.0101	0.0100	0.0081	0.0080
1355 (FTL Ferritin)	0.0107	0.0106	0.0086	0.0085
2083 (Azurocidin)	0.0107	0.0106	0.0086	0.0085
1867 (Protein MAD3)	0.0114	0.0113	0.0092	0.0091
6057 (PFC Properdin P factor)	0.0143	0.0142	0.0108	0.0107
3286 (Lysophospholipase homolog)	0.0168	0.0167	0.0126	0.0125
6487 (Lysozyme)	0.0170	0.0169	0.0127	0.0126
1510 (PPGB Protective protein)	0.0178	0.0177	0.0133	0.0132
6478 (LYZ Lysozyme)	0.0193	0.0191	0.0144	0.0142
6358 (HOX 2.2)	0.0210	0.0208	0.0160	0.0158
3733 (Catalase EC 1.11.1.6)	0.0216	0.0214	0.0162	0.0160
1075 (FTH1 Ferritin heavy chain)	0.0281	0.0279	0.0211	0.0209
6086 (CD36 CD36 antigen)	0.0300	0.0298	0.0224	0.0222
189 (ADM)	0.0350	0.0348	0.0260	0.0258
1948 (CDC25A Cell division cycle)	0.0356	0.0354	0.0263	0.0261
5722 (APLP2 Amyloid beta precursor-like protein)	0.0415	0.0413	0.0306	0.0304
5686 (TIMP2 Tissue inhibitor of metalloproteinase)	0.0425	0.0423	0.0314	0.0312
5453 (C-myb)	0.0461	0.0459	1.000	1.000
6059 (NF-IL6-beta protein mRNA)	0.0482	0.0480	0.0350	0.0348

Source: Golub, T.R. et al. 1999. *Science*, 286(15), 531–537 and Jung, S.H. et al. 2005. *Biostatistics*, 6(1), 157–169.

An interesting question is raised regarding the accuracy of the sample size formula when the gene expression data have distributions other than the multivariate normal distributions. We considered the setting $\alpha = 0.05$, $1 - \beta = 0.8$, $\delta = 1$, $D = 100$, $\rho = 0.1$ with CS structure, which results in the smallest sample size, $n = 34$, in the above sample size calculation. Gene expression data were generated from a correlated asymmetric distribution:

$$X_{kj} = \mu_{kj} + (e_{kj} - 2)\sqrt{\frac{\rho}{4}} + (e_{k0} - 2)\sqrt{\frac{(1-\rho)}{4}}$$

for $1 \leq j \leq m$ and $k = 1, 2$. Here, $\mu_{1j} = \delta_j$ and $\mu_{2j} = 0$, and $e_{k0}, e_{k1}, \dots, e_{km}$ are *i.i.d.* random variables from a χ^2 distribution with two degrees of freedom. Note that $(X_{k1}, \dots, X_{km})$ have means $(\mu_{k1}, \dots, \mu_{km})$, marginal variances 1, and a compound symmetry correlation structure with $\rho = 0.1$. In this case, we obtained an empirical *FWER* of 0.060 and an empirical global power of 0.832, which are close to the nominal $\alpha = 0.05$ and $1 - \beta = 0.8$, respectively, from a simulation with $B = N = 1000$.

12.4 Concluding Remarks

Microarray has been a major high-throughput assay method to display DNA or RNA abundance for a large number of genes concurrently. Discovery of the prognostic genes should be made taking multiplicity into account, but also with enough statistical power to identify important genes successfully. Owing to the costly nature of microarray experiments, however, often only a small sample size is available and the resulting data analysis does not provide reliable answers to the investigators. If the findings from a small study look promising, a large-scale study may be developed to confirm the findings using appropriate statistical tools. As a result, sample size calculation plays an important role in the design stage of such a confirmatory study. It can be used to check the statistical power, r_1/m_1, of a small-scale pilot study as well.

In recent years, many researchers have proposed the new concepts for controlling errors such as *FDR* and positive-*FDR* (i.e., *pFDR*), which control the expected proportion of type I error among the rejected hypotheses (Benjamini and Hochberg, 1995; Storey, 2002). Controlling these quantities relaxes the multiple testing criteria compared to controlling *FWER* in general and increase the number of declared significant genes. In particular, *pFDR* is motivated by Bayesian perspective and inherits the idea of single-step in constructing q-values, which are the counterpart of the adjusted p-values in this case (Ge, Dudoit, and Speed, 2003). In practice, it is of interest to compare sample sizes obtained using the methods for controlling *FDR*, *pFDR*, and *FWER*. *FWER* is important as a benchmark because the reexamination of Golub et al.'s data reveals that classical *FWER* control (along with global power) may not necessarily be as exceedingly conservative as many researchers thought and carries clear conceptual and practical interpretations.

The formula and/or procedure for the *FDR* approach described in this chapter is to calculate the sample size for a specified number of true rejections (or the expected number of true rejections given a sample size) while controlling the *FDR* at a given level. The input variables to be prespecified are total number of genes for testing m, projected number of prognostic genes m_1, allocation proportions a_k between groups, and effect sizes for the prognostic genes. When the effect sizes among the prognostic genes are the same, a closed form formula for sample size calculation is available.

It should be noted that although there are many research publications on sample size estimation in the microarray context, none examined the accuracy of their estimates. Most of them focused on exploratory and approximate relationships among statistical power, sample size (or the number of replicates), and effect size (often, in terms of fold-change), and used the most conservative Bonferroni adjustment without any attempt to incorporate the underlying correlation structure (Witte, Elston, and Cardon, 2000; Wolfinger et al., 2001; Black and Doerge, 2002; Lee and Whitmore, 2002; Pan, Lin, and Le, 2002; Simon, Radmacher, and Dobbin, 2002; Cui and Churchill, 2003). By comparing empirical power resulting from naive and modified methods, Jung, Bang, and Young (2005) showed that an ostensibly similar but incorrect choice of sample size ascertainment could cause considerable underestimation of required sample size. Thus, Jung, Bang, and Young recommended that the assessment of bias in empirical power (compared to nominal power) should be done as a conventional step in the research of sample size calculation.

13

Bayesian Sample Size Calculation

During the past decade, the approach for sample size determination originating from the Bayesian point of view has received much attention from academia, industry, and government. Although there are still debates between frequentist and Bayesian, Berger, Boukai, and Wang (1997, 1999) have successfully reconciled the merits from both frequentist and Bayesian approaches. However, no specific discussions regarding sample size determination for clinical trials at the planning stage are provided from the Bayesian point of view. Their work has stimulated research on sample size calculation using the Bayesian approach thereafter. The increasing popularity of sample size calculation using the Bayesian approach may be due to the following reason. The traditional sample size calculation based on the concept of frequentist assumes that the values of the true parameters under the alternative hypothesis are known. This is a strong assumption that can never be true in reality. In practice, these parameters are usually unknown and hence have to be estimated based on limited data from a pilot study. This raises an important question: how to control the uncertainty of the parameter from the pilot study (Wang, Chow, and Chen, 2005). Note that the relatively small pilot study may not be the only source of the parameter uncertainty. In some situations, the magnitude of the noncentrality parameter may be obtained simply from subjective clinical opinions (Spiegelhalter and Freedman, 1986). In such a situation, the true parameter specification uncertainty seems to be even severe. Some related works can be found in Joseph and Belisle (1997), Joseph, Wolfson, and du Berger (1995), Lindley (1997), and Pham-Gia (1997).

Many other researchers (e.g., Lee and Zelen, 2000) questioned the basic (frequentist) testing approach, which has been widely used in practice. More specifically, they argued that instead of using the frequentist type I and type II error rates for determining the needed sample size, it is more appropriate to use the posterior error rates from the Bayesian perspective. In other words, the Bayesian approach concerns "If the trial is significant, what is the probability that the treatment is effective?" It has been argued that by ignoring these fundamental considerations, the frequentist approach may result in positive harm due to inappropriate use of the error rates (Lee and Zelen, 2000). In this chapter, we summarize the current Bayesian sample size calculations into two categories. One category considers making use of the Bayesian framework to reflect the investigator's belief regarding the uncertainty of the true parameters, while the traditional frequentist testing procedure is still used for analyzing the data. The other category considers determining the required sample size when a Bayesian testing procedure is used. We consider both categories important and useful methodologies for biopharmaceutical research and development. Hence, in this chapter, we will focus on the research work done in these categories. On the other hand, we do believe the effort done thus far is far less than enough for providing a comprehensive overview of the Bayesian sample size calculations. Therefore, practice issues and possible future research topics will also be discussed whenever possible.

In Section 13.1, we introduce the procedure proposed by Joseph and Belisle (1997). In Section 13.2, the important work by Lee and Zelen (2000) is summarized. Lee and Zelen (2000) proposed a procedure for sample size calculation based on the concept for achieving a desired posterior error probability. The method is simple and yet very general. In Section 13.3, an alternative approach, which is referred to as the bootstrap-median approach, is proposed. This chapter concludes with a brief discussion, where some practical issues and possible future research topics are briefly outlined.

13.1 Posterior Credible Interval Approach

In their early work, Joseph and Belisle (1997) had studied the procedure for sample size calculation from the Bayesian point of view. As it can be seen from Chapter 3, for a one-sample two-sided hypotheses, the sample size needed to achieve the error rate of (α, β) is given by

$$n \geq \frac{4\sigma^2 z_{1-\alpha/2}}{l^2}. \tag{13.1}$$

Joseph and Belisle (1997) indicated that such a commonly used (frequentist) formula may suffer the following drawbacks:

1. The value of the standard deviation σ is usually unknown and yet it plays a critical role in determining the final sample size. Consequently, the resultant sample size estimate could be very sensitive to the choice of the σ value.

2. In practice, statistical inference is made based on the observed data at the end of the study regardless of the unknown σ value in Equation 13.1. At the planning stage, the investigator will have to determine the sample size with many uncertainties such as the unknown σ and the final observed data.

3. In some situations, prior information regarding the mean difference ε may be available. Ignoring this important prior information may lead to an unnecessarily large sample size, which could be a huge waste of the limited resources.

To overcome the above-mentioned limitations, Joseph and Belisle (1997) provided three elegant solutions from a Bayesian perspective. Specifically, three different criteria are proposed for sample size estimation. They are (i) the *average coverage criterion*, (ii) *average length criterion*, and (iii) *worst outcome criterion*, which will be discussed in detail in the following sections.

13.1.1 Three Selection Criteria

Under the Bayesian framework, let $\theta \in \Theta$ denote a generic parameter and its associated parameter space. Then, prior information regarding the value of θ is described by a prior distribution $f(\theta)$. Consider $x = (x_1, \ldots, x_n)$ a generic data set with n independent and identically distributed random observations. It is assumed that $x \in S$, where S is the associated sample space. Then, the marginal distribution of x is given by

$$f(x) = \int_{\Theta} f(x \mid \theta) f(\theta) d\theta,$$

where $f(x|\theta)$ denotes the conditional distribution of x given the parameter θ. Then, the posterior distribution of θ given x is given by

$$f(\theta\,|\,x) = \frac{f(x\,|\,\theta)f(\theta)}{f(x)}.$$

Based on the above Bayesian framework, the following three criteria can be used to select the optimal sample size.

13.1.1.1 Average Coverage Criterion

Consider the situation where a fixed posterior interval length l is prespecified for an acceptable precision of an estimate. Thus, the concept of the *average coverage criterion* (ACC) is to select the minimum sample size n such that the average coverage probability of such an interval is at least $(1 - \alpha)$, where α is a prespecified level of significance. More specifically, ACC selects the minimum sample size by solving the following inequality:

$$\int_S \left[\int_a^{a+l} f(\theta\,|\,x,n)d\theta\right] f(x)dx \geq 1-\alpha,$$

where a is some statistic to be determined by the data. Adcock (1988) first proposed to choose the interval $(a, a + l)$ so that it is symmetric about the mean. On the other hand, Joseph et al. (1995) proposed to select $(a, a + l)$ to be a highest posterior density interval. Note that for a symmetric distribution like normal, both methods of Adcock (1988) and Joseph et al. (1995) lead to the same solution. However, for a general asymmetric distribution, the two methods may lead to different results.

13.1.1.2 Average Length Criterion

Note that the ACC criterion fixed the posterior credible interval length l but optimize the sample size to achieve a desired coverage probability at the level of $(1 - \alpha)$. Following an idea similar to ACC, another possible solution is to fix the coverage probability of the posterior credible interval, then select the sample size so that the resulting posterior credible interval has a desired length l on average. Such a sample size selection criterion is referred to as *average length criterion* (ALC). More specifically, ALC selects the smallest sample size so that

$$\int_S l'(x,n)f(x)dx \leq l,$$

where $l'(x, n)$ is the length of the $100(1 - \alpha)\%$ posterior credible interval for data x, which is determined by solving

$$\int_a^{a+l'(x,n)} f(\theta\,|\,x,n)d\theta = 1-\alpha.$$

As before, different methods exist for the selection of *a*. For example, *a* can be chosen to be the highest posterior density interval or some meaningful symmetric intervals.

13.1.1.3 Worst Outcome Criterion

In practice, the investigators may not be satisfied with the average-based criterion due to its conservativeness. For example, the sample size selected by the ALC only ensures that the *average* of the posterior credible interval length will be no larger than *l*. On the other hand, one may wish to select a sample size such that the expected posterior credible interval length is as close to *l* as possible. Thus, we may expect about a 50% chance that the resultant interval length is larger than *l*, which may not be desirable due to its conservativeness.

In this situation, the following *worst outcome criterion* (WOC) may be useful. More specifically, WOC selects the smallest sample size by solving the following inequality:

$$\inf_{x \in S_0} \left\{ \int_{a}^{a+l(x,n)} f(\theta \mid x,n) d\theta \right\} \geq 1 - \alpha,$$

where S_0 is an appropriately selected subset of the original sample space S. For example, we may consider S_0 to be a region containing 95% of the sample S. Then, WOC ensures that the length of the posterior credible length will be at most *l* for any possible $x \in S_0$.

13.1.2 One Sample

We first consider a one-sample problem. Let $x = (x_1, \ldots, x_n)$ be n independent and identically distributed observations from $N(\mu, \sigma^2)$, where both μ and $\sigma^2 > 0$ are unknown parameters. Furthermore, we define precision $\lambda = \sigma^{-2}$, then we are able to utilize the following conjugate prior:

$$\lambda \sim \Gamma(v, \beta),$$

$$\mu \mid \lambda \sim N(\mu_0, n_0 \lambda),$$

where μ_0 and n_0 are hyperparameters. For a more detailed discussion for many standard Bayesian results used in this section, one may refer to Bernardo and Smith (1994).

13.1.2.1 Known Precision

When λ is known, it can be shown that the posterior distribution of μ given x is $N(\mu_n, \lambda_n)$, where

$$\lambda_n = (n + n_0)\lambda,$$

$$\mu_n = \frac{n_0 \mu_0 + n\bar{x}}{n_0 + n},$$

where $\bar{x}$ is the usual sample mean. Note that the posterior precision depends only on the sample size n and is independent of the value of the observed x. Consequently, all three criteria (ACC, ALC, and WOC) lead to the same solution, which is also equivalent to that of Adcock (1988):

$$n \geq \frac{4z^2_{1-\alpha/2}}{\lambda l^2} - n_0. \tag{13.2}$$

Note that if a noninformative prior with $n_0 = 0$ is used, then the above formula reduces to the usual formula for sample size calculation. Furthermore, by comparing Equation 13.2 with Equation 13.1, it can be seen that the sample size obtained from the Bayesian approach is smaller than that of the traditional frequentist estimate by a number n_0. This simply reflects the effect of the usefulness of the prior information.

13.1.2.2 Unknown Precision

If the precision parameter λ is unknown, then the posterior distribution of μ given the observed x is given by

$$\mu \mid x \sim t_{2v+n} \sqrt{\frac{\beta_n}{(n+n_0)(v+n/s)}} + \mu_n,$$

where

$$\mu_n = \frac{n_0\mu_0 + n\bar{x}}{n+n_0},$$

$$\beta_n = \beta + \frac{1}{2}ns^2 + \frac{nn_0}{2(n+n_0)}(\bar{x} - \mu_0)^2,$$

where $s^2 = n^{-1}\Sigma(x_i - \bar{x})^2$ and t_d represents a t-distribution with d degrees of freedom. As it can be seen, for such a situation, the posterior precision varies with the value of x. Hence, different selection criteria will lead to different sample size estimations.

1. *Average coverage criterion.* Adcock (1988) showed that the ACC sample size can be obtained as follows:

$$n = \frac{4\beta}{vl^2}t^2_{2v,1-\alpha/2} - n_0.$$

Note that the above formula is very similar to Equation 13.2 in the sense that (i) the precision λ is replaced by the mean precision v/β as specified in the prior and (ii) the normal quantile is replaced by an appropriate t-quantile. However, it should be noted that the degrees of freedom used in the t-quantile does not increase as the sample size increases. Consequently, the resultant sample size could be very different from the frequentist estimator (13.1).

2. *Average length criterion.* Consider the same Bayesian setup as for ACC. Then, the minimum sample size selected by ALC can be obtained by solving the following inequality:

$$2t_{n+2v,1-\alpha/2}\sqrt{\frac{2\beta}{(n+2v)(n+n_0)}}\,\frac{\Gamma\left(\frac{n+2v}{2}\right)\Gamma\left(\frac{2v-1}{2}\right)}{\Gamma\left(\frac{n+2v-1}{2}\right)\Gamma(v)}\le l. \tag{13.3}$$

Unfortunately, there exists no explicit solution for the above inequality. Since the left-hand side can be calculated with a given sample size n, a bisectional searching algorithm could be useful in finding the optimal sample size. However, in the case where the sample size n is large, the numerical evaluation of $\Gamma(0.5n + v)$ and $\Gamma(0.5n + v - 0.5)$ could be very unstable. According to Graham et al. (1994, MathWorld), we have

$$\frac{\Gamma(\frac{n+2v}{2})}{\Gamma(\frac{n+2v-1}{2})}=\sqrt{0.5n}\,\{1+o(1)\}.$$

Then, the inequality (13.3) can be approximated by the following:

$$\sqrt{2n}\,t_{n+2v,1-\alpha/2}\sqrt{\frac{2\beta}{(n+2v)(n+n_0)}}\,\frac{\Gamma(\frac{2v-1}{2})}{\Gamma(v)}\le l. \tag{13.4}$$

Note that the above formula (13.4) does provide an adequate approximation to Equation 13.3.

3. *Worst outcome criterion.* Let $S_0 \subset S$ be the subset of the sample space such that

$$\int_{S_0} f(x)dx = 1-w,$$

for some probability $w > 0$. Also, assume that $f(x) \ge f(y)$ for any $x \in S$ and $y \notin S$. Thus, the sample size needed can be approximated by

$$\frac{l^2(n+2v)(n+n_0)}{8\beta\{1+(n/2v)F_{n,2v,1-w}\}}\ge t^2_{n+2v,1-\alpha/2},$$

where $F_{d1,d2,1-\alpha}$ is the $(1-\alpha)$th quantile of the F-distribution with (d_1, d_2) degrees of freedom. Note that the subsample space S_0 cannot be exactly the same as S. In this case, we have $w = 0$, hence $F_{n,2v,1-w} = \infty$. As a result, the sample size is not well defined (Tables 13.1 and 13.2).

13.1.2.3 Mixed Bayesian-Likelihood

In practice, although many investigators may acknowledge the usefulness and importance of the prior information, they will still make the final inference based on the likelihood of the data. For example, the investigator may want to report the 95% confidence interval in

TABLE 13.1

One-Sample Sample Sizes with Unknown Precision and $v = 2$

β	n_0	1	α	FREQ	ACC	ALC	WOC (90%)	WOC (95%)
2	10	0.2	0.20	165	226	247	608	915
			0.10	271	445	414	1008	1513
			0.05	385	761	593	1436	2153
			0.01	664	2110	1034	2488	3727
		0.5	0.20	27	28	2	89	138
			0.10	44	63	57	154	234
			0.05	62	114	86	223	337
			0.01	107	330	158	392	590
	50	0.2	0.20	165	186	206	568	874
			0.10	271	405	374	968	1473
			0.05	385	721	553	1396	2113
			0.01	664	2070	994	2448	3687
		0.5	0.20	27	0	2	48	97
			0.10	44	23	2	113	194
			0.05	62	74	2	183	297
			0.01	107	290	118	352	550
5	10	0.2	0.20	165	578	634	1534	2301
			0.10	271	1127	1052	2535	3797
			0.05	385	1918	1498	3603	5396
			0.01	664	5290	2597	6231	9328
		0.5	0.20	27	85	92	237	360
			0.10	44	172	159	398	600
			0.05	62	299	231	569	856
			0.01	107	838	408	991	1486
	50	0.2	0.20	165	538	594	1494	2261
			0.10	271	1087	1012	2495	3757
			0.05	385	1878	1458	3563	5356
			0.01	664	5250	2557	6191	9288
		0.5	0.20	27	45	2	197	320
			0.10	44	132	118	358	560
			0.05	62	259	191	529	816
			0.01	107	798	368	951	1446

the final analysis. In such a situation, the sample size needed by ACC can also be found based on the 95% confidence interval. Consider the case where the confidence interval is symmetric about $\bar{x}$. It follows that

$$(a, a + l) = (\bar{x} - l/2, \bar{x} + l/2)$$

$$\mu \mid x \sim \bar{x} + t_{n-1} \sqrt{\frac{ns^2}{n(n-1)}}$$

$$\int_{\bar{x}-l/2}^{\bar{x}+l/2} f(\mu \mid x) d\mu = 2 p_t \left(\frac{l}{2} \sqrt{\frac{n(n-1)}{ns^2}}, n-1 \right),$$

TABLE 13.2

One-Sample Sample Sizes with Unknown Precision and $v = 5$

β	n_0	1	α	FREQ	ACC	ALC	WOC (90%)	WOC (95%)
2	10	0.2	0.20	165	66	2	124	155
			0.10	271	122	110	212	264
			0.05	385	189	164	306	380
			0.01	664	392	297	537	665
		0.5	0.20	27	3	2	10	15
			0.10	44	12	2	25	33
			0.05	62	22	2	41	52
			0.01	107	55	2	79	99
	50	0.2	0.20	165	26	2	83	114
			0.10	271	82	2	172	224
			0.05	385	149	2	266	340
			0.01	664	352	2	497	625
		0.5	0.20	27	0	2	2	2
			0.10	44	0	2	2	2
			0.05	62	0	2	2	2
			0.01	107	15	2	39	59
5	10	0.2	0.20	165	179	175	327	406
			0.10	271	319	301	546	676
			0.05	385	487	435	780	965
			0.01	664	995	764	1355	1675
		0.5	0.20	27	21	2	43	55
			0.10	44	43	2	79	99
			0.05	62	70	2	117	146
			0.01	107	151	109	210	261
	50	0.2	0.20	165	139	2	286	365
			0.10	271	279	2	506	636
			0.05	385	447	394	740	925
			0.01	664	955	724	1316	1635
		0.5	0.20	27	0	2	2	2
			0.10	44	3	2	37	57
			0.05	62	30	2	76	105
			0.01	107	111	2	170	221

where $p_t(c, d)$ is the area between 0 and c under a t-density with d degrees of freedom. Note that the area only depends on the data via ns^2. Hence, the following algorithm can be used to obtain an approximate solution:

a. Select an initial estimate of the sample size n.

b. Generate m values of the random variables ns^2. Note that $ns^2|\lambda \sim \Gamma((n-1)/2, \lambda/2)$ and λ from $\Gamma(v, \beta)$. Then, marginally, ns^2 follows a gamma–gamma distribution with parameter $(v, 2\beta, (n-1)/2)$. Note that a random variable x follows a gamma–gamma distribution if its density is given by

$$f(x \mid v, \beta, n) = \frac{\Gamma(v+n)\beta^v}{\Gamma(v)\Gamma(n)} \frac{x^{n-1}}{(\beta+x)^{v+n}}$$

for $x > 0$, $v > 0$, $\beta > 0$, and $n > 0$. For more details, one can refer to Bernardo and Smith (1994), pp. 430.

c. For each of the m values of ns_i^2, $i = 1, 2, \ldots, m$ in step (b), calculate

$$\text{coverage}\left(ns_i^2\right) = 2p_i\left(\frac{l}{2}\sqrt{\frac{n(n-1)}{ns_i^2}}, n-1\right).$$

d. Then, compute

$$\frac{1}{m}\sum_{i=1}^{m}\text{coverage}\left(ns_i^2\right)$$

as an approximate to the average coverage probability.

Repeating Steps (b) through (d) for values of n in conjunction with a bisectional search procedure will lead to an approximate sample size.

Similarly, the WOC sample size can be obtained by solving the following inequality:

$$2p_t\left(\frac{l}{2}\sqrt{\frac{n(n-1)}{ns_i^2}}, n-1\right) \geq 1-\alpha$$

for about 95% or 99% (for example) of the ns_i^2 values with a fixed sample size n. On the other hand, the sample size needed by ALC is given by

$$2t_{n-1,1-\alpha/2}\sqrt{\frac{2\beta}{n(n-1)}\frac{\Gamma(0.5n)}{\Gamma(0.5n-0.5)}\frac{\Gamma(v-0.5)}{\Gamma(v)}} \leq l.$$

Therefore, finding the smallest sample size n satisfying the above inequality provides an estimate for the sample size selected by ALC.

13.1.3 Two-Sample with Common Precision

Similarly, we can apply the idea of ACC, ALC, and WOC to the most commonly used two-arm parallel-group design. More specifically, let $x_1 = (x_{11}, \ldots, x_{n11})$ and $x_2 = (x_{12}, \ldots, x_{n22})$ be two independent samples obtained under two different treatments. Furthermore, it is assumed that x_{ij}, $i = 1, \ldots, n_j$ are independent and identically distributed normal random variables with mean μ_j and a common precision λ within each treatment. Similar to the one-sample problem, we assume a conjugate prior distribution for the common precision, that is,

$$\lambda \sim \Gamma(v, \beta),$$

for some v and β. That is, given λ, it is assumed that

$$\mu_j \mid \lambda \sim N(\mu_{0j}, n_{0j}\lambda) \text{ with } j = 1, 2.$$

Then, by treating the mean treatment difference $\theta = \mu_1 - \mu_2$ as the parameter of interest, the methods of ACC, ALC, and WOC can be carried out in a similar manner as in the one-sample situation.

13.1.3.1 Known Common Precision

If the common precision λ is known, then by conditioning on the observed data x_1 and x_2, the posterior distribution of the treatment difference θ is given by

$$\theta \mid x_1, x_2 \sim N \left\{ \mu_{n_2 2} - \mu_{n_1 1}, \frac{\lambda(n_{01} + n_1)(n_{02} + n_2)}{n_1 + n_2 + n_{01} + n_{02}} \right\}, \tag{13.5}$$

where

$$\mu_{n_j j} = \frac{n_{0j} + \mu_{0j} + n_j \bar{x}_j}{n_{0j} + n_j}, \quad j = 1, 2.$$

As it can be seen, the posterior variance of θ only depends on the data via the sample size n_1 and n_2. This situation is very similar to the one-sample case with known precision. Consequently, all of the three methods (i.e., ACC, ALC, and WOC) can be applied to obtain the desired sample size. For the case with equal sample size allocation (i.e., $n_1 = n_2$), the smallest sample size can be estimated by

$$n_1 = n_2 \geq \frac{-B + \sqrt{B^2 - 4AC}}{2A}, \tag{13.6}$$

where $A = \lambda^2$ and

$$B = \lambda^2(n_{01} + n_{02}) - 28 z_{1-\alpha/2}^2 / l^2$$

$$C = n_{01} n_{02} \lambda^2 - \frac{4(n_{01} + n_{02})\lambda z_{1-\alpha/2}^2}{l^2}.$$

In practice, it happens that $B^2 - 4AC \leq 0$, which implies that the prior information is sufficient and no additional sampling is needed.

As it can be seen from Equation 13.5, if we fix the total information $n_1 + n_{01} + n_2 + n_{02}$, then the optimal sample size allocation, which can minimize the posterior variance, is given by $n_1 + n_{01} = n_2 + n_{02}$. In such a situation, the minimum sample size can be obtained by solving the following inequality:

$$n_1 \geq \frac{8}{\lambda l^2} z_{1-\alpha/2}^2 - n_{01}, \tag{13.7}$$

with $n_2 = n_1 + n_{01} - n_{02}$. If $n_{01} = n_{02}$, then the sample size estimated by Equations 13.6 and 13.7 reduces to the sample quantity.

13.1.3.2 Unknown Common Precision

In the situation where the two treatment groups share the same but unknown precision, the posterior distribution of θ can be derived based on the normal-gamma prior family as described above. More specifically,

$$\theta \mid x_1, x_2 \sim A + t_{2v}\sqrt{\frac{B}{2CD}},$$

where

$$A = E(\theta \mid x_1, x_2) = \frac{n_2 \bar{x}_2 + n_{02}\mu_{02}}{n_2 + n_{02}} + \frac{n_1 \bar{x}_1 + n_{01}\mu_{01}}{n_1 + n_{01}},$$

$$B = 2\beta + n_1 s_1^2 + n_2 s_2^2 + \frac{n_1 n_{01}}{n_1 + n_{01}}(\bar{x}_1 - \mu_{01})^2 + \frac{n_2 n_{02}}{n_2 + n_{02}}(\bar{x}_2 - \mu_{02})^2,$$

$$C = \frac{n_1 + n_2}{2} + v, \text{ and } D = \frac{(n_1 + n_{01})(n_2 + n_{02})}{n_1 + n_{01} + n_2 + n_{02}}.$$

Based on such a posterior distribution, the method of ACC, ALC, and WOC can be easily implemented, which are outlined below.

Average coverage criterion. As it has been shown by Joseph and Belisle (1997), the sample size needed must satisfy the following inequality:

$$\frac{(n_1 + n_{01})(n_2 + n_{02})}{n_1 + n_2 + n_{01} + n_{02}} \geq \frac{4\beta}{vl^2}t_{2v,1-\alpha/2}^2.$$

The above inequality can have an explicit solution if we adopt an equal sample size allocation strategy (i.e., $n_1 = n_2 = n$). In such a situation, the optimal sample size is given by

$$n \geq \frac{-B + \sqrt{B^2 - 4AC}}{2A},$$

where

$$A = \frac{vl^2}{4},$$

$$B = \frac{vl^2}{4}(n_{01} + n_{02}) - 2\beta t_{2v,1-\alpha/2}^2,$$

$$C = \frac{n_{01}n_{02}vl^2}{4} - \beta t_{2v,1-\alpha/2}^2(n_{01} + n_{02}).$$

In the case where unequal sample allocation is allowed, how to minimize the expected posterior variance would be a reasonable criterion for an effective allocation of the sample sizes between treatment groups. Note that

$$\bar{x}_i \sim \mu_{0i} + t_{2v_i} \sqrt{\frac{\beta_i(n_i + n_{0i})}{v_i n_i n_{0i}}},$$

$$n_i s_i^2 \sim \Gamma - \Gamma\left(v_i, 2\beta_i, \frac{n_i - 1}{2}\right).$$

It can be verified that the expected variance of θ given the data is given by

$$\frac{1}{(n_1 + n_{01})(n_2 + n_{02})} \frac{(n_1 + n_2 + n_{01} + n_{02})}{n_1 + n_2 + 2v - 2}$$

$$\times \left\{2\beta + \frac{(n_1 + n_2 - 2)\Gamma(v-1)}{\Gamma(v)} + \frac{2\beta}{v - 1}\right\},$$

which implies that for a fixed total sample size $n_1 + n_2$, the minimum expected posterior variance is achieved if $n_1 + n_{01} = n_2 + n_{02}$.

Average length criterion. Similarly, the sample size selected by ALC can be obtained by solving the following inequality:

$$2t_{n_1+n_2+2v,1-\alpha/2} \sqrt{\frac{2\beta(n_1 + n_{01} + n_2 + n_{02})}{(n_1 + n_2 + 2v)(n_1 + n_{01})(n_2 + n_{02})}}$$

$$\times \frac{\Gamma\left(\frac{n_1+n_2+2v}{2}\right)\Gamma\left(\frac{2v-1}{2}\right)}{\Gamma\left(\frac{n_1+n_2+2v-1}{2}\right)\gamma\left(\frac{2v}{2}\right)} \leq l.$$

To solve the above inequality, the standard bisectional search can be used. For unequal sample size situation, the constraint $n_1 + n_{10} = n_2 + n_{20}$ can be used.

Worst outcome criterion. The WOC selects the sample size n_1 and n_2 according to the following inequality:

$$\frac{l^2(n_1 + n_{01})(n_2 + n_{02})}{8\beta(n_1 + n_{01} + n_2 + n_{02})} \frac{n_1 + n_2 + 2v}{1 + \{(n_1 + n_2)/2v\} F_{n_1+n_2,2v,1-w}}$$

$$\geq t_{n_1+n_2+2v,1-\alpha/2}^2.$$

Again, the simple bisectional search algorithm can be used to find the solution effectively.

Mixed Bayesian-likelihood. For the mixed Bayesian-likelihood method, there exists no exact solution. Therefore, an appropriate simulation method has to be employed to estimate the desired sample size. Simply speaking, one can first generate $(\bar{x}_1, \bar{x}_2, n_1 s_1^2, n_2 s_2^2)$ vectors, then ACC, ALC, and WOC can be estimated via a standard bisectional search algorithm. Then, by averaging these estimated sample sizes or taking appropriate infimum, the desired sample size can be obtained (Tables 13.3 and 13.4). For a more detailed discussion, one can refer to Section 4.3 of Joseph and Belisle (1997).

13.1.4 Two-Sample with Unequal Precisions

In this subsection, we consider the problem of two samples with unequal precisions. Similarly, we denote $x_1 = (x_{11}, \ldots, x_{n_i1})$ and $x_2 = (x_{21}, \ldots, x_{n_{21}})$ independent samples obtained from the two different treatments. Furthermore, it is assumed that within each treatment group, the observation x_{ij} is independent and identically distributed as a normal random variable with mean μ_j and precision λ_j. Once again, the normal-gamma prior is used to reflect the prior information about the mean and the variance, that is,

$$\lambda_j \sim \Gamma(v_j, \beta_j)$$
$$\mu_j \mid \lambda_j \sim N(\mu_{0j}, n_{0j}\lambda_j).$$

TABLE 13.3

Two-Sample Sample Sizes with Unknown Common Precision ($n = n_1 = n_2$, $n_0 = n_{01} = n_{02}$, and $v = 2$)

β	n_0	1	α	FREQ	ACC	ALC	WOC (90%)	WOC (95%)
2	10	0.2	0.20	329	461	506	1226	1840
			0.10	542	899	840	2026	3037
			0.05	769	1532	1198	2881	4315
			0.01	1327	4230	2076	4983	7460
		0.5	0.20	53	66	73	189	287
			0.10	87	136	126	317	478
			0.05	123	237	184	454	683
			0.01	213	669	325	791	1187
	50	0.2	0.20	329	421	466	1186	1800
			0.10	542	859	800	1986	2996
			0.05	769	1492	1158	2841	4275
			0.01	1327	4190	2036	4943	7420
		0.5	0.20	53	26	2	148	246
			0.10	87	96	86	277	438
			0.05	123	197	144	414	643
			0.01	213	629	285	751	1147
5	10	0.2	0.20	329	1166	1280	3079	4613
			0.10	542	2263	2115	5079	7605
			0.05	769	3845	3008	7215	10801
			0.01	1327	10589	5202	12468	18663
		0.5	0.20	53	179	196	485	730
			0.10	87	354	330	805	1209
			0.05	123	607	473	1147	1721
			0.01	213	1686	825	1988	2979
	50	0.2	0.20	329	1126	1240	3039	4573
			0.10	542	2223	2075	5039	7565
			0.05	769	3805	2968	7175	10761
			0.01	1327	10549	5162	12428	18623
		0.5	0.20	53	139	156	445	690
			0.10	87	314	290	765	1169
			0.05	123	567	433	1107	1681
			0.01	213	1646	785	1948	2939

TABLE 13.4

Two-Sample Sample Sizes with Unknown Common Precision ($n = n_1 = n_2$, $n_0 = n_{01} = n_{02}$, and $v = 2$)

β	n_0	1	α	FREQ	ACC	ALC	WOC (90%)	WOC (95%)
2	10	0.2	0.20	329	141	141	260	324
			0.10	542	253	241	436	540
			0.05	769	388	349	623	771
			0.01	1327	794	612	1083	1338
		0.5	0.20	53	15	2	33	43
			0.10	87	33	2	62	78
			0.05	123	54	45	92	115
			0.01	213	119	88	166	207
	50	0.2	0.20	329	101	2	220	283
			0.10	542	213	2	395	500
			0.05	769	348	308	583	731
			0.01	1327	754	572	1043	1298
		0.5	0.20	53	0	2	2	2
			0.10	87	0	2	20	37
			0.05	123	14	2	51	75
			0.01	213	79	2	126	167
5	10	0.2	0.20	329	367	373	666	824
			0.10	542	648	623	1103	1364
			0.05	769	983	890	1570	1941
			0.01	1327	1999	1547	2719	3359
		0.5	0.20	53	51	48	98	123
			0.10	87	96	89	169	210
			0.05	123	149	132	244	303
			0.01	213	312	238	428	530
	50	0.2	0.20	329	327	332	625	784
			0.10	542	608	583	1063	1324
			0.05	769	943	850	1530	1901
			0.01	1327	1959	1507	2679	3319
		0.5	0.20	53	11	2	57	83
			0.10	87	56	2	128	170
			0.05	123	109	2	203	263
			0.01	213	272	197	388	490

13.1.4.1 Known Precision

If the precisions λ_1 and λ_2 are known, then the posterior distribution of the treatment mean difference θ is given by

$$\theta \mid x_1, x_2 \sim N\left(\mu_{n_2 2} - \mu_{n_1 1}, \frac{\lambda_{n_1 1}\lambda_{n_2 2}}{\lambda_{n_1 1} + \lambda_{n_2 2}}\right),$$

where $\lambda_{ni} = \lambda_i(n_{0i} + n_i)$ and

$$\mu_{n_i i} = \frac{\lambda_i(n_{0i}\mu_{0i} + n_i\bar{x}_i)}{\lambda_{n_i i}} = \frac{n_{0i}\mu_{0i} + n_i\bar{x}_i}{n_{0i} + n_i}.$$

If we assume an equal sample size allocation (i.e., $n = n_1 = n_2$), then the sample sizes selected by ACC, ALC, and WOC are all given by

$$n \geq \frac{-B + \sqrt{B^2 - 4AC}}{2A},$$

where $A = \lambda_1\lambda_2$ and

$$B = \lambda_1 n_{02}\lambda_2 + \lambda_2 n_{01}\lambda_1 - \frac{4z_{1-\alpha/2}^2}{l^2}(\lambda_1 + \lambda_2),$$

$$C = n_{01}\lambda_1 n_{02}\lambda_2 - \frac{4z_{1-\alpha/2}^2}{l^2}(n_{01}\lambda_1 + n_{02}\lambda_2).$$

If unequal sample size allocation is considered, then the optimal sample size is given by

$$n_1 + n_{01} \geq \frac{4}{l^2} z_{1-\alpha/2}^2 \left\{ \frac{1}{\sqrt{\lambda_1\lambda_2}} + \frac{1}{\lambda_1} \right\},$$

$$n_2 + n_{02} = \sqrt{\frac{\lambda_1}{\lambda_2}}(n_1 + n_{01}).$$

13.1.4.2 Unknown Precisions

When $\lambda_1 \neq \lambda_2$, the exact posterior distribution of θ could be complicated. It, however, can be approximated by

$$\theta \,|\, x_1, x_2 \approx N(\mu_{n1} - \mu_{n2}, \lambda^*),$$

where

$$\mu_{nj} = \frac{n_{0j}\mu_{0j} + n_j \bar{x}_j}{n_j + n_{0j}},$$

$$\lambda^* = \left\{ \frac{2\beta_{n_1 1}}{(n_1 + n_{01})(2v_1 + n_1 - 2)} + \frac{2\beta_{n_2 2}}{(n_2 + n_{02})(2v_2 + n_2 - 2)} \right\}^{-1}.$$

Owing to the fact that the posterior distribution of θ does not have a standard form, appropriate numerical methods are necessarily employed for obtaining the required sample size. In most applications, however, normal approximation usually provides adequate sample sizes. Note that

$$\bar{x}_j \sim \mu_{0j} + \sqrt{\frac{\beta_i(n_j + n_{0j})}{v_j n_j n_{0j}}} t_{2v_j}$$

$$n_j s_j^2 \sim \Gamma\left(v_j, 2\beta_j, \frac{n_j - 1}{2}\right),$$

which is minimized if and only if

$$n_2 + n_{02} = \sqrt{\frac{\beta_2(v_1 - 1)}{\beta_1(v_2 - 1)}}(n_1 + n_{01}).$$

As a result, the optimal sample size, which minimizes the posterior variance, can be obtained.

13.2 Posterior Error Approach

Recently, Lee and Zelen (2000) developed a simple and yet general Bayesian sample size determination theory. It is very similar but different from the traditional frequentist approach. More specifically, the traditional frequentist approach selects the minimum sample size so that type I and type II error rates can be controlled at prespecified levels of significance, that is, α and β. Lee and Zelen (2000), on the other hand, proposed to select the minimum sample size so that the posterior error rate is controlled. Consequently, it is essential to understand the posterior error rate and its relationship with the traditional type I and type II error rates. Under this framework, the Bayesian sample size calculation for comparing means and survival rates are illustrated.

13.2.1 Posterior Error Rate

Following the notations of Lee and Zelen (2000), denote by δ the noncentrality parameter, which is usually a function of the clinically meaningful difference and the population standard deviation. From a traditional frequentist point of view, the objective of a clinical trial is to differentiate between the following two hypotheses:

$$H_0 : \delta = 0 \quad \text{versus} \quad H_a : \delta \neq 0.$$

To utilize the Bayesian framework, we assign a nondegenerate probability $\theta > 0$ to the joint even $\delta > 0$ or $\delta < 0$. Then, the prior probability for the null hypothesis (i.e., no treatment effect) would be $(1-\theta)$ by definition.

Consequently, the parameter θ summarizes the prior confidence regarding which hypothesis is more likely to be true. When there is no clear preference between H_0 and H_1, $\theta = 0.5$ seems to be a reasonable choice. However, if one wishes to be more conservative in concluding the treatment effect, then one may consider a $\theta < 0.5$ (e.g., $\theta = 0.25$). Similarly, one may also use a $\theta > 0.5$ (e.g., $\theta = 0.75$) to reflect a better confidence about the treatment effect. Furthermore, equal probability $\theta/2$ is assigned to each possible alternative $\delta > 0$ and $\delta < 0$. The reason to assign equal probability to each possible alternative is that if the physician has prior belief that one treatment is better or worse than the other, then the physician cannot ethically enter the patient into the trial (Lee and Zelen, 2000). Consequently, it seems that equal probability $\theta/2$ for $\delta > 0$ and $\delta < 0$ is a natural choice.

To facilitate discussion, we further classify the outcomes of a clinical trial into positive or negative ones. A positive outcome refers to a clinical trial with conclusion $\delta \neq 0$, while a negative outcome refers to a trial with conclusion $\delta = 0$. For convenience purpose, we use a binary indicator $C = +$ or $C = -$ to represent positive or negative outcomes, respectively. On the other hand, we consider T as a similar indicator but it refers to the true status of hypothesis. In other words, $T = +$ refers to the situation where the alternative hypothesis is actually true, while $T = -$ refers to the situation where the null hypothesis is in fact true. It follows immediately that $P(T = +) = \theta$.

For a prespecified type I and type II error rates α and β, the traditional frequentist approach tries to find the minimum sample size so that

$$\alpha = P(C = + \mid T = -),$$

$$\beta = P(C = - \mid T = +).$$

Similarly, from a Bayesian perspective, Lee and Zelen (2000) suggest to find the minimum sample size so that the posterior error rates are controlled as

$$a^* = P(T = - \mid C = +),$$

$$\beta^* = P(T = + \mid C = -).$$

Simply speaking, they are the posterior probabilities that the true situation is opposite to the outcome of the trial. As long as these posterior error rates are well controlled at given levels of a^* and β^*, we are able to accept any trial outcome (either positive or negative) at an acceptable confidence level.

The merit of Lee and Zelen's method is that their posterior error rates α^* and β^* are closely related to the most commonly used type I and type II error rates α and β via the following simple formula:

$$P_1 = 1 - \alpha^* = P(T = - \mid C = -) = \frac{(1-\theta)(1-\alpha)}{(1-\theta)(1-\alpha) + \theta\beta},$$

$$P_2 = 1 - \beta^* = P(T = + \mid C = +) = \frac{\theta(1-\beta)}{(1-\beta)\theta + \alpha(1-\theta)},$$

where P_1 and P_2 define the posterior probability that the true status of the hypothesis is indeed consistent with the clinical trial outcome. Then, we have

$$\alpha = \frac{(1-P_2)(\theta + P_1 - 1)}{(1-\theta)(P_1 + P_2 - 1)}, \tag{13.8}$$

$$\beta = \frac{(1-P_1)(P_2 - \theta)}{\theta(P_1 + P_2 - 1)}. \tag{13.9}$$

Consequently, for a given θ and (P_1, P_2), we can first compute the value of α and β. Then, the traditional frequentist sample size formula can be used to determine the needed sample size. Note that such a procedure is not only very simple but also free of model and testing procedures. Hence, it can be directly applied to essentially any type of clinical trial

TABLE 13.5

Type I and II Errors for Various P_1, P_2, and θ Values

θ	P_1	P_2	α	β	P_1	P_2	α	β
0.25	0.80	0.80	0.0222	0.7333	0.90	0.80	0.0571	0.3143
		0.85	0.0154	0.7385		0.85	0.0400	0.3200
		0.90	0.0095	0.7429		0.90	0.0250	0.3250
		0.95	0.0044	0.7467		0.95	0.0118	0.3294
	0.85	0.80	0.0410	0.5077	0.95	0.80	0.0711	0.1467
		0.85	0.0286	0.5143		0.85	0.0500	0.1500
		0.90	0.0178	0.5200		0.90	0.0314	0.1529
		0.95	0.0083	0.5250		0.95	0.0148	0.1556
0.50	0.80	0.80	0.2000	0.2000	0.90	0.80	0.2286	0.0857
		0.85	0.1385	0.2154		0.85	0.1600	0.0933
		0.90	0.0857	0.2286		0.90	0.1000	0.1000
		0.95	0.0400	0.2400		0.95	0.0471	0.1059
	0.85	0.80	0.2154	0.1385	0.95	0.80	0.2400	0.0400
		0.85	0.1500	0.1500		0.85	0.1688	0.0438
		0.90	0.0933	0.1600		0.90	0.1059	0.0471
		0.95	0.0438	0.1688		0.95	0.0500	0.0500
0.75	0.80	0.80	0.7333	0.0222	0.90	0.80	0.7429	0.0095
		0.85	0.5077	0.0410		0.85	0.5200	0.0178
		0.90	0.3143	0.0571		0.90	0.3250	0.0250
		0.95	0.1467	0.0711		0.95	0.1529	0.0314
	0.85	0.80	0.7385	0.0154	0.95	0.80	0.7467	0.0044
		0.85	0.5143	0.0286		0.85	0.5250	0.0083
		0.90	0.3200	0.0400		0.90	0.3294	0.0118
		0.95	0.1500	0.0500		0.95	0.1556	0.0148

and any type of testing procedure. For illustration purposes, we consider its usage for comparing means and survival in Sections 13.2.2 and 13.2.3. Table 13.5 provides a number of most likely (θ, P_1, P_2) values and their associated (α, β) values.

13.2.2 Comparing Means

In this subsection, we show in detail how Lee and Zelen's Bayesian approach can be used to determine the sample size for comparing means. For illustration purpose, we consider only the most commonly used two-arm parallel-group design with equal sample size allocation. The situation with multiple arms and unequal sample size allocation can be similarly obtained.

Following the notations of Chapter 3, denote x_{ij} the response observed from the jth subject in the ith treatment group, $j = 1, \ldots, n_i$, $i = 1, 2$. It is assumed that x_{ij}, $j = 1, \ldots, n_i$, $i = 1, 2$, are independent normal random variables with mean μ_i and variance σ^2. In addition, define

$$\bar{x}_i = \frac{1}{n_i} \sum_{j=1}^{n_i} x_{ij} \quad \text{and} \quad s^2 = \frac{1}{n_1 + n_2 - 2} \sum_{i=1}^{2} \sum_{j=1}^{n_i} (x_{ij} - \bar{x}_i)^2.$$

Suppose that the objective of a clinical trial is to test whether there is a difference between the mean responses of the test drug and a placebo control or an active control agent. The following hypotheses are of interest:

$$H_0 : \varepsilon = 0 \quad \text{versus} \quad H_\alpha : \varepsilon \neq 0.$$

When σ^2 is unknown, the null hypothesis H_0 is rejected at the α level of significance if

$$\left| \frac{\bar{x}_1 - \bar{x}_2}{s\sqrt{\frac{1}{n_1} + \frac{1}{n_2}}} \right| > t_{\alpha/2, n_1 + n_2 - 2}.$$

For given error rates α and β, as shown in Chapter 3, the minimum sample size needed can be estimated by

$$n_1 = n_2 = \frac{2(z_{\alpha/2} + z_\beta)^2 \sigma^2}{\varepsilon^2}. \tag{13.10}$$

To implement the Bayesian approach, one needs to first specify the value of θ to reflect the prior knowledge about the null and alternative hypotheses. Furthermore, one needs to specify the value of (P_1, P_2), which controls the posterior error rate. Based on these specifications, one can obtain the values of (α, β) according to Equations 13.8 and 13.9. Then, use the resultant significance levels (α, β) and the formula (13.10) to obtain the desired sample size.

Consider Example 3.2.4 as given in Chapter 3, where a clinical trial is conducted to evaluate the effectiveness of a test drug on cholesterol in patients with coronary heart disease (CHD). It is assumed that the clinical meaning difference is given by $\varepsilon = 5\%$ and the standard deviation is $\sigma = 10\%$. It has been demonstrated in Example 3.2.4 that if we specify the type I and type II error rates to be $(\alpha, \beta) = (0.05, 0.20)$, respectively, then the resulting sample size from Equation 13.10 is given by 63. However, if we use the Bayesian approach and specify $(\alpha^*, \beta^*) = (0.05, 0.20)$ with $\theta = 0.50$, then according to Equations 13.8 and 13.9, we have $(\alpha, \beta) = (0.24, 0.04)$. It follows that the sample size needed is given by

$$n_1 = n_2 = \frac{2(z_{\alpha/2} + z_\beta)^2 \sigma^2}{\varepsilon^2} = \frac{2 \times 2.46^2 \times 0.10^2}{0.05^2} \approx 49.$$

Consequently, only 49 subjects per treatment group are needed.

It, however, should be noted that the resultant sample sizes 63 and 49 are not directly comparable, because they control different error rates. The sample size 63 per treatment group is selected to control the type I and type II error rates at the levels of 0.05 and 0.20, respectively. On the other hand, the 49 subjects per treatment group is selected to control the two posterior error rates α^* and β^* at the levels of 0.05 and 0.20, respectively, which correspond to the traditional type I and type II error rates at the levels of 0.04 and 0.24 according to our computation. Therefore, which sample size should be used depends on what statistical test will be used for testing the hypotheses. To provide a better understanding, various sample sizes needed at different posterior error configurations are summarized in Table 13.6.

TABLE 13.6

Sample Sizes under Different θ and $P = P_1 = P_2$ Specifications

θ	P	α	β	(ε/σ)	n	P	α	β	(ε/σ)	n
0.25	0.80	0.0222	0.7333	0.1	554	0.90	0.0250	0.3250	0.1	1453
				0.2	139				0.2	364
				0.3	62				0.3	162
				0.4	35				0.4	91
				0.5	23				0.5	59
0.25	0.85	0.0286	0.5143	0.1	928	0.95	0.0148	0.1556	0.1	2381
				0.2	232				0.2	596
				0.3	104				0.3	265
				0.4	58				0.4	149
				0.5	38				0.5	96
0.50	0.80	0.2000	0.2000	0.1	902	0.90	0.1000	0.1000	0.1	1713
				0.2	226				0.2	429
				0.3	101				0.3	191
				0.4	57				0.4	108
				0.5	37				0.5	69
0.50	0.85	0.1500	0.1500	0.1	1227	0.95	0.0500	0.0500	0.1	2599
				0.2	307				0.2	650
				0.3	137				0.3	289
				0.4	77				0.4	163
				0.5	50				0.5	104
0.75	0.80	0.7333	0.0222	0.1	1106	0.90	0.3250	0.0250	0.1	1734
				0.2	277				0.2	434
				0.3	123				0.3	193
				0.4	70				0.4	109
				0.5	45				0.5	70
0.75	0.85	0.5143	0.0286	0.1	1305	0.95	0.1556	0.0148	0.1	2586
				0.2	327				0.2	647
				0.3	145				0.3	288
				0.4	82				0.4	162
				0.5	53				0.5	104

13.3 Bootstrap-Median Approach

As indicated earlier, both methods based on the posterior credible interval and the posterior error rate are useful Bayesian approach for sample size calculation. The method of posterior credible interval approach explicitly takes into consideration the prior information. In practice, however, prior information is either not available or not reliable. In this case, the application of Joseph and Belisle's method is limited. On the other hand, the method of posterior error rate does not suffer from this limitation. In practice, it is suggested that the posterior error rates P_1 and P_2 be chosen in such a way that they correspond to the type I and type II error rates from the frequentist point of view. This method, however, may not be widely used and accepted given the fact that the traditional testing procedure for controlling both type I and type II error rates have been widely used and accepted in practice.

As a result, two important questions are raised. First, if there is no reliable prior information available, is there still a need for Bayesian sample size calculation method? Second, if there is a need for the Bayesian method for sample size calculation, then how does one incorporate it into the framework that is accepted from the frequentist point of view? With an attempt to address these two questions, we will introduce an alternative method.

13.3.1 Background

As indicated in Chapter 1, a prestudy power analysis for sample size calculation is usually performed at the planning stage of an intended clinical trial. Sample size calculation is done under certain assumptions regarding the parameters (e.g., clinically important difference and standard deviation) of the target patient population. In practice, since the standard deviation of the target patient population is usually unknown, a typical approach is to use estimates from some pilot studies as the surrogate for the true parameters. We then treat these estimates as the true parameters to justify the sample size needed for the intended clinical trial. It should be noted that the parameter estimate obtained from the pilot study inevitably suffers from the sampling error, which causes the uncertainty (variation) about the true parameter. Consequently, it is likely to yield an unstable estimate of sample size for the intended clinical trial.

To fix the idea, we conduct a simple simulation study to examine the effect of the pilot study uncertainty. For illustration purposes, we consider a simple one-sample design with the type I and type II error rates given by 5% and 10%, respectively. Furthermore, we assume that the population standard deviation $\sigma = 1$ and the true mean response is $\varepsilon = 25\%$. Then, according to Equation 3.2, the sample size needed, if the true parameters are known, is given by

$$n = \frac{(z_{\alpha/2} + z_{\beta})^2 \sigma^2}{\varepsilon^2} \approx 169.$$

However, in practice, the true parameters are never known. Hence, it has to be estimated from some pilot studies (with small sizes). In the simulation, 10,000 independent simulation runs are generated to examine the distribution of the sample size estimates, which are obtained based on the estimates of the parameters of the pilot study. The results are summarized in Table 13.7.

As it can be seen from Table 13.7, for a pilot study with sample size as large as 100, the resulting sample size estimates are still very unstable with a mean given by 10,302, which is much larger than the ideal sample size 169. To have a more reliable sample size estimate,

TABLE 13.7

Effect of the Pilot Study Sample Size

n_0	Mean	SD	Minimum	Median	Maximum
10	20,060,568	1,824,564,572	2	122	181,767,638,343
50	72,746	2,754,401	10	163	225,999,504
100	10,302	452,841	22	168	38,265,202
500	188	87	56	168	1575
1000	177	49	66	168	594

we need to increase the size of the pilot study up to 500, which is much larger than the planned larger-scale trial with only 169 subjects! Hence, we conclude that the uncertainty due to sampling error of the pilot study is indeed a very important source of risk for the planned clinical trial.

13.3.2 Bootstrap-Median Approach

Note that Table 13.7 reveals two important points. First, it clearly demonstrates that there is a serious effect of the sample size estimation uncertainty due to the sampling error of the pilot study. Second, it seems to suggest that although the mean of the estimated sample size is very unstable, its median is more robust. To some extent, this is not a surprising finding. Let $\hat{\varepsilon}$ and $\hat{\sigma}^2$ be the sample mean and variance obtained from the pilot study, respectively. Then, by applying Equations 3.2, the estimated sample size would be given by

$$\hat{n} = \frac{(z_{\alpha/2} + z_\beta)^2}{\hat{\varepsilon}^2} \hat{\sigma}^2.$$

If we assume that the sample is normally distributed, then we have $\hat{\varepsilon}$, which also follows some normal distribution. Hence, we know immediately that $E(\hat{n}) = \infty$, which naturally explains why naïve sample size estimates based on pilot data could be extremely unstable. However, the median of $\hat{n}$ is always well defined and according to Table 13.7, it seems be able to approximate the true sample size with quite a satisfactory precision.

All the above discussion seems to suggest that if we are able to generate an independent copy of the sample (i.e., bootstrap data set), then we are able to estimate the sample size based on each bootstrapped data set. Then, the median of those bootstrap sample size estimates may provide a much more reliable approximate to the true sample size needed. More specifically, let S be the original sample, then the following simple bootstrap procedure can be considered:

1. First generate the bootstrap data set S_h from S. It is generated by simple random sampling with replacement and has the same sample size as S.
2. Compute the sample size n_h based on the bootstrap data set S_h and appropriate objective and formula.
3. Repeat Steps (1) and (2) for a total of B times, then take the median of the n_h, $h = 1$, ..., B to be the final sample size estimate.

To evaluate the finite sample performance of the above bootstrap procedure, a simple simulation study is conducted. It is simulated according to the same parameter configuration used in Table 13.7. The number of bootstrap iteration B is fixed to be $B = 1000$ and the total number of simulation iterations is fixed to be 100. The results are summarized in Table 13.8. As it can be seen, although the final estimate may still suffer from instability to some extent, it represents a dramatical improvement as compared with the results in Table 13.7. For detailed discussion about the bootstrap sample size calculation, we refer to Lee, Wang, and Chow (2006).

TABLE 13.8

Bootstrap Sample Size Estimation

n_0	Mean	SD	Minimum	Median	Maximum
10	103	71	9	88	239
50	266	275	45	150	1175
100	363	448	28	192	2090
500	182	80	79	154	479
1000	180	51	84	174	408

13.4 Concluding Remarks

As indicated by Lee, Wang, and Chow (2006), power analysis for sample size calculation of a later-phase clinical study based on limited information collected from a pilot study could lead to an extremely unstable estimate of sample size. Lee, Wang, and Chow (2006) proposed the use of bootstrap-median approach and evaluated the instability of the estimated sample size by means of the population squared coefficient of variation, that is, $CV = \sigma/\xi$, where ξ and σ are the population mean and standard deviation, respectively. Lee, Wang, and Chow (2006) considered the parameter $\theta = (\sigma^2 + \xi^2)/\xi^2$ and showed that

$$n_{0.5} = 1.5n^{-1}\theta\{1 + o(1)\},$$

where $n_{0.5}$ satisfies

$$P\left(\hat{\theta} \leq \eta_{0.5}\right) = 0.5,$$

where $\eta_{0.5}$ is the squared median of sample coefficient of variation.

14

Nonparametrics

In clinical trials, a parametric procedure is often employed for the evaluation of clinical efficacy and safety of the test compound under investigation. A parametric procedure requires assumptions on the underlying population from which the data are obtained. A typical assumption on the underlying population is the normality assumption. Under the normality assumption, statistical inference regarding treatment effects can be obtained through appropriate parametric statistical methods such as the analysis of variance under valid study designs. In practice, however, the primary assumptions on the underlying population may not be met. As a result, parametric statistical methods may not be appropriate. In this case, alternatively, a nonparametric procedure is often considered. Nonparametric methods require few assumptions about the underlying populations, which are applicable in situations where the normal theory cannot be utilized. In this chapter, procedures for sample size calculation for testing hypotheses of interest are obtained under appropriate nonparametric statistical methods.

In Section 14.1, the loss in power due to the violation of the normality assumption is examined. Nonparametric methods for testing differences in location are discussed for one-sample and two-sample problems, respectively, in Sections 14.2 and 14.3. Included in these sections are the corresponding procedures for sample size calculation. Nonparametric tests for independence and the corresponding procedure for sample size calculation are given in Section 14.4. Some practical issues are presented in Section 14.5.

14.1 Violation of Assumptions

Under a parametric model, normality is probably the most commonly made assumption when analyzing data obtained from clinical trials. In practice, however, it is not uncommon that the observed data do not meet the normality assumption at the end of the trial. The most commonly seen violation of the normality assumption is that the distribution of the observed variable is skewed (either to the right or to the left). In this case, a log-transformation is usually recommended to remove the skewness before data analysis. For a fixed sample size selected based on the primary assumption of normality, it is then of interest to know how the power is affected if the primary assumption is seriously violated. For illustration purposes, in this section, we address this question for comparing means. Other situations when comparing proportions or time-to-event data can be addressed in a similar manner.

Consider a randomized, parallel-group clinical trial comparing a treatment group and an active control agent. Let x_{ij} be the observation from the jth subject in the ith treatment, $i = 1, 2, j = 1, \ldots, n$. It is assumed that $\log(x_{ij})$ follows a normal distribution with mean μ_i and variance σ^2. Let $\mu_i^* = E(x_{ij}) = e^{\mu_i + \sigma^2/2}$. The hypothesis of interest is to test

$$H_0 : \mu_1^* = \mu_2^* \quad \text{versus} \quad H_a : \mu_1^* \neq \mu_2^*$$

which is equivalent to

$$H_0 : \mu_1 = \mu_2 \quad \text{versus} \quad H_a : \mu_1 \neq \mu_2.$$

At the planning stage of the clinical trial, sample size calculation is performed under the assumption of normality and the assumption that a two-sample t-test statistic will be employed. More specifically, the test statistic is given by

$$T_1 = \frac{\sqrt{n}(\bar{x}_1 - \bar{x}_2)}{\sqrt{2}s},$$

where $\bar{x}_i$ is the sample mean of the ith treatment group and s^2 is the pooled sample variance of x_{ij}'s. We reject the null hypothesis at the α level of significance if

$$|T_1| > z_{\alpha/2}.$$

Under the alternative hypothesis that $\mu_1 \neq \mu_2$, the power of the above testing procedure is given by

$$\Phi\left(\frac{\sqrt{n}\,|e^{\mu_1} - e^{\mu_2}|}{\sqrt{(e^{2\mu_1} + e^{2\mu_2})(e^{\sigma^2} - 1)}} - z_{\alpha/2} \right). \tag{14.1}$$

At end of the trial, it is found that the observed data are highly skewed and hence a log-transformation is applied. After the log-transformation, the data appear to be normally distributed. As a result, it is of interest to compare the power of the two-sample t-test based on either the untransformed (raw) data or the log-transformed data to determine the impact of the violation of normality assumption on power with the fixed sample size n selected under the normality assumption of the untransformed data. Let $y_{ij} = \log(x_{ij})$. The test statistic is given by

$$T_2 = \frac{\sqrt{n}(\bar{y}_1 - \bar{y}_2)}{2s_y} \tag{14.2}$$

where $\bar{y}_i$ is the sample mean of the log-transformed response from the ith treatment group and s_y^2 the pooled sample variance based on the log-transformed data. The power of the above test is given by

$$\Phi\left(\frac{\sqrt{n}\,|\mu_1 - \mu_2|}{\sqrt{2}\sigma} - z_{\alpha/2} \right). \tag{14.3}$$

From Equations 14.1 and 14.3, the loss in power is

$$\Phi\left(\frac{\sqrt{n}\,|e^{\mu_1} - e^{\mu_2}|}{\sqrt{(e^{2\mu_1} + e^{2\mu_2})(e^{\sigma^2} - 1)}} - z_{\alpha/2} \right) - \Phi\left(\frac{\sqrt{n}\,|\mu_1 - \mu_2|}{\sqrt{2}\sigma} - z_{\alpha/2} \right).$$

It can be seen that the violation of the model assumption can certainly have an impact on the power of a trial with a fixed sample size selected under the model assumption. If the true power is below the designed power, the trial may fail to detect a clinically meaningful difference, when it truly exists. If the true power is above the designed power, then the trial is not cost-effective. This leads to a conclusion that incorrectly applying a parametric procedure to a data set, which does not meet the parametric assumption, may result in a significant loss in power and efficiency of the trial. As an alternative, a nonparametric method is suggested. In what follows, sample size calculations based on nonparametric methods for comparing means are provided.

14.2 One-Sample Location Problem

As discussed in Chapter 3, one-sample location problem concerns two types of data. The first type of data consists of paired replicates. In clinical research, it may represent pairs of pretreatment and posttreatment observations. The primary interest is whether there is a shift in location due to the application of the treatment. The second type of data consists of observations from a single population. Statistical inference is made on the location of this population. For illustration purposes, in this section, we focus on nonparametric methods for paired replicates. Let x_i and y_i be the paired observations obtained from the ith subject before and after the application of treatment, $i = 1, \ldots, n$. Let $z_i = y_i - x_i, i = 1, \ldots, n$. Then, z_i can be described by the following model:

$$z_i = \theta + e_i, \quad i = 1, \ldots, n,$$

where θ is the unknown location parameter (or treatment effect) of interest and the e_i's are unobserved random errors having mean 0. It is assumed that (i) each e_i has a continuous population (not necessarily the same one) that is symmetric about zero and (ii) the e_i's are mutually independent. The hypotheses regarding the location parameter of interest are given by

$$H_0: \theta = 0 \quad versus \quad H_a: \theta \neq 0.$$

To test the above hypotheses, a commonly employed nonparametric test is the Wilcoxon signed rank test. Consider the absolute differences $|z_i|, i = 1, \ldots, n$. Let R_i denote the rank of $|z_i|$ in the joint ranking from least to greatest. Define

$$\psi_i = \begin{cases} 1 & \text{if } z_i > 0 \\ 0 & \text{if } z_i < 0 \end{cases} \quad i = 1, \ldots, n.$$

The statistic

$$T^+ = \sum_{i=1}^{n} R_i \psi_i$$

is the sum of the positive signed ranks. Based on T^+, the Wilcoxon signed rank test rejects the null hypothesis at the α level of significance if

$$T^+ \geq t(\alpha_2, n)$$

or

$$T^+ \leq \frac{n(n+1)}{2} - t(\alpha_1, n),$$

where $t(\alpha, n)$ satisfies

$$P(T^+ \geq t(\alpha, n)) = \alpha$$

under the null hypothesis and $\alpha = \alpha_1 + \alpha_2$. Values of $t(\alpha, n)$ are given in the most standard nonparametric references (e.g., Hollander and Wolfe, 1973). It should be noted that under the null hypothesis, the statistic

$$T^* = \frac{T^+ - E(T^+)}{\sqrt{\mathrm{var}(T^+)}} = \frac{T^+ - n(n+1)/4}{\sqrt{n(n+1)(2n+1)/24}}$$

has an asymptotic standard normal distribution. In other words, we may reject the null hypothesis at the α level of significance for large n if

$$|T^*| \geq z_{\alpha/2}.$$

To derive a formula for sample size calculation, we note that

$$T^+ = \sum_{i=1}^{n} \sum_{j=1}^{n} I\{|z_i| \geq |z_j|\} \psi_i$$

$$= \sum_{i=1}^{n} \psi_i + \sum_{i \neq j} I\{|z_i| \geq |z_j|\} \psi_i$$

$$= \sum_{i=1}^{n} \psi_i + \sum_{i<j} (I\{|z_i| \geq |z_j|\} \psi_i + I\{|z_j| \geq |z_i|\} \psi_j).$$

Hence, the variance of T^+ can be obtained as

$$\begin{aligned}
\mathrm{var}(T^+) = {} & n \, \mathrm{var}(\psi_i) \\
& + \frac{n(n-1)}{2} \mathrm{var}(I\{|z_i| \geq |z_j|\} \psi_i + I\{|z_j| \geq |z_i|\} \psi_j) \\
& + 2n(n-1) \mathrm{cov}(\psi_i, I\{|z_i| \geq |z_j|\} \psi_i + I\{|z_j| \geq |z_i|\} \psi_j) \\
& + n(n-1)(n-2) \mathrm{cov}(I\{|z_i| \geq |z_{j_1}|\} \psi_i + I\{|z_{j_1}| \geq |z_i|\} \psi_{j_1}, \\
& \quad I\{|z_i| \geq |z_{j_2}|\} \psi_i + I\{|z_{j_2}| \geq |z_i|\} \psi_{j_2}) \\
= {} & np_1(1-p_1) + n(n-1)\left(p_1^2 - 4p_1p_2 + 3p_2 - 2p_2^2\right) \\
& + n(n-1)(n-2)\left(p_3 + 4p_4 - 4p_2^2\right),
\end{aligned}$$

where

$$p_1 = P(z_i > 0),$$
$$p_2 = P(|z_i| \geq |z_j|, z_i > 0),$$
$$p_3 = P(|z_i| \geq |z_{j1}|, |z_i| \geq |z_{j2}|, z_i > 0),$$
$$p_4 = P(|z_{j1}| \geq |z_i| \geq |z_{j2}|, z_{j1} > 0, z_i > 0).$$

It should be noted that the above quantities can be readily estimated based on data from pilot studies. More specifically, suppose that $z_1, \ldots, z_n$ are data from a pilot study. Then, the corresponding estimators can be obtained as

$$\hat{p}_1 = \frac{1}{n} \sum_{i=1}^{n} I\{z_i > 0\},$$

$$\hat{p}_2 = \frac{1}{n(n-1)} \sum_{i \neq j} I\{|z_i| \geq |z_j|, z_i > 0\},$$

$$\hat{p}_3 = \frac{1}{n(n-1)(n-2)} \sum_{i \neq j_1 \neq j_2} I\{|z_i| \geq |z_{j_1}|, |z_i| \geq |z_{j_2}|, z_i > 0\},$$

$$\hat{p}_4 = \frac{1}{n(n-1)(n-2)} \sum_{i \neq j_1 \neq j_2} I\{|z_{j_1}| \geq |z_i| \geq |z_{j_2}|, z_{j_1} > 0, z_i > 0\}.$$

Under the alternative hypothesis, $E(T^+) \neq n(n+1)/4$. T^+ can be approximated by a normal random variable with mean $E(T^+) = np_1 + n(n-1)p_2$ and variance $\mathrm{var}(T^+)$. Without loss of generality, assume that $E(T^+) > n(n+1)/4$. Thus, the power of the test can be approximated by

$$\text{Power} = P(|T^*| > z_{\alpha/2})$$

$$\approx P(T^* > z_{\alpha/2})$$

$$= P\left(T^+ > z_{\alpha/2}\sqrt{n(n+1)(2n+1)/24} + \frac{n(n+1)}{4}\right)$$

$$\approx 1 - \Phi\left(\frac{z_{\alpha/2}/\sqrt{12} + \sqrt{n}(1/4 - p_2)}{\sqrt{p_3 + 4p_4 - 4p_2^2}}\right).$$

The last approximation in the above equation is obtained by ignoring the lower-order terms of n. Hence, the sample size required for achieving the desired power of $1 - \beta$ can be obtained by solving the following equation:

$$\frac{z_{\alpha/2}/\sqrt{12} + \sqrt{n}(1/4 - p_2)}{\sqrt{p_3 + 4p_4 - 4p_2^2}} = z_\beta.$$

This leads to

$$n = \frac{\left(z_{\alpha/2}/\sqrt{12} + z_{\beta}\sqrt{p_3 + 4p_4 - 4p_2^2}\right)^2}{(1/4 - p_2)^2}.$$

14.2.1 Remark

As indicated before, when there are no ties,

$$\text{var}(T^+) = \frac{1}{24}[n(n+1)(2n+1)].$$

When there are ties, var (T^+) is given by

$$\text{var}(T^+) = \frac{1}{24}\left[n(n+1)(2n+1) - \frac{1}{2}\sum_{j=1}^{g} t_j(t_j - 1)(t_j - 1)\right],$$

where g is the number of tied groups and t_j is the size of tied group j. In this case, the above formula for sample size calculation is necessarily modified.

14.2.2 An Example

To illustrate the use of sample size formula derived above, we consider the same example concerning a study of osteoporosis in postmenopausal women described in Chapter 3 for testing one-sample location problem. Suppose a clinical trial is planned to investigate the effect of a test drug on the prevention of the progression to osteoporosis in women with osteopenia. Suppose that a pilot study with five subjects was conducted. According to the data from the pilot study, it was estimated that $p_2 = 0.30$, $p_3 = 0.40$, and $p_4 = 0.05$. Hence, the sample size needed to achieve an 80% power for the detection of such a clinically meaningful improvement can be estimated by

$$n = \frac{\left(z_{\alpha/2}/\sqrt{12} + z_{\beta}\sqrt{p_3 + 4p_4 - 4p_2^2}\right)^2}{(1/4 - p_2)^2}$$

$$= \frac{\left(1.96/\sqrt{12} + 0.84\sqrt{0.4 + 4 \times 0.05 - 4 \times 0.3^2}\right)^2}{(0.25 - 0.3)^2}$$

$$\approx 383.$$

Thus, a total of 383 subjects are needed to have an 80% power to confirm the observed posttreatment improvement.

14.3 Two-Sample Location Problem

Let x_i, $i = 1, \ldots, n_1$, and y_j, $j = 1, \ldots, n_2$, be two independent random samples, which are, respectively, from a control population and a treatment population in a clinical trial. Suppose that the primary objective is to investigate whether there is a shift of location, which indicates the presence of the treatment effect. Similar to the one-sample location problem, the hypotheses of interest are given by

$H_0: \theta = 0$ versus $H_a: \theta \neq 0$,

where θ represents the treatment effect. Consider the following model:

$$x_i = e_i, \quad i = 1, \ldots, n_1,$$

and

$$y_j = \theta + e_{n_1+j}, \quad j = 1, \ldots, n_2,$$

where the e_i's are random errors having mean 0. It is assumed that (i) each e_i comes from the same continuous population and (ii) the $n_1 + n_2$ e_i's are mutually independent. To test the above hypotheses, the Wilcoxon rank sum test is probably the most commonly used nonparametric test (Wilcoxon, 1945; Wilcoxon and Bradley, 1964; Hollander and Wolfe, 1973). To obtain the Wilcoxon rank sum test, we first order the $N = n_1 + n_2$ observations from least to greatest and let R_j denote the rank of y_j in this ordering. Let

$$W = \sum_{j=1}^{n} R_j,$$

which is the sum of the ranks assigned to the y_j's. We then reject the null hypothesis at the α level of significance if

$$W \geq w(\alpha_2, n_2, n_1)$$

or

$$W \leq n_1(n_2 + n_1 + 1) - w(\alpha_1, n_2, n_1),$$

where $\alpha = \alpha_1 + \alpha_2$ and $w(\alpha, n_2, n_1)$ satisfies

$$P(W \geq w(\alpha, n_2, n_1)) = \alpha$$

under the null hypothesis. Values of $w(\alpha, n_2, n_1)$ are given in the most standard nonparametric references (e.g., Hollander and Wolfe, 1973). Under the null hypothesis, the test statistic

$$W^* = \frac{W - E(W)}{\sqrt{\operatorname{var}(W)}} = \frac{W - \frac{1}{2}n_2(n_2 + n_1 + 1)}{\sqrt{\frac{1}{12}n_1 n_2(n_1 + n_2 + 1)}} \tag{14.4}$$

is asymptotically distributed as a standard normal distribution. Thus, by normal theory approximation, we reject the null hypothesis at the α level of significance if $|W^*| \geq z_{\alpha/2}$.

Note that W can be written as

$$W = \sum_{i=1}^{n_2}\left(\sum_{j=1}^{n_2} I\{y_i \geq y_j\} + \sum_{j=1}^{n_1} I\{y_i \geq x_j\}\right)$$

$$= \frac{n_2(n_2+1)}{2} + \sum_{i=1}^{n_2}\sum_{j=1}^{n_1} I\{y_i \geq x_j\}.$$

Hence, the variance of W is given by

$$\text{var}(W) = \text{var}\left(\frac{n_2(n_2-1)}{2} + \sum_{i=1}^{n_2}\sum_{j=1}^{n_1} I\{y_i \geq x_j\}\right)$$

$$= \text{var}\left(\sum_{i=1}^{n_2}\sum_{j=1}^{n_1} I\{y_i \geq x_j\}\right)$$

$$= n_1 n_2 \,\text{var}(I\{y_i \geq x_j\}) + n_1 n_2 (n_1-1)\text{cov}(I\{y_i \geq x_{j1}\},$$
$$I\{y_i \geq x_{j_2}\}) + n_1 n_2 (n_2-1)\text{cov}(I\{y_{i_1} \geq x_j\}, I\{y_{i_2} \geq x_j\})$$
$$= n_1 n_2 p_1(1-p_1) + n_1 n_2 (n_1-1)(p_2 - p_1^2)$$
$$+ n_1 n_2 (n_2-1)(p_3 - p_1^2)$$

where

$$p_1 = P(y_i \geq x_j),$$
$$p_2 = P(y_i \geq x_{j_1} \text{ and } y_i \geq x_{j_2}),$$
$$p_3 = P(y_{i1} \geq x_j \text{ and } y_{i_2} \geq x_j).$$

The above quantities can be estimated readily based on data from pilot studies. More specifically, assume that $x_1, \ldots, x_{n_1}$, and $y_1, \ldots, y_{n_2}$ are the data from a pilot study. The corresponding estimators can be obtained as

$$\hat{p}_1 = \frac{1}{n_1 n_2}\sum_{i=1}^{n_2}\sum_{j=1}^{n_1} I\{y_i \geq x_j\},$$

$$\hat{p}_2 = \frac{1}{n_1 n_2 (n_1-1)}\sum_{i=1}^{n_2}\sum_{j_1 \neq j_2} I\{y_i \geq x_{j_1} \quad \text{and} \quad y_i \geq x_{j_2}\},$$

$$\hat{p}_3 = \frac{1}{n_1 n_2 (n_2-1)}\sum_{i_1 \neq j_2}\sum_{j=1}^{n_1} I\{y_{i_1} \geq x_j \quad \text{and} \quad y_{i_2} \geq x_j\}.$$

Under the alternative hypothesis that $\theta \neq 0$, it can be shown that $p_1 \neq 1/2$,

$$E(W) = \frac{n_2(n_2+1)}{2} + n_1 n_2 p_1,$$

and that W can be approximated by a normal random variable with mean

$$\mu_W = \frac{n_2(n_2+1)}{2} + n_1 n_2 p_1$$

and variance

$$\sigma_W^2 = n_1 n_2 p_1 (1-p_1) + n_1 n_2 (n_1-1)(p_2 - p_1^2) + n_1 n_2 (n_2-1)(p_3 - p_1^2).$$

Without loss of generality, we assume that $p_1 > 1/2$. The power of the test can be approximated by

$$
\begin{aligned}
\text{Power} &= P(|W^*| > z_{\alpha/2}) \\
&\approx P(W^* > z_{\alpha/2}) \\
&= P\left(\frac{W - n_2(n_2+1)/2 - n_1 n_2 p_1}{\sigma_W} \right. \\
&\qquad \left. > \frac{z_{\alpha/2}\sqrt{n_1 n_2(n_1 + n_2 + 1)/12} + n_1 n_2(1/2 - p_1)}{\sigma_W} \right).
\end{aligned}
$$

Under the assumption that $n_1/n_2 \to \kappa$, the above equation can be further approximated by

$$\text{Power} = 1 - \Phi\left(\frac{z_{\alpha/2}{}^* \sqrt{\kappa(1+\kappa)/12} + \sqrt{n_2}\kappa(1/2 - p_1)}{\sqrt{\kappa^2(p_2 - p_1^2) + \kappa(p_3 - p_1^2)}} \right).$$

As a result, the sample size needed to achieve a desired power of $1 - \beta$ can be obtained by solving

$$\frac{z_{\alpha/2}{}^* \sqrt{\kappa(1+\kappa)/12} + \sqrt{n_2}\kappa(1/2 - p_1)}{\sqrt{\kappa^2\left(p_2 - p_1^2\right) + \kappa\left(p_3 - p_1^2\right)}} = -z_\beta,$$

which leads to $n_1 = \kappa n_2$ and

$$n_2 = \frac{\left(z_{\alpha/2}\sqrt{\kappa(\kappa+1)/12} + z_\beta \sqrt{\kappa^2\left(p_2 - p_1^2\right) + \kappa\left(p_3 - p_1^2\right)} \right)^2}{\kappa^2(1/2 - p_1)^2}.$$

14.3.1 Remark

As indicated in Equation 14.4, when there are no ties,

$$\text{var}(W) = \frac{n_1 + n_2 + 1}{12}.$$

When there are ties among the N observations,

$$\text{var}(W) = \frac{n_1 n_2}{12} \left[n_1 + n_2 + \frac{\sum_{j=1}^{g} t_j(t_j^2 - 1)}{(n_1 + n_2)(n_1 + n_2 + 1)} \right],$$

where g is the number of tied groups and t_j is the size of tied group j. In this case, the above formula for sample size calculation is necessarily modified.

14.3.2 An Example

To illustrate the use of the sample size formula derived above, we consider the same example concerning a clinical trial for the evaluation of the effect of a test drug on cholesterol in patients with coronary heart disease (CHD). Suppose the investigator is interested in comparing two cholesterol-lowering agents for treatment of patients with CHD through a parallel design. The primary efficacy parameter is the LDL. The null hypothesis of interest is the one of no treatment difference. Suppose that a two-arm parallel pilot study was conducted. According to the data given in the pilot study, it was estimated that $p_2 = 0.70$, $p_3 = 0.80$, and $p_4 = 0.80$. Hence, the sample size needed to achieve an 80% power for the detection of a clinically meaningful difference between the treatment groups can be estimated by

$$n = \frac{\left(z_{\alpha/2}/\sqrt{6} + z_\beta \sqrt{p_2 + p_3 - p_1^2} \right)^2}{(1/2 - p_1)^2}$$

$$= \frac{\left(1.96/\sqrt{6} + 0.84\sqrt{0.80 + 0.08 - 2 \times 0.70^2} \right)^2}{(0.50 - 0.70)^2}$$

$$\approx 54.$$

Hence, a total of 54 subjects is needed to have an 80% power to confirm the observed difference between the two treatment groups when such a difference truly exists.

14.4 Test for Independence

In many clinical trials, data collected may consist of a random sample from a bivariate population, for example, the baseline value and the posttreatment value. For such a data set, it is of interest to determine whether there is an association between the two variates (say x and y) involved in the bivariate structure. In other words, it is of interest to test for independence between x and y. Let (x_i, y_i), $i = 1, \ldots, n$, be the n bivariate observation from the n subjects involved in a clinical trial. It is assumed that (i) (x_i, y_i), $i = 1, \ldots, n$, are mutually independent and (ii) each (x_i, y_i) comes from the same continuous bivariate population of (x, y). To obtain a nonparametric test for independence between x and y, define

$$\tau = 2P\{(x_1 - x_2)(y_1 - y_2) > 0\} - 1,$$

which is the so-called Kendall coefficient. Testing the hypothesis that x and y are independent, that is,

H_0: $P(x \le a \text{ and } y \le b) = P(x \le a)P(y \le b)$ for all a and b is equivalent to testing the hypothesis that $\tau = 0$. A nonparametric test can be obtained as follows. First, for $1 \le i < j \le n$, calculate $\zeta(x_i, x_j, y_i, y_j)$, where

$$\zeta(a,b,c,d) = \begin{cases} 1 & \text{if } (a-b)(c-d) > 0 \\ -1 & \text{if } (a-b)(c-d) < 0. \end{cases}$$

For each pair of subscripts (i, j) with $i < j$, $\zeta(x_i, x_j, y_i, y_j) = 1$ indicates that $(x_i - x_j)$ $(y_i - y_j)$ is positive while $\zeta(x_i, x_j, y_i, y_j) = -1$ indicates that $(x_i - x_j)(y_i - y_j)$ is negative. Consider

$$K = \sum_{i=1}^{n-1} \sum_{j=i+1}^{n} \zeta(x_i, x_j, y_i, y_j).$$

We then reject the null hypothesis that $\tau = 0$ at the α level of significance if

$$K \ge k(\alpha_2, n) \quad or \quad K \le -k(\alpha_1, n),$$

where $k(\alpha, n)$ satisfies

$$P(K \ge k(\alpha, n)) = \alpha$$

and $\alpha = \alpha_1 + \alpha_2$. Values of $k(\alpha, n)$ are given in the most standard nonparametric references (e.g., Hollander and Wolfe, 1973). Under the null hypothesis,

$$K^* = \frac{K - E(K)}{\sqrt{var(K)}} = K \left[\frac{n(n-1)(2n+5)}{18} \right]^{-1/2} \tag{14.5}$$

is asymptotically distributed as a standard normal. Hence, we would reject the null hypothesis at the α level of significance for large samples if $|K^*| \ge z_{\alpha/2}$. It should be noted that when there are ties among the n x observations or among the n y observations, $\zeta(a, b, c, d)$ should be replaced with

$$\zeta^*(a,b,c,d) = \begin{cases} 1 & \text{if } (a-b)(c-d) > 0 \\ 0 & \text{if } (a-b)(c-d) = 0. \\ -1 & \text{if } (a-b)(c-d) < 0 \end{cases}$$

As a result, under H_0, var (K) becomes

$$\text{var}(K) = \frac{1}{18}\left[n(n-1)(2n+5) - \sum_{i=1}^{g} t_i(t_i - 1)(2t_i + 5) \right.$$

$$\left. - \sum_{j=1}^{h} u_j(u_j - 1)(2u_j + 5) \right]$$

$$+ \frac{1}{9n(n-1)(n-2)}\left[\sum_{i=1}^{g} t_i(t_i - 1)(t_i - 2) \right]$$

$$\times \left[\sum_{j=1}^{h} u_j(u_j - 1)(u_j + 2) \right]$$

$$+ \frac{1}{2n(n-1)}\left[\sum_{i=1}^{g} t_i(t_i - 1) \right]\left[\sum_{j=1}^{h} u_j(u_j - 1) \right],$$

where g is the number of tied x groups, t_i is the size of the tied x group i, h is the number of tied y groups, and u_j is the size of the tied y group j.

A formula for sample size calculation can be derived based on test (14.5). Define

$$\zeta_{i,j} = \zeta(x_i, x_j, y_i, y_j).$$

It follows that

$$\text{var}(K) = \text{var}\left(\sum_{i=1}^{n-1} \sum_{j=i+1}^{n} \zeta_{i,j} \right)$$

$$= \frac{n(n-1)}{2}\text{var}(\zeta_{i,j}) + n(n-1)(n-2)\text{cov}(\zeta_{i,j_1}, \zeta_{i,j_2})$$

$$= \frac{n(n-1)}{2}[1 - (1 - 2p_1)^2]$$

$$+ n(n-1)(n-2)[2p_2 - 1 - (1 - 2p_1)^2],$$

where

$$p_1 = P((x_1 - x_2)(y_1 - y_2) > 0),$$
$$p_2 = P((x_1 - x_2)(y_1 - y_2)(x_1 - x_3)(y_1 - y_3) > 0).$$

The above quantities can be readily estimated based on data from pilot studies. More specifically, let $(x_1, y_1), \dots, (x_n, y_n)$ be the data from a pilot study, the corresponding estimators can be obtained by

$$\hat{p}_1 = \frac{1}{n(n-1)} \sum_{i \neq j} I\{(x_i - x_j)(y_i - y_j) > 0\},$$

$$\hat{p}_2 = \frac{1}{n(n-1)(n-2)} \sum_{i \neq j_1 \neq j_2} I\{(x_i - x_{j_1})(y_i - y_{j_1})(x_i - x_{j_2})(y_i - y_{j_2}) > 0\}.$$

Under the alternative hypothesis, K is approximately distributed as a normal random variable with mean

$$\mu_K = \frac{n(n-1)}{2}(2p_1 - 1)$$

and variance

$$\sigma_K^2 = \frac{n(n-1)}{2}[1 - (1-2p_1)^2] + n(n-1)(n-2)[2p_2 - 1 - (1-2p_1)^2].$$

Without loss of generality, we assume $p_1 > 1/2$. The power of test (14.5) can be approximated by

$$
\begin{aligned}
\text{Power} &= P(|K^*| > z_{\alpha/2}) \\
&\approx P(K^* > z_{\alpha/2}) \\
&= P\left(\frac{K - n(n-1)(2p_1 - 1)/2}{\sigma_K} \right. \\
&\qquad \left. > \frac{z_{\alpha/2}\sqrt{n(n-1)(2n+5)/18} - n(n-1)(p_1 - 1/2)}{\sigma_K} \right) \\
&\approx 1 - \Phi\left(\frac{z_{\alpha/2}/3 - \sqrt{n}(p_1 - 1/2)}{\sqrt{2p_2 - 1 - (2p_1 - 1)^2}} \right).
\end{aligned}
$$

Hence, the sample size needed to achieve a desired power of $1 - \beta$ can be obtained by solving the following equation:

$$\frac{z_{\alpha/2}/3 - \sqrt{n}(p_1 - 1/2)}{\sqrt{2p_2 - 1 - (2p_1 - 1)^2}} = -z_\beta.$$

This leads to

$$n = \frac{4\left(z_{\alpha/2}/3 + z_\beta\sqrt{2p_2 - 1 - (2p_1 - 1)^2}\right)^2}{(2p_1 - 1)^2}.$$

14.4.1 An Example

In a pilot study, it is observed that a larger x value resulted in a larger value of y. Thus, it is of interest to conduct a clinical trial to confirm such an association between two primary responses, x and y, truly exists. Suppose that a two-arm parallel pilot study was conducted. Based on the data from the pilot study, it was estimated that $p_1 = 0.60$ and $p_2 = 0.70$. Hence, the sample size required for achieving an 80% power is

$$n = \frac{\left(z_{\alpha/2}/3 + z_\beta\sqrt{2p_2 - 1 - (2p_1 - 1)^2}\right)^2}{(p_1 - 0.5)^2}$$

$$= \frac{\left(1.96/3 + 0.84\sqrt{2 \times 0.70 - 1 - (1.20 - 1.00)^2}\right)^2}{(0.6 - 0.5)^2}$$

$$\approx 135.$$

Thus, a total of 135 subjects is needed to achieve an 80% power to confirm the observed association in the pilot study.

14.5 Practical Issues

14.5.1 Bootstrapping

When a nonparametric method is used, a formula for sample size calculation may not be available or may not exist as a closed form, especially when the study design/objective is rather complicated. In this case, the technique of bootstrapping may be applied. For more details, see Shao and Tu (1999).

14.5.2 Comparing Variabilities

In practice, it is often of interest to compare variabilities between treatments observed from the trials. Parametric methods for comparing variabilities are examined in Chapter 9. Nonparametric methods for comparing variabilities between treatment groups, however, are much more complicated and require further research.

14.5.3 Multiple-Sample Location Problem

When there are more than two treatment groups, the method of analysis of variance is usually considered. The primary hypothesis is that there are no treatment differences across the treatment groups. Let x_{ij} be the observation from the ith subject receiving the jth treatment, where $i = 1, \ldots, n_j$ and $j = 1, \ldots, k$. Similar to the analysis of variance model for the parametric case, we consider the following model:

$$x_{ij} = \mu + \tau_j + e_{ij}, \quad i = 1,\ldots,n_j, \quad j = 1,\ldots,k,$$

where μ is the unknown overall mean, τ_j is the unknown jth treatment effect, and $\sum_{j=1}^{k} \tau_j = 0$. It is assumed that (i) each e_i comes from the same continuous population with mean 0 and (ii) the e_i's are mutually independent. The hypotheses of interest are

$$H_0: \tau_1 = \ldots = \tau_k \quad \text{versus} \quad H_a: \tau_i \neq \tau_j \text{ for some } i \neq j.$$

To test the above hypotheses, the following Kruskal–Wallis test is useful (Kruskal and Wallis, 1952). We first rank all $N = \sum_{j=1}^{k} n_j$ observations jointly from least to greatest. Let R_{ij} denote the rank of x_{ij} in this joint ranking, $R_j = \sum_{j=1}^{n_j} R_{ij}, R._j = R_j/n_j$, and $R.. = (N+1)/2$, $j = 1,\ldots,k$. Note that R_j is the sum of the ranks received by treatment j and $R._j$ is the average rank obtained by treatment j. Based on R_j, $R._j$, and $R..$, the Kruskal–Wallis test statistic for the above hypotheses can be obtained as

$$H = \frac{12}{N(N+1)} \sum_{j=1}^{k} n_j (R._j - R..)^2$$

$$= \left(\frac{12}{N(N+1)} \sum_{j=1}^{k} \frac{R_j^2}{n_j} \right) - 3(N+1).$$

We reject the null hypothesis at the α level of significance if

$$H \geq h(\alpha, k, n_1, \ldots, n_k),$$

where $h(\alpha, k, n_1, \ldots, n_k)$ satisfies

$$P\left(H \geq h(\alpha, k, n_1, \ldots, n_k)\right) = \alpha$$

under the null hypothesis. Values of $h(\alpha, k, (n_1, \ldots, n_k))$ are given in the most standard non-parametric references (e.g., Hollander and Wolfe, 1973). Note that under the null hypothesis, H has an asymptotic chi-square distribution with $k-1$ degrees of freedom (Hollander and Wolfe, 1973). Thus, we may reject the null hypothesis at the α level of significance for large samples if $H \geq \chi^2_{\alpha,k-1}$, where $\chi^2_{\alpha,k-1}$ is the upper αth percentile of a chi-square distribution with $k-1$ degrees of freedom.

Unlike the parametric approach, formulas or procedures for sample size calculation for testing difference in multiple-sample locations using nonparametric methods are much more complicated. Further research is needed.

14.5.4 Testing Scale Parameters

In clinical trials, the reproducibility of subjects' medical status in terms of intrasubject variability is often assessed. If the intrasubject variability is much larger than that of the standard therapy (or control), safety of the test product could be a concern. In practice, a replicate crossover design or a parallel-group design with replicates is usually recommended for comparing intrasubject variability. Although nonparametric methods for testing scale parameters are available in the literature (see, e.g., Hollander and Wolfe, 1973), powers of these tests under the alternative hypothesis are not fully studied. As a result, further research in this area is necessary.

15

Sample Size Calculations for Cluster Randomized Trials

Cluster randomized trials (CRTs), also known as group randomized trials, have become increasingly common in certain areas of public health research, such as evaluation of life-style interventions, vaccine field trials, and studies of the impact of hospital guidelines on patient health. In such studies, the unit of randomization is a group of subjects instead of an independent individual, for example, hospital, clinical practice, household, or village. Compared with individually randomized studies, CRTs are generally more complex and require the investigators to consider issues such as possible lack of blinding, selection of the unit of randomization and the unit of inference, matching or stratification to improve treatment balance across clusters, and additional analytical challenges. It is also well known that CRTs need more subjects than individually randomized trials to be adequately powered. Because of correlation between subjects from the same cluster, the effective sample size is smaller than the total number of subjects. If N is the total number of subjects needed to achieve power $(1 - \beta)$ in an individually randomized study, the number of subjects needed for the same power in a CRT is N multiplied by the *design effect* $DE = [1 + (m - 1)\rho]$, where m is the average cluster size and ρ is a measure of within-cluster correlation, known as the intracluster (intraclass) correlation coefficient (ICC). Despite the relative inefficiency and methodological complexities, researchers often choose cluster randomized designs over individually randomized trials for their logistical and practical convenience, or to avoid the possibility of treatment group contamination, or because the interventions are naturally delivered at the cluster level, among other reasons. CRT methodology is generally well developed (e.g., see Murray, 1998; Donner and Klar, 2000; Hayes and Moulton, 2009).

Cluster randomized designs usually fall into one of the three main categories: unmatched (completely randomized), matched (matched-pair), or stratified. For each of these three design types, we present sample size formulas for continuous outcomes, proportions, and incidence rates.

Depending on the primary question of interest and the chosen unit of inference, CRTs can be analyzed using cluster-level summaries or at the individual level using mixed models or generalized estimating equations. Here, we focus on sample size requirements for studies that use cluster-level analyses. Throughout this chapter, we will consider test statistics of the form $|\hat{d}|/\sqrt{\widehat{\mathrm{Var}(\hat{d})}}$, which have approximately standard normal distribution under the null hypothesis $H_0 : d = 0$, where d is the true value of the parameter of interest and $\hat{d}$ is its estimate. We will reject the null hypothesis in favor of the alternative $H_a : d \neq 0$ at the two-sided level of significance α if

$$\frac{|\hat{d}|}{\sqrt{\widehat{\mathrm{Var}(\hat{d})}}} > Z_{\alpha/2}.$$

Therefore, to obtain the sample size required to achieve power $(1 - \beta)$, we would need to solve the equation

$$\frac{|\hat{d}|}{\sqrt{\mathrm{Var}(\hat{d})}} = Z_{\alpha/2} + Z_{\beta}. \tag{15.1}$$

Sample size in cluster randomized studies can be increased by either adding more clusters or by recruiting more subjects per cluster. To ensure good replication and reliable conclusions, a study must enroll a sufficient number of clusters. As a rule of thumb, Hayes and Moulton (2009) recommend that unmatched studies should have at least four clusters per arm, while matched studies should include a minimum of six matched pairs. However, after a certain point, increasing the number of subjects per cluster does not produce any noticeable increase in power. Campbell et al. (2007) recommend that it is not worth recruiting more than $(1/\rho)$ subjects per cluster. Sample size formulas in this chapter are presented in terms of the number of clusters (or cluster pairs), assuming a common cluster size m.

15.1 Unmatched Trials

In this section, we will assume that $2c$ clusters are randomized with equal probability to either group 1 or group 2. The outcome of subject k from cluster j assigned to group i $(i = 1, 2)$ will be denoted by X_{ijk}.

We will use a general definition of the ICC ρ as

$$\rho = \frac{\mathrm{Cov}(X_{ijk}, X_{ijl})}{\sqrt{\mathrm{Var}(X_{ijk})\mathrm{Var}(X_{ijl})}},$$

where X_{ijk} and X_{ijl} are outcomes of subjects k and l from the same cluster j assigned to group i $(i = 1, 2)$. For unmatched cluster randomized design with continuous outcomes, this definition is equivalent to the frequently used expression $\rho = \sigma_B^2/(\sigma_W^2 + \sigma_B^2)$, where σ_B^2 denotes the variance due to differences between clusters (between-cluster variance) and σ_W^2 denotes the variance due to differences between subjects within the same cluster (within-cluster variance). For binary outcomes, several alternative expressions exist for the ICC. The one we adopt here (Fleiss, 1981; Donald and Donner, 1987; Mak, 1988) is based on a hierarchical random model, which assumes that X_{ijk} have Bernoulli distribution with probability p_{ij}, where p_{ij} themselves are independent and identically distributed with mean p_i and variance σ_p^2. Under this model, it is easy to show using the law of total variance that

$$\mathrm{Var}(X_{ijk}) = E[\mathrm{Var}(X_{ijk} \mid p_{ij})] + \mathrm{Var}[E(X_{ijk} \mid p_{ij})] = p_i(1 - p_i),$$

and for $k \neq l$

$$\mathrm{Cov}(X_{ijk}, X_{ijl}) = E[E(X_{ijk}X_{ijl} \mid p_{ij})] - \{E[E(X_{ijk} \mid p_{ij})]\}^2 = \mathrm{Var}(p_{ij}) = \sigma_p^2.$$

Therefore, for binary outcomes, ICC can be expressed as

$$\rho = \frac{\text{Cov}(X_{ijk}, X_{ijl})}{\sqrt{\text{Var}(X_{ijk})\text{Var}(X_{ijl})}} = \frac{\sigma_p^2}{p_i(1-p_i)}.$$

A useful overview of ICC in cluster randomized studies can be found in Eldridge et al. (2009). The methods for the estimation of ICC for different types of CRTs are described in Hayes and Bennett (1999), Campbell et al. (2001), and Hayes and Moulton (2009). Sample size formulas presented in this section have been reported by Cornfield (1978), Donner et al. (1981), Hsieh (1988), and Hayes and Bennett (1999).

15.1.1 Comparison of Means

Suppose we want to test the null hypothesis $H_0 : \mu_1 = \mu_2$ versus the alternative $H_a : \mu_1 \neq \mu_2$ in a study with a continuous primary outcome X that is normally distributed with mean μ_i and variance $\sigma_i^2 = \sigma_{Bi}^2 + \sigma_{Wi}^2$ in group i ($i = 1, 2$), where σ_{Bi}^2 and σ_{Wi}^2 are between-cluster and within-cluster variance components. Unbiased estimates of the group means are the usual sample means $\bar{X}_1$ and $\bar{X}_2$, calculated across all subjects in clusters randomized to groups 1 and 2, respectively. The variance of $\bar{X}_i$ can be obtained as follows:

$$\text{Var}(\bar{X}_i) = \text{Var}\left(\frac{1}{cm}\sum_{j=1}^{c}\sum_{k=1}^{m}X_{ijk}\right) = \frac{1}{c^2m^2}\sum_{j=1}^{c}\left[m\text{Var}(X_{ijk}) + \sum_{k\neq l}\text{Cov}(X_{ijk}, X_{ijl})\right]$$

$$= \frac{\sigma_i^2}{mc}[1 + (m-1)\rho_i].$$

We will reject the null hypothesis at the level of significance α if

$$\frac{|\bar{X}_1 - \bar{X}_2|}{\sqrt{\widehat{\text{Var}}(\bar{X}_1) + \widehat{\text{Var}}(\bar{X}_2)}} > Z_{\alpha/2}.$$

Solving Equation 15.1, we find that the number of clusters needed per group is equal to

$$c = \frac{(Z_{\alpha/2} + Z_\beta)^2}{(\mu_1 - \mu_2)^2}\left[\frac{\sigma_1^2 + \sigma_2^2}{m} + \frac{m-1}{m}(\sigma_1^2\rho_1 + \sigma_2^2\rho_2)\right].$$

When $\rho_1 = \rho_2 = 0$, this result is reduced to a standard sample size formula for individually randomized studies. When we can assume $\sigma_1 = \sigma_2$ and $\rho_1 = \rho_2$, the formula for c becomes

$$c = \frac{2\sigma^2(Z_{\alpha/2} + Z_\beta)^2[1 + (m-1)\rho]}{m(\mu_1 - \mu_2)^2}. \tag{15.2}$$

Example 15.1

To illustrate the implementation of formula (15.2), consider a study of the effect of a comprehensive school-level physical activity program on bone mineral density (BMD) in prepubertal children. In this study, schools are randomized to either the intervention or the control group, and the primary parameter of interest is the average change in whole-body BMD, where BMD measurements at baseline and postintervention will be z-transformed using means and standard deviations calculated from the entire study sample. Suppose the researchers plan to perform BMD measurements in $m = 20$ randomly selected prepubertal children in each school and they consider a value of 0.2 z score units for the difference in mean BMD changes between the groups as clinically meaningful. They also assume, based on prior studies, that the variance of BMD change is the same in each group and equal to 0.6. Then, assuming a common value of ICC ρ for each group, the number of schools that the researchers need to randomize to each group to have 80% power for two-sided $\alpha = 0.05$ is equal to

$$c = \frac{2 \times 0.6(1.96 + 0.84)^2[1 + 19\rho]}{20(0.2)^2}.$$

From prior studies, the researchers conservatively estimate the value of ρ to be 0.2, yielding the required number of schools per group to be equal to 57. Note that the study assumptions mean that within-school variance of BMD change is expected to be $(1 - \rho) \times 0.6 = 0.48$, while between-school variance is $0.2 \times 0.6 = 0.12$. We will need these numbers when we consider this example again in Chapter 16.

15.1.2 Comparison of Proportions

Consider an unmatched cluster randomized study where the primary outcome is a binary response X with response rate $P(X = 1)$ equal to p_i in group i ($i = 1, 2$). Suppose we want to test the null hypothesis $H_0 : p_1 = p_2$ versus the alternative $H_a : p_1 \neq p_2$. We will use the test statistics

$$\frac{|\hat{p}_1 - \hat{p}_2|}{\sqrt{\widehat{\text{Var}}(\hat{p}_1) + \widehat{\text{Var}}(\hat{p}_2)}},$$

where $\hat{p}_i = (1/cm)\sum_{j=1}^{cm} X_{ij}$ is the usual average response rate in group i ($i = 1, 2$) with the variance

$$\text{Var}(\hat{p}_i) = \frac{p_i(1 - p_i)}{cm}[1 + (m - 1)\rho_i].$$

Solving Equation 15.1, we obtain the number of clusters needed per group as

$$c = \frac{(Z_{\alpha/2} + Z_\beta)^2}{m(p_1 - p_2)^2}\{p_1(1 - p_1)(1 + \rho_1[m - 1]) + p_2(1 - p_2)(1 + \rho_2[m - 1])\}. \tag{15.3}$$

Example 15.2

Suppose the investigators are planning a study of the hospital-level intervention to reduce surgical site infections (SSIs). Hospitals will be randomized to either intervention or control groups and the outcome will be the proportion of surgeries where SSI occurred. From previously available data, researchers estimate the SSI rate to be 1% and they hope that intervention will reduce it to 0.75%, which they intend to be able to detect with 90% power at a two-sided 0.05 level of significance. From the same data, between-hospital standard deviation is estimated to be 1%, so that the ICC in both groups is expected to be $0.01^2/[0.01 \times 0.99] = 0.01$. The investigators also expect that on average 2750 procedures will be performed in each hospital during the year when the study is conducted. Then, using formula (15.3), the number of hospitals needed per group is

$$c = \frac{(1.96 + 1.28)^2[0.01 \times 0.99 + 0.0075 \times 0.9925][1 + 0.01 \times (2750 - 1)]}{2750 \times 0.0025^2},$$

or approximately 302 hospitals per group.

15.1.3 Comparison of Incidence Rates

Consider an unmatched CRT with a survival outcome where each cluster is followed for a total of t person-years. Suppose we want to compare incidence rates λ_1 and λ_2 in two groups. We will assume that the true incidence rate λ_{ij} in cluster j in group i ($i = 1, 2$) is sampled randomly from a distribution with mean $E(\lambda_{ij}) = \lambda_i$ and variance $\mathrm{Var}(\lambda_{ij}) = \sigma_i^2$. Denote by r_{ij} the number of events observed in the jth cluster in group i. We will test the null hypothesis $H_0 : \lambda_1 = \lambda_2$ using test statistic

$$\frac{|\hat{\lambda}_1 - \hat{\lambda}_2|}{\sqrt{\widehat{\mathrm{Var}}(\hat{\lambda}_1) + \widehat{\mathrm{Var}}(\hat{\lambda}_2)}},$$

where $\hat{\lambda}_i = (1/c)\sum_{j=1}^c \hat{\lambda}_{ij}$ and $\hat{\lambda}_{ij} = (r_{ij}/t)$ is the maximum likelihood estimate of λ_{ij}. First, we need to calculate the variance of $\hat{\lambda}_{ij}$. Using the law of total variance,

$$\mathrm{Var}(\hat{\lambda}_{ij}) = E[\mathrm{Var}(\hat{\lambda}_{ij} \mid \lambda_{ij})] + \mathrm{Var}[E(\hat{\lambda}_{ij} \mid \lambda_{ij})] = E\left[\frac{\lambda_{ij}}{t}\right] + \mathrm{Var}[\lambda_{ij}] = \frac{\lambda_i}{t} + \sigma_i^2.$$

For comparison of incidence rates, it is common to express the sample size in terms of the coefficient of variation $k = \sigma/\lambda$. Then,

$$\mathrm{Var}(\hat{\lambda}_i) = \frac{1}{c}\left(\frac{\lambda_i}{t} + \sigma_i^2\right) = \frac{1}{c}\left(\frac{\lambda_i}{t} + k_i^2\lambda_i^2\right).$$

Substituting this expression and solving Equation 15.1, we obtain the number of clusters needed per group as

$$c = \frac{(Z_{\alpha/2} + Z_\beta)^2(\lambda_1 + \lambda_2)}{t(\lambda_1 - \lambda_2)^2}\left[1 + t\frac{k_1^2\lambda_1^2 + k_2^2\lambda_2^2}{\lambda_1 + \lambda_2}\right]. \tag{15.4}$$

Example 15.3

In a study of the impact of enhanced terminal (i.e., after the previous patient has been discharged and before the next patient entered) room cleaning strategy on the incidence of healthcare-associated infections (HAI), the investigators plan to enroll c hospitals and randomize them with equal probability to either enhanced or standard room disinfection protocols. They assume that under standard disinfection strategy, the HAI incidence rate will be $\lambda_1 = 4.5$ per 1000 patient-days and they hope to detect a clinically meaningful decrease of 20% (i.e., $\lambda_2 = 3.6$ per 1000 patient-days) with 80% power. For the purpose of sample size calculations, they further assume that each hospital will have an average total of 50,000 patient-days of care. Then, assuming a common value of the coefficient of variation k for each group, the number of hospitals that the researchers need to randomize to each group to have 80% power for two-sided $\alpha = 0.05$ is equal to

$$c = \frac{(1.96 + 0.84)^2 (4.5 + 3.6)}{50(4.5 - 3.6)^2}\left(1 + 50k^2 \frac{4.5^2 + 3.6^2}{4.5 + 3.6}\right).$$

From previous limited data, the researchers estimated the coefficient of variation to be 0.4. Using these parameters, the investigators need to recruit a total of 106 hospitals into the study, or 53 per group.

15.1.4 Further Remarks

The sample size formulas in this chapter include a common cluster size m or the number of years of follow-up t, while in practice, clusters are rarely expected to have the same size. From a practical perspective, when variation in the cluster size or the follow-up is not considerable, m and t can be replaced by their average values for the purpose of sample size estimation.

  The results presented above are based on a standard normal approximation of a t-distribution and will underestimate the required sample size when the number of the degrees of freedom is relatively small. As a simple correction, one cluster per group can be added to c when $\alpha = 0.05$ and two clusters per group when $\alpha = 0.1$ (Snedecor and Cochran, 1989). When c is approximately 30 or greater, the number of degrees of freedom is sufficiently large and such adjustment is not necessary (Johnson and Kotz, 1972). In particular, for Examples 15.1 through 15.3 considered in this section, the addition of one cluster per group is not needed.

15.2 Matched Trials

Typical CRTs involve a relatively small number of clusters and for that reason, unlike in individually randomized studies, simple randomization cannot guarantee that intervention groups will be balanced. Pair matching can be used to ensure better balance between the groups and potentially to gain power by decreasing between-cluster variation. In matched, or matched-pair, CRTs, pairs of similar clusters are formed and then one cluster from the pair is randomized to group 1 while the other is assigned to group 2. Sample size formulas for matched CRTs presented in this section are based on the work by Hsieh (1988), Shipley et al. (1989), Byar (1988), Freedman et al. (1990), and Hayes and Bennett

(1999). As in the case of unmatched design, it is recommended that the numbers of cluster pairs shown below are increased by two pairs to adjust for using the critical values from standard normal distribution instead of t-distribution.

15.2.1 Comparison of Means

Suppose we want to compare group means of a continuous normally distributed outcome X in a matched CRT with c pairs of clusters. Following Thompson et al. (1997), we can write the model for the kth observation in a cluster assigned to group i within cluster pair j as

$$X_{ijk} = \alpha_i + \beta_j + (\alpha\beta)_{ij} + \varepsilon_{ijk}, \tag{15.5}$$

where α_i is the fixed effect for group i, β_j is the effect of cluster pair j, and $(\alpha\beta)_{ij}$ is the random cluster effect within cluster pair ($i = 1, 2; j = 1, \ldots, c; k = 1, \ldots, m_{ij}$). Here, random variables ε_{ijk} are i.i.d. with $E[\varepsilon_{ijk}] = 0$ and $(\alpha\beta)_{ij}$ are i.i.d. with $E[(\alpha\beta)_{ij}] = 0$; ε_{ijk} and $(\alpha\beta)_{ij}$ are independent for all i, j. Denote the group difference as $d = \alpha_1 - \alpha_2$, within-cluster variability as $\mathrm{Var}(\varepsilon_{ijk}) = \sigma_{ij}^2$ and between-cluster variability within matched cluster pair as $\mathrm{Var}[(\alpha\beta)_{ij}] = \sigma_{bm}^2$. Usually, it is reasonable to assume that within-cluster variability does not change across clusters, that is, $\sigma_{ij}^2 = \sigma_w^2$. We are interested in testing the null hypothesis $H_0 : d = 0$ versus the alternative $H_a : d \neq 0$.

Sample mean difference $\hat{d}_j$, calculated within cluster pair j as

$$\hat{d}_j = \bar{X}_{1j} - \bar{X}_{2j},$$

is an unbiased estimate of d. Assuming for simplicity that the cluster size is the same across all clusters, that is, $m_{ij} = m$ for all $j = 1, \ldots, c$ and $i = 1,2$, the variance of $\hat{d}_j$ can be calculated based on model (15.5) as

$$\mathrm{Var}\left(\hat{d}_j\right) = 2\left(\frac{\sigma_w^2}{m} + \sigma_{bm}^2\right).$$

We will use test statistic

$$\frac{|\bar{d}|}{\sqrt{\widehat{\mathrm{Var}(\bar{d})}}},$$

where $\bar{d} = c^{-1}\sum_{j=1}^{c} \hat{d}_j$ is an average difference. Then, from Equation 15.1, we can obtain the number of cluster pairs needed to achieve power $(1-\beta)$ as

$$c = \frac{2\left((\sigma_w^2/m) + \sigma_{bm}^2\right)(Z_{\alpha/2} + Z_\beta)^2}{d^2}. \tag{15.6}$$

Note that in the extreme scenario where matching is ineffective and σ_{bm}^2 is the same as between-cluster variance σ_B^2 in the unmatched design, formula (15.6) is the same as formula (15.2).

Example 15.4

We will revisit Example 15.1 of a study of the effect of a comprehensive school-level physical activity program on BMD in prepubertal children, but now we assume that schools in the study are paired based on characteristics such as rural versus city location, average income level in the school area, proportion of girls versus boys, and average weight. One school within a pair is then randomized to intervention and the other one to the control group. Within-school variance of BMD change is assumed to be $\sigma_w^2 = 0.48$, just like in the unpaired design, but now investigators expect that pairing will result in between-school, within-pair variance to be $\sigma_{bm}^2 = 0.05$, reduced compared to between-school variance of 0.12 in unpaired design. Then the number of school pairs needed to detect a group difference of 0.2 z score units with 80% power at two-sided 0.05 level of significance is

$$c = \frac{2((0.48/20) + 0.05)(1.96 + 0.84)^2}{(0.2)^2} = 29,$$

almost half the number of schools required per group in the unmatched design in Example 15.1. The investigators may want to increase this number to 31 to account for using standard normal approximation of the t-distribution.

15.2.2 Comparison of Proportions

Consider a matched cluster randomized study with c pairs of clusters where the primary outcome is a binary response X. As in the previous section, we will denote by m_{ij} the size of a cluster assigned to group i ($i = 1, 2$) in a cluster pair j. We will regard the jth pair of clusters ($j = 1, \ldots, c$) as randomly sampled from a stratum of similar clusters (which we can also index by j), where, if this stratum j was exposed to treatment i ($i = 1, 2$), cluster-level true response rates would be distributed with mean p_{ij} and variance σ_{ij}^2. We will additionally assume that response rates $p_{i1}, \ldots, p_{ic}$ are a random sample from a distribution with mean p_i.

We will consider observed response rate differences $\hat{d}_j = \hat{p}_{1j} - \hat{p}_{2j}$, where $\hat{p}_{ij} = (1/m_{ij})\sum_{k=1}^{m_{ij}} X_{ijk}$ is an unbiased estimate of p_{ij}. The distribution of $\hat{d}_j$ can be described in terms of the hierarchical model:

$$\hat{d}_j \,|\, (p_{1jA}, p_{2jB}) \sim \text{mean}(\, p_{1jA} - p_{2jB}), \text{ variance} \left[\frac{p_{1jA}(1 - p_{1jA})}{m_{1j}} + \frac{p_{2jB}(1 - p_{2jB})}{m_{2j}} \right].$$

We can calculate the variance of $\hat{d}_j$ using the law of total variance as

$$\text{Var}(\hat{d}_j) = E[\text{Var}(\hat{d}_j \,|\, p_{1jA}, p_{2jB})] + \text{Var}[E(\hat{d}_j \,|\, p_{1jA}, p_{2jB})]$$

$$= E\left[\frac{p_{1jA}(1 - p_{1jA})}{m_{1j}} + \frac{p_{2jB}(1 - p_{2jB})}{m_{2j}} \right] + \text{Var}[p_{1jA} - p_{2jB}]$$

$$= \frac{p_{1j}}{m_{1j}} - \frac{E\left(p_{1jA}^2\right)}{m_{1j}} + \frac{p_{2j}}{m_{2j}} - \frac{E\left(p_{2jB}^2\right)}{m_{2j}} + \left(\sigma_{1j}^2 + \sigma_{2j}^2\right).$$

Noting that $E(p_{1jA}^2) = \text{Var}(p_{1jA}) + [E(p_{1jA})]^2 = \sigma_{1j}^2 + p_{1j}^2$, we can write

$$\text{Var}(\hat{d}_j) = \frac{p_{1j}(1-p_{1j})}{m_{1j}} + \frac{p_{2j}(1-p_{2j})}{m_{2j}} + \sigma_{1j}^2 \frac{m_{1j}-1}{m_{1j}} + \sigma_{2j}^2 \frac{m_{2j}-1}{m_{2j}}.$$

To test the null hypothesis $H_0: p_1 - p_2 = 0$ versus the alternative $H_a: p_1 - p_2 \neq 0$, we can use the test statistic $|\bar{d}| / \sqrt{\widehat{\text{Var}}(\bar{d})}$, where $\bar{d} = c^{-1}\sum_{j=1}^c \hat{d}_j$ with variance $\text{Var}(\bar{d}) = c^{-2}\sum_{j=1}^c \text{Var}(\hat{d}_j)$. Assuming $m_{1j} = m_{2j} = m$ and $\sigma_{1j}^2 = \sigma_{2j}^2 = \sigma_{bm}^2$ for all $j = 1, \ldots, c$ and substituting p_{1j}, p_{2j} in the formula for $\text{Var}(\hat{d}_j)$ by their expected values, we can write

$$\text{Var}(\bar{d}) = \frac{p_1(1-p_1)}{cm} + \frac{p_2(1-p_2)}{cm} + 2\sigma_{bm}^2 \frac{m-1}{cm},$$

and therefore the required number of cluster pairs is

$$c = \frac{(Z_{\alpha/2} + Z_\beta)^2}{m(p_1-p_2)^2}\{p_1(1-p_1) + p_2(1-p_2) + 2\sigma_{bm}^2(m-1)\}. \tag{15.7}$$

In case matching has failed to reduce between-cluster variance within matched pairs and σ_{bm}^2 is the same as between-cluster variance σ_p^2 in the unmatched design, formula (15.7) is the same as result (15.3).

Example 15.5

Consider again the study described in Example 15.2. The researchers decide that 300 hospitals per group, required for the unmatched design, is not feasible and hope to achieve better efficiency by pairing hospitals based on characteristics such as bed size, tertiary care versus community hospital, average number of procedures per surgeon per year, and prevalence of surgical procedures with greater complexity. They assume that pairing will reduce between-hospital variation from $(1\%)^2$ in unmatched design to between-hospital, within-pair variance $\sigma_{bm}^2 = (0.75\%)^2$ in a matched study. Then, assuming as before an average of 2750 procedures per hospital, the number of hospital pairs needed to detect a group difference of 0.25% in SSI rate with 90% power at two-sided 0.05 level of significance is

$$c = \frac{(1.96+1.28)^2\left[0.01\times0.99 + 0.0075\times0.9925 + 2\times0.0075^2(2750-1)\right]}{2750\times0.0025^2},$$

or approximately 200 hospital pairs. Since c is greater than 30, the correction of adding two more pairs is not needed. While the number of hospitals needed for matched design is still large, it reduced the number of hospitals needed for unmatched design by a third. To reduce between-cluster variation even further, researchers may want to consider grouping surgeons performing similar types of procedures within each hospital into clusters and conducting a paired study with such clusters as units of randomization.

15.2.3 Comparison of Incidence Rates

Consider a matched cluster randomized study with c pairs of clusters, where the outcome is survival and where each cluster is followed for a total of t person-years. We will regard jth pair of clusters ($j = 1, \ldots, c$) as randomly sampled from a stratum of similar clusters (which we can also index by j), where, if this stratum j was exposed to treatment i ($i = 1, 2$), cluster-level true incidence rates would be distributed with mean λ_{ij} and variance σ_{ij}^2. We will additionally assume that incidence rates $\lambda_{i1}, \ldots, \lambda_{ic}$ are a random sample from a distribution with mean λ_i. Suppose we want to compare incidence rates λ_1 and λ_2. Within cluster pair j, denote by r_{ij} the number of events observed in the cluster assigned to group i and consider the observed difference $\hat{d}_j = \hat{\lambda}_{1j} - \hat{\lambda}_{2j}$, where $\hat{\lambda}_{ij} = (r_{ij}/t)$. Using a similar logic as in Section 15.1.3, it is easy to show that

$$\mathrm{Var}(\hat{\lambda}_{ij}) = \frac{\lambda_{ij}}{t} + \sigma_{ij}^2,$$

and therefore

$$\mathrm{Var}(\hat{d}_j) = \frac{\lambda_{1j} + \lambda_{2j}}{t} + \sigma_{1j}^2 + \sigma_{2j}^2 = \frac{\lambda_{1j} + \lambda_{2j}}{t} + k_{1j}^2 \lambda_{1j}^2 + k_{2j}^2 \lambda_{2j}^2,$$

where k_{1j}, k_{2j} are within-stratum j coefficients of variation under treatments 1 and 2, respectively.

To test the null hypothesis $H_0 : \lambda_1 - \lambda_2 = 0$ versus the alternative $H_a : \lambda_1 - \lambda_2 \neq 0$, we will use the test statistic $|\bar{d}| / \sqrt{\widehat{\mathrm{Var}(\bar{d})}}$, where an average difference $\bar{d} = c^{-1} \sum_{j=1}^{c} \hat{d}_j$ has variance $\mathrm{Var}(\bar{d}) = c^{-2} \sum_{j=1}^{c} \mathrm{Var}(\hat{d}_j)$.

For the purpose of sample size calculations, we would usually assume that $k_{1j} = k_{2j} = k_{bm}$ and would substitute λ_{1j} and λ_{2j} in the formula for $\mathrm{Var}(\hat{d}_j)$ by their expected values. Then, the number of clusters needed can be calculated as

$$c = \frac{(Z_{\alpha/2} + Z_\beta)^2 (\lambda_1 + \lambda_2)}{t(\lambda_1 - \lambda_2)^2} \left[1 + t \frac{k_{bm}^2 (\lambda_1^2 + \lambda_2^2)}{\lambda_1 + \lambda_2} \right], \tag{15.8}$$

which is the same expression as in Equation 15.4 but with k_1 and k_2 substituted by a between-cluster coefficient of variation k_{bm}.

Example 15.6

Recall the study described in Example 15.3. Suppose the researchers now want to design a more efficient crossover study, where each hospital is randomized to either {enhanced disinfection, standard disinfection} or {standard disinfection, enhanced disinfection} protocol sequence, with a "washout" period between the two study periods where different disinfection strategies were employed. The washout period needs to be sufficiently long to avoid the potential carryover effects. In this study design, study periods for each hospital can be regarded as a matched pair, so that formula (15.8) can be applied. The investigators still assume that true mean HAI incidence rates are $\lambda_1 = 4.5$ per 1000 patient-days under standard disinfection strategy and $\lambda_2 = 3.6$ per 1000 patient-days under enhanced disinfection strategy, and that for each study period, the hospital will have an average total of 50,000 patient-days of care. However, now they assume that between-study period, within-hospital coefficient of variation k_{bm} is equal to 0.3, reduced from between-hospital coefficient of variation $k = 0.4$ in the unmatched design. Then we can find the number of hospitals needed for the study to have 80% power for two-sided $\alpha = 0.05$ as

$$c = \frac{(1.96 + 0.84)^2 (4.5 + 3.6)}{50(4.5 - 3.6)^2} \left(1 + 50 \times 0.3^2 \frac{4.5^2 + 3.6^2}{4.5 + 3.6}\right),$$

or 31 hospitals. Again, c is greater than 30, so the correction of adding two more hospitals may not be necessary. Thus, in this crossover study, less than a third of the number of hospitals required for an unmatched design needs to be enrolled; however, the study will also continue more than twice as long. If the increased length of the study is unacceptable (e.g., for budgetary reasons), the investigators can consider recruiting more hospitals with shorter study periods.

15.3 Stratified Trials

In unmatched designs with relatively small number of clusters, it is possible that a chance of imbalance can occur, leading to large values of between-cluster variation and subsequent loss of power. On the other hand, in matched designs, better balance can be offset by a decrease in the number of degrees of freedom. Stratified cluster randomized designs can be viewed as a compromise between unmatched and matched designs. Relatively little work has been done to derive sample size formulas specifically for stratified CRTs (Donner, 1992). As a practical approach, formulas (15.6) through (15.8) can be used to calculate sample size in stratified CRTs, where between-cluster variance v and coefficient of variation k_{bm} have to be calculated within strata instead of within matched pair.

.

16

Test for Homogeneity of Two Zero-Inflated Poisson Population

In clinical trials, it is not uncommon to observe data of the occurrence of certain events such as the number of tumors in cancer trials and the number of erosions in gastrointestinal (GI) safety studies. These counts are usually considered as markers for the prediction of clinical outcomes if they are not directly related to clinical outcomes. For example, a large number of erosions will most likely lead to gastroduodenal ulcers. These counts are usually assumed to follow a Poisson distribution and test for the null hypothesis that there is no difference in mean counts between treatment groups. In practice, these counts could include a large proportion of zeros. In this case, it is suggested that a zero-inflated Poisson (ZIP) be used for an accurate and reliable assessment of the treatment effect. Under a zero-inflated Poisson distribution, however, there are three possible scenarios: (i) there is no difference between treatments in the group of zeros; (ii) there is no difference between treatments in the group of nonzeros; and (iii) there are no differences between treatments in both groups of zeros and nonzeros. It should be noted that the proportion of zeros could be as high as 60% in practice.

A Poisson model is often used to model count data. However, the Poisson model may not be able to provide good fit to the data if excess zeros occur in count data. Neyman (1939) and Feller (1943) first introduced the concept of zero inflation to address the problem of excess zeros. Since then, there have been extensive studies related to the development of ZIP modeling. Van den Broek (1995) discussed the tests of zero inflation and proposed to use the Rao's score test for hypotheses testing. A simulation study was conducted to show that the score test was appropriate. Properties and inference of the ZIP distribution, including the MLE, can be found in Gupta, Gupta, and Tripathi (1996). Janakul and Hinde (2002) modified the score test to the regression situation where the zero-mixing probability is dependent on covariates. A review of the related literature was given in Böhning (1998) and examples from different disciplines were also discussed in the study. Further examples of the ZIP regression model can be found in Böhning et al. (1999) and Lee, Wang, and Yau (2001). However, most of these works were conducted in one population case while relatively little works can be found in comparing two ZIP models. Tse et al. (2009) developed statistical methodology for testing homogeneity of the model parameters associated with two different ZIP models.

The purpose of this chapter is then to follow the idea of Tse et al. (2009) to derive likelihood ratio tests for homogeneity of two ZIP populations under the null hypotheses that (1) there is no difference in the group of zeros, (2) there is no difference in the group of nonzeros, and (3) there is no difference in both groups of zeros and nonzeros. Under appropriate alternative hypotheses, approximate formulas for sample size calculation are also obtained for achieving a desired power for detecting a clinically meaningful difference at a prespecified level of significance.

The remaining part of this chapter is organized as follows. In the next section, likelihood ratio tests under a randomized parallel-group design comparing a test treatment with a control assuming that the study endpoint follows a zero-inflated Poisson are derived in Sections 16.1 and 16.2. Under appropriate alternative hypotheses, approximate formulas for sample size calculation are obtained for achieving a desired power for detecting a clinically meaningful difference at a prespecified level of significance in Section 16.3. Also included in this section are the procedure for sample size calculation and an example concerning the assessment of the gastrointestinal (GI) safety in terms of the number of erosion counts of a newly developed compound for the treatment of osteoarthritis (OA) and rheumatoid arthritis (RA). Section 16.4 considers the test for the treatment effect under multivariate (bivariate) ZIP model. Some concluding remarks are given in Section 16.5.

16.1 Zero-Inflated Poisson Distribution

A random variable X is said to have a zero-inflated Poisson (ZIP) distribution with parameters θ and λ, denoted as ZIP (θ, λ) if the probability function is given by

$$P(X = 0) = \theta + (1-\theta)e^{-\lambda} \quad \text{and}$$

$$P(X = x) = (1-\theta)\frac{e^{-\lambda}\lambda^x}{x!}, \quad x = 1, 2, \dots .$$

Suppose that a clinical study is conducted to compare the effect between a treatment and a control and n subjects are randomly selected to participate in each of the two groups. Let X_{ij} be the erosion count of the ith patient in the jth treatment group. Assume that X_{ij} follows a ZIP model with parameters θ_j and λ_j. The corresponding probability functions are

$$P(x_{ij}|\theta_j, \lambda_j) = \theta_j I(x_{ij} = 0) + (1-\theta_j)\frac{e^{-\lambda_j}\lambda_j^{x_{ij}}}{x_{ij}!} \quad i = 1, 2, \dots, n, j = 1, 2. \tag{16.1}$$

Define $K_1 = \#\{x_{i1} = 0\}$, $K_2 = \#\{x_{i2} = 0\}$. Without loss of generality, we can denote the observations such that $x_{11} = x_{21} = \dots = x_{K_11} = x_{12} = x_{22} = \dots = x_{K_22} = 0$. Define $\theta = (\theta_1, \theta_2, \lambda_1, \lambda_2)$. In particular, $L = \prod_{j=1}^{2}\prod_{i=1}^{n} P(x_{ij}|\theta_j, \lambda_j)$ is the likelihood function and the log-likelihood function is $l(\theta) = \log L = l_1 + l_2$, where for $j = 1, 2$,

$$l_j = \sum_{i=1}^{n}\log f(x_{ij}|\theta_j, \lambda_j)$$

$$= K_1 \log(\theta_j + (1-\theta_j)e^{-\lambda_j})$$

$$+ \sum_{i=K_j+1}^{n}[\log(1-\theta_j) - \lambda_j + x_{ij}\log \lambda_j - \log(x_{ij}!)]$$

Consider the reparameterization $\alpha_1 = \lambda_1 - \lambda_2$, $\alpha_2 = \lambda_2$, $\beta_1 = \theta_1 - \theta_2$, and $\beta_2 = \theta_2$. The corresponding partial derivatives are

$$\frac{\partial l(\theta)}{\partial \alpha_1} = \frac{\partial l(\theta)}{\partial \lambda_1}\frac{\partial \lambda_1}{\partial \alpha_1} = \frac{1}{\lambda_1}\sum_{i=K_1+1}^{n}x_{i1} - \frac{K_1(1-\theta_1)e^{-\lambda_1}}{\theta_1 + (1-\theta_1)e^{-\lambda_1}} - (n - K_1),$$

$$\frac{\partial l(\theta)}{\partial \alpha_2} = \frac{\partial l(\theta)}{\partial \lambda_1}\frac{\partial \lambda_1}{\partial \alpha_2} + \frac{\partial l(\theta)}{\partial \lambda_2}\frac{\partial \lambda_2}{\partial \alpha_2}$$

$$= \frac{\partial l(\theta)}{\partial \alpha_1} + \frac{1}{\lambda_2}\sum_{i=K_2+1}^{n} x_{i2} - \frac{K_2(1-\theta_2)e^{-\lambda_2}}{\theta_2 + (1-\theta_2)e^{-\lambda_2}} - (n - K_2),$$

$$\frac{\partial l(\theta)}{\partial \beta_1} = \frac{\partial l(\theta)}{\partial \theta_1}\frac{\partial \theta_1}{\partial \beta_1} = \frac{K_1(1-e^{-\lambda_1})}{\theta_1 + (1-\theta_1)e^{-\lambda_1}} - \frac{n - K_1}{1 - \theta_1},$$

$$\frac{\partial l(\theta)}{\partial \beta_2} = \frac{\partial l(\theta)}{\partial \theta_1}\frac{\partial \theta_1}{\partial \beta_2} + \frac{\partial l(\theta)}{\partial \theta_2}\frac{\partial \theta_2}{\partial \beta_2}$$

$$= \frac{\partial l(\theta)}{\partial \beta_1} + \frac{K_2(1-e^{-\lambda_2})}{\theta_2 + (1-\theta_2)e^{-\lambda_2}} - \frac{n - K_2}{1 - \theta_2},$$

Then, the score function is

$$U(\theta) = \frac{\partial l(\theta)}{\partial \theta} = \left(\frac{\partial l(\theta)}{\partial \alpha_1}, \frac{\partial l(\theta)}{\partial \beta_1}, \frac{\partial l(\theta)}{\partial \alpha_1}, \frac{\partial l(\theta)}{\partial \beta_2}\right)^T$$

Furthermore, denote the Fisher information matrix by

$$i(\theta) = E(I(\theta)) \quad \text{with} \quad I(\theta) = \left[-\frac{\partial^2 l(\theta)}{\partial \theta \, \partial \theta^T}\right]; \tag{16.2}$$

the detailed expressions of the entries in the Fisher information matrix, which are the expected values of the second-order partial derivatives of the log-likelihood function with respect to the parameters, are given in the Appendix.

16.2 Testing Differences between Treatment Groups

Tse et al. (2009) studied the problem of testing treatment difference in the occurrence of a safety parameter in a randomized parallel-group comparative clinical trial under the assumption that the number of occurrence follows a zero-inflated Poisson (ZIP) distribution. Tse et al. (2009) derived likelihood ratio tests for homogeneity of two ZIP populations under the hypotheses that (1) there is no difference in the group of zeros, (2) there is no difference in the group of nonzeros, and (3) there is no difference in both groups of zeros and nonzeros. In addition, statistical properties of the derived likelihood ratio tests are examined for extreme cases when the proportion of zeros increases to 1 and decreases to 0. Under each scenario, approximate formulas for sample size calculation were also obtained for achieving a desired power for detecting a clinically meaningful difference under the corresponding alternative hypotheses. These tests and the corresponding approximate formulas for sample size requirements are described, respectively, below.

16.2.1 Testing the Difference in Both Groups of Zeros and Nonzeros

Suppose that the primary interest is to test whether there is any difference in the groups of zeros between treatments. In particular, the problem is formulated as the testing of hypotheses $H_0 : \lambda_1 = \lambda_2$, $\theta_1 = \theta_2$ versus H_1: not both are equal. However, this is equivalent to test

$$H_{01} : \alpha_1 = 0, \beta_1 = 0 \quad \text{versus} \quad H_{11} : \text{not both are equal.} \tag{16.3}$$

Likelihood ratio test method is used to develop the test procedure for these hypotheses. Let $\hat{\theta} = (\hat{\alpha}_1, \hat{\beta}_1, \hat{\alpha}_2, \hat{\beta}_2)$ be the unrestricted MLE of θ. Denote $\theta_0 = (0, 0, \alpha_2, \beta_2)$ and $\hat{\theta}_0 = (0, 0, \hat{\alpha}_{2,0}, \hat{\beta}_{2,0})$, where $\hat{\alpha}_{2,0}$ and $\hat{\beta}_{2,0}$ are the MLE of α_2 and β_2 under H_{01}. In particular, $\hat{\alpha}_{2,0}$ and $\hat{\beta}_{2,0}$ are given by the likelihood equations

$$\left. \frac{\partial l(\theta)}{\partial \alpha_2} \right|_{\theta = \hat{\theta}_0} = \frac{1}{\hat{\alpha}_{20}} \sum_{i=1}^{n} (x_{i1} + x_{i2}) + \frac{\hat{\beta}_{20}(K_1 + K_2)}{\hat{\beta}_{20} + (1 - \hat{\beta}_{20})e^{-\hat{\alpha}_{20}}} - 2n = 0$$

$$\left. \frac{\partial l(\theta)}{\partial \beta_2} \right|_{\theta = \hat{\theta}_0} = \frac{(K_1 + K_2)(1 - e^{-\hat{\alpha}_{20}})}{\hat{\beta}_{20} + (1 - \hat{\beta}_{20})e^{-\hat{\alpha}_{20}}} - \frac{2n - (K_1 + K_2)}{1 - \hat{\beta}_{20}} = 0.$$

After some algebras, the MLE are given by solving the following equations:

$$2n - (K_1 + K_2) - \frac{1 - e^{-\hat{\alpha}_{20}}}{\hat{\alpha}_{20}} \sum_{i=1}^{n} (x_{i1} + x_{i2}) = 0$$

and

$$\hat{\beta}_{20} = 1 - \frac{1}{2n\hat{\alpha}_{20}} \sum_{i=1}^{n} (x_{i1} + x_{i2}).$$

Partition the Fisher information matrix $i(\theta)$, given in Equation 16.2, into

$$i(\theta) = \begin{pmatrix} i_{aa}(\theta) & i_{ab}(\theta) \\ i_{ba}(\theta) & i_{bb}(\theta) \end{pmatrix},$$

where

$$
i_{aa}(\theta) = -E \begin{pmatrix} \dfrac{\partial^2 l(\theta)}{\partial \alpha_1^2} & \dfrac{\partial^2 l(\theta)}{\partial \alpha_1 \partial \beta_1} \\[2ex] \dfrac{\partial^2 l(\theta)}{\partial \alpha_1 \partial \beta_1} & \dfrac{\partial^2 l(\theta)}{\partial \beta_1^2} \end{pmatrix},
$$

$$
i_{bb}(\theta) = -E \begin{pmatrix} \dfrac{\partial^2 l(\theta)}{\partial \alpha_2^2} & \dfrac{\partial^2 l(\theta)}{\partial \alpha_2 \partial \beta_2} \\[2ex] \dfrac{\partial^2 l(\theta)}{\partial \alpha_2 \partial \beta_2} & \dfrac{\partial^2 l(\theta)}{\partial \beta_2^2} \end{pmatrix},
$$

and

$$
i_{ba}(\theta) = i_{ab}^T(\theta) = -E \begin{pmatrix} \dfrac{\partial^2 l(\theta)}{\partial \alpha_1 \partial \alpha_2} & \dfrac{\partial^2 l(\theta)}{\partial \beta_1 \partial \alpha_2} \\[2ex] \dfrac{\partial^2 l(\theta)}{\partial \alpha_1 \partial \beta_2} & \dfrac{\partial^2 l(\theta)}{\partial \beta_1 \partial \beta_2} \end{pmatrix}.
$$

Using the results in Cox and Hinkley (1974), the likelihood ratio test statistic for the hypotheses (16.3) is

$$
W_1 = 2\left[l(\hat{\theta}) - l(\hat{\theta}_0)\right]
$$

$$
= (\hat{\alpha}_1, \hat{\beta}_1)\left[i_{aa}(\theta_0) - i_{ab}(\theta_0)i_{bb}^{-1}(\theta_0)i_{ba}(\theta_0)\right](\hat{\alpha}_1, \hat{\beta}_1)^T + o_p(1).
$$

It is easy to see that

$$
i_{aa}(\theta_0) = i_{ab}(\theta_0) = i_{ba}(\theta_0) = \frac{1}{2}i_{bb}(\theta_0)
$$

$$
= n \begin{pmatrix} \dfrac{(1-\beta_2)}{\alpha_2} - \dfrac{\beta_2(1-\beta_2)e^{-\alpha_2}}{\beta_2 + (1-\beta_2)e^{-\alpha_2}} & -\dfrac{e^{-\alpha_2}}{\beta_2 + (1-\beta_2)e^{-\alpha_2}} \\[3ex] -\dfrac{e^{-\alpha_2}}{\beta_2 + (1-\beta_2)e^{-\alpha_2}} & \dfrac{1-e^{-\alpha_2}}{(1-\beta_2)\left[\beta_2 + (1-\beta_2)e^{-\alpha_2}\right]} \end{pmatrix} \tag{16.4}
$$

Then,

$$
i_{aa}(\theta_0) - i_{ab}(\theta_0)i_{bb}^{-1}(\theta_0)i_{ba}(\theta_0) = \frac{1}{2}i_{aa}(\theta_0)
$$

and

$$
W_1 = \frac{1}{2}(\hat{\alpha}_1, \hat{\beta}_1)i_{aa}(\theta_0)(\hat{\alpha}_1, \hat{\beta}_1)^T + o_p(1) \tag{16.5}
$$

From the limiting marginal normal distribution of $(\hat{\alpha}_1, \hat{\beta}_1)$, the likelihood ratio statistic W_1 asymptotically follows a central χ_2^2 under the null hypothesis and a noncentral chi-squared distribution $\chi_2^2(\delta_1)$ with a noncentrality parameter δ_1, where

$$\delta_1^2 = \frac{1}{2}(\alpha_1, \beta_1)i_{aa}(\theta_0)(\alpha_1, \beta_1)^T$$

$$= \frac{n}{2}\left[\frac{\alpha_1^2(1-\beta_2)}{\alpha_2} + \frac{\beta_1^2(1-e^{-\alpha_2})}{(1-\beta_2)\left[\beta_2 + (1-\beta_2)e^{-\alpha_2}\right]} - \frac{\left[\alpha_1^2\beta_2(1-\beta_2) + 2\alpha_1\beta_1\right]e^{-\alpha_2}}{\beta_2 + (1-\beta_2)e^{-\alpha_2}}\right]$$

under the alternative hypothesis. Thus, the null hypothesis H_{01} is rejected at a significant level α if $W_1 > \chi_2^2(\alpha)$, where $\chi_2^2(\alpha)$ is the upper α-quantile of a χ_2^2 distribution. Based on this result, for a given level α, the critical region is given by

$$\left\{(x_{11}, ..., x_{n_1}, x_{12}, ..., x_{n_2}) \,\middle|\, W_1 > \chi_2^2(\alpha)\right\}$$

and the power function of the test is given as follows:

$$1 - \beta = P\left\{W_1 > \chi_2^2(\alpha) \,\middle|\, H_{11}\right\}$$

$$= P\left\{\left(W_1 - \frac{1}{3}\right)^{\frac{1}{2}} - \left(\frac{5 + 3\delta_1^2}{3}\right)^{\frac{1}{2}} > \left(\chi_2^2(\alpha) - \frac{1}{3}\right)^{\frac{1}{2}} - \left(\frac{5 + 3\delta_1^2}{3}\right)^{\frac{1}{2}}\,\middle|\, H_{11}\right\}. \tag{16.6}$$

16.2.2 Testing the Difference in the Groups of Zeros

If the difference in the groups of nonzeros between treatments is of primary interest, the corresponding hypotheses are

$$H_0 : \theta_1 = \theta_2 \quad \text{versus} \quad H_1 : \theta_1 \neq \theta_2,$$

which is equivalent to

$$H_{02} : \beta_1 = 0 \quad \text{versus} \quad H_{12} : \beta_1 \neq 0. \tag{16.7}$$

Denote

$$i_{\beta_1\beta_1}(\theta) = -E\left(\frac{\partial^2 l(\theta)}{\partial \beta_1^2}\right), \quad i_{\beta_1\bar{\beta}_1}(\theta) = -E\left(\frac{\partial^2 l(\theta)}{\partial \beta_1 \partial \alpha_1}, \frac{\partial^2 l(\theta)}{\partial \beta_1 \partial \alpha_2}, \frac{\partial^2 l(\theta)}{\partial \beta_1 \partial \beta_2}\right),$$

and

$$i_{\bar{\beta}_1\bar{\beta}_1}(\theta) = -E\begin{vmatrix} \dfrac{\partial^2 l(\theta)}{\partial \alpha_1^2} & \dfrac{\partial^2 l(\theta)}{\partial \alpha_1 \partial \alpha_2} & \dfrac{\partial^2 l(\theta)}{\partial \alpha_1 \partial \beta_2} \\[2mm] \dfrac{\partial^2 l(\theta)}{\partial \alpha_1 \partial \alpha_2} & \dfrac{\partial^2 l(\theta)}{\partial \alpha_2^2} & \dfrac{\partial^2 l(\theta)}{\partial \alpha_2 \partial \beta_2} \\[2mm] \dfrac{\partial^2 l(\theta)}{\partial \alpha_1 \partial \beta_2} & \dfrac{\partial^2 l(\theta)}{\partial \alpha_2 \partial \beta_2} & \dfrac{\partial^2 l(\theta)}{\partial \beta_2^2} \end{vmatrix}.$$

Let $\theta_{\beta_1} = (\alpha_1, 0, \alpha_2, \beta_2)$ and $\hat{\theta}_{\beta_1} = (\hat{\alpha}_{11}, 0, \hat{\alpha}_{21}, \hat{\beta}_{21})$, where $\hat{\alpha}_{11}, \hat{\alpha}_{21}$, and $\hat{\beta}_{21}$ are the MLE of α_1, α_2 and β_2 subject to $\beta_1 = 0$, respectively. Following the similar ideas used in Section 16.2.1, we obtain the likelihood radio test statistic W_2 for hypotheses (16.7):

$$W_2 = 2\left(l(\hat{\theta}) - l(\hat{\theta}_{\beta_1})\right)$$
$$= \hat{\beta}_1^2 [i_{\beta_1\beta_1}(\theta_{\beta_1}) - i_{\beta_1\bar{\beta}_1}(\theta_{\beta_1})i_{\bar{\beta}_1\bar{\beta}_1}^{-1}(\theta_{\beta_1})i_{\bar{\beta}_1\beta_1}^T(\theta_{\beta_1})] + o_p(1). \tag{16.8}$$

From the limiting marginal normal distribution of $\hat{\beta}_1$, the likelihood ratio statistic W_2 asymptotically follows a central χ_1^2 under the null hypothesis H_{02} and a noncentral chi-squared distribution $\chi_1^2(\delta_2)$ with

$$\delta_2^2 = \beta_1^2 [i_{\beta_1\beta_1}(\theta_{\beta_1}) - i_{\beta_1\bar{\beta}_1}(\theta_{\beta_1})i_{\bar{\beta}_1\bar{\beta}_1}^{-1}(\theta_{\beta_1})i_{\bar{\beta}_1\beta_1}^T(\theta_{\beta_1})]$$

under the local alternative. For a given significant level α, the critical region is

$$\left\{ (x_{11}, ..., x_{n1}, x_{12}, ..., x_{n2}) \;\middle|\; W_2 > \chi_1^2(\alpha) \right\}$$

and the power function of the test is as follows:

$$1 - \beta = P\left\{ W_2 > \chi_1^2(\alpha) \;\middle|\; H_{12} \right\}. \tag{16.9}$$

16.2.3 Testing the Difference of the Groups of Nonzeros

The set of hypotheses of interests in this case is

$$H_0 : \lambda_1 = \lambda_2 \quad \text{versus} \quad H_1 : \lambda_1 \neq \lambda_2,$$

which is equivalent to

$$H_{03} : \quad \alpha_1 = 0 \quad \text{versus} \quad H_{13} : \quad \alpha_1 \neq 0. \tag{16.10}$$

Denote

$$i_{\alpha_1\alpha_1}(\theta) = -E\left(\frac{\partial^2 l(\theta)}{\partial \alpha_1^2}\right),$$

$$i_{\alpha_1\bar{\alpha}_1}(\theta) = -E\left(\frac{\partial^2 l(\theta)}{\partial \alpha_1 \partial \beta_1}, \frac{\partial^2 l(\theta)}{\partial \alpha_1 \partial \alpha_2}, \frac{\partial^2 l(\theta)}{\partial \alpha_1 \partial \beta_2}\right),$$

and

$$i_{\bar{\alpha}_1\bar{\alpha}_1}(\theta) = -E \begin{vmatrix} \dfrac{\partial^2 l(\theta)}{\partial \beta_1^2} & \dfrac{\partial^2 l(\theta)}{\partial \beta_1 \partial \alpha_2} & \dfrac{\partial^2 l(\theta)}{\partial \beta_1 \partial \beta_2} \\[2ex] \dfrac{\partial^2 l(\theta)}{\partial \beta_1 \partial \alpha_2} & \dfrac{\partial^2 l(\theta)}{\partial \alpha_2^2} & \dfrac{\partial^2 l(\theta)}{\partial \alpha_2 \partial \beta_2} \\[2ex] \dfrac{\partial^2 l(\theta)}{\partial \beta_1 \partial \beta_2} & \dfrac{\partial^2 l(\theta)}{\partial \alpha_2 \partial \beta_2} & \dfrac{\partial^2 l(\theta)}{\partial \beta_2^2} \end{vmatrix}.$$

Let $\theta_{\alpha_1} = (0, \beta_1, \alpha_2, \beta_2)$ and $\hat{\theta}_{\alpha_1} = (0, \hat{\beta}_{12}, \hat{\alpha}_{22}, \hat{\beta}_{22})$, where $\hat{\beta}_{12}, \hat{\alpha}_{22}$, and $\hat{\beta}_{22}$ are the MLE of β_1, α_2 and β_2 subject to $\alpha_1 = 0$, respectively. Using similar arguments as those in the previous section, we can derive the likelihood ratio test statistic W_3 and is given as

$$W_3 = 2\left(l(\hat{\theta}) - l(\hat{\theta}_{\alpha_1})\right)$$
$$= \hat{\alpha}_1^2\left[i_{\alpha_1\alpha_1}(\theta_{\alpha_1}) - i_{\alpha_1\bar{\alpha}_1}(\theta_{\alpha_1})i_{\bar{\alpha}_1\bar{\alpha}_1}^{-1}(\theta_{\alpha_1})i_{\alpha_1\bar{\alpha}_1}^T(\theta_{\alpha_1})\right] + o_p(1), \tag{16.11}$$

and the critical region for a given significant level α is given as

$$\left\{ (x_{11}, \ldots, x_{n1}, x_{12}, \ldots, x_{n2}) \mid W_3 > \chi_1^2(\alpha) \right\}.$$

From the limiting marginal normal distribution of $\hat{\alpha}_1$, the likelihood ratio statistic W_3 asymptotically follows a central χ_1^2 under the null hypothesis H_{03} and a noncentral chi-squared distribution $\chi_1^2(\delta_3)$ with

$$\delta_3^2 = \alpha_1^2[i_{\alpha_1\alpha_1}(\theta_{\alpha_1}) - i_{\alpha_1\bar{\alpha}_1}(\theta_{\alpha_1})i_{\bar{\alpha}_1\bar{\alpha}_1}^{-1}(\theta_{\alpha_1})i_{\alpha_1\bar{\alpha}_1}^T(\theta_{\alpha_1})]$$

under the local alternative.

16.3 Sample Size Calculation

16.3.1 Testing the Difference in the Groups of Both Zeros and Nonzeros

Based on the power function given in Equation 16.8, we can determine the required sample size so that the resulting sample would achieve a given power. Using normal approximation to the probability, we note that

$$\left(2 + \delta_1^2\right)^{\frac{1}{2}}\left[\left(\frac{W_1 - 1/3}{2 + \delta_1^2}\right)^{\frac{1}{2}} - \left(1 - \frac{1}{3(2 + \delta_1^2)}\right)^{\frac{1}{2}}\right],$$

or equivalently,

$$\left(W_1 - \frac{1}{3}\right)^{\frac{1}{2}} - \left(\frac{5 + 3\delta_1^2}{3}\right)^{\frac{1}{2}}$$

approximately follows a standard normal distribution. Therefore,

$$\left(\chi_2^2(\alpha) - \frac{1}{3}\right)^{\frac{1}{2}} - \left(\frac{5 + 3\delta_1^2}{3}\right)^{\frac{1}{2}} = -z(\beta), \tag{16.12}$$

where $z(\beta)$ is the upper β-quantile of the standard normal distribution. Simplification of Equation 16.12 gives a formula of the sample size for achieving a $(1 - \beta)$ power as follows:

$$n \geq C_1 \left\{ 2 \left[\left[\chi_2^2(\alpha) - \frac{1}{3} \right]^{\frac{1}{2}} + z(\beta) \right]^2 - \frac{10}{3} \right\}, \tag{16.13}$$

where

$$C_1 = \left\{ \frac{\alpha_1^2(1 - \beta_2)}{\alpha_2} + \frac{\beta_1^2(1 - e^{-\alpha_2})}{(1 - \beta_2)\left[\beta_2 + (1 - \beta_2)e^{-\alpha_2}\right]} - \frac{\left[\alpha_1^2 \beta_2 (1 - \beta_2) + 2\alpha_1 \beta_1\right] e^{-\alpha_2}}{\beta_2 + (1 - \beta_2)e^{-\alpha_2}} \right\}^{-1}.$$

Some values of C_1 are presented in Table 16.1.

16.3.2 Testing the Difference in the Groups of Zeros between Treatments

To determine the required sample for this test, we follow the similar idea outlined in the previous section. In particular, applying normal approximation to the likelihood ratio statistic W_2 given Equation 16.6, we can find that the required sample size to achieve $(1 - \beta)$ power for the test is

$$n \geq C_2 \left\{ \left[(\chi_1^2(\alpha))^{\frac{1}{2}} + z(\beta) \right]^2 - 1 \right\}. \tag{16.14}$$

where

$$C_2 = \frac{1}{n} \beta_1^2 [i_{\beta_1 \beta_1}(\theta_{\beta_1}) - i_{\beta_1 \bar{\beta}_1}(\theta_{\beta_1}) i_{\bar{\beta}_1 \bar{\beta}_1}^{-1}(\theta_{\beta_1}) i_{\bar{\beta}_1 \bar{\beta}}^{\tau}(\theta_{\beta_1})].$$

Some values of C_2 are listed in Table 16.2.

16.3.3 Testing the Difference in the Groups of Nonzeros between Treatments

Following similar arguments as described above, we apply normal approximation to the likelihood ratio statistic W_3 given Equation 16.8. The required sample size to give power $(1 - \beta)$ is given by

$$n \geq C_3 \left\{ \left[(\chi_1^2(\alpha))^{\frac{1}{2}} + z(\beta) \right]^2 - 1 \right\}, \tag{16.15}$$

where

$$C_3 = \frac{1}{n} \alpha_1^2 [i_{\alpha_1 \alpha_1}(\theta_{\alpha_1}) - i_{\alpha_1 \bar{\alpha}_1}(\theta_{\alpha_1}) i_{\bar{\alpha}_1 \bar{\alpha}_1}^{-1}(\theta_{\alpha_1}) i_{\bar{\alpha}_1 \bar{\alpha}_1}^{\tau}(\theta_{\alpha_1})].$$

Some values of C_3 are listed in Table 16.3.

TABLE 16.1

Values of C_1

	$\lambda_1 - \lambda_2$	$\lambda_2 = 10$					$\lambda_2 = 2.0$					$\lambda_2 = 3.0$				
		0.15	0.25	0.50	0.75	1.00	0.15	0.25	0.50	0.75	1.00	0.15	0.25	0.50	0.75	1.00
$\theta_2 = 0.10$	$\theta_1 - \theta_2 = -0.05$	28.25	13.01	3.96	1.88	1.10	34.64	19.69	7.14	3.59	2.14	33.97	22.92	9.74	5.12	3.11
	0.00	53.99	19.44	4.86	2.16	1.21	112.49	40.50	10.12	4.50	2.53	165.19	59.47	14.87	6.61	3.72
	0.10	108.54	39.88	7.31	2.85	1.50	29.50	26.67	12.33	5.75	3.16	14.57	13.79	9.45	5.79	3.66
	0.15	59.60	41.48	8.74	3.25	1.66	12.68	13.09	9.55	5.48	3.23	6.47	6.45	5.56	4.20	3.03
	0.20	30.78	31.97	9.97	3.67	1.83	6.87	7.30	6.67	4.71	3.08	3.61	3.65	3.45	2.94	2.36
$\theta_2 = 0.15$	$\theta_1 - \theta_2 = -0.10$	17.62	9.75	3.54	1.80	1.08	16.18	11.57	5.57	3.15	2.00	13.98	11.53	6.82	4.21	2.78
	0.00	59.37	21.37	5.34	2.37	1.34	123.49	44.46	11.11	4.94	2.78	177.55	63.92	15.98	7.10	3.99
	0.10	110.32	43.28	8.08	3.15	1.65	32.09	28.29	12.94	6.09	3.38	17.86	16.44	10.55	6.25	3.90
	0.15	58.04	42.88	9.61	3.59	1.83	13.99	14.17	10.03	5.75	3.41	8.01	7.87	6.47	4.69	3.30
	0.20	29.91	31.67	10.82	4.05	2.02	7.63	8.00	7.07	4.93	3.23	4.49	4.50	4.11	3.38	2.64
$\theta_2 = 0.20$	$\theta_1 - \theta_2 = -0.15$	11.80	7.44	3.15	1.71	1.06	9.43	7.55	4.40	2.75	1.85	7.93	7.06	4.98	3.45	2.46
	0.00	65.27	23.50	5.87	2.61	1.47	134.78	48.52	12.13	5.39	3.03	190.38	68.54	17.13	7.62	4.28
	0.10	111.37	46.88	8.95	3.48	1.82	34.12	29.66	13.59	6.46	3.61	20.60	18.61	11.48	6.68	4.14
	0.15	56.19	44.04	10.58	3.98	2.03	15.03	15.02	10.47	6.04	3.61	9.32	9.05	7.21	5.10	3.54
	0.20	28.90	31.18	11.72	4.47	2.24	8.24	8.55	7.41	5.15	3.40	5.24	5.21	4.65	3.74	2.87
$\theta_2 = 0.30$	$\theta_1 - \theta_2 = -0.20$	8.55	6.03	2.97	1.73	1.12	6.58	5.66	3.80	2.60	1.85	5.69	5.27	4.13	3.12	2.36
	-0.10	20.57	11.87	4.49	2.32	1.41	20.91	15.20	7.40	4.18	2.65	20.18	16.50	9.44	5.66	3.67
	0.00	79.16	28.50	7.12	3.17	1.78	159.87	57.55	14.39	6.39	3.60	219.90	79.16	19.79	8.80	4.95
	0.10	110.69	54.66	11.04	4.29	2.24	36.68	31.72	15.04	7.34	4.15	24.61	21.86	13.12	7.58	4.70
	0.20	26.43	29.50	13.67	5.48	2.76	9.02	9.24	7.93	5.60	3.76	6.34	6.25	5.47	4.32	3.28

TABLE 16.2

Values of C_2

$\lambda_1 - \lambda_2$	$\lambda_2 = 2.0$					$\lambda_2 = 3.0$					$\lambda_2 = 4.0$				
	0.15	0.25	0.50	0.75	1.00	0.15	0.25	0.50	0.75	1.00	0.15	0.25	0.50	0.75	1.00
$\theta_2 = 0.10$ $\theta_1 - \theta_2 = -0.10$	55.07	52.93	48.87	46.07	44.10	28.29	27.77	26.74	25.98	25.41	21.36	21.20	20.88	20.63	20.45
0.15	24.47	23.53	21.72	20.48	19.60	12.57	12.34	11.88	11.55	11.29	9.49	9.42	9.28	9.17	9.09
0.20	13.77	13.23	12.22	11.52	11.03	7.07	6.94	6.69	6.49	6.35	5.34	5.30	5.22	5.16	5.11
0.25	8.81	8.47	7.82	7.37	7.06	4.53	4.44	4.28	4.16	4.07	3.42	3.39	3.34	3.30	3.27
0.30	6.12	5.88	5.43	5.12	4.90	3.14	3.09	2.97	2.89	2.82	2.37	2.36	2.32	2.29	2.27
$\theta_2 = 0.15$ $\theta_1 - \theta_2 = \pm 0.10$	60.51	58.49	54.66	52.01	50.15	35.21	34.73	33.75	33.03	32.50	28.67	28.52	28.22	27.99	27.81
0.15	26.89	26.00	24.29	23.12	22.29	15.65	15.44	15.00	14.68	14.44	12.74	12.68	12.54	12.44	12.36
0.20	15.13	14.62	13.66	13.00	12.54	8.80	8.68	8.44	8.26	8.12	7.17	7.13	7.05	7.00	6.95
0.25	9.68	9.36	8.74	8.32	8.02	5.63	5.56	5.40	5.29	5.20	4.59	4.56	4.52	4.48	4.45
0.30	6.72	6.50	6.07	5.78	5.57	3.91	3.86	3.75	3.67	3.61	3.19	3.17	3.14	3.11	3.09
$\theta_2 = 0.20$ $\theta_1 - \theta_2 = \pm 0.10$	64.95	63.05	59.44	56.95	55.20	41.14	40.69	39.77	39.09	38.59	34.99	34.85	34.56	34.34	34.17
± 0.15	28.87	28.02	26.42	25.31	24.53	18.29	18.08	17.68	17.37	17.15	15.55	15.49	15.36	15.26	15.19
0.20	16.24	15.76	14.86	14.24	13.80	10.29	10.17	9.94	9.77	9.65	8.75	8.71	8.64	8.59	8.54
0.25	10.39	10.09	9.51	9.11	8.83	6.58	6.51	6.36	6.25	6.17	5.60	5.58	5.53	5.49	5.47
0.30	7.22	7.01	6.60	6.33	6.13	4.57	4.52	4.42	4.34	4.29	3.89	3.87	3.84	3.82	3.80
$\theta_2 = 0.30$ $\theta_1 - \theta_2 = \pm 0.10$	70.83	69.17	66.01	63.83	62.30	50.00	49.60	48.80	48.21	47.76	44.61	44.49	44.24	44.05	43.90
± 0.15	31.48	30.74	29.34	28.37	27.69	22.22	22.05	21.69	21.42	21.23	19.83	19.77	19.66	19.58	19.51
± 0.20	17.71	17.29	16.50	15.96	15.58	12.50	12.40	12.20	12.05	11.94	11.15	11.12	11.06	11.01	10.98
± 0.25	11.33	11.07	10.56	10.21	9.97	8.00	7.94	7.81	7.71	7.64	7.14	7.12	7.08	7.05	7.02
0.30	7.87	7.69	7.33	7.09	6.92	5.56	5.51	5.42	5.36	5.31	4.96	4.94	4.92	4.89	4.88

TABLE 16.3

Values of C_3

$\lambda_1 - \lambda_2$	$\lambda_2 = 1$					$\lambda_2 = 2.0$					$\lambda_2 = 3.0$				
	0.25	0.50	0.75	1.00	1.50	0.25	0.50	0.75	1.00	1.50	0.25	0.50	0.75	1.00	1.50
$\theta_2 = 0.10$															
$\theta_1 - \theta_2 = -0.05$	82.82	20.70	9.20	5.18	2.30	100.79	25.20	11.20	6.30	2.80	123.23	30.81	13.69	7.70	3.42
0.00	85.06	21.26	9.45	5.32	2.36	103.51	25.88	11.50	6.47	2.88	126.56	31.64	14.06	7.91	3.52
0.10	90.37	22.59	10.04	5.65	2.51	109.98	27.50	12.22	6.87	3.06	134.47	33.62	14.94	8.40	3.74
0.15	93.56	23.39	10.40	5.85	2.60	113.87	28.47	12.65	7.12	3.16	139.22	34.80	15.47	8.70	3.87
0.20	97.21	24.30	10.80	6.08	2.70	118.30	29.58	13.14	7.39	3.29	144.64	36.16	16.07	9.04	4.02
$\theta_2 = 0.15$															
$\theta_1 - \theta_2 = -0.10$	85.32	21.33	9.48	5.33	2.37	103.84	25.96	11.54	6.49	2.88	126.95	31.74	14.11	7.93	3.53
0.00	90.06	22.51	10.01	5.63	2.50	109.60	27.40	12.18	6.85	3.04	134.01	33.50	14.89	8.38	3.72
0.10	96.06	24.02	10.67	6.00	2.67	116.91	29.23	12.99	7.31	3.25	142.94	35.73	15.88	8.93	3.97
0.15	99.71	24.93	11.08	6.23	2.77	121.35	30.34	13.48	7.58	3.37	148.36	37.09	16.48	9.27	4.12
0.20	103.92	25.98	11.55	6.49	2.89	126.47	31.62	14.05	7.90	3.51	154.62	38.66	17.18	9.66	4.30
$\theta_2 = 0.20$															
$\theta_1 - \theta_2 = -0.15$	88.13	22.03	9.79	5.51	2.45	107.26	26.82	11.92	6.70	2.98	131.14	32.78	14.57	8.20	3.64
0.00	95.69	23.92	10.63	5.98	2.66	116.45	29.11	12.94	7.28	3.23	142.38	35.60	15.82	8.90	3.96
0.10	102.52	25.63	11.39	6.41	2.85	124.77	31.19	13.86	7.80	3.47	152.55	38.14	16.95	9.53	4.24
0.15	106.73	26.68	11.86	6.67	2.96	129.89	32.47	14.43	8.12	3.61	158.81	39.70	17.65	9.93	4.41
0.20	111.64	27.91	12.40	6.98	3.10	135.86	33.97	15.10	8.49	3.77	166.11	41.53	18.46	10.38	4.61
$\theta_2 = 0.30$															
$\theta_1 - \theta_2 = -0.20$	97.21	24.30	10.80	6.08	2.70	118.30	29.58	13.14	7.39	3.29	144.64	36.16	16.07	9.04	4.02
−0.10	102.52	25.63	11.39	6.41	2.85	124.77	31.19	13.86	7.80	3.47	152.55	38.14	16.95	9.53	4.24
0.00	109.36	27.34	12.15	6.83	3.04	133.09	33.27	14.79	8.32	3.70	162.72	40.68	18.08	10.17	4.52
0.10	118.47	29.62	13.16	7.40	3.29	144.18	36.05	16.02	9.01	4.01	176.28	44.07	19.59	11.02	4.90
0.20	131.23	32.81	14.58	8.20	3.65	159.71	39.93	17.75	9.98	4.44	195.26	48.82	21.70	12.20	5.42

16.3.4 An Example

For the management of acute pain and relief of signs and symptoms associated with osteo-arthritis (OA) and rheumatoid arthritis (RA), traditional nonsteroidal anti-inflammatory drugs (NSAIDs) are commonly used as standard therapy. However, it is recognized that the use of NSAIDs is frequently associated with an increased risk of gastroduodenal ulcers and hemorrhages. A pharmaceutical company is interested in conducting a clinical trial to assess the gastrointestinal (GI) safety of its newly developed compound as compared to the NSAID class in the treatment of OA and RA. Since it is believed that occurrence of gastroduodenal erosions could lead to gastroduodenal ulcers, one of the primary study endpoints for GI safety is the number of erosion counts. In practice, many patients may have zero occurrences of gastroduodenal erosions. Thus, it is reasonable to assume that the erosion counts follow a zero-inflated Poisson distribution with parameters θ and λ.

Case 1: Difference in both groups of zeros and nonzeros between treatments

Suppose that for independent samples of size $n = 30$, the numbers of subjects with erosion counts from 0 to 4 are given below.

The hypotheses of interests are H_0: $\lambda_1 = \lambda_2$, $\theta_1 = \theta_2$ versus H_1: not both are equal, which is equivalent to H_{01}: $\alpha_1 = 0$, $\beta_1 = 0$ versus H_{11}: not both are equal. Based on the observed counts, we have

$$\text{Treatment A: } K_1 = 4 \text{ and } \sum_{i=1}^{n} x_{i1} = 4 \times 0 + 2 \times 1 + 5 \times 2 + 11 \times 3 + 8 \times 4 = 77;$$

$$\text{Treatment B: } K_2 = 15 \text{ and } \sum_{i=1}^{n} x_{i2} = 15 \times 0 + 4 \times 1 + 6 \times 2 + 4 \times 3 + 1 \times 4 = 32.$$

Thus, the likelihood equations are given as

$$\hat{\beta}_2 = 1 - \frac{32}{30\hat{\alpha}_2}, \quad \frac{\hat{\alpha}_2}{1 - e^{-\hat{\alpha}_2}} = \frac{32}{15}, \quad \hat{\beta}_1 + \hat{\beta}_2 = 1 - \frac{77}{30(\hat{\alpha}_1 + \hat{\alpha}_2)}, \quad \text{and} \quad \frac{\hat{\alpha}_1 + \hat{\alpha}_2}{1 - e^{-\hat{\alpha}_1 - \hat{\alpha}_2}} = \frac{77}{26}$$

Solving these equations by numerical method, the MLEs of the parameters are $\hat{\alpha}_1 = 1.0074$, $\hat{\beta}_1 = -0.3275$ $\hat{\alpha}_2 = 1.7699$, and $\hat{\beta}_2 = 0.3973$. For the constrained MLEs $\hat{\alpha}_{20}$ and $\hat{\beta}_{20}$ of α_2 and β_2 under H_{01}, the likelihood equations are given as $\hat{\beta}_{20} = 1 - (109/60\hat{\alpha}_{20})$ and $\hat{\alpha}_{20}/1 - e^{-\hat{\alpha}_{20}} = 109/41$, which gives $\hat{\alpha}_{20} = 2.4228$ and $\hat{\beta}_{20} = 0.2502$. Thus, $\hat{\theta}_0 = (0, 0, 2.4228, 0.2502)$.

The test statistic $W_1 = 13.1852 > \chi_2^2(0.05) = 5.9915$. Therefore, we reject H_{01} at a significant level of 0.05. The results indicate that there is a significant difference in the number of erosions in (1) either group of zeros or group of nonzeros between treatments or (2) both groups of zeros and nonzeros between treatments.

Note that the relative changes are $\hat{\alpha}_1/\hat{\alpha}_2 = 0.5691$ and $|\hat{\beta}_1/\hat{\beta}_2| = 0.8092$. Based on the results in Table 16.4, suppose an experimenter intends to plan a future study that aims at

TABLE 16.4

Erosion Data

	Counts	0	1	2	3	4
Number of subjects	Treatment A	4	2	5	11	8
	Treatment B	15	4	6	4	1

detecting the clinical difference of 20% relative changes of parameters between the two populations. The sample size required for detecting such differences to achieve an 80% power at 5% level of significance should be at least greater than 23.

Case 2: Difference in the groups of zeros between treatments

Suppose that the hypotheses of interests are $H_0: \theta_1 = \theta_2$ versus $H_1: \theta_1 \neq \theta_2$, which is equivalent to $H_{02}: \beta_1 = 0$ versus $H_{12}: \beta_1 \neq 0$. Then, under H_{12}, the constrained likelihood equations for $\hat{\theta}_{\beta_1} = (\hat{\alpha}_{11}, 0, \hat{\alpha}_{21}, \hat{\beta}_{21})$ are given by

$$\frac{77}{\hat{\alpha}_{11} + \hat{\alpha}_{21}} + \frac{4\hat{\beta}_{21}}{\hat{\beta}_{21} + (1 - \hat{\beta}_{21})e^{-(\hat{\alpha}_{11} + \hat{\alpha}_{21})}} - 30 = 0,$$

$$\frac{32}{\hat{\alpha}_{21}} + \frac{15\hat{\beta}_{21}}{\hat{\beta}_{21} + (1 - \hat{\beta}_{21})e^{-\hat{\alpha}_{21}}} - 30 = 0,$$

$$\frac{4(1 - e^{-\hat{\alpha}_{11} - \hat{\alpha}_{21}})}{\hat{\beta}_{21} + (1 - \hat{\beta}_{21})e^{-\hat{\alpha}_{11} - \hat{\alpha}_{21}}} + \frac{15(1 - e^{-\hat{\alpha}_{21}})}{\hat{\beta}_{21} + (1 - \hat{\beta}_{21})e^{-\hat{\alpha}_{21}}} - \frac{41}{1 - \hat{\beta}_{21}} = 0$$

Thus, the MLEs are given by $\hat{\alpha}_{11} = 1.5166$, $\hat{\alpha}_{21} = 1.3380$, and $\hat{\beta}_{21} = 0.1518$.

The test statistic $W_2 = 4.3596 > \chi_1^2(0.05) = 3.8415$, which implies that there is significant difference in the number of erosions between the groups of zeros between treatment A and treatment B. Therefore, we reject H_{02} at a significant level of 0.05.

Suppose that an experimenter intends to plan a future study that aims at detecting the clinical difference at 20% relative changes of parameters between the two populations. The sample size required for detecting such differences to achieve an 80% power at 5% level of significance should be at least greater than 48.

Case 3: Difference in the groups of nonzeros between treatments

Suppose that the set of hypotheses of interests is

$$H_0: \lambda_1 = \lambda_2 \quad \text{versus} \quad H_1: \lambda_1 \neq \lambda_2,$$

which is equivalent to

$$H_{03}: \alpha_1 = 0 \quad \text{versus} \quad H_{13}: \alpha_1 \neq 0.$$

Under H_{03}, the constrained likelihood equations subject to $\alpha_1 = 0$ are

$$\frac{15(1 - e^{-\hat{\alpha}_{22}})}{\hat{\beta}_{22} + (1 - \hat{\beta}_{22})e^{-\hat{\alpha}_{22}}} - \frac{15}{1 - \hat{\beta}_{22}} = 0,$$

$$\frac{4(1 - e^{-\hat{\alpha}_{22}})}{\hat{\beta}_{12} + \hat{\beta}_{22} + (1 - \hat{\beta}_{12} - \hat{\beta}_{22})e^{-\hat{\alpha}_{22}}} - \frac{26}{1 - \hat{\beta}_{12} - \hat{\beta}_{22}} = 0,$$

and

$$\frac{109}{\hat{\alpha}_{22}} + \frac{4(\hat{\beta}_{12} + \hat{\beta}_{22})}{\hat{\beta}_{12} + \hat{\beta}_{22} + (1 - \hat{\beta}_{12} - \hat{\beta}_{22})e^{-\hat{\alpha}_{22}}} + \frac{15\hat{\beta}_{22}}{\hat{\beta}_{22} + (1 - \hat{\beta}_{22})e^{-\hat{\alpha}_{22}}} - 60 = 0.$$

Thus, the constrained MLEs under H_{03} are

$$\hat{\alpha}_{22} = 2.4228, \quad \hat{\beta}_{12} = -0.4023, \quad \text{and} \quad \hat{\beta}_{22} = 0.4513.$$

The test statistic $W_3 = 3.4147 < \chi_1^2(0.05) = 3.8415$, suggests that there is not enough evidence to reject the null hypothesis. In other words, no significant difference in the number of erosions between the groups of nonzeros between treatment A and treatment B was observed.

Suppose that an experimenter intends to plan a future study that aims at detecting the clinical difference at 20% relative changes of parameters between the two populations. The sample size required for detecting such differences to achieve an 80% power at 5% level of significance should be at least greater than 51.

16.4 Multivariate ZIP

16.4.1 Bivariate ZIP

In many clinical applications, outcomes with unusually large number of zeros may be taken repeatedly on the same individual. These repeated outcomes certainly possess a certain correlated structure, and analysis using univariate ZIP models would produce a misleading conclusion since the correlated structure is not properly addressed in the analysis process. Li et al. (1999) discussed several ways to construct multivariate ZIP (MZIP) distribution and presented real-life applications of the MZIP in equipment-fault detection. However, as the dimension increases, the number of model parameters would increase rapidly to describe the covariance structure of the multivariate observations. Large number of model parameters not only requires a relatively large number of observations but also imposes difficulty in the estimation of the model parameters. Therefore, it is desirable to choose an MZIP model such that the corresponding number of model parameters is in a manageable size. Following a similar idea of Li et al. (1999), Yuen, Chow, and Tse (2015) considered a bivariate ZIP model with five parameters as follows. A bivariate ZIP (Y_1, Y_2) is constructed from a mixture of a point mass at $(0, 0)$, two univariate Poisson distributions with parameters λ_1 and λ_2, and a bivariate Poisson distribution with parameters $(\lambda_{10}, \lambda_{20}, \lambda_{00})$, where

$$
\begin{array}{lll}
(Y_1, Y_2) & \sim (0, 0) & \text{with probability } p_0 \\
& \sim (\text{Poisson}(\lambda_1), 0) & \text{with probability } p_1 \\
& \sim (0, \text{Poisson}(\lambda_2)) & \text{with probability } p_2 \\
& \sim \text{bivariate Poisson}(\lambda_{10}, \lambda_{20}, \lambda_{00}) & \text{with probability } p_3 = 1 - p_0 - p_1 - p_2
\end{array}
$$

Assume that $\lambda_1 = \lambda_{10} + \lambda_{00}$ and $\lambda_2 = \lambda_{20} + \lambda_{00}$. The probability function of the bivariate ZIP (Y_1, Y_2) is then given by

$$P(Y_1 = 0, Y_2 = 0) = p_0 + p_1 \exp(-\lambda_1) + p_2 \exp(-\lambda_2) + p_3 \exp(-\lambda_{10} - \lambda_{20} - \lambda_{00})$$

$$P(Y_1 = y_1, Y_2 = 0) = \frac{[p_1 \lambda_1^{y_1} \exp(-\lambda_1) + p_3 \lambda_{10}^{y_1} \exp(-\lambda_{10} - \lambda_{20} - \lambda_{00})]}{y_1!}$$

$$P(Y_1 = 0, Y_2 = y_2) = \frac{[p_2 \lambda_2^{y_2} \exp(-\lambda_2) + p_3 \lambda_{20}^{y_2} \exp(-\lambda_{10} - \lambda_{20} - \lambda_{00})]}{y_2!} \qquad (16.16)$$

$$P(Y_1 = y_1, Y_2 = y_2) = p_3 \sum_{j=0}^{\min(y_1, y_2)} \lambda_{10}^{y_1 - j} \lambda_{20}^{y_2 - j} \lambda_{00}^{j} \frac{\exp(-\lambda_{10} - \lambda_{20} - \lambda_{00})}{(y_1 - j)!(y_2 - j)!j!}$$

The marginal distributions of Y_1 is a univariate ZIP, which is a mixture of point mass at 0 with probability $p_0 + p_2$ and a Poisson distribution with parameter λ_1 with probability $1 - (p_0 + p_2)$; similarly, Y_2 is a mixture of point mass at 0 with probability $p_0 + p_1$ and a Poisson distribution with parameter with probability $1 - (p_0 + p_2)$. Yuen, Chow, and Tse (2015) assume that Y_1 and Y_2 are identically distributed, that is, $\lambda_1 = \lambda_2$. To simplify notations, let $\lambda_1 = \lambda_2 = \lambda_0$, then $\lambda_{10} = \lambda_{20} = \lambda_0$ and $\lambda_{00} = \lambda - \lambda_0$. Thus, the parameters are λ, λ_0, p_0, p_1 and p_2.

Suppose that a clinical study is conducted to compare the effect between a test treatment (T) and a control treatment (C). For $j = T, C$, assume that n_j subjects are randomly assigned to the two groups. Let (Y_{1ij}, Y_{2ij}) be the observed responses corresponding to the ith subject in the jth treatment group. Furthermore, assume that (Y_{1ij}, Y_{2ij}) follows a bivariate ZIP model with parameters $\lambda_C, \lambda_{0C}, p_{0C}, p_{1C}, p_{2C}$ for the control treatment, and $\lambda_T, \lambda_{0T}, p_{0T}, p_{1T}, p_{2T}$ for the test treatment. Thus, the corresponding probability functions are given by Equation 16.16.

To facilitate the comparison of the two treatments, consider the following transformation of parameters:

$$
\begin{aligned}
\theta_1 &= \lambda_{0T} - \lambda_{0C} \quad \text{and} \quad \theta_6 = \lambda_{0C} \Rightarrow \lambda_{0T} = \theta_1 + \theta_6 \\
\theta_2 &= \lambda_T - \lambda_C \quad \text{and} \quad \theta_7 = \lambda_C \Rightarrow \lambda_T = \theta_2 + \theta_7 \\
\theta_3 &= p_{0T} - p_{0C} \quad \text{and} \quad \theta_8 = p_{0c} \Rightarrow p_{0T} = \theta_3 + \theta_8 \\
\theta_4 &= p_{1T} - p_{1C} \quad \text{and} \quad \theta_9 = p_{1c} \Rightarrow p_{1T} = \theta_4 + \theta_9 \\
\theta_5 &= p_{2T} - p_{2C} \quad \text{and} \quad \theta_{10} = p_{2C} \Rightarrow p_{2T} = \theta_5 + \theta_{10}
\end{aligned}
\qquad (16.17)
$$

Define the index sets

$$
\begin{aligned}
I_{0j} &= \{i : y_{1ij} = 0 \quad \text{and} \quad y_{2ij} = 0; i = 1, 2, \ldots, n_j\} \\
I_{1j} &= \{i : y_{1ij} \neq 0 \quad \text{and} \quad y_{2ij} = 0; i = 1, 2, \ldots, n_j\} \\
I_{2j} &= \{i : y_{1ij} = 0 \quad \text{and} \quad y_{2ij} \neq 0; i = 1, 2, \ldots, n_j\} \\
I_{3j} &= \{i : y_{1ij} \neq 0 \quad \text{and} \quad y_{2ij} \neq 0; i = 1, 2, \ldots, n_j\}
\end{aligned}
$$

Then, given observations (y_{1ij}, y_{2ij}), $i = 1, 2, \ldots, n_j$; $j = T, C$, the log-likelihood function $\ln L$ is given by

$$L = \prod_{k=0}^{3} \prod_{j=C,T} \prod_{i \in I_{kj}} P_{ijk} \Rightarrow \ln L = \sum_{k=0}^{3} \sum_{j=C,T} \sum_{i \in I_{kj}} \ln P_{ijk} \qquad (16.18)$$

where

$$P_{iC0} = \theta_8 + (\theta_9 + \theta_{10})\exp(-\theta_7) + (1 - \theta_8 - \theta_9 - \theta_{10})\exp(-(\theta_6 + \theta_7))$$

$$P_{iC1} = \frac{\left[\theta_9\theta_7^{y_{iC1}}\exp(-\theta_7) + (1 - \theta_8 - \theta_9 - \theta_{10})\theta_6^{y_{iC1}}\exp(-(\theta_6 + \theta_7))\right]}{y_{iC1}!}$$

$$P_{iC2} = \frac{\left[\theta_{10}\theta_7^{y_{iC2}}\exp(-\theta_7) + (1 - \theta_8 - \theta_9 - \theta_{10})\theta_6^{y_{iC2}}\exp(-(\theta_6 + \theta_7))\right]}{y_{iC2}!}$$

$$P_{iC3} = (1 - \theta_8 - \theta_9 - \theta_{10})\sum_{j=0}^{\min(y_{iC1}, y_{iC2})}\theta_1^{(y_{iC1} + y_{iC2} - 2j)}(\theta_7 - \theta_6)^j\frac{\exp(-(\theta_6 + \theta_7))}{(y_{iC1} - j)!(y_{iC2} - j)!j!}$$

and

$$P_{iT0} = (\theta_3 + \theta_8) + (\theta_4 + \theta_5 + \theta_9 + \theta_{10})\exp(-(\theta_2 + \theta_7)) + (1 - u)\exp(-v)$$

$$P_{iT1} = \frac{\left[(\theta_4 + \theta_9)(\theta_2 + \theta_7)^{y_{iT1}}\exp(-(\theta_2 + \theta_7)) + (1 - u)(\theta_1 + \theta_6)^{y_{iT1}}\exp(-v)\right]}{y_{iT1}!}$$

$$P_{iT2} = \frac{\left[(\theta_5 + \theta_{10})(\theta_2 + \theta_7)^{y_{iT2}}\exp(-(\theta_2 + \theta_7)) + (1 - u)(\theta_1 + \theta_6)^{y_{iT2}}\exp(-v)\right]}{y_{iT2}!}$$

$$P_{iT3} = (1 - u)\sum_{j=0}^{\min(y_{iT1}, y_{iT2})}(\theta_1 + \theta_6)^{(y_{iT1} + y_{iT2} - 2j)}(\theta_2 + \theta_7 - \theta_1 - \theta_6)^j\frac{\exp(-v)}{(y_{iT1} - j)!(y_{iT2} - j)!j!}$$

with $u = \theta_3 + \theta_4 + \theta_5 + \theta_8 + \theta_9 + \theta_{10}$ and $v = \theta_1 + \theta_2 + \theta_6 + \theta_7$.

Denote $\theta = (\theta_1, \theta_2, \theta_3, \theta_4, \theta_5, \theta_6, \theta_7, \theta_8, \theta_9, \theta_{10})$. Then, the maximum likelihood estimates (MLE) $\hat{\theta} = (\hat{\theta}_1, \hat{\theta}_2, \hat{\theta}_3, \hat{\theta}_4, \hat{\theta}_5, \hat{\theta}_6, \hat{\theta}_7, \hat{\theta}_8, \hat{\theta}_9, \hat{\theta}_{10})$ of θ can be found by solving the following likelihood equations:

$$\frac{\partial \ln L}{\partial \theta_i} = 0; \quad i = 1, \ldots, 10. \tag{16.19}$$

Thus, the MLEs of the model parameters $\lambda_C, \lambda_{0C}, p_{0C}, p_{1C}, p_{2C}$ for the control and $\lambda_T, \lambda_{0T}, p_{0T}, p_{1T}, p_{2T}$ for the test treatment can be found easily using the invariance principle. Assume that both the control and test treatments have the same sample size, that is, $n_C = n_T = n$. Denote the Fisher information matrix by

$$I(\theta) = E\left[-\frac{\partial^2 \ln L}{\partial \theta \, \partial \theta^T}\right]. \tag{16.20}$$

Let

$$I_1(\theta) = \frac{1}{n}I(\theta), \tag{16.21}$$

then $I_1(\theta)$ is the Fisher information per unit. The inverse of the Fisher information matrix $I(\theta)$ is the asymptotic covariance matrix of the MLE $\hat{\theta}$ of the model parameters θ. Thus, the corresponding diagonal elements of $I^{-1}(\theta)$ would be the asymptotic variances of the MLE $\hat{\theta}$. Statistical inference of the individual parameters can be drawn based on this information.

16.4.2 Comparing the Effects of Control and Test Treatment

Suppose that the interest is to test whether there is any difference between a control treatment and a test treatment. If the responses are bivariate ZIP distributed, the problem is formulated as the testing of the following hypotheses:

$$H_0 : \lambda_C = \lambda_T, \lambda_{0C} = \lambda_{0T}, p_{0C} = p_{0T}, p_{1C} = p_{1T}, p_{2C} = p_{2T} \quad \text{versus} \quad H_A : \text{not all are equal}$$

However, using the transformation given in Equation 16.17, this is equivalent to test

$$H_0 : \theta_1 = \theta_2 = \theta_3 = \theta_4 = \theta_5 = 0 \quad \text{versus} \quad H_A : \text{Some of these } \theta_i \text{ are not equal to 0.} \tag{16.22}$$

Likelihood ratio test method is used to develop the test procedure for these hypotheses. Denote $\theta_0 = (0,0,0,0,0,\theta_6,\theta_7,\theta_8,\theta_9,\theta_{10})$ and $\hat{\theta}_0 = (0,0,0,0,0,\hat{\theta}_{06},\hat{\theta}_{07},\hat{\theta}_{08},\hat{\theta}_{09},\hat{\theta}_{0,10})$, where $\hat{\theta}_{0i}, i = 6,...,10$ are the MLE of $\theta_i, i = 6, ..., 10$ under H_0.

Partition the matrix $i(\theta)$ given in Equation 16.21 into

$$i(\theta) = \begin{pmatrix} i_{aa}(\theta) & i_{ab}(\theta) \\ i_{ba}(\theta) & i_{bb}(\theta) \end{pmatrix}_{10 \times 10}$$

where

$$i_{aa}(\theta) = -\frac{1}{n} E \left(\frac{\partial^2 \ln L}{\partial \theta_i \partial \theta_j} \right)_{i=1,...,5; j=1,...,5},$$

$$i_{bb}(\theta) = -\frac{1}{n} E \left(\frac{\partial^2 \ln L}{\partial \theta_i \partial \theta_j} \right)_{i=6,...,10; j=6,...,10},$$

and

$$i_{ab}(\theta) = -\frac{1}{n} E \left(\frac{\partial^2 \ln L}{\partial \theta_i \partial \theta_j} \right)_{i=1,...,5; j=6,...,10}.$$

The likelihood ratio test statistic for the hypotheses (16.22) can be obtained as

$$W = 2 \left[\ln L(\hat{\theta}) - \ln L(\hat{\theta}_0) \right]. \tag{16.23}$$

It can be verified that the likelihood ratio statistic W asymptotically follows a central χ^2_5 under the null hypothesis H_0 and a noncentral chi-squared distribution with a noncentrality parameter δ under the alternative hypothesis H_A, where

$$\delta^2 = n(\theta_1,\theta_2,\theta_3,\theta_4,\theta_5)\left[i_{aa}(\theta) - i_{ab}(\theta)i_{bb}^{-1}(\theta)i_{ba}(\theta)\right](\theta_1,\theta_2,\theta_3,\theta_4,\theta_5)^T.$$

Thus, the null hypothesis H_0 is rejected at a significance level α if $W > \chi^2_5(\alpha)$, where $\chi^2_5(\alpha)$ is the upper α-quantile of a χ^2_5 distribution. Based on this result, for a given level α, the critical region is given by

$$\left\{(y_{1ij}, y_{2ij}), \quad i = 1, 2, \ldots, n_j; \quad j = T, C, \middle| W > \chi^2_5(\alpha)\right\}$$

and the power function of the test is given by

$$1 - \beta = P\left\{W > \chi^2_5(\alpha)\middle|H_A\right\}. \tag{16.24}$$

16.4.3 Sample Size Calculation

Based on the power function given in Equation 16.24, we can determine the required sample size n so that the resulting sample would achieve a given power $(1 - \beta)$. Using normal approximation to the probability, we have

$$(2+\delta^2)^{\frac{1}{2}}\left[\left(\frac{W-1/3}{2+\delta^2}\right)^{\frac{1}{2}} - \left(1 - \frac{1}{3(2+\delta^2)}\right)^{\frac{1}{2}}\right],$$

or equivalently,

$$\left(W - \frac{1}{3}\right)^{\frac{1}{2}} - \left(\frac{5+3\delta^2}{3}\right)^{\frac{1}{2}}$$

approximately follows a standard normal distribution. Therefore, based on Equation 16.24, we have

$$\left(\chi^2_5(\alpha) - \frac{1}{3}\right)^{\frac{1}{2}} - \left(\frac{5+3\delta^2}{3}\right)^{\frac{1}{2}} = -z(\beta), \tag{16.25}$$

where $z(\beta)$ is the upper β-quantile of a standard normal distribution. Simplification of Equation 16.25 gives a formula of the sample size for achieving a $(1 - \beta)$ power as follows:

$$n \geq C^{-1}\left\{\left[\left(\chi^2_5(\alpha) - \frac{1}{3}\right)^{\frac{1}{2}} + z(\beta)\right]^2 - \frac{5}{3}\right\}, \tag{16.26}$$

where

$$C = (\theta_1, \theta_2, \theta_3, \theta_4, \theta_5) \left[i_{aa}(\theta) - i_{ab}(\theta) i_{bb}^{-1}(\theta) i_{ba}(\theta) \right] (\theta_1, \theta_2, \theta_3, \theta_4, \theta_5)^T.$$

For illustration purpose, using $\alpha = 0.05$, $1 - \beta = 0.8$, and selected combinations of parameter values, the required sample n is listed in Table 16.4. Note that the number of model parameters involved in the study is large even in the bivariate cases. In general, the model parameters of a bivariate ZIP can be classified as the location parameters (λ, λ_0) and the weights (p_0, p_1, p_2). Part A of Table 16.5 compares cases with fixed choices of (λ, λ_0) for different patterns of (p_0, p_1, p_2) for the control and test distributions, while Part B of Table 16.5 considers cases for different choices of $(\lambda_T, \lambda_{0T})$ against $(\lambda_C, \lambda_{0C}) = (2, 4)$ with the same weight pattern.

In general, the results suggest that the difference of λ_C and λ_T plays an important role in determining the required sample size to discriminate the two bivariate ZIP models. When λ_C and λ_T are close, it requires a very large sample size to achieve the nominal power level. Since the results presented in Table 16.5 are based on approximation, a simulation study is conducted to assess the accuracy of these results. For the cases considered in Table 16.5, 1000 trails based on the suggested sample sizes n were simulated. The empirical power, which is computed as the proportion of the trails out of 1000 with $W > \chi_5^2(\alpha)$, was computed. The corresponding results are also presented in Table 16.5. All simulations were done in Excel 2010 using Visual Basic for Applications. To determine the accuracy of the approximated power, a bivariate sample of size n was simulated 1000 times. For each simulated bivariate sample, the set of restricted parameters under the null hypothesis and the set of unrestricted parameters under the alternative hypothesis were estimated by maximum likelihood estimation. The method-of-moments estimates as given in Li et al. (1999) were used as the initial estimates for finding the MLEs. As the likelihood function is very complicated, the generalized reduced gradient nonlinear method was used to obtain the estimates. The method is available in Solver of Excel 2010. All computations were done in Excel 2010 with Visual Basic for Applications.

It should be noted that the simulated powers are higher than the nominal power level 0.8 for cases with sample sizes larger than 20. This suggests that the required sample sizes derived by the approximation results would give the desired power level, albeit in a somewhat conservative manner. However, for cases with sample sizes less than 20, the simulated power is much lower than the nominal value. Thus, these results suggest that the approximation results only work for cases with relatively large sample size.

16.4.4 An Example

To illustrate the test statistic described in the previous subsection, Yuen, Chow, and Tse (2015) considered a clinical trial for the evaluation of a test treatment for patients with addictive drug or substance abuse. Addiction is a chronic, often relapsing, brain disease that causes compulsive drug seeking and use, despite harmful consequences to the addicted individuals and those around him or her. Drug or substance abuse and addiction have negative consequences for individuals and for society such as family disintegration, loss of employment, failure in school, domestic violence, and child abuse.

In this clinical trial, 226 qualified subjects were randomly assigned to receive a test treatment and another 207 subjects were assigned to receive an active control treatment in conjunction with behavioral therapy. The treatment duration was for 6 months. Clinical

TABLE 16.5

Sample Size and Simulated Power for Testing Homogeneity of Two Bivariate ZIPs with $\alpha = 0.05$, $\beta = 0.8$

		Required Sample Size					Simulated Power				
		$*W_{1C}$	W_{2C}	W_{3C}	W_{4C}	W_{5C}	W_{1C}	W_{2C}	W_{3C}	W_{4C}	W_{5C}

Case A: $\lambda_{0C} = 2$ and $\lambda_C = 4$ For $j = T, C$, $W_{1j} = (p_0, p_1, p_2) = (0.1, 0.4, 0.4)$; $W_{2j} = (p_0, p_1, p_2) = (0.2, 0.1, 0.1)$; $W_{3j} = (p_0, p_1, p_2) = (0.25, 0.25, 0.25)$; $W_{4j} = (p_0, p_1, p_2) = (0.4, 0.1, 0.1)$, and $W_{5j} = (p_0, p_1, p_2) = (0.6, 0.1, 0.1)$.

$\lambda_{0T} = 3, \lambda_T = 5$	W_{1T}	152	13	49	14	13	0.875	0.433	0.832	0.490	0.409
	W_{2T}	13	127	37	70	27	0.409	0.860	0.841	0.870	0.831
	W_{3T}	52	38	164	51	39	0.843	0.828	0.874	0.851	0.849
	W_{4T}	14	75	54	173	83	0.427	0.869	0.870	0.852	0.864
	W_{5T}	14	28	42	90	282	0.483	0.816	0.845	0.863	0.888
$\lambda_{0T} = 4, \lambda_T = 5$	W_{1T}	141	13	48	14	13	0.858	0.476	0.858	0.472	0.441
	W_{2T}	13	95	35	61	26	0.407	0.858	0.852	0.849	0.797
	W_{3T}	50	36	138	48	38	0.849	0.814	0.846	0.817	0.843
	W_{4T}	14	66	49	137	76	0.441	0.888	0.853	0.865	0.850
	W_{5T}	13	28	41	83	229	0.389	0.825	0.841	0.878	0.867

	Required Sample Size					Simulated Power				
	$*I$	II	III	IV	V	I	II	III	IV	V

Case B: $\lambda_{0C} = 2$ and $\lambda_C = 4$

$\lambda_{0T} = 2, \lambda_T = 5$	154	144	173	194	312	0.860	0.864	0.861	0.880	0.895
$\lambda_{0T} = 3, \lambda_T = 5$	151	127	163	177	289	0.871	0.893	0.877	0.868	0.883
$\lambda_{0T} = 3, \lambda_T = 6$	42	39	47	54	82	0.856	0.855	0.859	0.870	0.851
$\lambda_{0T} = 4, \lambda_T = 5$	143	96	140	137	226	0.866	0.845	0.877	0.886	0.855
$\lambda_{0T} = 4, \lambda_T = 6$	42	36	45	49	77	0.848	0.867	0.850	0.858	0.863

$*$I: $(p_{0T}, p_{1T}, p_{2T}) = (p_{0C}, p_{1C}, p_{2C}) = (0.1, 0.4, 0.4)$
II: $(p_{0T}, p_{1T}, p_{2T}) = (p_{0C}, p_{1C}, p_{2C}) = (0.2, 0.1, 0.1)$
III: $(p_{0T}, p_{1T}, p_{2T}) = (p_{0C}, p_{1C}, p_{2C}) = (0.25, 0.25, 0.25)$
IV: $(p_{0T}, p_{1T}, p_{2T}) = (p_{0C}, p_{1C}, p_{2C}) = (0.4, 0.1, 0.1)$
V: $(p_{0T}, p_{1T}, p_{2T}) = (p_{0C}, p_{1C}, p_{2C}) = (0.6, 0.1, 0.1)$

assessment such as measurements for social behavior and safety implications of drug abuse and addiction was done at three months and six months (endpoint visit). In addition, individual patient diary was used to capture the days of illicit drug use at three months and six months. Bivariate ZIP models were used to model the data observed based on the two treatments, namely, control and test treatments. Assessment of the therapy effect is done by comparison of the two bivariate ZIP models. In particular, following the model formulation outlined in Section 16.4.1, the hypotheses of interests are

$$H_0 : \lambda_C = \lambda_T, \lambda_{0C} = \lambda_{0T}, p_{0C} = p_{0T}, p_{1C} = p_{1T},$$
$$p_{2C} = p_{2T} \quad \text{versus} \quad H_A : \text{not all are equal,}$$

which is equivalent to

$$H_0 : \theta_1 = \theta_2 = \theta_3 = \theta_4 = \theta_5 = 0 \quad \text{versus}$$
$$H_A : \text{Some of these } \theta_i \text{ are not equal to 0}$$

Based on the data, the unrestricted MLEs of the parameters are

$$\hat{\theta} = (\hat{\theta}_1, \hat{\theta}_2, \hat{\theta}_3, \hat{\theta}_4, \hat{\theta}_5, \hat{\theta}_6, \hat{\theta}_7, \hat{\theta}_8, \hat{\theta}_9, \hat{\theta}_{10})$$
$$= (-0.478, -1.164, 0.000, -0.101, 0.000, 2.980, 5.073, 0.317, 0.204, 0.131).$$

Consequently, the MLEs of the control treatment model parameters $(\lambda_{0C}, \lambda_C, p_{0C}, p_{1C}, p_{2C})$ are (2.980, 5.073, 0.317, 0.204, 0.131) and the MLEs of the test treatment model parameters $(\lambda_{0T}, \lambda_T, p_{0T}, p_{1T}, p_{2T})$ are (2.502, 3.909, 0.317, 0.103, 0.131). Furthermore, under H_0, the MLEs are

$$\hat{\theta}_0 = (0, 0, 0, 0, 0, \hat{\theta}_{06}, \hat{\theta}_{07}, \hat{\theta}_{08}, \hat{\theta}_{09}, \hat{\theta}_{0,10})$$
$$= (0, 0, 0, 0, 0, 2.647, 4.445, 0.318, 0.151, 0.131).$$

The test statistic is then given by $W = 2 [-1320.56 - (-1332.98)] = 24.84$ which is larger than $\chi_5^2(0.05) = 11.07$. Therefore, we reject H_0 at a 5% level of significance. The results indicate that there is a significant difference between the test treatment and the control treatment.

16.5 Concluding Remarks

Tse et al. (2009) studied the problem of comparing two ZIP populations and derived likelihood ratio tests to test the homogeneity of the two populations under different scenarios. In the previous sections, although we only focused on testing hypotheses of equality, similar idea can be applied for testing noninferiority/equivalence and superiority hypotheses.

It should be noted that the use of ZIP is motivated by the fact that the counts could include a large proportion of zeros such that they can be modeled by a Poisson model, that is, the proportion k_1/n_1 is unusually large with n_1 being the sample size and k_1 being the number of zero counts. For a given sample of size n, the estimated distribution F_n is given by

$$F_n(x) = \hat{\theta}_n I(x = 0) + (1 - \hat{\theta}_n) \frac{\exp(-\hat{\lambda}_n)\hat{\lambda}_n^x}{x!},$$

where $\hat{\theta}_n$ and $\hat{\lambda}_n$ are the MLE of θ and λ based on the observed data. Suppose that the ratio of the number of zero counts k_1/n_1 tends to 1, then it can be shown that F_n converge to a degenerate distribution with point mass 1 at $x = 0$. On the other hand, if k_1/n_1 tends to 0, then $\hat{\theta}_n$ converges to a value θ and $\theta < 0$. Thus, the probability of having zero count is smaller than that given by a Poisson model. The corresponding distribution is the so-called zero-deflated Poisson model. Some discussions of this model can be found in Johnson, Kotz, and Kemp (1992). In general, the results developed in this study can be applied to the general situation that the probability of zero is being modified. Whether it is zero-inflated or zero-deflated is pending on the pattern of the data.

Yuen, Chow, and Tse (2015) considered the problem of testing treatment difference in the occurrence of a study endpoint in a randomized parallel-group comparative clinical trial with repeated responses. These occurrence counts usually could include a large proportion of zeros and are usually modeled by a zero-inflated Poisson (ZIP) model to give an accurate assessment of the treatment effect. However, there is a need to extend a univariate ZIP model to the multivariate case since repeated responses are observed in the study. Following the idea of Li et al. (1999), Yuen, Chow, and Tse (2015) proposed a bivariate ZIP model to compare the effect of two treatments in a two-armed clinical trial. In particular, likelihood ratio test is derived to test the homogeneity of the two bivariate zero-inflated Poisson populations. Yuen, Chow, and Tse (2015) also derived an approximate formula for sample size determination to achieve a desired power. It should be noted that although the discussion of the example and the results given are based on only two repeated responses, the model can be readily extended to more than two repeated responses. However, the number of parameters would inevitably increase, which would require a relatively large data set and impose considerable amount of difficulty on the estimation of the model parameters. In any event, the proposed model provides a viable tool for practitioners to assess whether a new treatment is more effective than another treatment when the responses are repeated counts with usually large proportion of zeros.

Appendix

Define $\theta = (\theta_1, \theta_2, \lambda_1, \lambda_2)$. In particular, $L = \Pi_{j=1}^{2}\Pi_{i=1}^{n}P(x_{ij}\,|\,\theta_j,\lambda_j)$ is the likelihood function and the log-likelihood function is $l(\theta) = \log L = l_1 + l_2$, where for $j = 1, 2$;

$$l_j = \sum_{i=1}^{n} \log f(x_{ij}|\theta_j,\lambda_j)$$
$$= K_1 \log\left(\theta_j + (1-\theta_j)e^{-\lambda_j}\right)$$
$$+ \sum_{i=K_j+1}^{n}\left[\log(1-\theta_j) - \lambda_j + x_{ij}\log\lambda_j - \log(x_{ij}!)\right].$$

Note that $\alpha_1 = \lambda_1 - \lambda_2$, $\alpha_2 = \lambda_2$, $\beta_1 = \theta_1 - \theta_2$, and $\beta_2 = \theta_2$. Taking second-order partial derivative of the $l(\theta)$ with respect to the parameters, we have

$$\frac{\partial^2 l(\theta)}{\partial \alpha_1^2} = \frac{\partial^2 l(\theta)}{\partial \alpha_1 \partial \alpha_1} = -\frac{1}{\lambda_1^2}\sum_{i=1}^{n} x_{i1} + \frac{K_1\theta_1(1-\theta_1)e^{-\lambda_1}}{\left[\theta_1 + (1-\theta_1)e^{-\lambda_1}\right]^2}, \tag{A1}$$

$$\frac{\partial^2 l(\theta)}{\partial \beta_1^2} = \frac{\partial^2 l(\theta)}{\partial \beta_1 \partial \beta_2} = -\frac{K_1(1-e^{-\lambda_1})^2}{\left[\theta_1 + (1-\theta_1)e^{-\lambda_1}\right]^2} - \frac{n-K_1}{(1-\theta_1)^2}, \tag{A2}$$

$$\frac{\partial^2 l(\theta)}{\partial \alpha_1 \partial \beta_1} = \frac{\partial^2 l(\theta)}{\partial \alpha_1 \partial \beta_2} = \frac{\partial^2 l(\theta)}{\partial \alpha_2 \partial \beta_1} = \frac{K_1 e^{-\lambda_1}}{\left[\theta_1 + (1-\theta_1)e^{-\lambda_1}\right]^2}, \tag{A3}$$

$$\frac{\partial^2 l(\theta)}{\partial \alpha_2^2} = \frac{\partial^2 l(\theta)}{\partial \alpha_1^2} - \frac{1}{\lambda_2^2} \sum_{i=1}^{n} x_{i2} + \frac{K_2 \theta_2 (1-\theta_2) e^{-\lambda_2}}{\left[\theta_2 + (1-\theta_2) e^{-\lambda_2}\right]^2} \tag{A4}$$

$$\frac{\partial^2 l(\theta)}{\partial \beta_2^2} = \frac{\partial^2 l(\theta)}{\partial \beta_1^2} - \frac{K_2 (1-e^{-\lambda_2})^2}{\left[\theta_2 + (1-\theta_2) e^{-\lambda_2}\right]^2} - \frac{n-K_2}{(1-\theta_2)^2}, \tag{A5}$$

and

$$\frac{\partial^2 l(\theta)}{\partial \alpha_2 \partial \beta_2} = \frac{\partial^2 l(\theta)}{\partial \alpha_1 \partial \beta_1} + \frac{K_2 e^{-\lambda_2}}{\left[\theta_2 + (1-\theta_2) e^{-\lambda_2}\right]^2}. \tag{A6}$$

Note that $E(X_{ij}) = (1-\theta_j)\lambda_j$ and $E(K_j) = n[\theta_j + (1-\theta_j)e^{-\lambda_j}]$ for $j = 1, 2$. It can be shown that the expected values of the expressions A1 through A6 are

$$E\left[\frac{\partial^2 l(\theta)}{\partial \alpha_1^2}\right] = E\left[\frac{\partial^2 l(\theta)}{\partial \alpha_1 \partial \alpha_2}\right] = -\frac{n(1-\theta_1)}{\lambda_1} + \frac{n\theta_1 (1-\theta_1) e^{-\lambda_1}}{\theta_1 + (1-\theta_1) e^{-\lambda_1}},$$

$$E\left[\frac{\partial^2 l(\theta)}{\partial \beta_1^2}\right] = E\left[\frac{\partial^2 l(\theta)}{\partial \beta_1 \partial \beta_2}\right] = -\frac{n(1-e^{-\lambda_1})}{(1-\theta_1)\left[\theta_1 + (1-\theta_1) e^{-\lambda_1}\right]},$$

$$E\left[\frac{\partial^2 l(\theta)}{\partial \alpha_1 \partial \beta_1}\right] = E\left[\frac{\partial^2 l(\theta)}{\partial \alpha_1 \partial \beta_2}\right] = E\left[\frac{\partial^2 l(\theta)}{\partial \beta_1 \partial \alpha_2}\right] = \frac{n e^{-\lambda_1}}{\theta_1 + (1-\theta_1) e^{-\lambda_1}},$$

$$E\left[\frac{\partial^2 l(\theta)}{\partial \alpha_2^2}\right] = E\left[\frac{\partial^2 l(\theta)}{\partial \alpha_1^2}\right] - \frac{n(1-\theta_2)}{\lambda_2} + \frac{n\theta_2 (1-\theta_2) e^{-\lambda_2}}{\theta_2 + (1-\theta_2) e^{-\lambda_2}},$$

$$E\left[\frac{\partial^2 l(\theta)}{\partial \beta_2^2}\right] = E\left[\frac{\partial^2 l(\theta)}{\partial \beta_1^2}\right] - \frac{n(1-e^{-\lambda_2})}{(1-\theta_2)\left[\theta_2 + (1-\theta_2) e^{-\lambda_2}\right]},$$

and

$$E\left[\frac{\partial^2 l(\theta)}{\partial \alpha_2 \partial \beta_2}\right] = E\left[\frac{\partial^2 l(\theta)}{\partial \alpha_1 \partial \beta_1}\right] + \frac{n e^{-\lambda_2}}{\theta_2 + (1-\theta_2) e^{-\lambda_2}}.$$

17

Sample Size for Clinical Trials with Extremely Low Incidence Rate

As indicated by Himabindu (2012), one of the major challenges when conducting a clinical trial for the investigation of treatments for rare disease is probably the selection of study endpoint and power calculation for sample size based on the selected study endpoint. In some clinical trials, the incidence rate of certain events such as adverse events, immune responses, and infections are commonly considered clinical study endpoints for the evaluation of safety and efficacy of the test treatment under investigation (O'Neill, 1988). In epidemiological/clinical studies, the incidence rate expresses the number of new cases of disease that occur in a defined population of disease-free individuals. The observed incidence rate provides a direct estimate of the probability or risk of illness. Boyle and Parkin (1991) introduced the methods required for using incidence rates in comparative studies. For example, one may compare incidence rates from different time periods or from different geographical areas in the studies. On the other hand, one may compare incidence rates (e.g., adverse events, infections postsurgery, or immune responses for immunogenicity) in clinical trials (FDA, 2002; Chow and Liu, 2004). In practice, however, there are only few references available in the literature regarding sample size required for achieving certain statistical inference (e.g., in terms of power calculation or precision analysis) for the studies with extremely low incidence rates (Chow and Chiu, 2013).

In clinical trials, a prestudy power analysis for sample size calculation is usually performed for determining an appropriate sample size (usually minimum sample size required) for achieving a desired power (e.g., 80% or 90%) for detecting a clinically meaningful difference at a prespecified level of significance (e.g., 1% or 5%) (see, e.g., Chow and Liu, 2004; Chow, Shao, and Wang, 2008). In practice, a much larger sample size is expected for detecting a relatively smaller difference, especially for clinical trials with extremely low incidence rate. For example, the incidence rate for hemoglobin A_{1C} (H_bA_{1C}) in diabetic studies and the immune responses for immunogenicity in biosimilar studies and/or vaccine clinical trials are usually extremely low. In the case of clinical trials with extremely low incidence rate (the primary endpoint of the study), the obtained sample size required for achieving a desired power and/or precision is often too large to be of practical consideration. Thus, alternative methods for sample size determination are necessarily developed.

In Section 7.1, an example concerning clinical trials with extremely low incidence rate is described to illustrate the issue of infeasible sample size requirement as a result of power calculation. Section 17.2 outlines classical methods for sample size determination. Also included in this section is the application of a Bayesian approach with a noninformative uniform prior. An alternative method for sample size determination for clinical studies with extremely low incidence rate and a statistical procedure for data safety monitoring based on probability statement during the conduct of a clinical trial with extremely low incidence rate proposed by Chow and Chiu (2013) are outlined in Sections 17.3 and 17.4, respectively. Section 17.5 gives some concluding remarks.

17.1 Clinical Studies with Extremely Low Incidence Rate

A typical example for clinical studies with extremely low incidence rate is clinical trials conducted for preventive HIV vaccine development. In 2008, it was estimated that the total number of people living with HIV was 33.4 million, with 97% living in low- and mid-income countries (UNAIDS, 2009). As a result, the development of a safe and efficacious preventive HIV vaccine had become the top priority in global health for the control of the HIV-1 in the long term. In their excellent review article, Kim et al. (2010) indicated that the immune response elicited by a successful vaccine will likely require both antibodies and T cells that recognize, neutralize, and/or inactivate diverse strains of HIV and that reach the site of infection before the infection becomes irreversibly established (see also Haynes and Shattock, 2008). Basically, the development of HIV vaccine focuses on evaluating vaccines capable of reducing viral replication after infection as the control of viral replication could prevent the transmission of HIV in heterosexual population (Excler, Tomaras, and Russell, 2013), and/or conceivably slow the rate of disease progression as suggested by nonhuman primate (NHP) challenge studies (see, e.g., Mattapallil et al., 2006; Gupta et al., 2007; Watkins et al., 2008).

The goal of a preventive HIV vaccine is to induce cell-mediated immune (CMI) responses and subsequently to reduce the plasma viral load at set point and preserve memory CD4+ lymphocytes. As a result, clinical efforts have mainly focused on CMI-inducing vaccines such as DNA and vectors alone or in prime-boost regimens (Belyakov et al., 2008; Esteban, 2009). In a recent Thai efficacy trial (RV144), the first evidence that HIV-1 vaccine protection against HIV-1 acquisition could be achieved was revealed. The results of RV144 indicated that patients with the lowest risk (yearly incidence of 0.23/100 person years) had an apparent efficacy of 40%, while those with the highest risk (incidence of 0.36/100 person years) had an efficacy of 3.7%. This finding suggested that clinical meaningful difference in vaccine efficacy can be detected by means of the difference in the incidence of risk rate. In addition, the vaccine efficacy appeared to decrease with time (e.g., at 12 months, the vaccine efficacy was about 60% and fell to 29% by 42 months). As a result, at a specific time point, sample size required for achieving a desired vaccine efficacy can be obtained by detecting a clinically meaningful difference in the incidence of risk rate at baseline.

17.2 Classical Methods for Sample Size Determination

In clinical trials, a prestudy power analysis for sample size calculation is often performed to ensure that an intended clinical trial will achieve a desired power for correctly detecting a clinically meaningful treatment effect at a prespecified level of significance. For clinical trials with extremely low incidence rate, sample size calculation based on power analysis may not be feasible. Alternatively, it is suggested that sample size calculation be done based on precision analysis. In this section, prestudy power analysis and precision analysis for sample size calculation are briefly described.

17.2.1 Power Analysis

Under a two-sample parallel group design, let x_{ij} be a binary response (e.g., adverse events immune responses, or infection rate postsurgery) from the jth subject in the ith group, $j = 1, \ldots, n, i = T$ (test), R (reference or control). Then,

$$\hat{p}_i = \frac{1}{n}\sum_{i=1}^{n} x_{ij}$$

are the infection rates for the test group and the control group, respectively. Let $\delta = p_R - p_T$ be the difference in response rate between the test group and the control group. For simplicity, consider the following hypotheses for testing equality between p_R and p_T:

H_0: $\delta = 0$ versus H_a: $\delta \neq 0$

Thus, under the alternative hypothesis, the power of $1 - \beta$ can be approximately obtained by the following equation (see, e.g., Chow, Shao, and Wang, 2008):

$$\Phi\left(\frac{|\hat{\delta}|}{\sqrt{\dfrac{\hat{p}_R(1-\hat{p}_R)+\hat{p}_T(1-\hat{p}_T)}{n}}} - Z_{1-(\alpha/2)}\right),$$

where Φ is the cumulative standard normal distribution function and $Z_{1-\alpha/2}$ is the upper $\alpha/2$th quantile of the standard normal distribution. As a result, the sample size needed for achieving a desired power of $1 - \beta$ at the α level of significance can be obtained by the following equation:

$$n_{power} = \frac{(Z_{1-\alpha/2} + Z_{\beta})^2}{\hat{\delta}^2}\hat{\sigma}^2 \tag{17.1}$$

where $\hat{\delta} = \hat{p}_R - \hat{p}_T$, $\hat{\sigma}^2 = \hat{p}_R(1-\hat{p}_R) + \hat{p}_T(1-\hat{p}_T)$.

17.2.2 Precision Analysis

On the other hand, the $(1 - \alpha) \times 100\%$ confidence interval (CI) for $\delta = p_R - p_T$ based on large sample normal approximation is given by

$$\hat{\delta} \pm Z_{1-\alpha/2}\frac{\hat{\sigma}}{\sqrt{n}},$$

where $\hat{\delta} = \hat{p}_R - \hat{p}_T$, $\hat{\sigma} = \sqrt{\hat{\sigma}_R^2 + \hat{\sigma}_T^2}$, $\hat{\sigma}_R^2 = \hat{p}_R(1-\hat{p}_R)$, $\hat{\sigma}_T^2 = \hat{p}_T(1-\hat{p}_T)$, and $Z_{1-\alpha/2}$ is the upper $\alpha/2$th quantile of the standard normal distribution.

Denote half of the width of the confidence interval by $w = Z_{1-\alpha/2}\hat{\sigma}$, which is usually referred to as the *maximum error margin allowed* for a given sample size n. In practice, the maximum error margin allowed represents the precision that one would expect for the selected sample size. The precision analysis for sample size determination is to consider the maximum error margin allowed. In other words, we are confident that the true difference $\delta = p_R - p_T$ would fall within the margin of $w = Z_{1-\alpha/2}\hat{\sigma}$ for a given sample size of n. Thus, the sample size required for achieving the desired precision can be chosen as

$$n_{precision} = \frac{Z_{1-\alpha/2}^2\hat{\sigma}^2}{w^2}, \tag{17.2}$$

where $\hat{\sigma}^2 = \hat{p}_R(1-\hat{p}_R) + \hat{p}_T(1-\hat{p}_T)$.

This approach, based on the interest of type I error, is only to specify precision while estimating true δ for selecting n.

By formulas (17.1) and (17.2), we can also get the relationship between the sample size based on power analysis and precision analysis:

$$R = \frac{n_{power}}{n_{precision}} = \left(1 + \frac{Z_\beta}{Z_{1-\alpha/2}}\right)^2 \frac{w^2}{\delta^2}.$$

Thus, R is proportional to $1/\delta^2$ or w^2.

Under a fixed power and significant level, the sample size based on power analysis is much larger than the sample size based on precision analysis with extremely low infection rate difference or large error margin allowed.

Without loss of generality, $(1 + (Z_\beta/(Z_{1-\alpha/2})))$ is always much larger than 1 (e.g., power $= 80\%$, significant level $= 5\%$ then, $(1 + (Z_\beta/(Z_{1-\alpha/2})))^2 = 2.04$). It means that if $(w/\delta) > 0.7$, the proposed sample size based on power analysis will be larger than the one based on precision analysis. The sample size determined by power analysis will be large when the difference between the test group and the control group is extremely small. Table 17.1 shows the comparison of sample sizes determined by power analysis and precision analysis. The power is fixed at 80% and the significance level is 5%. When $(\hat{p}_R, \hat{p}_R) = (2\%, 1\%)$, compare the sample size calculated by two methods. The sample sizes determined by precision analysis are much smaller than the sample sizes determined by power analysis.

17.2.3 Remarks

In previous sections, formulas for sample size calculation were derived based on the concept of frequentist. Information obtained from previous studies or small pilot

TABLE 17.1

Sample Size Based on Power Analysis and Precision Analysis

ω		δ		n_{power}	$n_{precision}$	R
$0.08\hat{\sigma}$	1.37%	$0.04\hat{\sigma}$	0.69%	4906	600	8.2
	1.37%	$0.05\hat{\sigma}$	0.86%	3140	600	5.2
	1.37%	$0.06\hat{\sigma}$	1.03%	2180	600	3.6
	1.37%	$0.07\hat{\sigma}$	1.20%	1602	600	2.7
	1.37%	$0.08\hat{\sigma}$	1.37%	1226	600	2.0
$0.10\hat{\sigma}$	1.72%	$0.04\hat{\sigma}$	0.69%	4906	384	12.8
	1.72%	$0.05\hat{\sigma}$	0.86%	3140	384	8.2
	1.72%	$0.06\hat{\sigma}$	1.03%	2180	384	5.7
	1.72%	$0.07\hat{\sigma}$	1.20%	1602	384	4.2
	1.72%	$0.08\hat{\sigma}$	1.37%	1226	384	3.2
$0.12\hat{\sigma}$	2.06%	$0.04\hat{\sigma}$	0.69%	4906	267	18.4
	2.06%	$0.05\hat{\sigma}$	0.86%	3140	267	11.8
	2.06%	$0.06\hat{\sigma}$	1.03%	2180	267	8.2
	2.06%	$0.07\hat{\sigma}$	1.20%	1602	267	6.0
	2.06%	$0.08\hat{\sigma}$	1.37%	1226	267	4.6

studies is often used to estimate the parameters required for sample size calculation. In practice, sample size required for achieving the desired precision may be further improved by taking Bayesian approach into consideration. For the purpose of illustration, sample size calculation based on precision analysis in conjunction with Bayesian approach with a noninformative uniform prior is performed based on the following assumptions:

1. Since the primary endpoint x_{ij} are binary response, it follows Bernoulli distribution.
2. $p_R > p_T$.
3. Let $\delta_i | \theta, \sigma^2 \sim N(\theta, \sigma^2)$ and $\sigma^2 \sim Uniform(0, 1)$.

We would like to estimate the likelihood of the data, which is normal distribution with a known mean θ and unknown variance σ^2.
Thus,

$$f(\delta_i | \theta, \sigma^2) = \frac{1}{\sqrt{2\pi\sigma^2}} exp\left\{\frac{-(\delta_i - \theta)^2}{2\sigma^2}\right\} \quad \text{and} \quad \pi(\sigma^2) \equiv 1,$$

$$L(\delta_i | \theta, \sigma^2) = \prod_{i=1}^{n} f(\delta_i | \theta, \sigma^2) = [2\pi\sigma^2]^{-n/2} exp\left\{\frac{-\sum_{i=1}^{n}(\delta_i - \theta)^2}{2\sigma^2}\right\}$$

As a result, the posterior distribution can be obtained as follows:

$$\pi(\sigma^2 | \theta, \delta_i) \propto (\sigma^2)^{-n/2} exp\left\{\frac{-\sum_{i=1}^{n}(\delta_i - \theta)^2}{2\sigma^2}\right\},$$

$$\pi(\sigma^2 | \theta, \delta_i) \sim inverse\text{-}gamma\left(\alpha = \frac{n}{2} - 1, \beta = \frac{-\sum_{i=1}^{n}(\delta_i - \theta)^2}{2}\right)$$

As a result, the sample size required for achieving a desired precision can be obtained by following iterative steps:

Step 1: Start with an initial guess for n_0.
Step 2: Generate σ^2 from inverse-gamma

$$\left(\alpha = \frac{n}{2} - 1, \beta = \frac{-\sum_{i=1}^{n}(\delta_i - \theta)^2}{2}\right).$$

Step 3: Calculate the required sample size n with σ^2 (generated from step 2) by formula (17.2).

Step 4: If $n \neq n_0$, then let $n_0 = n$ and repeat steps 1 through 4. If $n = n_0$, then let $n_b = n$.

The sample size based on Bayesian approach n_b can be obtained, which will converge in probability to n after several iterations.

17.3 Chow and Chiu's Procedure for Sample Size Estimation

17.3.1 Basic Idea of Chow and Chiu's Procedure

Chow and Chiu (2013) proposed a procedure based on precision analysis for sample size calculation for clinical studies with extremely low incidence rate. The idea of Chow and Chiu's procedure is to justify a selected sample size based on a precision analysis and a sensitivity analysis in conjunction with a power analysis. As indicated earlier, for clinical trials with extremely low incidence rates, the sample size required for achieving a desired power for detecting a small difference may not be feasible. Sample size justification based on a small difference (absolute change) may not be of practical interest. Alternatively, sample size justification based on relative change is often considered. For example, suppose the postsurgery infection rate for the control group is 2% and the incidence rate for the test group is 1%. The absolute change in infection rate is 1% = 2%–1%. This small difference may not be of any clinical or practical meaning. However, if we consider relative change, then the difference becomes appealing. In other words, there is 50% relative reduction in infection rate from 2% to 1%. In this section, we propose a procedure based on precision analysis for selecting an appropriate sample size for clinical trials with an extremely low incidence rate.

Suppose p_R and p_T are infection rates for the control group and the test group, respectively. Define relative improvement (or % improvement) as follows:

$$\%\text{improvement} = \frac{p_R - p_T}{p_R} \times 100\%$$

Note that in case when, $p_R < p_T$ the above measurement becomes a measure of % worsening. Based on the precision analysis and considering the relative improvement at the same time, the following step-by-step procedure for choosing an appropriate sample size is recommended:

Step 1: Determine the maximum error margin allowed. Choose a maximum error margin that one feels comfortable with. In other words, we are 95% confident that the true difference in incidence rate between the two groups is within the maximum error margin.

Step 2: Select highest % improvement. Since it is expected that the relative improvement in infection rate is somewhere within the range, we may choose the combination of incidence rates that gives the highest % improvement.

Step 3: Select sample size that reaches statistical significance. We then select the sample size for achieving statistical significance (i.e., those confidence intervals that do not cover 0). In other words, the observed difference is not by chance above and it is reproducible if we shall repeat the study under similar experimental conditions.

Note that with a selected sample size (based on the above procedure), we can also evaluate the corresponding power. If one feels uncomfortable, one may increase the sample size. In practice, it is suggested that the selected sample size should have at least 50% power at a prespecified level of significance.

17.3.2 Sensitivity Analysis

Since $n_{precision}$ in Equation 17.2 is very sensitive to a small change in p_R and p_T, the following sensitivity analysis is studied to evaluate the impact of small deviations from the true incidence rates. The true infection rate for the reference and test groups has a small shift, then,

$$p'_R = p_R + \varepsilon_R \quad \text{and} \quad p'_T = p_T + \varepsilon_T.$$

Thus, the sample size required can be chosen as

$$n_s = \frac{Z^2_{1-\alpha/2}\hat{\sigma}'^2}{w^2}, \tag{17.3}$$

where

$$\hat{\sigma}' = \sqrt{\sigma'^2_R + \sigma'^2_T}, \quad \sigma'^2_R = (\hat{p}_R + \varepsilon_R)(1 - \hat{p}_R - \varepsilon_R), \quad \sigma'^2_T = (\hat{p}_T + \varepsilon_T)(1 - \hat{p}_T - \varepsilon_T),$$

and $Z_{1-\alpha/2}$ is the upper $(\alpha/2)$th quantile of the standard normal distribution.

If we let the shift be the same for the reference and test groups, $\varepsilon_R = \varepsilon_T = \varepsilon$. By formulas (17.2) and (17.3), we can also get the relationship between the sample size based on precision analysis and the sample size adjusted by sensitivity analysis.

$$\frac{n_s}{n_{precision}} = \frac{(p_R + \varepsilon)(1 - p_R - \varepsilon) + (p_T + \varepsilon)(1 - p_T - \varepsilon)}{p_R(1 - p_R) + p_T(1 - p_T)}$$

$$= \frac{p_R + p_T - p_R^2 - p_T^2 - 2(p_R + p_T)\varepsilon + 2\varepsilon - 2\varepsilon^2}{p_R + p_T - p_R^2 - p_T^2}$$

Obviously, the ratio above is independent of $Z_{1-\alpha/2}$ and w. It becomes a ratio of variances before and after shift. Moreover,

$$\frac{n_s}{n_{precision}} < 1 \quad \text{if} \quad \varepsilon > 0,$$

$$\frac{n_s}{n_{precision}} > 1 \quad \text{if} \quad \varepsilon < 0,$$

and

$$\frac{n_s}{n_{precision}} = 1 \quad \text{if} \quad \varepsilon = 0.$$

It means that when $\varepsilon > 0$, the sample size adjusted by sensitivity analysis n_s will be smaller than the one $n_{precision}$ proposed by precision analysis. On the contrary, n_s will be larger than $n_{precision}$ if $\varepsilon < 0$.

17.3.3 An Example

A pharmaceutical company is conducting a clinical trial for developing a preventive vaccine for HIV. The incidence rate for immune responses is extremely low, which ranges from 1.7% to 2.1% with a mean incidence rate of 1.9%. The sponsor expects the incidence rate for the vaccine candidate (test) is about 1.0% and targets for an at least 50% improvement in the incidence rate postvaccination. Based on prestudy power analysis for sample size calculation, a total sample size of 6532 (3266 per group) is required for achieving a 90% power for detecting a difference in incidence rate of 0.95% if such a difference truly exists at the 5% level of significance, assuming that the true incidence rate for the control group is 1.9%. With this huge sample size, not only that the sponsor cannot afford to support the study, it may not be of any practical use. Alternatively, Chow and Chiu (2013) suggested the following steps for choosing an appropriate sample size for the proposed vaccine clinical trial using Tables 17.2 (95% CIs for $p_R = 1.90\%$) and 17.3 (95% CIs for $p_R = 2.0\%$):

Step 1: Assuming that the true incidence rate for the control group is 1.95%, we expect there is a 50% relative reduction for the test group. In other words, the true incidence rate for the test group is 0.95%. Now suppose the sponsor is willing to tolerate a 0.5% error margin. Thus, we choose a maximum error margin allowed to be 0.05%.

Step 2: We then use Table 17.2 to select the combination of (p_R, p_T) with the highest possible % improvement. Table 17.2 suggests the second column with 53% relative improvement be considered.

Step 3: We then select the sample size that reaches statistical significance. As it can be seen from Table 17.2, $n = 1100$ per group will reach statistical significance (i.e., the observed difference is *not* by chance alone and it is reproducible).

Thus, total sample size required for the proposed vaccine trial for achieving the desired precision (i.e., maximum error margin allowed) and the relative improvement of 53% is 2200 (1100 per group) assuming that the true incidence rate for the control group is 1.9%. With the selected sample size of $N = 2n = 2200$, the corresponding power for correctly detecting a difference of $\delta = p_R - p_T = 1.9\% - 0.9\% = 1.0\%$ is 53.37%. Note that the selected sample size does not account for possible dropout of the proposed study.

17.4 Data Safety Monitoring Procedure

For vaccine clinical trials with extremely low incidence rate, it will take a large sample to observe a few responses. The time and cost are a great concern to the sponsor of the trial.

TABLE 17.2

95% CIs for $p_R = 1.90\%$

p_T	0.85%		0.90%		0.95%		1.00%		1.05%	
% improvement	55%		53%		50%		47%		45%	
	Lower	Upper	Lower	Upper	Lower	Upper	Lower	Upper	Lower	Upper
n	Power		Power		Power		Power		Power	
200	−1.23%	3.33%	−1.30%	3.30%	−1.37%	3.27%	−1.44%	3.24%	−1.51%	3.21%
	14.52%		13.39%		12.35%		11.38%		10.48%	
300	−0.81%	2.91%	−0.88%	2.88%	−0.95%	2.85%	−1.01%	2.81%	−1.08%	2.78%
	19.64%		17.97%		16.42%		14.98%		13.66%	
400	−0.56%	2.66%	−0.63%	2.63%	−0.69%	2.59%	−0.76%	2.56%	−0.82%	2.52%
	24.71%		22.51%		20.45%		18.55%		16.80%	
500	−0.39%	2.49%	−0.46%	2.46%	−0.52%	2.42%	−0.58%	2.38%	−0.64%	2.34%
	29.71%		26.99%		24.46%		22.10%		19.92%	
600	−0.27%	2.37%	−0.33%	2.33%	−0.39%	2.29%	−0.45%	2.25%	−0.51%	2.21%
	34.58%		31.40%		28.42%		25.62%		23.03%	
700	−0.17%	2.27%	−0.23%	2.23%	−0.29%	2.19%	−0.35%	2.15%	−0.41%	2.11%
	39.30%		35.71%		32.30%		29.10%		26.11%	
800	−0.09%	2.19%	−0.15%	2.15%	−0.21%	2.11%	−0.27%	2.07%	−0.33%	2.03%
	43.85%		39.89%		36.11%		32.52%		29.15%	
900	−0.02%	2.12%	−0.08%	2.08%	−0.14%	2.04%	−0.20%	2.00%	−0.26%	1.96%
	48.19%		43.93%		39.81%		35.88%		32.16%	
1000	0.03%	2.07%	−0.03%	2.03%	−0.09%	1.99%	−0.15%	1.95%	−0.21%	1.91%
	52.32%		47.81%		43.40%		39.16%		35.11%	
1100	0.08%	2.02%	0.02%	1.98%	−0.04%	1.94%	−0.10%	1.90%	−0.16%	1.86%
	56.23%		51.51%		46.87%		42.35%		38.01%	
1200	0.12%	1.98%	0.06%	1.94%	0.00%	1.90%	−0.06%	1.86%	−0.11%	1.81%
	59.91%		55.04%		50.20%		45.44%		40.84%	
1300	0.16%	1.94%	0.10%	1.90%	0.04%	1.86%	−0.02%	1.82%	−0.08%	1.78%
	63.35%		58.39%		53.40%		48.44%		43.60%	
1400	0.19%	1.91%	0.13%	1.87%	0.07%	1.83%	0.02%	1.78%	−0.04%	1.74%
	66.57%		61.56%		56.45%		51.33%		46.28%	
1500	0.22%	1.88%	0.16%	1.84%	0.10%	1.80%	0.05%	1.75%	−0.01%	1.71%
	69.56%		64.55%		59.37%		54.12%		48.89%	

In practice, it is then of particular interest to stop the trial early if the candidate vaccine will not achieve the study objectives. In this section, a statistical data safety monitoring procedure based on probability statement is developed to assist the sponsor in making decision as to whether the trial should stop at interim.

Assume that an interim look is to take place when we reach the sample size of $N' = N/2 = n$, where $N = 2n$ is the total sample size for the trial. At interim, suppose $p\%$

TABLE 17.3

95% CIs for $p_R = 2.00\%$

p_T	0.90%		0.95%		1.00%		1.05%		1.10%	
% improvement	55%		53%		50%		48%		45%	
	Lower	Upper	Lower	Upper	Lower	Upper	Lower	Upper	Lower	Upper
n	Power		Power		Power		Power		Power	
200	−1.24%	3.44%	−1.31%	3.41%	−1.38%	3.38%	−1.45%	3.35%	−1.52%	3.32%
	14.95%		13.83%		12.79%		11.82%		10.92%	
300	−0.81%	3.01%	−0.88%	2.98%	−0.94%	2.94%	−1.01%	2.91%	−1.08%	2.88%
	20.28%		18.61%		17.07%		15.63%		14.30%	
400	−0.55%	2.75%	−0.62%	2.72%	−0.68%	2.68%	−0.75%	2.65%	−0.81%	2.61%
	25.55%		23.36%		21.32%		19.41%		17.65%	
500	−0.38%	2.58%	−0.44%	2.54%	−0.51%	2.51%	−0.57%	2.47%	−0.63%	2.43%
	30.73%		28.05%		25.52%		23.17%		20.98%	
600	−0.25%	2.45%	−0.31%	2.41%	−0.37%	2.37%	−0.44%	2.34%	−0.50%	2.30%
	35.78%		32.64%		29.67%		26.89%		24.28%	
700	−0.15%	2.35%	−0.21%	2.31%	−0.27%	2.27%	−0.33%	2.23%	−0.39%	2.19%
	40.65%		37.11%		33.74%		30.55%		27.56%	
800	−0.07%	2.27%	−0.13%	2.23%	−0.19%	2.19%	−0.25%	2.15%	−0.31%	2.11%
	45.32%		41.44%		37.71%		34.15%		30.79%	
900	0.00%	2.20%	−0.06%	2.16%	−0.12%	2.12%	−0.18%	2.08%	−0.24%	2.04%
	49.77%		45.60%		41.55%		37.67%		33.97%	
1000	0.05%	2.15%	−0.01%	2.11%	−0.06%	2.06%	−0.12%	2.02%	−0.18%	1.98%
	53.98%		49.58%		45.27%		41.09%		37.08%	
1100	0.10%	2.10%	0.04%	2.06%	−0.01%	2.01%	−0.07%	1.97%	−0.13%	1.93%
	57.94%		53.37%		48.85%		44.41%		40.12%	
1200	0.14%	2.06%	0.09%	2.01%	0.03%	1.97%	−0.03%	1.93%	−0.09%	1.89%
	61.66%		56.97%		52.27%		47.62%		43.09%	
1300	0.18%	2.02%	0.12%	1.98%	0.07%	1.93%	0.01%	1.89%	−0.05%	1.85%
	65.12%		60.36%		55.54%		50.72%		45.97%	
1400	0.22%	1.98%	0.16%	1.94%	0.10%	1.90%	0.04%	1.86%	−0.01%	1.81%
	68.34%		63.56%		58.65%		53.69%		48.76%	
1500	0.25%	1.95%	0.19%	1.91%	0.13%	1.87%	0.07%	1.83%	0.02%	1.78%
	71.32%		66.55%		61.60%		56.54%		51.46%	

(where $p = p_R + p_T/2$) incidences are expected to be observed from the N' samples. Then one can follow the procedure described below for data safety monitoring (Figure 17.1).

For the purpose of illustration, consider the example described in Section 17.4. Assume that the incidence rate of the control group is 1.9% and that for the test group is 0.95%. Thus, the blinded expected total incidence rate will be 1.425% and the expected mean is $15.68 \approx 16$. We can also find the 95% upper and lower limits of the expected observed

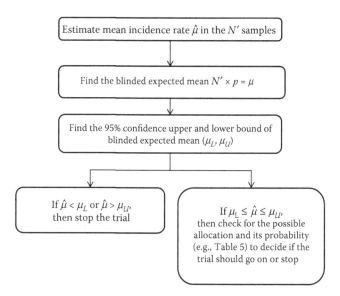

FIGURE 17.1
Data safety monitoring procedure.

number for the whole trial is (8, 23). If the observed incidence number is 23 ($\hat{\mu} = 23$, which is within the 95% confidence interval) at intern, one can consider to continue the trial by the probability of each case. The expected incidence numbers in each case are shown in Table 17.4 for reference. For the whole trial, the estimated expected mean and 95% confidence limit in the reference group are 28 and (16, 40); in the test group, they are 14 and (6, 23) if $\hat{\mu} = \mu_U$. For the possible combination of sample sizes, we provide the probabilities $P\left(\sum_{i=1}^n x_{Rj} \mid p_R\right) \times P\left(\sum_{i=1}^n x_{Tj} \mid p_T\right)$ in each case in Table 17.5.

17.5 Concluding Remarks

For vaccine clinical trials with extremely low incidence rates, the sample size required for achieving a desired power of correctly detecting a small clinically meaningful difference

TABLE 17.4

Expected Incidence Numbers and the Corresponding 95% Confidence Limit for Each Group for *Whole Trail* at Intern

	Intern	Target	Reference			Test		
	$N' = 1100$	$N = 2200$	$n_R = 1100$			$n_T = 1100$		
Sample Size	Expected incidence numbers		Lower	Mean	Upper	Lower	Mean	Upper
Upper	23	47	16	28	40	6	14	23
Mean	16	31	12	21	30	4	10	17
Lower	8	16	8	14	19	3	7	11

TABLE 17.5

Probabilities of Each Sample Size Combinations Occur

Incidence Numbers			Incidence Numbers			Incidence Numbers			Incidence Numbers			Incidence Numbers		
Reference	Test	Probabilities (%)	Reference	Test	Probabilities (%)	Reference	Test	Probabilities (%)	Reference	Test	Probabilities (%)	Reference	Test	Probabilities (%)
16	6	1.71	21	6	5.84	26	6	9.17	31	6	10.17	36	6	10.30
16	7	3.00	21	7	10.26	26	7	16.10	31	7	17.87	36	7	18.10
16	8	4.69	21	8	16.05	26	8	25.19	31	8	27.95	36	8	28.31
16	9	6.66	21	9	22.78	26	9	35.76	31	9	39.68	36	9	40.19
16	10	8.73	21	10	29.83	26	10	46.83	31	10	51.96	36	10	52.62
16	11	10.68	21	11	36.53	26	11	57.34	31	11	63.62	36	11	64.44
16	12	12.39	21	12	42.36	26	12	66.50	31	12	73.78	36	12	74.73
16	13	13.76	21	13	47.04	26	13	73.84	31	13	81.93	36	13	82.98
16	14	14.78	21	14	50.53	26	14	79.31	31	14	88.00	36	14	89.13
16	15	15.49	21	15	52.95	26	15	83.11	31	15	92.22	36	15	93.40
16	16	15.95	21	16	54.52	26	16	85.58	31	16	94.96	36	16	96.18
16	17	16.23	21	17	55.49	26	17	87.10	31	17	96.63	36	17	97.87
16	18	16.39	21	18	56.04	26	18	87.97	31	18	97.60	36	18	98.85
16	19	16.48	21	19	56.34	26	19	88.44	31	19	98.13	36	19	99.39
16	20	16.53	21	20	56.50	26	20	88.69	31	20	98.40	36	20	99.67
16	21	16.55	21	21	56.58	26	21	88.81	31	21	98.54	36	21	99.80
16	22	16.56	21	22	56.62	26	22	88.87	31	22	98.60	36	22	99.87
16	23	16.56	21	23	56.63	26	23	88.90	31	23	98.63	36	23	99.90
17	6	2.38	22	6	6.70	27	6	9.52	32	6	10.23	37	6	10.31
17	7	4.18	22	7	11.76	27	7	16.71	32	7	17.96	37	7	18.10
17	8	6.54	22	8	18.40	27	8	26.14	32	8	28.10	37	8	28.32
17	9	9.29	22	9	26.13	27	9	37.12	32	9	39.90	37	9	40.20
17	10	12.17	22	10	34.21	27	10	48.60	32	10	52.24	37	10	52.64

(*Continued*)

TABLE 17.5

Probabilities of Each Sample Size Combinations Occur

Incidence Numbers			Incidence Numbers			Incidence Numbers			Incidence Numbers			Incidence Numbers		
Reference	Test	Probabilities (%)	Reference	Test	Probabilities (%)	Reference	Test	Probabilities (%)	Reference	Test	Probabilities (%)	Reference	Test	Probabilities (%)
17	11	14.90	22	11	41.89	27	11	59.52	32	11	63.97	37	11	64.47
17	12	17.27	22	12	48.58	27	12	69.02	32	12	74.18	37	12	74.75
17	13	19.18	22	13	53.95	27	13	76.64	32	13	82.38	37	13	83.01
17	14	20.60	22	14	57.95	27	14	82.32	32	14	88.48	37	14	89.16
17	15	21.59	22	15	60.72	27	15	86.27	32	15	92.72	37	15	93.44
17	16	22.23	22	16	62.53	27	16	88.83	32	16	95.47	37	16	96.21
17	17	22.63	22	17	63.63	27	17	90.40	32	17	97.16	37	17	97.91
17	18	22.85	22	18	64.27	27	18	91.31	32	18	98.13	37	18	98.89
17	19	22.98	22	19	64.62	27	19	91.80	32	19	98.66	37	19	99.43
17	20	23.04	22	20	64.80	27	20	92.06	32	20	98.94	37	20	99.71
17	21	23.07	22	21	64.88	27	21	92.18	32	21	99.08	37	21	99.84
17	22	23.09	22	22	64.93	27	22	92.24	32	22	99.14	37	22	99.91
17	23	23.09	22	23	64.95	27	23	92.27	32	23	99.17	37	23	99.94
18	6	3.17	23	6	7.48	28	6	9.77	33	6	10.26	38	6	10.31
18	7	5.56	23	7	13.13	28	7	17.17	33	7	18.02	38	7	18.11
18	8	8.70	23	8	20.54	28	8	26.85	33	8	28.19	38	8	28.32
18	9	12.35	23	9	29.16	28	9	38.13	33	9	40.03	38	9	40.21
18	10	16.17	23	10	38.19	28	10	49.92	33	10	52.42	38	10	52.65
18	11	19.81	23	11	46.76	28	11	61.13	33	11	64.19	38	11	64.48
18	12	22.97	23	12	54.23	28	12	70.89	33	12	74.43	38	12	74.77
18	13	25.50	23	13	60.22	28	13	78.72	33	13	82.66	38	13	83.03
18	14	27.39	23	14	64.68	28	14	84.56	33	14	88.78	38	14	89.18

(*Continued*)

TABLE 17.5

Probabilities of Each Sample Size Combinations Occur

Incidence Numbers			Incidence Numbers			Incidence Numbers			Incidence Numbers			Incidence Numbers		
Reference	Test	Probabilities (%)	Reference	Test	Probabilities (%)	Reference	Test	Probabilities (%)	Reference	Test	Probabilities (%)	Reference	Test	Probabilities (%)
18	15	28.71	23	15	67.78	28	15	88.61	33	15	93.03	38	15	93.46
18	16	29.56	23	16	69.79	28	16	91.24	33	16	95.80	38	16	96.23
18	17	30.08	23	17	71.02	28	17	92.85	33	17	97.49	38	17	97.93
18	18	30.38	23	18	71.73	28	18	93.78	33	18	98.47	38	18	98.91
18	19	30.55	23	19	72.12	28	19	94.29	33	19	99.00	38	19	99.45
18	20	30.63	23	20	72.32	28	20	94.55	33	20	99.27	38	20	99.73
18	21	30.67	23	21	72.42	28	21	94.68	33	21	99.41	38	21	99.86
18	22	30.69	23	22	72.47	28	22	94.74	33	22	99.48	38	22	99.93
18	23	30.70	23	23	72.49	28	23	94.77	33	23	99.50	38	23	99.96
19	6	4.03	24	6	8.15	29	6	9.96	34	6	10.28	39	6	10.31
19	7	7.08	24	7	14.32	29	7	17.49	34	7	18.06	39	7	18.11
19	8	11.08	24	8	22.40	29	8	27.36	34	8	28.25	39	8	28.33
19	9	15.73	24	9	31.80	29	9	38.85	34	9	40.11	39	9	40.22
19	10	20.59	24	10	41.64	29	10	50.87	34	10	52.52	39	10	52.66
19	11	25.22	24	11	50.99	29	11	62.29	34	11	64.32	39	11	64.49
19	12	29.24	24	12	59.13	29	12	72.23	34	12	74.58	39	12	74.78
19	13	32.48	24	13	65.66	29	13	80.21	34	13	82.82	39	13	83.04
19	14	34.88	24	14	70.53	29	14	86.16	34	14	88.96	39	14	89.19
19	15	36.55	24	15	73.91	29	15	90.28	34	15	93.22	39	15	93.47
19	16	37.64	24	16	76.11	29	16	92.97	34	16	95.99	39	16	96.24
19	17	38.30	24	17	77.45	29	17	94.61	34	17	97.69	39	17	97.94
19	18	38.69	24	18	78.22	29	18	95.56	34	18	98.67	39	18	98.92
19	19	38.90	24	19	78.65	29	19	96.07	34	19	99.20	39	19	99.46

(Continued)

TABLE 17.5

Probabilities of Each Sample Size Combinations Occur

Incidence Numbers			Incidence Numbers			Incidence Numbers			Incidence Numbers			Incidence Numbers		
Reference	Test	Probabilities (%)	Reference	Test	Probabilities (%)	Reference	Test	Probabilities (%)	Reference	Test	Probabilities (%)	Reference	Test	Probabilities (%)
19	20	39.01	24	20	78.87	29	20	96.34	34	20	99.48	39	20	99.74
19	21	39.06	24	21	78.98	29	21	96.47	34	21	99.62	39	21	99.87
19	22	39.08	24	22	79.03	29	22	96.54	34	22	99.68	39	22	99.94
19	23	39.10	24	23	79.05	29	23	96.56	34	23	99.71	39	23	99.97
20	6	4.94	25	6	8.72	30	6	10.09	35	6	10.30	40	6	10.31
20	7	8.67	25	7	15.31	30	7	17.72	35	7	18.08	40	7	18.11
20	8	13.57	25	8	23.95	30	8	27.71	35	8	28.29	40	8	28.33
20	9	19.26	25	9	34.00	30	9	39.35	35	9	40.16	40	9	40.22
20	10	25.22	25	10	44.52	30	10	51.52	35	10	52.59	40	10	52.66
20	11	30.89	25	11	54.52	30	11	63.09	35	11	64.40	40	11	64.49
20	12	35.82	25	12	63.22	30	12	73.16	35	12	74.67	40	12	74.78
20	13	39.77	25	13	70.21	30	13	81.24	35	13	82.92	40	13	83.05
20	14	42.72	25	14	75.41	30	14	87.26	35	14	89.07	40	14	89.20
20	15	44.77	25	15	79.02	30	15	91.44	35	15	93.34	40	15	93.47
20	16	46.10	25	16	81.37	30	16	94.16	35	16	96.11	40	16	96.25
20	17	46.91	25	17	82.81	30	17	95.82	35	17	97.81	40	17	97.95
20	18	47.38	25	18	83.63	30	18	96.78	35	18	98.79	40	18	98.93
20	19	47.64	25	19	84.09	30	19	97.31	35	19	99.32	40	19	99.47
20	20	47.77	25	20	84.32	30	20	97.58	35	20	99.60	40	20	99.74
20	21	47.84	25	21	84.44	30	21	97.71	35	21	99.74	40	21	99.88
20	22	47.87	25	22	84.49	30	22	97.78	35	22	99.80	40	22	99.94
20	23	47.88	25	23	84.52	30	23	97.80	35	23	99.83	40	23	99.97

if such a difference truly exists is often huge. This huge sample size may not be of practical use. In this chapter, an alternative approach based on precision analysis and sensitivity analysis in conjunction with power analysis proposed by Chow and Chiu (2013) was discussed. Chow and Chiu's proposed method reduces the sample size required for achieving a desired precision with certain statistical assurance.

Most importantly, the statistical data safety monitoring procedure proposed by Chow and Chiu (2013), which was developed based on probability statement, is useful in assisting the investigator or sponsor in making decisions as to whether the trial should stop due to safety, futility, and/or efficacy early at interim.

18

Sample Size Calculation for Two-Stage Adaptive Trial Design

In the past decade, adaptive design methods in clinical research have attracted much attention due to (1) their flexibility for identifying clinical benefit of a test treatment under investigation and (2) their efficiency for shortening (or speeding up) the development process. To assist the sponsors, the U.S. Food and Drug Administration (FDA) published a draft guidance on adaptive clinical trial design in 2010. The FDA draft guidance defines an adaptive design as a clinical study that includes a prospectively planned opportunity for the modification of one or more specified aspects of the study design and hypotheses based on the analysis of data (usually interim data) from subjects in the study (FDA, 2010a). The use of adaptive design methods in clinical trials may allow researchers to correct assumptions made at the planning stage and select the most promising option early. In addition, adaptive trial designs make use of cumulative information of the ongoing trial, which provide the investigator an opportunity to react earlier to surprises regardless of positive or negative results. As a result, adaptive design approaches may shorten (or speed up) the drug development process.

Despite the possible benefits for having a second chance to modify the trial at interim when utilizing an adaptive design, it could be more problematic operationally due to potential biases that may have been introduced to the conduct of the trial especially after the review of interim data. As indicated by the FDA draft guidance, operational biases may occur when adaptations in trial and/or statistical procedures are applied after the review of interim (unblinded) data. As a result, it is a concern whether scientific integrity and validity of trial are warranted. Chow and Chang (2011) indicated that trial procedures include, but are not limited to, inclusion/exclusion criteria, dose/dose regimen and treatment duration, endpoint selection and assessment, and/or laboratory testing procedures employed. On the other hand, statistical procedures are referred to as study design, statistical hypotheses (which can reflect study objectives), endpoint selection, power analysis for sample size calculation, sample size reestimation, and/or sample size adjustment, randomization schedules, and statistical analysis plan (SAP). With respect to these trial and statistical procedures, commonly employed adaptations at interim include, but are not limited to, (1) sample size reestimation at interim analysis, (2) adaptive randomization with unequal treatment allocation (e.g., change from 1:1 ratio to 2:1 ratio), (3) deleting, adding, or modifying treatment arms after the review of interim data, (4) shifting in patient population due to protocol amendment, (5) different statistical methods, (6) changing study endpoints (e.g., change response rate and/or survival to time-to-disease progression in cancer trials), and (7) changing hypotheses/objectives (e.g., switch a superiority hypothesis to a noninferiority hypothesis). In practice, although the use of the adaptive design methods in clinical trials seems promising, major adaptations may have an impact on the integrity and validity of the clinical trials, which may raise some critical concerns to the accurate and reliable evaluation of the test treatment under investigation. These concerns include (1) the control of the overall type I error rate at a prespecified level of significance,

(2) the correctness of the obtained *p*-values, and (3) the reliability of the obtained confidence interval. Most importantly, major (significant) adaptations may have resulted in a totally different trial that is unable to address the scientific/medical questions the original study intended to answer.

Chow (2011) indicated that a seamless trial design is one that combines two (independent) trials into a single study, which can address study objectives from individual studies. An adaptive seamless design is referred to as a seamless trial design that would use data collected before and after the adaptation in the final analysis. In practice, a two-stage seamless adaptive design typically consists of two stages (phases): a learning (or exploratory) phase (stage 1) and a confirmatory phase (stage 2). The objective of the learning phase is not only to obtain information regarding the uncertainty of the test treatment under investigation but also to provide the investigator the opportunity to stop the trial early due to safety and/or futility/efficacy based on accrued data or to apply some adaptations such as adaptive randomization at the end of stage 1. The objective of the second stage is to confirm the findings observed from the first stage. A two-stage seamless adaptive trial design has the following advantages: (1) it may reduce lead time between studies (the traditional approach) and (2) it provides the investigator the second chance to redesign the trial after the review of accumulated data at the end of stage 1. Most importantly, data collected from both stages are combined for a final analysis to fully utilize all data collected from the trial for a more accurate and reliable assessment of the test treatment under investigation.

18.1 Types of Two-Stage Adaptive Designs

Chow and Tu (2008) and Chow (2011) classified two-stage seamless adaptive trial designs into four categories, depending upon study objectives and study endpoints at different stages.

Table 18.1 indicates that there are four different types of two-stage adaptive trial designs depending upon whether study objectives and/or study endpoints at different stages are the same. For example, Category I designs (i.e., SS designs) include those designs with the same study objectives and the same study endpoints, while Category II and Category III designs (i.e., SD and DS designs) are referred to those designs with the same study objectives but different study endpoints and different study objectives but the same study endpoints respectively. Category IV designs (i.e., DD designs) are those designs with different study objectives and different study endpoints. In practice, different study objectives could be treatment selection for stage 1 and efficacy confirmation for stage 2. On the other hand,

TABLE 18.1

Types of Two-Stage Adaptive Designs

	Study Endpoint	
Study Objectives	Same (S)	Different (D)
Same (S)	I = SS	II = SD
Different (D)	III = DS	IV = DD

Source: Chow, S.C. 2011. *Controversial Issues in Clinical Trials.* Chapman & Hall/CRC, Taylor & Francis, New York, NY.

different study endpoints could be biomarker, surrogate endpoints, or a clinical endpoint with a shorter duration at the first stage versus clinical endpoint at the second stage. Note that a group sequential design with one planned interim analysis is often considered an SS design.

In practice, typical examples for a two-stage adaptive seamless design include a two-stage adaptive seamless phase I/II design and a two-stage adaptive seamless phase II/III design. For the two-stage adaptive seamless phase I/II design, the objective at the first stage may be for biomarker development and the study objective for the second stage is usually to establish early efficacy. For a two-stage adaptive seamless phase II/III design, the study objective is often for treatment selection (or dose finding) while the study objective at the second stage is for efficacy confirmation. In this chapter, our focus will be on Category II designs. The results can be similarly applied to Category III and Category IV designs.

It should be noted that the terms "seamless" and "phase II/III" were not used in the FDA draft guidance as they have sometimes been adopted to describe various design features (FDA, 2010a,b). In this chapter, a two-stage adaptive seamless phase II/III design only refers to a study containing an exploratory phase II stage (stage 1) and a confirmatory phase III stage (stage 2), while data collected at both phases (stages) will be used for final analysis.

One of the questions that is commonly asked when applying a two-stage adaptive seamless design in clinical trials is on sample size calculation/allocation. For the first kind (i.e., Category I, SS) of two-stage seamless designs, the methods based on individual *p*-values as described in Chow and Chang (2011) can be applied. However, for other kinds (i.e., Category II through Category IV) of two-stage seamless trial designs, standard statistical methods for group sequential design are not appropriate and hence should not be applied directly. For Category II–IV trial designs, power analysis and/or statistical methods for data analysis are challenging to the biostatistician. For example, a commonly asked question is, "How do we control the overall type I error rate at a prespecified level of significance?" In the interest of stopping trial early, the question of "How to determine stopping boundaries?" is a challenge to the investigator and the biostatistician. In practice, it is often of interest to determine whether the typical O'Brien–Fleming type of boundaries are feasible. Another challenge is, "How to perform a valid analysis that combines data collected from different stages?" To address these questions, Cheng and Chow (2015) proposed the concept of a multiple-stage transitional seamless adaptive design, which takes into consideration different study objectives and study endpoints.

18.2 Analysis and Sample Size for Category SS Adaptive Designs

Category I design with the same study objectives and the same study endpoints at different stages is considered similar to a typical group sequential design with one planned interim analysis. Thus, standard statistical methods for group sequential design are often employed. It, however, should be noted that with various adaptations applied, these standard statistical methods may not be appropriate. In practice, many interesting methods for Category I designs are available in the literature. These include (1) Fisher's criterion for combining independent *p*-values (Bauer and Kohne, 1994; Bauer and Rohmel, 1995; Posch and Bauer, 2000), (2) weighted test statistics (Cui, Hung, and Wang, 1999), (3) the conditional error function approach (Proschan and Hunsberger, 1995; Liu and Chi, 2001), and (4) conditional power approaches (Li, Shih, and Wang, 2005).

Among these methods, Fisher's method for combining p-values provides great flexibility in selecting statistical tests for individual hypotheses based on subsamples. Fisher's method, however, lacks flexibility in the choice of boundaries (Müller and Schafer, 2001). For Category I adaptive designs, many related issues have been studied. For example, Rosenberger and Lachin (2003) explored the potential use of response-adaptive randomization. Chow, Chang, and Pong (2005) examined the impact of population shift due to protocol amendments. Li, Shih, and Wang (2005) studied a two-stage adaptive design with a survival endpoint, while Hommel, Lindig, and Faldum (2005) studied a two-stage adaptive design with correlated data. An adaptive design with a bivariate endpoint was studied by Todd (2003). Tsiatis and Mehta (2003) showed that there exists a more powerful group sequential design for any adaptive design with sample size adjustment.

For illustration purpose, in what follows, we will introduce the method based on sum of p-values (MSP) by Chang (2007) and Chow and Chang (2011). The MSP follows the idea of considering a linear combination of the p-values from different stages.

18.2.1 Theoretical Framework

Consider a clinical trial utilizing a multiple-stage (say K-stage) design. This is similar to a clinical trial with K interim analyses, while the final analysis is the Kth interim (final) analysis. Suppose that at each interim analysis, a hypothesis test is performed. The objective of the trial can be formulated as the following intersection of the individual hypothesis tests from the interim analyses:

$$H_0 : H_{01} \cap \cdots \cap H_{0K},$$

where $H_{0i}, i = 1,...,K$ is the null hypothesis to be tested at the ith interim analysis. Note that there are some restrictions on H_{0i}, that is, rejection of any $H_{0i}, i = 1,...,K$ will lead to the same clinical implication (e.g., drug is efficacious); hence all $H_{0i}, i = 1,...,K$ are constructed for testing the same endpoint within a trial. Otherwise the global hypothesis cannot be interpreted.

In practice, H_{0i} is tested based on a subsample from each stage, and without loss of generality, assume H_{0i} is a test for the efficacy of a test treatment under investigation, which can be written as

$$H_{0i} : \eta_{i1} \geq \eta_{i2} \quad \text{versus} \quad H_{ai} : \eta_{i1} < \eta_{i2},$$

where η_{i1} and η_{i2} are the responses of the two treatment groups at the ith stage and we assume bigger values are better. It is often the case that when $\eta_{i1} = \eta_{i2}$, the p-value p_i for the subsample at the ith stage is uniformly distributed on [0, 1] under H_0. Under the null hypothesis, Bauer and Kohne (1994) used Fisher's combination of the p-values to construct a test statistic for multiple-stage adaptive designs. Following similar idea, Chang (2007) and Chow and Chang (2011) considered a linear combination of the p-values as follows:

$$T_k = \sum_{i=1}^{K} w_{ki} p_i, \quad i = 1,...,K, \tag{18.1}$$

where $w_{ki} > 0$ and K are the number of interim analyses planned. If $w_{ki} = 1$, this leads to

$$T_k = \sum_{i=1}^{K} p_i, \quad i = 1, \ldots, K. \tag{18.2}$$

T_k can be viewed as cumulative evidence against H_0. Thus, the smaller the T_k, the stronger the evidence. Alternatively, we can consider $T_k = \Sigma_{i=1}^{k} p_i / K$, which is an average of the evidence against H_0. Intuitively, one may consider the stopping rules

$$\begin{cases} \text{Stop for efficacy} & \text{if } T_k \leq \alpha_k \\ \text{Stop for futility} & \text{if } T_k \geq \beta_k, \\ \text{Continue} & \text{otherwise} \end{cases} \tag{18.3}$$

where T_k, α_k, and β_k are monotonic increasing functions of k, $\alpha_k < \beta_k$, $k = 1, \ldots, K - 1$ and $\alpha_k = \beta_k$. Note that α_k and β_k are referred to as the efficacy and futility boundaries, respectively. To reach the kth stage, a trial has to pass 1 to $(k - 1)$th stages. Therefore, a so-called proceeding probability can be defined as the following unconditional probability:

$$\psi_k(t) = P(T_k < t, \alpha_1 < T_1 < \beta_1, \ldots, \alpha_{k-1} < T_{k-1} < \beta_{k-1})$$

$$= \int_{\alpha_1}^{\beta_1} \cdots \int_{\alpha_{k-1}}^{\beta_{k-1}} \int_{-\infty}^{t} f_{T_1 \cdots T_k}(t_1, \ldots, t_k) dt_k dt_{k-1} \cdots dt_1, \tag{18.4}$$

where $t \geq 0$, t_i, $i = 1, \ldots, k$ is the test statistic at the ith stage, and $f_{T_1 \cdots T_k}$ is the joint probability density function. Thus, the error rate at the kth stage can be obtained as

$$\pi_k = \psi_k(\alpha_k). \tag{18.5}$$

Since the type I error rates at different stages are mutually exclusive, the experiment-wise type I error rate is sum of π_k, $k = 1, \ldots, K$. Thus, we have

$$\alpha = \sum_{k=1}^{K} \pi_k. \tag{18.6}$$

Note that stopping boundaries can be determined with appropriate choices of α_k. The adjusted p-value calculation is the same as the one in a classic group sequential design (Jennison and Turnbull, 2000). The key idea is that when the test statistic at the kth stage $T_k = t = \alpha_k$ (i.e., just on the efficacy stopping boundary), the p-value is equal to alpha spent $\Sigma_{i=1}^{k} \pi_i$. This is true regardless of which error spending function is used and consistent with the p-value definition of the traditional design. As indicated in Chang (2007), the adjusted p-value corresponding to an observed test statistic $T_k = t$ at the kth stage can be defined as

$$p(t; k) = \sum_{i=1}^{k-1} \pi_i + \psi_k(t), \quad k = 1, \ldots, K. \tag{18.7}$$

Note that p_i in Equation 18.1 is the stage-wise (unadjusted) p-value from a subsample at the ith stage, while $p(t;k)$ are adjusted p-values calculated from the test statistic, which are based on the cumulative sample up to the kth stage where the trial stops; Equations 18.6 and 18.7 are valid regardless of how p_i is calculated.

18.2.2 Two-Stage Design

In this section, for simplicity, we will consider the method of sum of p-values (MSP) and apply the general framework to the two-stage design as outlined in Chang (2007) and Chow and Chang (2011), which are suitable for the following adaptive designs that allow (1) early efficacy stopping, (2) early stopping for both efficacy and futility, and (3) early futility stopping. These adaptive designs are briefly described below.

Early efficacy stopping—For simplicity, consider $K = 2$ (i.e., a two-stage design), which allows for early efficacy stopping (i.e., $\beta_1 = 1$). By Equation 18.5, the type I error rates to spend at stage 1 and stage 2 are given by

$$\pi_1 = \psi_1(\alpha_1) = \int_0^{\alpha_1} dt_1 = \alpha_1, \tag{18.8}$$

and

$$\pi_2 = \psi_2(\alpha_2) = \int_{\alpha_1}^{\alpha_2} \int_{t_1}^{\alpha_2} dt_2 dt_1 = \frac{1}{2}(\alpha_2 - \alpha_1)^2, \tag{18.9}$$

respectively. Using Equations 18.8 and 18.9, Equation 18.6 becomes

$$\alpha = \alpha_1 + \frac{1}{2}(\alpha_2 - \alpha_1)^2. \tag{18.10}$$

Solving for α_2, we obtain

$$\alpha_2 = \sqrt{2(\alpha - \alpha_1)} + \alpha_1. \tag{18.11}$$

α_1 is the stopping probability (error spent) at the first stage under the null hypothesis condition and $\alpha - \alpha_1$ is the error spent at the second stage. As a result, if the test statistic $t_1 = p_1 > \alpha_2$, it is certain that $t_2 = p_1 + p_2 > \alpha_2$. Therefore, the trial should stop when $p_1 > \alpha_2$ for futility.

Based on relationship among α_1, α_2, and α as given in Equation 18.10, various stopping boundaries can be considered with appropriate choices of α_1, α_2, and α. For illustration purpose, Table 18.2 provides some examples of the stopping boundaries from Equations 18.10 and 18.11.

By Equations 18.7 through 18.11, the adjusted p-value is given by

$$p(t;k) = \begin{cases} t & \text{if } k = 1 \\ \alpha_1 + \dfrac{1}{2}(t - \alpha_1)^2 & \text{if } k = 2 \end{cases}, \tag{18.12}$$

TABLE 18.2

Stopping Boundaries for Two-Stage Efficacy Designs

One-Sided α		0.005	0.010	0.015	0.020	0.025	0.030
0.025	α_1	0.2050	0.1832	0.1564	0.1200	0.0250	–
0.05	α_2	0.3050	0.2928	0.2796	0.2649	0.2486	0.2300

Source: Chang, M. 2007. *Statistics in Medicine*, 26, 2772–2784.

where $t = p_1$ if the trial stops at stage 1 and $t = p_1 + p_2$ if the trial stops at stage 2.

Early efficacy or futility stopping—For this case, it is obvious that if $\beta_1 \geq \alpha_2$ the stopping boundary is the same as it is for the design with early efficacy stopping. However, futility boundary β_1 when $\beta_1 \geq \alpha_2$ is expected to affect the power of the hypothesis testing. Therefore,

$$\pi_1 = \int_0^{\alpha_1} dt_1 = \alpha_1, \tag{18.13}$$

and

$$\pi_2 = \begin{cases} \displaystyle\int_{\alpha_1}^{\beta_1} \int_{t_1}^{\alpha_2} dt_2 dt_1 & \text{for } \beta_1 \leq \alpha_2 \\[2ex] \displaystyle\int_{\alpha_1}^{\alpha_2} \int_{t_1}^{\alpha_2} dt_2 dt_1 & \text{for } \beta_1 > \alpha_2, \end{cases} \tag{18.14}$$

Thus, it can be verified that

$$\alpha = \begin{cases} \alpha_1 + \alpha_2(\beta_1 - \alpha_1) - \dfrac{1}{2}(\beta_1^2 - \alpha_1^2) & \text{for } \beta_1 \leq \alpha_2 \\[2ex] \alpha_1 + \dfrac{1}{2}(\alpha_2 - \alpha_1)^2 & \text{for } \beta_1 > \alpha_2, \end{cases} \tag{18.15}$$

Similarly, under Equation 18.15, various boundaries can be obtained with appropriate choices of α_1, α_2, β_1, and α (Table 18.3). The adjusted *p*-value is given by

$$p(t;k) = \begin{cases} t & \text{if } k = 1 \\[1ex] \alpha_1 + t(\beta_1 - \alpha_1) - \dfrac{1}{2}(\beta_1^2 - \alpha_1^2) & \text{if } k = 2 \text{ and } \beta_1 < \alpha_2, \\[2ex] \alpha_1 + \dfrac{1}{2}(t - \alpha_1)^2 & \text{if } k = 2 \; \beta_1 \geq \alpha_2 \end{cases} \tag{18.16}$$

TABLE 18.3

Stopping Boundaries for Two-Stage Efficacy and Futility Designs

One-sided α				$\beta_1 = 0.15$		
0.025	α_1	0.005	0.010	0.015	0.020	0.025
	α_2	0.2154	0.1871	0.1566	0.1200	0.0250
0.05	α_1	0.005	0.010	0.015	0.020	0.025
	α_2	0.3333	0.3155	0.2967	0.2767	0.2554

Source: Chang, M. 2007. *Statistics in Medicine*, 26, 2772–2784.

where $t = p_1$ if the trial stops at stage 1 and $t = p_1 + p_2$ if the trial stops at stage 2.

For a trial design with early futility stopping, it is a special case of the previous design, where $\alpha_1 = 0$ in Equation 18.15. Hence, we have

$$\alpha = \begin{cases} \alpha_2 \beta_1 - \dfrac{1}{2}\beta_1^2 & \text{for } \beta_1 < \alpha_2 \\ \dfrac{1}{2}\alpha_2^2 & \text{for } \beta_1 \geq \alpha_2 \end{cases}, \tag{18.17}$$

Solving for α_2, we have

$$\alpha_2 = \begin{cases} \dfrac{\alpha}{\beta_1} + \dfrac{1}{2}\beta_1 & \text{for } \beta_1 < \sqrt{2\alpha} \\ \sqrt{2\alpha} & \text{for } \beta_1 \geq \alpha_2 \end{cases}, \tag{18.18}$$

Table 18.4 gives examples of the stopping boundaries generated using Equation 18.18. The adjusted p-value can be obtained from Equation 18.16, where $\alpha_1 = 0$, that is,

$$p(t;k) = \begin{cases} t & \text{if } k = 1 \\ t\beta_1 - \dfrac{1}{2}\beta_1^2 & \text{if } k = 2 \text{ and } \beta_1 < \alpha_2, \\ \dfrac{1}{2}t^2 & \text{if } k = 2 \; \beta_1 \geq \alpha_2 \end{cases} \tag{18.19}$$

TABLE 18.4

Stopping Boundaries for Two-Stage Futility Design

One-sided α	β_1	0.1	0.2	0.3	≥ 0.4
0.025	α_2	0.3000	0.2250	0.2236	0.2236
0.05	α_2	0.5500	0.3500	0.3167	0.3162

Source: Chang, M. 2007. *Statistics in Medicine*, 26, 2772–2784.

18.2.3 Conditional Power

Conditional power with or without clinical trial simulation is often considered for sample size reestimation in adaptive trial designs. As discussed earlier, since the stopping boundaries for the most existing methods are either based on z-scale or p-value, to link a z-scale and a p-value, we will consider $p_k = 1 - \Phi(z_k)$ or inversely, $z_k = \Phi^{-1}(1-p_k)$, where z_k and p_k are the normal z-score and the p-value from the subsample at the kth stage, respectively. It should be noted that z_2 has asymptotically normal distribution with $N(\delta/se(\hat{\delta}_2), 1)$ under the alternative hypothesis, where $\hat{\delta}_2$ is the estimation of treatment difference in the second stage and

$$se(\hat{\delta}_2) = \sqrt{2\hat{\sigma}^2/n_2} \approx \sqrt{2\sigma^2/n_2}.$$

The conditional power can be evaluated under the alternative hypothesis when rejecting the null hypothesis H_0. That is,

$$z_2 \geq B(\alpha_2, p_1). \tag{18.20}$$

Thus, the conditional probability given the first stage naïve p-value p_1, at the second stage is given by

$$P_C(p_1, \delta) = 1 - \Phi\left(B(\alpha_2, p_1) - \frac{\delta}{\sigma}\sqrt{\frac{n_2}{2}}\right), \quad \alpha_1 < p_1 \leq \beta_1. \tag{18.21}$$

As an example, for the method based on the product of stage-wise p-values (MPP), the rejection criterion for the second stage is

$$p_1 p_2 \leq \alpha_2, \quad \text{i.e., } z_2 \geq \Phi^{-1}\left(\frac{1-\alpha_2}{p_1}\right).$$

Therefore, $B(\alpha_2, p_1) = \Phi^{-1}(1-\alpha_2/p_1)$.

Similarly, for the method based on the sum of stage-wise p-values (MSP), the rejection criterion for the second stage is

$$p_1 + p_2 \leq \alpha_2, \quad \text{i.e., } z_2 = B(\alpha_2, p_1) = \Phi^{-1}(1-\max(0, \alpha_2 - p_1)).$$

On the other hand, for the inverse-normal method (Lehmacher and Wassmer, 1999), the rejection criterion for the second stage is

$$w_1 z_1 + w_2 z_2 \geq \Phi^{-1}(1-\alpha_2),$$

that is, $z_2 \geq (\Phi^{-1}(1-\alpha_2) - w_1\Phi^{-1}(1-p_1))/w_2$, where w_1 and w_2 are prefixed weights satisfying the condition of $w_1^2 + w_2^2 = 1$. Note that the group sequential design and CHW method (Cui, Hung, and Wang, 1999) are special cases of the inverse-normal method. Since the inverse-normal method requires two additional parameters (w_1 and w_2), for simplicity, we will only compare the conditional powers of MPP and MSP. For a valid comparison, the same α_1 is used for both methods. As it can be seen from Equation 18.21, the comparison of the conditional power is equivalent to the comparison of function $B(\alpha_2, p_1)$. Equating the two $B(\alpha_2, p_1)$, we have

$$\frac{\hat{\alpha}_2}{p_1} = \tilde{\alpha}_2 - p_1, \tag{18.22}$$

where $\hat{\alpha}_2$ and $\tilde{\alpha}_2$ are the final rejection boundaries for MPP and MSP, respectively. Solving Equation 18.22 for p_1, we obtain the critical point for p_1

$$\eta = \frac{\tilde{\alpha}_2 \mp \sqrt{\tilde{\alpha}_2^2 - 4\hat{\alpha}_2}}{2}. \tag{18.23}$$

Equation 18.23 indicates that when $p_1 < \eta_1$ or $p_2 > \eta_2$, MPP has a higher conditional power than that of MSP. When $\eta_1 < p_1 < \eta_2$, MSP has a higher conditional power than MPP. As an example, for one-sided test at $\alpha = 0.025$, if we choose $\alpha_1 = 0.01$ and $\beta_1 = 0.3$, then $\hat{\alpha}_2 = 0.0044$, and $\tilde{\alpha}_2 = 0.2236$, which results in $\eta_1 = 0.0218$, $\eta_2 = 0.2018$ by Equation 18.23.

Note that the unconditional power P_w is nothing but the expectation of conditional power, that is,

$$P_w = E_\delta[P_C(p_1, \delta)]. \tag{18.24}$$

Therefore, the difference in unconditional power between MSP and MPP is dependent on the distribution of p_1, and consequently, dependent on the true difference δ, and the stopping boundaries at the first stage (α_1, β_1).

Note that in Bauer and Kohne's method using Fisher's combination (Bauer and Kohne, 1994), which leads to the equation $\alpha_1 + \ln(\beta_1/\alpha_1)e^{-(1/2)\chi_{4,1-\alpha}^2} = \alpha$, it is obvious that the determination of β_1 leads to a unique α_1, and consequently α_2. This is a nonflexible approach. However, it can be verified that the method can be generalized to $\alpha_1 + \alpha_2 \ln \beta_1/\alpha_1 = \alpha$, where α_2 does not have to be $e^{-(1/2)\chi_{4,1-\alpha}^2}$.

18.3 Analysis and Sample Size for Category II SD Adaptive Designs

For illustration purpose, consider a two-stage phase II/III seamless adaptive designs, which have the same study objectives but different study endpoints. In what follows, we will consider the cases of continuous, binary responses, and time-to-event endpoints, respectively.

18.3.1 Continuous Endpoints

Let x_i be the observed value of the study endpoint (e.g., a biomarker) from the ith subject in phase II (stage 1), $i = 1,...,n$ and y_j be the observed value of the study endpoint (i.e., the primary clinical endpoint) from the jth subject in phase III (stage 2), $j = 1,...,m$. Suppose that $x_i's$ and $y_j's$ are independently and identically distributed with $E(x_i) = \nu$ and $\text{Var}(x_i) = \tau^2$, and $E(y_j) = \mu$ and $\text{Var}(y_j) = \sigma^2$, respectively. Chow, Lu, and Tse (2007) proposed obtaining predicted values of the clinical endpoint based on data collected from the biomarker (or surrogate endpoint) under an established relationship between the biomarker and the clinical endpoint. These predicted values are then combined with the data collected at the confirmatory phase (stage 2) to derive a statistical inference on the treatment effect under investigation. For simplicity, suppose that x and y can be correlated in the following straight-line relationship:

$$y = \beta_0 + \beta_1 x + \varepsilon, \tag{18.25}$$

where ε is the random error with zero mean and variance ς^2. ε is assumed to be independent of x. In practice, we assume that this relationship is well established. In other words, the parameters β_0 and β_1 are assumed to be known. Based on Equation 18.25, the observations x_i observed in the first stage can then be transformed into $\beta_0 + \beta_1 x_i$ (denoted by $\hat{y}_i$). $\hat{y}_i$ is then considered as the observation of the clinical endpoint and combined with those observations y_i collected in the second stage to estimate the treatment mean μ. Chow, Lu, and Tse (2007) proposed the following weighted mean estimator:

$$\hat{\mu} = w\overline{\hat{y}} + (1 - w)\overline{y}, \tag{18.26}$$

where $\overline{\hat{y}} = 1/n\Sigma_{i=1}^{n}\hat{y}_i$, $\overline{y} = 1/m\Sigma_{j=1}^{m}y_j$, and $0 \le w \le 1$. It should be noted that $\hat{\mu}$ is the minimum variance unbiased estimator among all weighted mean estimators when the weight is given by

$$w = \frac{n/(\beta_1^2 \tau^2)}{n/(\beta_1^2 \tau^2) + m/\sigma^2} \tag{18.27}$$

if β_1, τ^2, and σ^2 are known. In practice, τ^2 and σ^2 are usually unknown and w is commonly estimated by

$$\hat{w} = \frac{n/s_1^2}{n/s_1^2 + m/s_2^2}, \tag{18.28}$$

where s_1^2 and s_2^2 are the sample variances of $\hat{y}_i$'s and y_j's, respectively. The corresponding estimator of μ, is denoted by

$$\hat{\mu}_{GD} = \hat{w}\overline{\hat{y}} + (1 - \hat{w})\overline{y}, \tag{18.29}$$

and is referred to as the Graybill–Deal (GD) estimator of μ. Note that Meier (1953) proposed an approximate unbiased estimator of the variance of the GD estimator, which has bias of order $O(n^{-2} + m^{-2})$. Khatri and Shah (1974) gave an exact expression of the variance of this estimator in the form of an infinite series, which is given as

$$\widehat{\text{Var}}(\hat{\mu}_{GD}) = \frac{1}{n/S_1^2 + m/S_2^2}\left[1 + 4\hat{w}(1 - \hat{w})\left(\frac{1}{n-1} + \frac{1}{m-1}\right)\right].$$

Based on the GD estimator, the comparison of the two treatments can be made by testing the following hypotheses:

$$H_0 : \mu_1 = \mu_2 \quad \text{versus} \quad H_a : \mu_1 \neq \mu_2. \tag{18.30}$$

Let $\hat{y}_{ij}$ be the predicted value (based on $\beta_0 + \beta_1 x_{ij}$), which is used as the prediction of y for the jth subject under the ith treatment in phase II (stage 1). From Equation 18.29, the GD estimator of μ_i is given by

$$\hat{\mu}_{GDi} = \hat{\omega}_i \bar{\hat{y}}_i + (1 - \hat{\omega}_i) \bar{y}_i, \tag{18.31}$$

where $\bar{\hat{y}}_i = 1/n_i \Sigma_{j=1}^{n_i} \hat{y}_{ij}$, $\bar{y}_i = 1/m_i \Sigma_{j=1}^{m_i} y_{ij}$, and $\hat{\omega}_i = n_i/S_{1i}^2 / (n_i/S_{1i}^2 + m_i/S_{2i}^2)$ with S_{1i}^2 and S_{2i}^2 being the sample variances of $(\hat{y}_{i1}, \ldots, \hat{y}_{in_i})$ and $(y_{i1}, \ldots, y_{im_i})$, respectively. For hypotheses (18.30), consider the following test statistic:

$$\tilde{T}_1 = \frac{\hat{\mu}_{GD1} - \hat{\mu}_{GD2}}{\sqrt{\widehat{\text{Var}}(\hat{\mu}_{GD1}) + \widehat{\text{Var}}(\hat{\mu}_{GD2})}}, \tag{18.32}$$

where

$$\widehat{\text{Var}}(\hat{\mu}_{GDi}) = \frac{1}{n_i/S_{1i}^2 + m_i/S_{2i}^2} \left[1 + 4\hat{\omega}_i(1 - \hat{\omega}_i)\left(\frac{1}{n_i - 1} + \frac{1}{m_i - 1} \right) \right]$$

is an estimator of $\text{Var}(\hat{\mu}_{GDi})$, $i = 1, 2$. Consequently, an approximate $100(1 - \alpha)\%$ confidence interval of $\mu_1 - \mu_2$ is given as

$$\left(\hat{\mu}_{GD1} - \hat{\mu}_{GD2} - z_{\alpha/2}\sqrt{V_T}, \quad \hat{\mu}_{GD1} - \hat{\mu}_{GD2} + z_{\alpha/2}\sqrt{V_T} \right), \tag{18.33}$$

where $V_T = \text{Var}(\hat{\mu}_{GD1}) + \text{Var}(\hat{\mu}_{GD2})$. As a result, the null hypothesis H_0 is rejected if the above confidence interval does not contain 0. Thus, under the local alternative hypothesis that $H_a : \mu_1 - \mu_2 = \delta \neq 0$, the required sample size to achieve a $1 - \beta$ power satisfies

$$-z_{\alpha/2} + |\delta| / \sqrt{\text{Var}(\hat{\mu}_{GD1}) + \text{Var}(\hat{\mu}_{GD2})} = z_\beta.$$

Thus, if we let $m_i = \rho n_i$ and $n_2 = \gamma n_1$, then N_T, the total sample size required for achieving a desired power for detecting a clinically meaningful difference between the two treatments, is $(1 + \rho)(1 + \gamma)n_1$, which is given by

$$n_1 = \frac{1}{2} AB\left(1 + \sqrt{1 + 8(1 + \rho)A^{-1}C} \right), \tag{18.34}$$

where $A = (z_{\alpha/2} + z_\beta)^2/\delta^2$, $B = \sigma_1^2/\rho + r_1^{-1} + \sigma_2^2/\gamma(\rho + r_2^{-1})$, and $C = B^{-2}\{\sigma_1^2/[r_1(\rho + r_1^{-1})^3 + \sigma_2^2/\gamma^2 r_2(\rho + r_2^{-1})^3]\}$ with $r_i = \beta_1^2 \tau_i^2/\sigma_i^2$, $i = 1, 2$.

If one wishes to test for the following superiority hypotheses:

$$H_1 : \mu_1 - \mu_2 = \delta_1 > \delta,$$

the required sample size for achieving $1 - \beta$ power satisfies

$$-z_\alpha + (\delta_1 - \delta)/\sqrt{\text{Var}(\hat{\mu}_{GD1}) + \text{Var}(\hat{\mu}_{GD2})} = z_\beta.$$

This gives

$$n_1 = \frac{1}{2} DB \left(1 + \sqrt{1 + 8(1 + \rho)D^{-1}C} \right), \tag{18.35}$$

where $D = (z_\alpha + z_\beta)^2/(\delta_1 - \delta)^2$. For the case of testing for equivalence with a significance level α, consider the local alternative hypothesis that H_1: $\mu_1 - \mu_2 = \delta_1$ with $|\delta_1| < \delta$. The required sample size to achieve $1 - \beta$ power satisfies

$$-z_\alpha + (\delta - \delta_1)/\sqrt{\text{Var}(\hat{\mu}_{GD1}) + \text{Var}(\hat{\mu}_{GD2})} = z_\beta.$$

Thus, the total sample size for two treatment groups is $(1 + \rho)(1 + \gamma)n_1$ with n_1 given as

$$n_1 = \frac{1}{2} EB \left(1 + \sqrt{1 + 8(1 + \rho)E^{-1}C} \right), \tag{18.36}$$

where

$$E = \frac{(z_\alpha + z_{\beta/2})^2}{(\delta - |\delta_1|)^2}.$$

Table 18.5 provides a summary of formulas for sample size requirements for testing equality, superiority, noninferiority, and equivalence under a Category II SD two-stage adaptive trial designs with different continuous endpoints at different stages.

TABLE 18.5

Summary of Sample Size Requirements for Category II SD Two-Stage Adaptive Designs

Hypothesis Testing	Continuous Endpoint	Binary Response
Equality	$n_1 = \frac{1}{2} AB \left(1 + \sqrt{1 + 8(1 + \rho)A^{-1}C} \right)$	$n_1 = \dfrac{(z_{\alpha/2} + z_\beta)^2(\tilde{\sigma}_1^2(\lambda_1) + \tilde{\sigma}_2^2(\lambda_2))}{(\lambda_2 - \lambda_1)^2}$
Noninferiority $(\delta < 0)$	$n_1 = \frac{1}{2} BD \left(1 + \sqrt{1 + 8(1 + \rho)D^{-1}C} \right)$	$n_1 = \dfrac{(z_\alpha + z_\beta)^2(\tilde{\sigma}_1^2(\lambda_1) + \tilde{\sigma}_2^2(\lambda_2))}{(\lambda_2 - \lambda_1 - \delta)^2}$
Superiority $(\delta > 0)$	$n_1 = \frac{1}{2} DB \left(1 + \sqrt{1 + 8(1 + \rho)D^{-1}C} \right)$	$n_1 = \dfrac{(z_\alpha + z_\beta)^2(\tilde{\sigma}_1^2(\lambda_1) + \tilde{\sigma}_2^2(\lambda_2))}{(\lambda_2 - \lambda_1 - \delta)^2}$
Equivalence	$n_1 = \frac{1}{2} EB \left(1 + \sqrt{1 + 8(1 + \rho)E^{-1}C} \right)$	$n_1 = \dfrac{(z_\alpha + z_\beta)^2(\tilde{\sigma}_1^2(\lambda_1) + \tilde{\sigma}_2^2(\lambda_2))}{(\delta - |\lambda_2 - \lambda_1|)^2}$

18.3.2 Binary Responses

Consider the case where the primary study endpoint is a binary response with different treatment durations at different stages. Suppose that the study duration of the first stage is L, while the study duration of the second stage is CL with $C > 1$. Assume that the response is determined by the lifetime t, and the corresponding lifetime distribution for the test treatment is $G_1(t,\theta_1)$, while for the control it is $G_2(t,\theta_2)$. Denote by r_i the number of responders among n_i individuals in the ith stage for the test treatment, $i = 1, 2$. Similarly, denote by s_i the number of responders among m_i individuals in the ith stage for the control treatment. Based on the observed data, suppose $G_1(t,\theta_1) = G(t,\lambda_1)$ and $G_2(t,\theta_2) = G(t,\lambda_2)$, then the likelihood functions become

$$L(\lambda_i) = \left(1 - e^{-\lambda_i cL}\right)^{r_i} e^{-(n_i - r_i)\lambda_i cL} \left(1 - e^{-\lambda_i L}\right)^{s_i} e^{-(m_i - s_i)\lambda_i L}. \tag{18.37}$$

Let $\hat{\lambda}_i$ be the maximum likelihood estimate (MLE) of λ_i. Utilizing numerical methods such as Newton–Raphson method, $\hat{\lambda}_i$ can be found by solving the following equation:

$$\frac{r_i c}{e^{\lambda_i cL} - 1} + \frac{s_i}{e^{\lambda_i L} - 1} - (n_i - r_i)c - (m_i - s_i) = 0, \tag{18.38}$$

which is obtained by setting the first-order partial derivative with respect to the parameter to zero. Note that the MLE of λ_i exist if and only if r_i/n_i and s_i/m_i do not equal 0 or 1 at the same time.

Based on asymptotic normality of MLE, $\hat{\lambda}_i$ asymptotically follows a normal distribution. In particular, as n_i and m_i tend to infinity, $(\hat{\lambda}_i - \lambda_i)/\sigma_i(\lambda_i)$ follows the standard normal distribution where

$$\sigma_i(\lambda_i) = L^{-1}\left(n_i c^2 (e^{\lambda_i cL} - 1)^{-1} + m_i (e^{\lambda_i L} - 1)^{-1}\right)^{-1/2}.$$

Let $\sigma_i(\hat{\lambda}_i)$ be the MLE of $\sigma_i(\lambda_i)$. Then, based on the consistency of MLE, by the Slutsky's theorem $(\hat{\lambda}_i - \lambda_i)/\sigma_i(\hat{\lambda}_i)$ asymptotically follows the standard normal distribution. Consequently, an approximated $(1 - \alpha)$ confidence interval of λ_i is given as $(\hat{\lambda}_i - z_{\alpha/2}\sigma_i(\hat{\lambda}_i), \quad \hat{\lambda}_i + z_{\alpha/2}\sigma_i(\hat{\lambda}_i))$, where z_u is the upper u-quantile of the standard normal distribution. Under the exponential model, comparison of two treatments usually focuses on the hazard rate λ_i. In what follows, hypothesis testing and sample size calculation for different types of comparison are considered.

Test for equality—For equality testing, the hypotheses are formulated as

$$H_0 : \lambda_1 = \lambda_2 \quad \text{versus} \quad H_1 : \lambda_1 \neq \lambda_2. \tag{18.39}$$

Since $\hat{\lambda}_i$ is asymptotically normally distributed, and $\hat{\lambda}_1$ and $\hat{\lambda}_2$ are independent, it follows that under the null hypothesis, $(\hat{\lambda}_1 - \hat{\lambda}_2)/\sqrt{\sigma_1^2(\hat{\lambda}_1) + \sigma_2^2(\hat{\lambda}_2)}$ asymptotically follows the standard normal distribution. Thus, the null hypothesis in Equation 18.39 is rejected at approximate α level of significance if

$$|\hat{\lambda}_1 - \hat{\lambda}_2| / \sqrt{\sigma_1^2(\hat{\lambda}_1) + \sigma_2^2(\hat{\lambda}_2)} > z_{\alpha/2}.$$

Under the alternative hypothesis that $\lambda_1 \neq \lambda_2$, the approximate power of the above test is given as

$$\Phi\left(\left|\lambda_1 - \lambda_2\right| / \sqrt{\sigma_1^2(\lambda_1) + \sigma_2^2(\lambda_2)} - z_{\alpha/2}\right),$$

where $\Phi(t)$ is the cumulative function of the standard normal distribution. Thus, to achieve a prespecific power of $1 - \beta$, the required sample size satisfies

$$\left|\lambda_1 - \lambda_2\right| / \sqrt{\sigma_1^2(\lambda_1) + \sigma_2^2(\lambda_2)} - z_{\alpha/2} = z_\beta.$$

Let $m_i = \rho n_i$ and $n_2 = \gamma n_1$, $i = 1, 2$. Then, the total sample size N_T for two treatments is $(1 + \rho)(1 + \gamma)n_1$ with n_1 given as

$$n_1 = \frac{(z_{\alpha/2} + z_\beta)^2(\tilde{\sigma}_1^2(\lambda_1) + \tilde{\sigma}_2^2(\lambda_2))}{(\lambda_1 - \lambda_2)^2}, \tag{18.40}$$

where

$$\tilde{\sigma}_1^2(\lambda_1) = L^{-2}\left(c^2(e^{\lambda_1 cL} - 1)^{-1} + \rho(e^{\lambda_1 L} - 1)^{-1}\right)^{-1}$$

and

$$\tilde{\sigma}_2^2(\lambda_2) = L^{-2}\gamma^{-1}\left(c^2(e^{\lambda_2 cL} - 1)^{-1} + \rho(e^{\lambda_2 L} - 1)^{-1}\right)^{-1}.$$

Test for superiority—Under the exponential model, a smaller hazard rate indicates a better performance of the treatment. As a result, to identify superiority of the new treatment over the control, the following hypotheses are considered:

$$H_0 : \lambda_2 - \lambda_1 \leq \delta \quad \text{versus} \quad H_1 : \lambda_2 - \lambda_1 > \delta, \tag{18.41}$$

where $\delta > 0$ is a difference of clinical importance. Obviously, the null hypothesis should be rejected for a large value of $(\hat{\lambda}_2 - \hat{\lambda}_1 - \delta)/\sqrt{\sigma_1^2(\hat{\lambda}_1) + \sigma_2^2(\hat{\lambda}_2)}$. Under the null hypothesis, $(\hat{\lambda}_2 - \hat{\lambda}_1 - \delta)/\sqrt{\sigma_1^2(\hat{\lambda}_1) + \sigma_2^2(\hat{\lambda}_2)}$ is asymptotically normally distributed. Thus, the null hypothesis is rejected at approximately α level of significance if

$$(\hat{\lambda}_2 - \hat{\lambda}_1 - \delta)/\sqrt{\sigma_1^2(\hat{\lambda}_1) + \sigma_2^2(\hat{\lambda}_2)} > z_\alpha.$$

Under the alternative hypothesis that $\lambda_2 - \lambda_1 > \delta$, the approximate power of the above test is given as

$$\Phi\left((\lambda_2 - \lambda_1 - \delta)/\sqrt{\sigma_1^2(\lambda_1) + \sigma_2^2(\lambda_2)} - z_{\alpha/2}\right).$$

Thus, to achieve a power of $1 - \beta$, the required sample size satisfies

$$(\lambda_2 - \lambda_1 - \delta)/\sqrt{\sigma_1^2(\lambda_1) + \sigma_2^2(\lambda_2)} - z_{\alpha/2} = z_\beta.$$

Thus, the total sample size N_T for two treatments is $(1 + \rho)(1 + \gamma)n_1$ with n_1 given as follows:

$$n_1 = \frac{(z_\alpha + z_\beta)^2(\tilde{\sigma}_1^2(\lambda_1) + \tilde{\sigma}_2^2(\lambda_2))}{(\lambda_2 - \lambda_1 - \delta)^2}. \qquad (18.42)$$

Test for noninferiority—To show that the new treatment is not worse than the control, we may consider the following hypotheses:

$$H_0 : \lambda_1 - \lambda_2 \geq \delta \quad \text{versus} \quad H_1 : \lambda_1 - \lambda_2 < \delta,$$

which are equivalent to

$$H_0 : \lambda_2 - \lambda_1 \leq -\delta \quad \text{versus} \quad H_1 : \lambda_2 - \lambda_1 > -\delta, \qquad (18.43)$$

where $\delta > 0$ is a difference of clinical importance. The hypotheses in Equation 18.43 are of similar form as those for superiority testing. Therefore, the null hypothesis is rejected at approximate α level of significance if

$$(\hat{\lambda}_2 - \hat{\lambda}_1 + \delta)/\sqrt{\sigma_1^2(\hat{\lambda}_1) + \sigma_2^2(\hat{\lambda}_2)} > z_\alpha,$$

and the total sample size N_T for two treatments to achieve a power of $1 - \beta$ is $(1 + \rho)$ $(1 + \gamma)n_1$ with n_1 given as

$$n_1 = \frac{(z_\alpha + z_\beta)^2(\tilde{\sigma}_1^2(\lambda_1) + \tilde{\sigma}_2^2(\lambda_2))}{(\lambda_2 - \lambda_1 + \delta)^2}. \qquad (18.44)$$

Test for equivalence—In clinical trial, it is commonly unknown whether the performance of a new treatment is better than the (active) control, especially when prior knowledge of the new treatment is not available. In this case, it is more appropriate to consider the following hypotheses for therapeutic equivalence:

$$H_0 : |\lambda_1 - \lambda_2| \geq \delta \quad \text{versus} \quad H_1 : |\lambda_1 - \lambda_2| < \delta. \qquad (18.45)$$

The above hypotheses can be tested by constructing the confidence interval of $\lambda_2 - \lambda_1$. It can be verified that the null hypothesis is rejected at a significance level α if and only if the $100(1 - 2\alpha)\%$ confidence interval $\hat{\lambda}_2 - \hat{\lambda}_1 \pm z_\alpha \sqrt{\sigma_1^2(\hat{\lambda}_1) + \sigma_2^2(\hat{\lambda}_2)}$ falls within $(-\delta, \delta)$. In other words, the test treatment is concluded to be equivalent to the control if

$$(\hat{\lambda}_2 - \hat{\lambda}_1 - \delta)/\sqrt{\sigma_1^2(\hat{\lambda}_1) + \sigma_2^2(\hat{\lambda}_2)} < -z_\alpha,$$

and

$$(\hat{\lambda}_2 - \hat{\lambda}_1 + \delta)/\sqrt{\sigma_1^2(\hat{\lambda}_1) + \sigma_2^2(\hat{\lambda}_2)} > z_\alpha.$$

Under the alternative hypothesis that $|\lambda_1 - \lambda_2| < \delta$, the approximate power of the above test is given by

$$2\Phi\left((\delta - |\lambda_2 - \lambda_1|)/\sqrt{\sigma_1^2(\lambda_1) + \sigma_2^2(\lambda_2)} - z_\alpha\right) - 1.$$

Thus, to achieve a power of $1 - \beta$, the required sample size satisfies

$$(\delta - |\lambda_2 - \lambda_1|)/\sqrt{\sigma_1^2(\lambda_1) + \sigma_2^2(\lambda_2)} - z_\alpha = z_{\beta/2}.$$

Then, with the same notations in Section 3.1, the total sample size N_T for two treatments is $(1 + \rho)(1 + \gamma)n_1$ with n_1 given as follows:

$$n_1 = \frac{(z_\alpha + z_{\beta/2})^2 (\tilde{\sigma}_1^2(\lambda_1) + \tilde{\sigma}_2^2(\lambda_2))}{(\delta - |\lambda_2 - \lambda_1|)^2}. \tag{18.46}$$

Table 18.5 summarizes formulas for sample size requirements for testing equality, superiority, noninferiority, and equivalence under a Category II SD two-stage adaptive trial designs with different binary endpoints at different stages.

18.3.3 Time-to-Event Endpoints

Weibull distribution—Consider a two-stage adaptive clinical trial design for comparing two treatment groups, that is, a test (T) treatment and a control or reference (R) treatment. Let t_{ijk} denote the length of time of a patient from entering the trial to the occurrence of a defined event of interest for the kth subject at the jth stage in the ith treatment, where $k = 1, 2, \ldots, n_{ij}, j = 1, 2, i = T$ and R. Assume that the study durations for the first stage and the second stage are different, which are given by cL and L, respectively, where $c < 1$. Furthermore, assume that t_{ijk} follows a distribution with $G(t,\theta_i)$ and $g(t,\theta_i)$ as the cumulative distribution function (cdf) and probability density function (pdf) with parameter vector θ_i, respectively. Then, the data collected from the study can be represented by (x_{ijk},δ_{ijk}), where $\delta_{ijk} = 1$ indicates that the event of interest is observed and that $x_{ijk} = t_{ijk}$, while $\delta_{ijk} = 0$ means that the event is not observed during the study, that is, x_{ijk} is censored and $x_{ijk} < t_{ijk}$.

Given the observed data, the likelihood function for the test treatment and the control treatment is

$$L(\theta_i) = \prod_{j=1}^{2} \prod_{k=1}^{n_{ij}} g^{\delta_{ijk}}(x_{ijk},\theta_i)[1 - G(x_{ijk},\theta_i)]^{1-\delta_{ijk}} \tag{18.47}$$

for $i = T, R$. In particular, suppose that the observed time-to-event data are assumed to follow a Weibull distribution. Denote the cumulative distribution function of a

Weibull distribution with $\lambda, \beta > 0$ by $G(t; \lambda, \beta)$, where $G(t; \lambda, \beta) = 1 - e^{-(t/\lambda)^{\beta}}$. Suppose that $G(t; \theta_T) = G(t; \lambda_T, \beta_T)$ and $G(t; \theta_R) = G(t; \lambda_R, \beta_R)$, that is, t_{ijk} follows a Weibull distribution with cdf $G(t; \lambda_i, \beta_i)$. Then the likelihood function in Equation 18.47 becomes

$$L(\lambda_i, \beta_i) = \left(\beta_i \lambda_i^{-\beta_i} \right)^{\sum_{j=1}^{2} \sum_{k=1}^{n_{ij}} \delta_{ijk}} e^{-\sum_{j=1}^{2} \sum_{k=1}^{n_{ij}} \tilde{x}_{ijk}} \prod_{j=1}^{2} \prod_{k=1}^{n_{ij}} x_{ijk}^{(\beta_i - 1)\delta_{ijk}}, \tag{18.48}$$

where $\tilde{x}_{ijk} = (x_{ijk}/\lambda_i)^{\beta_i}$. Let $l(\lambda_i, \beta_i) = \log(L(\lambda_i, \beta_i))$ be the log-likelihood function. The following equations can be obtained by setting the first-order partial derivatives of $l(\lambda_i, \beta_i)$ with respect to λ_i and β_i to zero:

$$\sum_{j=1}^{2} \sum_{k=1}^{n_{ij}} (\tilde{x}_{ijk} - \delta_{ijk}) = 0,$$

$$\sum_{j=1}^{2} \sum_{k=1}^{n_{ij}} \delta_{ijk} + \sum_{j=1}^{2} \sum_{k=1}^{n_{ij}} (\delta_{ijk} - \tilde{x}_{ijk}) \log(\tilde{x}_{ijk}) = 0.$$

Denote the MLEs of β_i and λ_i by $\hat{\beta}_i$ and $\hat{\lambda}_i$, respectively. Then, $\hat{\beta}_i$ can be obtained by solving the following equation using numerical method:

$$\beta_i^{-1} + d_i^{-1} \sum_{j=1}^{2} \sum_{k=1}^{n_{ij}} \delta_{ijk} \log x_{ijk} - \frac{1}{\sum_{j=1}^{2} \sum_{k=1}^{n_{ij}} x_{ijk}^{\beta_i}} \sum_{j=1}^{2} \sum_{k=1}^{n_{ij}} x_{ijk}^{\beta_i} \log x_{ijk} = 0, \tag{18.49}$$

where $d_i = \sum_{j=1}^{2} \sum_{k=1}^{n_{ij}} \delta_{ijk}$. Consequently, $\hat{\lambda}_i$ is given by

$$\hat{\lambda}_i = \left(d^{-1} \sum_{j=1}^{2} \sum_{k=1}^{n_{ij}} x_{ijk}^{\hat{\beta}_i} \right)^{\hat{\beta}_i^{-1}}. \tag{18.50}$$

Note that δ_{ijk} is the indicator of the event of interest and x_{ijk} is the observed time, which may be censored. Then, $E[\delta_{ijk}] = 1 - e^{-\pi_{ij}}$,

$$E\left[\tilde{x}_{ijk} \log^2(\tilde{x}_{ijk}) \right] = \pi_{ij} \log^2 \pi_{ij} e^{-\pi_{ij}} + \int_0^{\pi_{ij}} x e^{-x} \log^2 x \, dx,$$

and

$$E[\delta_{ijk} \beta_i \log(\tilde{x}_{ijk})] = \int_0^{\pi_{ij}} e^{-x} \log x \, dx,$$

where $\pi_{ij} = (L_j / \lambda_i)^{\beta_i}$, $L_1 = cL$, and $L_2 = L$, $i = T, R$ and $j = 1, 2$. Consequently, the entries of the Fisher information matrix correspond to the parameters (λ_i, β_i) and are given as

$$E\left[-\frac{\partial^2 l(\lambda_i, \beta_i)}{\partial \beta_i^2}\right] = \frac{1}{\beta_i^2} \sum_{j=1}^{2} n_{ij} \left[1 - e^{-\pi_{ij}} + \pi_{ij} e^{-\pi_{ij}} \log^2 \pi_{ij} + \int_0^{\pi_{ij}} x e^{-x} \log^2 x\, dx\right],$$

$$E\left[-\frac{\partial^2 l(\lambda_i, \beta_i)}{\partial \lambda_i \partial \beta_i}\right] = -\frac{1}{\lambda_i} \sum_{j=1}^{2} n_{ij} \left[1 - e^{-\pi_{ij}} + \int_0^{\pi_{ij}} e^{-x} \log x\, dx\right],$$

and

$$E\left[-\frac{\partial^2 l(\lambda_i, \beta_i)}{\partial \lambda_i^2}\right] = \frac{\beta_i^2}{\lambda_i^2} \sum_{j=1}^{2} n_{ij} (1 - e^{-\pi_{ij}}).$$

Based on the asymptotic normality of MLE (Serfling, 1980), for $i = T, R, \sqrt{n_{i1}}(\hat{\lambda}_i - \lambda_i)/\sigma(\lambda_i)$ and $\sqrt{n_{i1}}(\hat{\beta}_i - \beta_i)/\sigma(\beta_i)$ converge in distribution to a standard normal distribution, where

$$\Delta_i = E\left[-\frac{\partial^2 l(\lambda_i, \beta_i)}{\partial \lambda_i^2}\right] E\left[-\frac{\partial^2 l(\lambda_i, \beta_i)}{\partial \beta_i^2}\right] - E^2\left[-\frac{\partial^2 l(\lambda_i, \beta_i)}{\partial \lambda_i \partial \beta_i}\right],$$

$$\sigma(\lambda_i) = \left[\frac{n_{i1}}{\Delta_i} E\left(-\frac{\partial^2 l(\lambda_i, \beta_i)}{\partial \beta_i^2}\right)\right]^{1/2},$$

and

$$\sigma(\beta_i) = \left[\frac{n_{i1}}{\Delta_i} E\left(-\frac{\partial^2 l(\lambda_i, \beta_i)}{\partial \lambda_i^2}\right)\right]^{1/2}.$$

Note that $\sigma(\lambda_i)$ and $\sigma(\beta_i)$ do not depend on n_{ij} for given n_{i2}/n_{i1}. Let $\sigma(\hat{\lambda}_i)$ and $\sigma(\hat{\beta}_i)$ be the MLEs of $\sigma(\lambda_i)$ and $\sigma(\beta_i)$, respectively. Thus, approximate $100(1 - \alpha)\%$ confidence intervals of λ_i and β_i can be constructed as follows:

$$\left(\hat{\lambda}_i - z_{\alpha/2}\sigma(\hat{\lambda}_i)/\sqrt{n_{i1}}, \quad \hat{\lambda}_i + z_{\alpha/2}\sigma(\hat{\lambda}_i)/\sqrt{n_{i1}}\right), \tag{18.51}$$

and

$$\left(\hat{\beta}_i - z_{\alpha/2}\sigma(\hat{\beta}_i)/\sqrt{n_{i1}}, \quad \hat{\beta}_i + z_{\alpha/2}\sigma(\hat{\beta}_i)/\sqrt{n_{i1}}\right), \tag{18.52}$$

where z_u is the $(1 - u)$-quantile of the standard normal distribution.

In clinical studies such as oncology trials, it is of interest to estimate the median (survival) lifetime. Thus, the comparison is often based on the medians of the lifetime distributions corresponding to the control and test treatments. Denote the median of a Weibull distribution with parameters (λ, β) by M, which is given as $\lambda(\log 2)^{1/\beta}$. In what follows, results for the hypotheses testing of equality, superiority, noninferiority, and equivalence of the medians of the two treatments are discussed.

Test for equality—For testing equality of two medians, the following hypotheses are usually considered:

$$H_0 : M_T = M_R \quad \text{versus} \quad H_1 : M_T \neq M_R, \tag{18.53}$$

where M_i is the median of $G(t;\lambda_i,\beta_i)$, $i = T, R$. Let $\hat{M}_i$ be the MLE of M_i. Based on the asymptotic normality of $(\hat{\lambda}_i, \hat{\beta}_i)$, $\hat{M}_i$ is approximately normally distributed for sufficiently large n_{ij}. In particular,

$$(\hat{M}_i - M_i)/\sqrt{n_{i1}^{-1} v_i} \xrightarrow{d} N(0,1),$$

where

$$v_i = M_i^2 \left[\lambda_i^{-2} \sigma^2(\lambda_i) + \beta_i^{-4} \log^2(\log(2)) \sigma^2(\beta_i) - 2\lambda_i^{-1} \beta_i^{-2} \log(\log(2)) \sigma(\lambda_i, \beta_i) \right]$$

and

$$\sigma(\lambda_i, \beta_i) = \frac{n_{i1}}{\Delta} E\left(\frac{\partial^2 l(\lambda_i, \beta_i)}{\partial \lambda_i \partial \beta_i} \right).$$

Note that $\hat{M}_T$ and $\hat{M}_R$ are independent. Thus, we have

$$[(\hat{M}_T - \hat{M}_R) - (M_T - M_R)]/\sqrt{n_{T1}^{-1} v_T + n_{R1}^{-1} v_R} \xrightarrow{d} N(0,1).$$

Let $\hat{v}_i$ be the MLE of v_i with λ_i and β_i replaced by the corresponding MLE. Then, according to the Slutsky's theorem, the above result also holds if v_i is replaced by $\hat{v}_i$. Consequently, $(\hat{M}_T - \hat{M}_R)/\sqrt{n_{T1}^{-1} \hat{v}_T + n_{R1}^{-1} \hat{v}_R}$ is asymptotically distributed as the standard normal distribution under the null hypothesis. Thus, we reject the null hypothesis at an approximate α level of significance if

$$|\hat{M}_T - \hat{M}_R| / \sqrt{n_{T1}^{-1} \hat{v}_T + n_{R1}^{-1} \hat{v}_R} > z_{\alpha/2}. \tag{18.54}$$

Since $(\hat{M}_T - M_T)/\sqrt{n_{T1}^{-1} \hat{v}_T}$ approximately follows the standard normal distribution, the power of the above test under H_1 can be approximated by $\Phi\left(|M_T - M_R| / \sqrt{n_{T1}^{-1} v_T} - z_{\alpha/2} \right)$, where Φ is the distribution function of the standard normal distribution. Hence, to achieve a power of $1 - \beta$, the required sample size satisfies $|M_T - M_R| / \sqrt{n_{T1}^{-1} v_T} - z_{\alpha/2} = z_\beta$. If the sample size of the two stages is related by $n_{T2} = \rho n_{T1}$ for a known constant ρ, then the required total sample size N for the two phases is $N = (1 + \rho)n_{T1}$ with

$$n_{T1} = \frac{(z_{\alpha/2} + z_\beta)^2 v_T}{(M_T - M_R)^2}. \tag{18.55}$$

Test for superiority—In general, a larger median survival time indicates better performance of the test treatment. Thus, it is of interest to test the superiority of the test treatment over the control. In this case, the following hypotheses are usually considered:

$$H_0 : M_T - M_R \leq \delta \quad \text{versus} \quad H_1 : M_T - M_R > \delta, \tag{18.56}$$

where $\delta > 0$ is a difference of clinical importance. Obviously, the null hypothesis should be rejected for a large value of $(\hat{M}_T - \hat{M}_R - \delta)/\sqrt{n_{T1}^{-1}\hat{v}_T + n_{R1}^{-1}\hat{v}_R}$. Under the null hypothesis, $(\hat{M}_T - \hat{M}_R - \delta)/\sqrt{n_{T1}^{-1}\hat{v}_T + n_{R1}^{-1}\hat{v}_R}$ asymptotically follows the standard normal distribution for sufficiently large n_{ij}. Thus, the null hypothesis in Equation 18.55 is rejected at an approximate α level of significance if

$$(\hat{M}_T - \hat{M}_R - \delta)/\sqrt{n_{T1}^{-1}\hat{v}_T + n_{R1}^{-1}\hat{v}_R} > z_\alpha. \tag{18.57}$$

Similarly, for testing of superiority, the sample size n_{T1} is given by

$$n_{T1} = \frac{(z_\alpha + z_\beta)^2 (v_T + \gamma^{-1} v_R)}{(M_T - M_R - \delta)^2}, \tag{18.58}$$

Test for noninferiority—To show that the test treatment is at least as effective as or not worse than the control, we may consider the following hypotheses for testing noninferiority:

$$H_0 : M_R - M_T \geq \delta \quad \text{versus} \quad H_1 : M_R - M_T < \delta,$$

which are equivalent to testing the following hypotheses:

$$H_0 : M_T - M_R \leq -\delta \quad \text{versus} \quad H_1 : M_T - M_R > -\delta, \tag{18.59}$$

where $\delta > 0$ is a difference of clinical importance. The hypotheses in Equation 18.59 are of the same form as those for superiority testing. Thus, the null hypothesis is rejected at an approximate α level of significance if

$$(\hat{M}_T - \hat{M}_R + \delta)/\sqrt{n_{T1}^{-1}\hat{v}_T + n_{R1}^{-1}\hat{v}_R} > z_\alpha. \tag{18.60}$$

Similarly, for the testing of noninferiority, the sample size n_{T1} is given as follows:

$$n_{T1} = \frac{(z_\alpha + z_\beta)^2 (v_T + \gamma^{-1} v_R)}{(M_T - M_R + \delta)^2}, \tag{18.61}$$

Test for equivalence—In clinical trial, it is commonly unknown whether the performance of the test treatment is better than the (active) control, especially when prior knowledge of the test treatment is not available. In this case, it is more appropriate to consider the following hypotheses for testing therapeutic equivalence:

$$H_0 : |M_T - M_R| \geq \delta \quad \text{versus} \quad H_1 : |M_T - M_R| < \delta. \tag{18.62}$$

The above hypotheses can be tested by constructing the confidence interval of $M_T - M_R$. It can be verified that the null hypothesis is rejected at an approximate significance level α if and only if the $100(1 - 2\alpha)\%$ confidence interval, which is $\hat{M}_T - \hat{M}_R \pm z_\alpha \sqrt{n_{T1}^{-1}\hat{v}_T + n_{R1}^{-1}\hat{v}_R}$, falls within $(-\delta, \delta)$. In other words, the test treatment is concluded to be equivalent to the control if

$$\frac{(\hat{M}_T - \hat{M}_R - \delta)}{\sqrt{n_{T1}^{-1}\hat{v}_T + n_{R1}^{-1}\hat{v}_R}} < -z_\alpha \quad \text{and} \quad \frac{(\hat{M}_T - \hat{M}_R + \delta)}{\sqrt{n_{T1}^{-1}\hat{v}_T + n_{R1}^{-1}\hat{v}_R}} > z_\alpha. \tag{18.63}$$

Similarly, for testing equivalence, the sample size n_{T1} is given by

$$n_{T1} = \frac{(z_\alpha + z_{\beta/2})^2(v_T + \gamma^{-1}v_R)}{(|M_T - M_R| - \delta)^2}. \tag{18.64}$$

Cox's proportional hazards model—The discussion in the previous section is based on a parametric approach. However, the corresponding results derived may not be robust against model misspecification. Thus, alternatively, hypotheses testing for equality, superiority, and noninferiority/equivalence and the corresponding formulas for sample size calculation and sample size allocation between the two stages under these hypotheses are explored using a semiparametric approach. In particular, the Cox's proportional hazard model is considered.

Let n_j be the total sample size for the two treatments in the jth stage, $j = 1, 2$ and d_j be the number of distinct failure times in the jth stage, which are denoted by $t_{j1} < t_{j2} < \ldots < t_{jd_j}$. Furthermore, denote the observation based on the kth subject in the jth stage by

$$\left(T_{jk}, \delta_{jk}, z_{jk}(t), \, 0 \leq t \leq T_{jk}\right) = \left(\min(\tilde{T}_{jk}, C_{jk}), \quad I(\tilde{T}_{jk} < C_{jk}), \quad z_{jk}(t), \, 0 \leq t \leq T_{jk}\right),$$

where, correspondingly, T_{jk} is the observed time, $\tilde{T}_{jk}$ is time to event, δ_{jk} is the indicator for the observed failure, C_{jk} is a censoring time that is assumed to be independent of $\tilde{T}_{jk}$, and $z_{jk}(t)$ is a covariate vector at time t. Let $h(t|z)$ be the hazard rate at time t for an individual with a covariate vector z. The Cox proportional hazard model (Cox, 1972) assumes

$$h(t \mid z(t)) = h(t \mid 0)e^{b'z(t)},$$

where the baseline $h(t|0)$ is unspecified and b is a coefficient vector with the same dimension as $z(t)$. Thus, the partial likelihood function is

$$L(b) = \prod_{j=1}^{2} \prod_{k=1}^{d_j} P(\text{observed failure at time } t_{jk} \mid R(t_{jk})),$$

$$= \prod_{j=1}^{2} \prod_{k=1}^{d_j} \frac{\exp(b'z_{(jk)}(t_{jk}))}{\sum_{l \in R(t_{jk})} \exp(b'z_l(t_{jk}))},$$

where the risk set $R(t_{jk}) = \{js : \tilde{T}_{js} \geq T_{jk}\}$ is the collection of subjects still on study just prior to t_{jk} in the jth stage. Furthermore, the first-derivative of the log partial likelihood equation is

$$U(b) = \sum_{j=1}^{2} \sum_{k=1}^{d_j} \left[z_{(jk)}(t_{jk}) - e(b, t_{jk}) \right], \tag{18.65}$$

where

$$e(b, t_{jk}) = \frac{\sum_{l \in R(t_{jk})} \exp(b' z_l(t_{jk})) z_l(t_{jk})}{\sum_{l \in R(t_{jk})} \exp(b' z_l(t_{jk}))}.$$

Based on Equation 18.65, the corresponding observed information matrix is

$$I(b) = \sum_{j=1}^{2} \sum_{k=1}^{d_j} \left[\frac{\sum_{l \in R(t_{jk})} \exp(b' z_l(t_{jk})) z_l(t_{jk}) z_l'(t_{jk})}{\sum_{l \in R(t_{jk})} \exp(b' z_l(t_{jk}))} - e(b, t_{jk}) e'(b, t_{jk}) \right]. \tag{18.66}$$

Test for equality—Based on the formulation of the Cox model, the testing of equality can be conducted through the comparison of the coefficient vector b. Thus, consider the following hypotheses:

$$H_0 : b = b_0 \quad \text{versus} \quad H_1 : b \neq b_0. \tag{18.67}$$

To test the above hypotheses, the score statistic $T_s = U'(b_0)I^{-1}(b_0)U(b_0)$ is considered. Under the null hypothesis H_0, T_s asymptotically follows a chi-squared distribution with p degrees of freedom where $p = \dim(b)$ (Cox and Hinkley, 1974). Thus, H_0 is rejected at an approximate α level of significance if $T_s > \chi_p^2(\alpha)$, where $\chi_p^2(\alpha)$ is the α-upper quantile of a chi-squared random variable with p degrees of freedom.

Consider the special case that the treatment indicator is the only covariate considered in the study. Let $z_{jk} = 1$ for the test treatment and $z_{jk} = 0$ for the control treatment. Then, the baseline $h(t|0)$ is the hazard in the control treatment and b is the log-relative risk, which measures the relative treatment effect. In particular, $b > 0$ (<0) implies that the test treatment increases (decreases) the risk of failure and $b = 0$ means no difference in risk between the two treatments. Define $P_{jk} = n_{Tjk}e^b/(n_{Tjk}e^b + n_{Rjk})$, where n_{Tjk} and n_{Rjk} denote the number of subjects at risk, that is, those who have not failed or censored just prior to the kth observed failure in the jth stage in the test and the control treatment, respectively. Consequently, the score function in Equation 18.65 and the observed Fisher information matrix in Equation 18.67 can be simplified to

$$U(b) = \sum_{j=1}^{2} \sum_{k=1}^{d_j} [z_{(jk)} - P_{jk}] \quad \text{and} \quad I(b) = \sum_{j=1}^{2} \sum_{k=1}^{d_j} P_{jk}(1 - P_{jk}),$$

respectively, where $z_{(jk)}$ is the treatment indicator for the kth observed failure in the jth stage. For the testing of the hypotheses for the equality of the two treatments defined in Equation 18.67, the corresponding score test statistic is

$$T_s = \frac{U^2(0)}{I(0)} = \frac{\left[\sum_{j=1}^{2} \sum_{k=1}^{d_j} \left(z_{(jk)} - n_{Tjk}/(n_{Rjk} + n_{Tjk})\right)\right]^2}{\sum_{j=1}^{2} \sum_{k=1}^{d_j} \left(n_{Rjk}n_{Tjk}/(n_{Rjk} + n_{Tjk})^2\right)}. \qquad (18.68)$$

Under the null hypothesis in Equation 18.67, T_s is asymptotically distributed as a chi-squared distribution with 1 degree of freedom. Equivalently, consider the statistic

$$T_z(b) = \frac{U(b)}{I^{1/2}(b)} = \frac{\sum_{j=1}^{2} \sum_{k=1}^{d_j} (z_{(jk)} - P_{jk})}{\sqrt{\sum_{j=1}^{2} \sum_{k=1}^{d_j} P_{jk}(1 - P_{jk})}}. \qquad (18.69)$$

$T_z(0)$ is asymptotic standard normally distributed. Therefore, the null hypothesis in Equation 18.67 is rejected at an approximate α level of significance if $T_z(0) > z_{\alpha/2}$, where z_u is the upper u-quantile of the standard normal distribution.

Let p_{ij} be the proportion of subjects allocated to the ith treatment in the jth stage, $H_i(t)$ be the distribution function of censoring, and $f_i(t)$ and $F_i(t)$ be the density and cumulative distribution function of survival for the ith treatment, respectively, $j = 1, 2; i = T, R$. Furthermore, define

$$V_j(t) = p_{Rj}f_R(t)(1 - H_R(t)) + p_{Tj}f_T(t)(1 - H_T(t))$$

and

$$\pi_j(t,b) = \frac{p_{Tj}(1 - F_T(t))(1 - H_T(t))e^b}{p_{Rj}(1 - F_R(t))(1 - H_R(t)) + p_{Tj}(1 - F_T(t))(1 - H_T(t))e^b}.$$

Similar to Schoenfeld (1981), it can be shown that $T_z(b)$ is asymptotically normally distributed with variance equals to 1 and the mean is given by

$$n_1^{1/2}b\left[\int_0^{cL} \pi_1(t,0)(1 - \pi_1(t,0))V_1(t)dt + \rho \int_0^{L} \pi_2(t,0)(1 - \pi_2(t,0))V_2(t)dt\right]^{1/2}. \qquad (18.70)$$

Suppose that $H_R(t) = H_T(t)$. Under the local alternative hypothesis, b is of order $O\left(n_1^{-1/2}\right)$ and $F_R(t) \approx F_T(t)$ for sufficiently large n_1 and n_2. Thus, $\pi_j(t,0) \approx p_{Tj}$ and $1 - \pi_j(t,0) \approx p_{Rj}$. Thus, Equation 18.70 can be approximated by $b\sqrt{n_1(p_{T1}p_{R1}u_1 + \rho p_{T2}p_{R2}u_2)}$, where $\rho = n_2/n_1$, $u_1 = \int_0^{cL} V_1(t)dt$, and $u_2 = \int_0^{L} V_2(t)dt$. Therefore, under the local alternative hypothesis, the power of the above test is approximated by $\Phi(b\sqrt{n_1(p_{T1}p_{R1}u_1 + \rho p_{T2}p_{R2}u_2)} - z_{\alpha/2})$. To achieve a power level of $1 - \beta$, the required total sample size for two treatments, which is denoted by N_T, is $(1 + \rho)n_1$ with

$$n_1 = \frac{(z_{\alpha/2} + z_\beta)^2}{b^2(p_{T1}p_{R1}u_1 + \rho p_{T2}p_{R2}u_2)}. \tag{18.71}$$

Test for superiority/noninferiority—Note that the log-relative risk $b > 0$ implies worse treatment effect (inferiority) of the test treatment and $b < 0$ indicates better treatment effect (superiority) of the test treatment. To demonstrate superiority/noninferiority, the following hypotheses are considered:

$$H_0 : b \geq \delta \quad \text{versus} \quad H_1 : b < \delta, \tag{18.72}$$

where δ is a given superiority or noninferiority margin. For $\delta < 0(>0)$, the rejection of the null hypothesis implies superiority (noninferiority) of the test treatment against the control. If $\delta - b$ is of order $O(n_1^{-1/2})$, then following similar arguments in Schoenfeld (1981), $T_z(b)$ is asymptotically normally distributed with unit variance and mean $\mu(b)$ given by

$$n_1^{1/2}(b - \delta) \left[\int_0^{cL} \pi_1(t, \delta)(1 - \pi_1(t, \delta))V_1(t)dt + \rho \int_0^L \pi_2(t, \delta)(1 - \pi_2(t, \delta))V_2(t)dt \right]^{1/2}. \tag{18.73}$$

Consequently when $b = \delta$, the test statistic $T_z(b)$ approximately follows a standard normal distribution for sufficiently large sample size. Thus, the null hypothesis H_0 is rejected at an approximate level α of significance if $T_z(\delta) < -z_\alpha$.

For the determination of the required sample size to achieve a given level of power, assume $H_R(t) = H_T(t)$. Note that $1 - F_T(t) = (1 - F_R(t))^{e^b}$. To obtain an idea of the required sample size, approximate $\pi_j(t, \delta)$ by $p_{Tj}e^\delta / (p_{Tj}e^\delta + p_{Rj})$. Then, $\mu(b)$, as defined in Equation 18.73, can be approximated by $n_1^{1/2}(b - \delta)e^{\delta/2}K^{1/2}$, where $K = p_{R1}p_{T1}u_1 / (p_{R1} + p_{T1}e^\delta)^2 + \rho p_{R2}p_{T2}u_2 / (p_{R2} + p_{T2}e^\delta)^2$.

Under the local alternative hypothesis defined in Equation 18.72, the power of the test is approximately equal to $\Phi(-z_\alpha - \mu(b))$. To achieve a power of $1 - \beta$, the required total sample size N_T for two treatments is $N_T = (1 + \rho)n_1$ with

$$n_1 = \frac{(z_\alpha + z_\beta)^2}{e^\delta(b - \delta)^2 K}. \tag{18.74}$$

Test for equivalence—If the question of interest is to assess whether the performance of the test treatment is better than the (active) control, especially when prior knowledge of the test treatment is not available, it is more appropriate to consider the following hypotheses for the testing of therapeutic equivalence:

$$H_0 : |b| > \delta \quad \text{versus} \quad H_1 : |b| < \delta. \tag{18.75}$$

Since $|b| > \delta$ is equivalent to $b > \delta$ or $b < -\delta$, the above hypotheses can be tested by two one-sided test procedures. In particular, the null hypothesis is rejected at an approximate

TABLE 18.6

Summary of Sample Size Requirements for Category II SD Two-Stage Adaptive Designs with Time-to-Event Endpoints

Hypothesis Testing	Weibull Distribution	Cox Proportional Hazard Model
Equality	$n_1 = \dfrac{(z_{\alpha/2}+z_\beta)^2(v_T+\gamma^{-1}v_R)}{(M_T-M_R)^2}$	$n_1 = \dfrac{(z_{\alpha/2}+z_\beta)^2}{b^2(p_{T_1}p_{R_1}u_1+\rho p_{T_1}p_{R_1}u_2)^2}$
Noninferiority ($\delta < 0$)	$n_1 = \dfrac{(z_{\alpha/2}+z_\beta)^2(v_T+\gamma^{-1}v_R)}{(M_T-M_R-\delta)^2}$	$n_1 = \dfrac{(z_{\alpha/2}+z_\beta)^2}{e^\delta(b-\delta)^2 K} \quad (\delta<0)$
Superiority ($\delta > 0$)	$n_1 = \dfrac{(z_{\alpha/2}+z_\beta)^2(v_T+\gamma^{-1}v_R)}{(M_T-M_R+\delta)^2}$	$n_1 = \dfrac{(z_\alpha+z_\beta)^2}{e^\delta(b-\delta)^2 K} \quad (\delta>0)$
Equivalence	$n_1 = \dfrac{(z_{\alpha/2}+z_\beta)^2(v_T+\gamma^{-1}v_R)}{(|M_T-M_R|-\delta)^2}$	$n_1 = \dfrac{(z_{\alpha/2}+z_{\beta/2})^2}{e^\delta(b-|\delta|)^2 K}$

α level of significance if $T_z(\delta) < -z_\alpha$ and $T_z(-\delta) > z_\alpha$. Based on the asymptotic normality of $T_z(b)$, under the local alternative hypothesis defined in Equation 18.75, the power of the test can be approximated by $2\Phi((\delta-|b|)\sqrt{n_1 e^\delta K}-z_\alpha)-1$, where

$$K = \frac{p_{R1}p_{T1}u_1}{(p_{R1}+p_{T1}e^\delta)^2} + \frac{\rho p_{R2}p_{T2}u_2}{(p_{R2}+p_{T2}e^\delta)^2}.$$

Thus, the required total sample size N_T for two treatments to achieve a power of $1-\beta$ is $N_T = (1+\rho)n_1$ with

$$n_1 = \frac{(z_\alpha+z_{\beta/2})^2}{e^\delta(b-|\delta|)^2 K}. \tag{18.76}$$

Table 18.6 summarizes formulas for sample size requirements for testing equality, superiority, noninferiority, and equivalence under a Category II SD two-stage adaptive trial designs with different time-to-event endpoints at different stages.

18.4 Analysis and Sample Size for Category III DS and IV DD Two-Stage Adaptive Designs

For a Category III DS two-stage adaptive design, the study objectives at different stages are different (e.g., dose selection vs. efficacy confirmation), but the study endpoints are same at different stages. For a Category IV design, both study objectives and endpoints at different stages are different (e.g., dose selection vs. efficacy confirmation with surrogate endpoint vs. clinical study endpoint).

As indicated earlier, how to control the overall type I error rate at a prespecified level is one of the major regulatory concerns when adaptive design methods are employed in confirmatory clinical trials. Another concern is how to perform power analysis for sample size calculation/allocation for achieving individual study objectives originally set by the two separate studies (different stages). In addition, how to combine data collected from both stages for a combined and valid final analysis is yet another concern. Under a Category III or IV phase II/III seamless adaptive design, in addition, the investigator plans to have an interim analysis at each stage. Thus, if we consider the initiation of the study, first interim analysis, end of stage 1 analysis, second interim analysis, and final analysis as critical milestones, the two-stage adaptive design becomes a four-stage transitional seamless trial design. In what follows, we will focus on the analysis of a four-stage transitional seamless design without (nonadaptive version) and with (adaptive version) adaptations, respectively.

18.4.1 Nonadaptive Version

For a given clinical trial comparing k treatment groups, $E_1, ..., E_k$ with a control group C, suppose a surrogate (biomarker) endpoint and a well-established clinical endpoint are available for the assessment of the treatment effect. Denoted by θ_i and $\psi_i, i = 1, ..., k$ the treatment effect comparing E_i with C assessed by the surrogate (biomarker) endpoint and the clinical endpoint, respectively. Under the surrogate and clinical endpoints, the treatment effect can be tested by the following hypotheses:

$$H_{0,2} : \psi_1 = \cdots = \psi_k, \tag{18.77}$$

which is for the clinical endpoint, while the hypothesis

$$H_{0,1} : \theta_1 = \cdots \theta_k, \tag{18.78}$$

is for the surrogate (biomarker) endpoint. Cheng and Chow (2015) assumed that ψ_i is a monotone increasing function of the corresponding θ_i and proposed to test the hypotheses (18.77) and (18.78) at three stages (i.e., stage 1, stage 2a, stage 2b, and stage 3) based on accrued data at four interim analyses. Their proposed tests are briefly described below. For simplicity, the variances of the surrogate (biomarker) endpoint and the clinical outcomes are denoted by σ^2 and τ^2, which are assumed known.

Stage 1—At this stage, $(k + 1)n_1$ subjects are randomly assigned to receive either one of the k treatments or the control at a 1:1 ratio. In this case, we have n_1 subjects in each group. At the first interim analysis, the most effective treatment will be selected based on the surrogate (biomarker) endpoint and proceed to subsequent stages. For pairwise comparison, consider test statistics $\hat{\theta}_{i,1}$, $i = 1, ..., k$ and $S = \text{argmax}_{1 \leq j \leq k} \hat{\theta}_{i,1}$. Thus, if $\hat{\theta}_{S,1} \leq c_1$ for some prespecified critical value c_1, then the trial is stopped and we are in favor of $H_{0,1}$. On the other hand, if $\hat{\theta}_{S,1} > c_{1,1}$, then we conclude that the treatment E_S is considered the most promising treatment and proceed to subsequent stages. Subjects who receive either the promising treatment or the control will be followed for the clinical endpoint. Treatment assessment for all other subjects will be terminated but will undergo necessary safety monitoring.

Stage 2a—At stage 2a, $2n_2$ additional subjects will be equally randomized to receive either the treatment E_S or the control C. The second interim analysis is scheduled when the short-term surrogate measures from these $2n_2$ stage 2 subjects and the primary endpoint measures from those $2n_1$ stage 1 subjects who receive either the treatment E_S or the control

C become available. Let $T_{1,1} = \hat{\theta}_{S,1}$ and $T_{1,2} = \hat{\psi}_{S,1}$ be the pairwise test statistics from stage 1 based on the surrogate endpoint and the primary endpoint, respectively, and $\hat{\theta}_{S,2}$ be the statistic from stage 2 based on the surrogate. If

$$T_{2,1} = \sqrt{\frac{n_1}{n_1 + n_2}}\hat{\theta}_{S,1} + \sqrt{\frac{n_2}{n_1 + n_2}}\hat{\theta}_{S,2} \leq c_{2,1},$$

then stop the trial and accept $H_{0,1}$. If $T_{2,1} > c_{2,1}$ and $T_{1,2} > c_{1,2}$, then stop the trial and reject both $H_{0,1}$ and $H_{0,2}$. Otherwise, if $T_{2,1} > c_{2,1}$ but $T_{1,2} \leq c_{1,2}$, then we will move on to stage 2b.

Stage 2b—At stage 2b, no additional subjects will be recruited. The third interim analysis will be performed when the subjects in stage 2a complete their primary endpoints. Let

$$T_{2,2} = \sqrt{\frac{n_1}{n_1 + n_2}}\hat{\psi}_{S,1} + \sqrt{\frac{n_2}{n_1 + n_2}}\hat{\psi}_{S,2},$$

where $\hat{\psi}_{S,2}$ is the pairwise test statistic from stage 2b. If $T_{2,2} > c_{2,2}$, then stop the trial and reject $H_{0,2}$. Otherwise, we move on to stage 3.

Stage 3—At stage 3, the final stage, $2n_3$ additional subjects will be recruited and followed till their primary endpoints. At the fourth interim analysis, define

$$T_3 = \sqrt{\frac{n_1}{n_1 + n_2 + n_3}}\hat{\psi}_{S,1} + \sqrt{\frac{n_2}{n_1 + n_2 + n_3}}\hat{\psi}_{S,2} + \sqrt{\frac{n_1}{n_1 + n_2 + n_3}}\hat{\psi}_{S,3},$$

where $\hat{\psi}_{S,3}$ is the pairwise test statistic from stage 3. If $T_3 > c_3$, then stop the trial and reject $H_{0,2}$; otherwise, accept $H_{0,2}$. The parameters in the above designs, n_1, n_2, n_3, $c_{1,1}, c_{1,2}, c_{2,1}, c_{2,2}$, and c_3 are determined such that the procedure will have a controlled type I error rate of α and a target power of $1 - \beta$.

In the above design, the surrogate data in the first stage are used to select the most promising treatment rather than assessing $H_{0,1}$. This means that upon completion of stage one, a dose does not need to be significant to be used in subsequent stages. In practice, it is recommended that the selection criterion be based on precision analysis (desired precision or maximum error allowed) rather than power analysis (desired power). This property is attractive to the investigator since it does not suffer from any lack of power because of limited sample sizes.

As discussed above, under the four-stage transitional seamless design, two sets of hypotheses, namely, $H_{0,1}$ and $H_{0,2}$ are to be tested. Since the rejection of $H_{0,2}$ leads to the claim of efficacy, it is considered the hypothesis of primary interest. However, in the interest of controlling the overall type I error rate at a prespecified level of significance, $H_{0,1}$ needs to be tested following the principle of closed testing procedure to avoid any statistical penalties.

In summary, the two-stage phase II/III seamless adaptive design is attractive due to its efficiency, such as potentially reducing the lead time between studies (i.e., a phase II trial and a phase III study) and flexibility, such as making an early decision and taking appropriate actions (e.g., stop the trial early or delete/add dose groups).

18.4.2 Adaptive Version

The approach for trial design with nonadaptive version discussed in the previous section is basically a group sequential procedure with treatment selection at interim. There are no additional adaptations involved. With additional adaptations (adaptive version), Tsiatis and Mehta (2003) and Jennison and Turnbull (2006) argue that adaptive designs typically suffer from loss of efficiency and hence are typically not recommended in regular practice. Proschan, Lan, and Wittes (2006), however, also indicated that in some scenarios, particularly when there is not enough primary outcome information available, it is appealing to use an adaptive procedure as long as it is statistically valid and justified. The transitional feature of the multiple-stage design enables us not only to verify whether the surrogate (biomarker) endpoint is predictive of the clinical outcome, but also to modify the design adaptively after the review of interim data. A possible modification is to adjust the treatment effect of the clinical outcome while validating the relationship between the surrogate (e.g., biomarker) endpoint and the clinical outcome. In practice, it is often assumed that there exists a local linear relationship between ψ and θ, which is a reasonable assumption if we focus only on the values at a neighborhood of the most promising treatment E_S. Thus, at the end of stage 2a, we can reestimate the treatment effect of the primary endpoint using

$$\hat{\delta}_S = \frac{\hat{\psi}_{S,1}}{\hat{\theta}_{S,1}} T_{2,1}.$$

Consequently, sample size can be reassessed at stage 3, based on a modified treatment effect of the primary endpoint $\delta = \max\{\delta_S, \delta_0\}$, where δ_0 is a minimally clinically relevant treatment effect. Suppose m is the reestimated stage 3 sample size based on δ. Then, there is no modification for the procedure if $m \leq n_3$. On the other hand, if $m > n_3$, then m (instead of n_3 as originally planned) subjects per arm will be recruited at stage 3. The detailed justification of the above adaptation can be found in Cheng and Chow (2015).

18.4.3 A Case Study of Hepatitis C Virus Infection

A pharmaceutical company is interested in conducting a clinical trial for the evaluation of safety, tolerability, and efficacy of a test treatment for patients with hepatitis C virus (HCV) infection. For this purpose, a two-stage seamless adaptive design is considered. The proposed trial design is to combine two independent studies (one phase IIb study for treatment selection and one phase III study for efficacy confirmation) into a single study. Thus, the study consists of two stages: treatment selection (stage 1) and efficacy confirmation (stage 2). The study objective at the first stage is for treatment selection, while the study objective at stage 2 is to establish the noninferiority of the treatment selected from the first stage as compared to the standard of care (SOC). Thus, this is a typical Category IV design (a two-stage adaptive design with different study objectives at different stages).

For genotype 1 HCV patients, the treatment duration is usually 48 weeks of treatment followed by a 24 weeks follow-up. The well-established clinical endpoint is the sustained virologic response (SVR) at week 72. The SVR is defined as an undetectable HCV RNA level (<10 IU/mL) at week 72. Thus, it will take a long time to observe a response. The pharmaceutical company is interested in considering a biomarker or a surrogate endpoint such as a regular clinical endpoint with short duration to make early decision for treatment selection of four active treatments under study at end of stage 1. As a result, the clinical endpoint of early virologic response (EVR) at week 12 is considered as a surrogate endpoint for treatment selection at stage 1. At this point, the trial design has become a typical Category IV adaptive trial design (i.e., a two-stage adaptive design with different study endpoints and different study objectives at different stages). The resultant Category IV adaptive design is briefly outlined below (Figure 18.1).

Stage 1—At this stage, the design begins with five arms (four active treatment arms and one control arm). Qualified subjects are randomly assigned to receive one of the five treatment arms at a 1:1:1:1:1 ratio. After all stage 1 subjects have completed week 12 of the study, an interim analysis will be performed based on EVR at week 12 for treatment selection. Treatment selection will be made under the assumption that the 12-week EVR is predictive of 72-week SVR. Under this assumption, the most promising treatment arm will be selected using precision analysis under some prespecified selection criteria. In other words, the treatment arm with the highest confidence level for achieving statistical significance (i.e., the observed difference as compared to the control is not by chance alone) will be selected. Stage 1 subjects who have not yet completed the study protocol will continue with their assigned therapies for the remainder of the planned 48 weeks, with final follow-up at week 72. The selected treatment arm will then proceed to stage 2.

Stage 2—At stage 2, the selected treatment arm from stage 1 will be tested for noninferiority against the control (SOC). A separate cohort of subjects will be randomized to receive either the selected treatment from stage 1 or the control (SOC) at a 1:1 ratio. A second interim analysis will be performed when all stage 2 subjects have completed week 12 and 50% of the subjects (stage 1 and stage 2 combined) have completed 48 weeks treatment and follow-up of 24 weeks. The purpose of this interim analysis is twofold. First, it is to validate the assumption that EVR at week 12 is predictive of SVR at week 72. Second, it is to perform sample size reestimation to determine whether the trial will achieve study

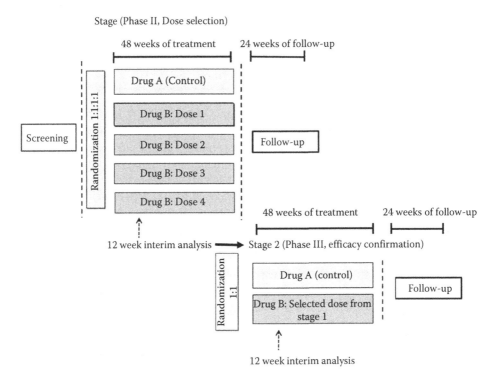

FIGURE 18.1
Diagram of four-stage transitional seamless trial design.

objective (establishing noninferiority) with the desired power if the observed treatment preserves till the end of the study.

Statistical tests as described in the previous section will be used to test noninferiority hypotheses at interim analyses and at end-of-stage analyses. For the two planned interim analyses, the incidence of EVR at week 12 as well as safety data will be reviewed by an independent data safety monitoring board (DSMB). The commonly used O'Brien–Fleming type of conservative boundaries will be applied for controlling the overall type I error rate at 5% (O'Brien and Fleming, 1979). Adaptations such as stopping the trial early, discontinuing selected treatment arms, and reestimating the sample size based on the prespecified criteria may be applied as recommended by the DSMB. Stopping rules for the study will be designated by the DSMB, based on their ongoing analyses of the data and as per their charter.

18.5 Concluding Remarks

Chow and Chang (2011) pointed out that the standard statistical methods for a group sequential trial (with one planned interim analysis) is often applied for planning and data analysis of a two-stage adaptive design regardless of whether the study objectives and/or the study endpoints are the same at different stages. As discussed earlier, two-stage seamless adaptive designs can be classified into four categories depending upon the study objectives and endpoints used at different stages. The direct application of standard statistical methods leads to the concern that the obtained p-value and confidence interval for the assessment of the treatment effect may not be correct or reliable. Most importantly, sample size required for achieving a desired power obtained under a standard group sequential trial design may not be sufficient for achieving the study objectives under the two-stage seamless adaptive trial design, especially when the study objectives and/or study endpoints at different stages are different.

As indicated in the 2010 FDA draft guidance on adaptive clinical trial design, adaptive designs were classified as either well-understood designs or less well-understood designs depending upon the availability of well-established statistical methods of specific designs (FDA, 2010a). In practice, most of the adaptive designs (including the two-stage seamless adaptive designs discussed in this chapter) are considered less well-understood designs. Thus, the major challenge is not only the development of valid statistical methods for those less well-understood designs, but also the development of a set of criteria for choosing an appropriate design among these less well-understood designs for valid and reliable assessment of test treatment under investigation.

.

19

Simulation-Based Sample Size and Power Analysis

Analytic sample size formulas have been developed for a range of standard study designs, many of which are considered in this book, and a variety of software packages for sample size and power analysis are available, both commercially and freely distributed. When closed-form expressions for sample size exist, calculations can be quickly performed and easily verified by the study reviewers. In practice, however, study designs can be quite sophisticated and may not be suitable for implementing any of the existing sample size formulas. Some examples of such studies include randomized trials with noncompliance, multilevel modeling, repeated measures analyses with complex correlation structure, longitudinal studies with dropout, and adaptive designs. In these situations, Monte Carlo simulations may be the only available approach for estimating the sample size and statistical power. Simulation studies can be equally useful for precision analysis, when the analytic expressions for the standard error or confidence limits are not available and these quantities instead have to be determined via complex analysis such as bootstrapping.

Even when analytic sample size solutions are available, they are often based on large sample theory or other restrictive assumptions that may not hold in practice. Simulations can be then used to verify that a study designed using existing sample size formula would have adequate type 1 error and planned statistical power or target precision. There are several additional reasons why simulation studies can be an important and powerful tool for every practicing statistician:

1. Flexibility: Simulation studies can be used to model very complex design features. While the true underlying population model and the analysis model have to be the same for sample size formulas to be applicable, this restriction is not required for simulation studies.

2. Intuitive approach: The simulation study mimics the actual study.

3. Ability to answer other study-related questions: In addition to power and sample size calculations, simulation studies can evaluate other study characteristics, such as type 1 error or conditional power. They can also be used to model the impact of changes in the study design, such as addition of a treatment arm, changes in the recruitment pattern, etc., on the statistical power and type 1 error.

4. Ease of implementation: Simulation studies can be performed using statistical software (SAS, Stata, R) or general-purpose programming languages (C++, Java). Once a simulation study has been designed, its parameters can be easily varied to investigate the study performance under various scenarios.

Numerous published articles reported the results of simulation studies to determine the sample size for specific examples (Taylor and Bosch, 1990; Gastañaga et al. 2006; Guimaraes and Palesch, 2007; Moineddin et al. 2007; Sutton et al. 2007; Abbas et al. 2008; Reynolds et al. 2008; Orloff et al. 2009; Sutton et al. 2009), and a discussion of the general principles of a well-designed simulation study for sample size and power estimation can be found in Feiveson, 2002; Burton et al. 2006; Arnold et al. 2011. For any study design, a researcher planning and executing a simulation study needs to consider the following components:

1. *Population model and the data-generating mechanism.* The underlying "true" model for the target population needs to be completely specified, with consideration given to potential issues such as dependence of the outcome on other covariates (that are not of primary interest), clustering, treatment noncompliance, study dropout, etc. Specified model should be sufficiently plausible for the simulation results to be applicable to the real data.

2. *Choice of the values for the model parameters.* They can be estimated from previous studies or available historical data, or solicited from the published literature. True effect size should ideally be chosen as the minimal important effect that is worth detecting ("clinically meaningful effect" in medical studies).

3. *Study design and sampling strategy.* The simulation study should reflect planned study design, including a range of possible sample sizes, and the method by which study subjects will be selected, including subpopulation or cluster sampling, if appropriate. Randomization mechanism should also be specified (e.g., simple randomization and stratified randomization).

4. *Choice of the analysis approach.* The analysis method that will be used to obtain inferences about the parameter of interest has to be clearly specified, including analysis model and statistical test used to reject the null hypothesis or a method used to obtain confidence limits.

5. *Choice of output metrics.* Here, we are primarily interested in evaluating the statistical power of the study (or precision of the confidence interval). Other metrics often used as an output of Monte Carlo simulations include type 1 error, bias, conditional power, and confidence interval coverage.

After all study parameters have been chosen, the simulation process mimics the actual study: the data are generated according to the population model, the study sample is chosen according to the experimental design and sampling strategy and is then analyzed using the selected method. This process is repeated a sufficiently large number of iterations (at least 1000) to determine the distribution of the test statistic of interest. Power can be determined as a proportion of times when the null hypothesis is rejected. Precision can be found as half-width of the interval between upper and lower $\alpha/2$ percentiles of the sample distribution of the parameter estimates across the iterations.

To find the sample size needed for the study, simulations over a grid of potential sample size values are often needed. The initial value of the sample size for a grid search could be set as a maximum possible sample size determined by the budgetary and logistical constraints of the study, or it could be determined by applying a formula available for a simplified study design or analysis.

We illustrate the process of a simulation study to determine the sample size and statistical power of a randomized trial on two examples.

19.1 Example: Survival Study with Nonconstant Treatment Effect

In this first example, we will consider a clinical trial where $N = 10,000$ patients will be randomized in 1:1 ratio to either new or standard treatment and the outcome of interest is time to event. The data will be analyzed using Cox regression model with treatment as the only covariate. The investigators believe that the effect of the new treatment may not fully manifest until the drug is taken for at least two months and they want to make sure the study has sufficient power to detect the average treatment effect even with this reduction. Specifically, they hypothesize that in the first two months of treatment, the hazard ratio will be 0.92, and after two months it will be 0.85. The standard sample size formula for Cox regression (Schoenfeld, 1983) assumes the proportional hazards assumption throughout time, so it cannot be used in this example. For $HR = 0.85$, Schoenfeld's formula yields 1591 events needed to be observed in the study to have 90% power at a two-sided 0.05 level; for $HR = 0.92$, this number is equal to 6045 events. Clearly, the number of events the investigators need lies somewhere between these two extremes, and it is probably closer to 1591 since $HR = 0.85$ is expected for the majority of the time on treatment. For simplicity, we will assume that there are no dropouts, that is, all subjects are followed to either the event of interest or the end of the study. In other situations researchers may want to specify the distribution of time to study dropout, typically, using exponential or Weibull model. In the simulation described below, we will evaluate the power of the study for different values D of the total number of events observed in the study. Then we can perform a grid search to find the smallest value of D that corresponds to the target power. The simulation study will include the following steps:

1. Evaluate the distribution of time to event in the standard arm from the prior data. Here, we will assume that it is exponential with a rate of 0.02 events per patient-month. We will also need to specify an enrollment pattern; based on the input from the investigators, we will assume that patients will be recruited at a uniform rate of 400 per month. Therefore, 10,000 subjects would be enrolled in the study over a 25-month period.

2. Create a sample of 10,000 subjects and generate random assignment with probability 0.5 to standard treatment ($X = 0$) or new treatment ($X = 1$). For each subject, generate the time of enrollment since the beginning of the study u from uniform distribution over 25 months.

3. For subjects in the control group, generate time to event T from the exponential distribution with rate $\lambda_0 = 0.02$ events per patient-month. For subjects in the new treatment group, generate time to event T from the distribution

$$F(t) = \begin{cases} 1 - \exp[-0.92\lambda_0 * t], & \text{if } t \leq 2 \text{ months,} \\ 1 - \exp[-0.92\lambda_0 * 2 - 0.85\lambda_0 * (t-2)], & \text{if } t > 2 \text{ months.} \end{cases}$$

4. Sort values of ($u + T$) in ascending order. Denote by L the Dth ordered value of ($u + T$)—this is the time since the beginning of the study where a total of D events have occurred.

5. For each subject, define observed time $V = \min(T, L - u)$ and the observed event indicator $\delta = 1$ if $T < L - u$ and $\delta = 0$ otherwise.

6. Fit Cox regression model with treatment group X as the only covariate to the observed data (V, δ), and store the two-sided p-value from the test of no treatment effect.

7. Repeat Steps 2–7 a sufficiently large number of times M and calculate the power as the proportion of times where the p-value is less than 0.05.

Results: From a simulation study with $M = 10,000$ repetitions, we have found that a trial continuing until a total of $D = 1930$ events were observed would have 90% power to reject the null hypothesis of no treatment effect at two-sided 0.05 level of significance, assuming a true hazard ratio of 0.92 in the first two months of treatment and a hazard ratio of 0.85 thereafter.

19.2 Example: Cluster Randomized Study with Stepped Wedge Design

Stepped wedge cluster randomized studies have recently started to gain wider popularity in various areas of research as an alternative to parallel arm cluster randomized trials (Mdege et al., 2011). In this approach, all clusters begin the study in the control group and gradually cross over to the intervention group, where they remain until the end of the study. Crossovers occur at regular intervals (steps) and the timing of the crossover is randomized before the start of the study. While stepped wedge studies typically take longer time to complete than traditional parallel arm designs, they can be considered advantageous in some situations for ethical or logistical reasons.

We will return to the Example 15.2 of a cluster randomized study of the hospital-level intervention to reduce surgical site infections (SSI). As we have shown in Chapter 15, under the given assumptions, both unmatched and matched cluster randomized designs would require a prohibitively large number of hospitals to be enrolled for the study to be adequately powered. To reduce between-cluster variation and therefore to lower sample size requirements, the researchers chose to perform a study where clusters are formed within each hospital by groups of surgeons performing similar types of procedures. A stepped wedge design was then selected so that all clusters could eventually get the intervention. The investigators can realistically enroll 18 hospitals with five clusters per hospital. They also assume that on average, 230 surgeries are performed by each cluster in a six-month period. The trial design includes a six-month baseline period when no intervention will be implemented and five six-month "steps," where in each step, one cluster in each hospital will be randomly selected to cross over to the intervention arm. Under stepped wedge design, the treatment schedule for each hospital can be represented by the following diagram (0 = control, 1 = intervention):

	Time					
	1	2	3	4	5	6
1	0	1	1	1	1	1
Cluster 2	0	0	1	1	1	1
3	0	0	0	1	1	1
4	0	0	0	0	1	1
5	0	0	0	0	0	1

The planned analysis is a Poisson regression of the cluster-level number of SSI using a generalized estimating equations (GEE) approach. The analysis model will include time step and current group as covariates and will account for within-cluster correlation over time and between-cluster correlation within each study hospital, both in the same time step and in different time steps. Using these assumptions, the investigators would like to evaluate the statistical power to detect a reduction in SSI rate from 1% under the control condition to 0.75% under the intervention.

Sample size and power analysis for stepped wedge designs is still a very active area of research. Following the seminal paper by Hussey and Hughes (2007), sample size formulas have been derived for the analyses of continuous and binary outcomes using linear mixed models; however, for GEE methods, closed-form sample size expressions are not yet available and sample size analysis has to be performed through simulation.

We will assume that the number of SSI in cluster k in hospital m under treatment i during time step j is generated from a binomial distribution with probability p_{ijkm}, where logit (p_{ijkm}) follows a model

$$\log\left(\frac{p_{ijkm}}{1-p_{ijkm}}\right) = \log\left(\frac{p_{ij}}{1-p_{ij}}\right) + \varepsilon_{ijkm}.$$

Here, $p_{ij} = p_i + \alpha_j$, where p_i is the average probability of SSI in group i across the three years of the study and α_j is the time effect in step j $\left(\sum \alpha_j = 0\right)$. We further assume that for each hospital, a vector of residuals ε_{ijkm} follows multivariate normal distribution with zero mean, between-cluster variance $\text{Var}(\varepsilon_{ijkm}) = \sigma^2$, within-cluster between-step correlation $\text{Corr}(\varepsilon_{ijkm}, \varepsilon_{ij'km}) = \rho_w$, between-cluster within-step correlation $\text{Corr}(\varepsilon_{ijkm}, \varepsilon_{ij'k'm}) = \rho_{bw}$, and between-cluster between-step correlation $\text{Corr}(\varepsilon_{ijkm}, \varepsilon_{ij'k'm}) = \rho_b$.

The simulation study was implemented following these steps.

1. From the available historical data, estimates were obtained for between-cluster variance ($\sigma^2 = 0.5$), within-cluster between-step correlation ($\rho_w = 0.3$), between-cluster within-step correlation ($\rho_{bw} = 0.1$), and between-cluster between-step correlation ($\rho_b = 0.05$).

2. We created a sample of 18 hospitals with five clusters per hospital. Clusters within every hospital were randomly matched with equal probability to one of the five treatment sequences: (0,1,1,1,1,1), (0,0,1,1,1,1), (0,0,0,1,1,1), (0,0,0,0,1,1), or (0,0,0,0,0,1), so that all treatment sequences were represented within each hospital.

3. The investigators believed that it was unlikely that time effect would be present, so we assumed that $p_{ij} = p_i$ ($i = 0,1$). For each hospital m, we generated a 30×1 vector w_m from a multivariate normal distribution with mean corresponding to the cluster treatment sequence and variance–covariance matrix as described above.

4. For each cluster, the number of SSIs were generated from the binomial distribution with parameters $n = 230$ and probability $p_{ijkm} = \exp(w_{ijkm})/(1 + \exp(w_{ijkm}))$, where w_{ijkm} is the element of vector w_m corresponding to time step j for cluster k in hospital m.

5. Next, a Poisson regression model was fitted to cluster counts of SSIs, with log (230) as an offset term and time step and group as covariates. The working correlation structure was assumed to be the same as in the data-generating model.

6. Steps 2–5 were repeated $M = 5000$ times.

Results: After performing a simulation study with 5000 repetitions, we determined that a planned stepped wedge cluster randomized study would have approximately 88% power to detect a reduction in SSI rate from 1% under the control condition to 0.75% under the intervention at two-sided 0.05 level of significance.

20

Sample Size Calculation in Other Areas

As indicated earlier, sample size calculation is an integral part of clinical research. It is undesirable to observe positive results with insufficient power. Sample size calculation should be performed based on the primary study endpoint using appropriate statistical methods under a valid study design with correct hypotheses, which can reflect the study objectives. In the previous chapters, we have examined formulas or procedures for sample size calculation based on various primary study endpoints for comparing means, proportions, variabilities, functions of means and variance components, and time-to-event data. In this chapter, in addition, we discuss several procedures for sample size calculation based on different study objectives and/or hypotheses using different statistical methods, which are not covered in the previous chapters.

In Section 20.1, sample size calculations for QT/QTc studies with time-dependent replicates are examined. Sample size calculation based on propensity score analysis for nonrandomized studies is given in Section 20.2. Section 20.3 discusses sample size calculation under an analysis of variance (ANOVA) model with repeated measures. Section 20.4 discusses sample size calculation for the assessment of quality of life (QOL) under a time series model. In Section 20.5, the concept of reproducibility and sensitivity index for bridging studies is introduced. Also included in this section is a proposed method for assessing the similarity of bridging studies, which is used for the derivation of a procedure for sample size calculation. Statistical methods and the corresponding procedure for sample size calculation for vaccine clinical trials are briefly outlined in Section 20.6.

20.1 QT/QTc Studies with Time-Dependent Replicates

In clinical trials, a 12-lead electrocardiogram (ECG) is usually conducted for the assessment of potential cardiotoxicity induced by the treatment under study. On an ECG tracing, the QT interval is measured from the beginning of the Q wave to the end of the T wave. QT interval is often used to indirectly assess the delay in cardiac repolarization, which can predispose to the development of life-threatening cardiac arrhythmias such as torsade de pointes (Moss, 1993). QTc interval is referred to as the QT interval corrected by heart rate. In clinical practice, it is recognized that the prolongation of the QT/QTc interval is related to increased risk of cardiotoxicity such as a life-threatening arrhythmia (Temple, 2003). Thus, it is suggested that a careful evaluation of potential QT/QTc prolongation be assessed for potential drug-induced cardiotoxicity.

For the development of a new pharmaceutical entity, most regulatory agencies such as the U.S. Food and Drug Administration (FDA) require the evaluation of proarrhythmic potential (see, e.g., CPMP, 1997; FDA/TPD, 2003; ICH, 2003). As a result, a draft guidance on the clinical evaluation of QT/QTc interval prolongation and proarrhythmic potential for nonantiarrhythmic drugs is being prepared by the International Conference on

Harmonization (ICH E14). This draft guidance calls for a placebo-controlled study in normal healthy volunteers with a positive control to assess cardiotoxicity by examining QT/QTc prolongation. Under a valid study design (e.g., a parallel-group design or a crossover design), ECGs will be collected at baseline and at several time points posttreatment for each subject. Malik and Camm (2001) recommend that it would be worthwhile to consider three to five replicate ECGs at each time point within two-to-five-min period. Replicate ECGs are then defined as single ECG recorded within several minutes of a nominal time (PhRMA QT Statistics Expert Working Team, 2003). Along this line, Strieter et al. (2003) studied the effect of replicate ECGs on QT variability in healthy subjects. In practice, it is then of interest to investigate the impact of recording replicates on power and sample size calculation in routine QT studies.

In clinical trials, a prestudy power analysis for sample size calculation is usually performed to ensure that the study achieves the desired power (or the probability of correctly detecting a clinically meaningful difference if such a difference truly exists). For QT studies, the following information is necessarily obtained prior to the conduct of the prestudy power analysis for sample size calculation. These information include (i) the variability associated with the primary study endpoint such as the QT intervals (or the QT interval endpoint change from baseline) and (ii) the clinically meaningful difference in QT interval between treatment groups. Under the above assumption, the procedures as described in Longford (1993a,b) and Chow, Shao, and Wang (2003c) can then be applied for sample size calculation under the study design (e.g., a parallel-group design or a crossover design).

In what follows, commonly used study designs such as a parallel-group design or a crossover design for routine QT studies with recording replicates are briefly described. Power analyses and the corresponding sample size calculations under a parallel-group design and a crossover design are derived. Extensions to the designs with covariates (pharmacokinetic [PK] responses) are also considered.

20.1.1 Study Designs and Models

A typical study design for QT studies is either a parallel-group design or a crossover design depending upon the primary objectives of the study. Statistical models under a parallel-group design and a crossover design are briefly outlined below.

Under a parallel-group design, qualified subjects will be randomly assigned to receive either treatment A or treatment B. ECGs will be collected at baseline and at several time points posttreatment. Subjects fast at least three hours and rest at least 10 minutes prior to scheduled ECG measurements. Identical lead placement and same ECG machine will be used for all measurements. As recommended by Malik and Camm (2001), 3 to 5 recording replicate ECGs at each time point will be obtained within a two-to-five-minute period.

Let y_{ijk} be the QT interval observed from the kth recording replicate of the jth subject who receives treatment i, where $i = 1, 2, j = 1, \ldots, n$, and $k = 1, \ldots, K$. Consider the following model:

$$y_{ijk} = \mu_i + e_{ij} + \varepsilon_{ijk}, \tag{20.1}$$

where e_{ij} are independent and identically distributed as a normal with mean 0 and variance σ_s^2 (between-subject or intersubject variability) and ε_{ijk} are independent and identically distributed as a normal with mean 0 and variance σ_e^2 (within-subject or intrasubject variability or measurement error variance). Thus, we have $\text{Var}(y_{ijk}) = \sigma_s^2 + \sigma_e^2$.

Under a crossover design, qualified subjects will be randomly assigned to receive one of the two sequences of test treatments under study. In other words, subjects who are randomly assigned to sequence 1 will receive treatment 1 first and then be crossed over to receive treatment 2 after a sufficient period of washout. Let y_{ijkl} be the QT interval observed from the kth recording replicate of the jth subject in the lth sequence who receives the ith treatment, where $i = 1, 2, j = 1, \ldots, n, k = 1, \ldots, K$, and $l = 1, 2$. We consider the following model:

$$y_{ijkl} = \mu_i + \beta_{il} + e_{ijl} + \varepsilon_{ijkl},$$ (20.2)

where β_{il} are independent and identically distributed normal random period effects (period uniquely determined by sequence l and treatment i) with mean 0 and variance σ_p^2, e_{ijl} are independent and identically distributed normal subject random effects with mean 0 and variance σ_s^2, and ε_{ijkl} are independent and identically distributed normal random errors with mean 0 and variance σ_e^2. Thus, $\mathrm{Var}(y_{ijkl}) = \sigma_p^2 + \sigma_s^2 + \sigma_e^2$.

To have a valid comparison between the parallel design and the crossover design, we assume that μ_i, σ_s^2, and σ_e^2 are the same as those given in Equations 20.1 and 20.2 and consider an extra variability σ_p^2, which is due to the random period effect for the crossover design.

20.1.2 Power and Sample Size Calculation

Parallel-Group Design

Under the parallel-group design as described in the previous section, to evaluate the impact of recording replicates on power and sample size calculation, for simplicity, we will consider only one time point posttreatment. The results for recording replicates at several posttreatment intervals can be similarly obtained. Under model (20.1), consider sample mean of QT intervals of the jth subject who receives the ith treatment, then $\mathrm{Var}(\bar{y}_{ij\cdot}) = \sigma_s^2 + \frac{\sigma_e^2}{K}$. The hypotheses of interest regarding treatment difference in QT interval are given by

$$H_0 : \mu_1 - \mu_2 = 0 \quad \text{versus} \quad H_1 : \mu_1 - \mu_2 = d,$$ (20.3)

where $d \neq 0$ is a difference of clinical importance. Under the null hypothesis of no treatment difference, the following statistic can be derived:

$$T = \frac{\bar{y}_{1\cdot\cdot} - \bar{y}_{2\cdot\cdot}}{\sqrt{\frac{2}{n}\left(\hat{\sigma}_s^2 + \frac{\hat{\sigma}_e^2}{K}\right)}},$$

where

$$\hat{\sigma}_e^2 = \frac{1}{2n(K-1)}\sum_{i=1}^{2}\sum_{j=1}^{n}\sum_{k=1}^{K}(y_{ijk} - \bar{y}_{ij\cdot})^2,$$

and

$$\hat{\sigma}_s^2 = \frac{1}{2(n-1)}\sum_{i=1}^{2}\sum_{j=1}^{n}(\bar{y}_{ij\cdot} - \bar{y}_{i\cdot\cdot})^2 - \frac{1}{2nK(K-1)}\sum_{i=1}^{2}\sum_{j=1}^{n}\sum_{k=1}^{K}(y_{ijk} - \bar{y}_{ij\cdot})^2.$$

Under the null hypothesis in Equation 20.3, T has a central t-distribution with $2n-2$ degrees of freedom. Let $\sigma^2 = \mathrm{Var}(y_{ijk}) = \sigma_s^2 + \sigma_e^2$ and $\rho = \sigma_s^2 / (\sigma_s^2 + \sigma_e^2)$, then under the alternative hypothesis in Equation 20.3, the power of the test can be obtained as follows:

$$1 - \beta \approx 1 - \Phi\left(z_{\alpha/2} - \frac{\delta}{\sqrt{\dfrac{2}{n}\left(\rho + \dfrac{1-\rho}{K}\right)}}\right) + \Phi\left(-z_{\alpha/2} - \frac{\delta}{\sqrt{\dfrac{2}{n}\left(\rho + \dfrac{1-\rho}{K}\right)}}\right), \qquad (20.4)$$

where $\delta = d/\sigma$ is the relative effect size and Φ is the cumulative distribution of a standard normal. To achieve the desired power of $1 - \beta$ at the α level of significance, the sample size needed per treatment is

$$n = \frac{2(z_{\alpha/2} + z_\beta)^2}{\delta^2}\left(\rho + \frac{1-\rho}{K}\right). \qquad (20.5)$$

Crossover Design

Under a crossover model (20.2), it can be verified that $\bar{y}_{i..}$ is an unbiased estimator of μ_i with variance $\frac{\sigma_p^2}{2} + \frac{\sigma_s^2}{2n} + \frac{\sigma_e^2}{2nK}$. Thus, we used the following test statistic to test the hypotheses in Equation 20.3:

$$T = \frac{\bar{y}_{1..} - \bar{y}_{2..}}{\sqrt{\hat{\sigma}_p^2 + \dfrac{1}{n}\left(\hat{\sigma}_s^2 + \dfrac{\hat{\sigma}_e^2}{K}\right)}},$$

where

$$\hat{\sigma}_e^2 = \frac{1}{4n(K-1)}\sum_{i=1}^{2}\sum_{j=1}^{n}\sum_{k=1}^{K}\sum_{l=1}^{2}(y_{ijkl} - \bar{y}_{ij.l})^2,$$

$$\hat{\sigma}_s^2 = \frac{1}{4(n-1)}\sum_{i=1}^{2}\sum_{j=1}^{n}\sum_{l=1}^{2}(y_{ij.l} - \bar{y}_{i..l})^2$$

$$- \frac{1}{4nK(K-1)}\sum_{i=1}^{2}\sum_{j=1}^{n}\sum_{k=1}^{K}\sum_{l=1}^{2}(y_{ijkl} - \bar{y}_{ij.l})^2,$$

and

$$\hat{\sigma}_p^2 = \frac{1}{2}\sum_{i=1}^{2}\sum_{l=1}^{2}(\bar{y}_{i..l} - \bar{y}_{....})^2 - \frac{1}{4n(n-1)}\sum_{i=1}^{2}\sum_{j=1}^{n}\sum_{l=1}^{2}(\bar{y}_{ij.l} - \bar{y}_{i..l})^2.$$

Under the null hypothesis in Equation 20.3, T has a central t-distribution with $2n - 4$ degrees of freedom. Let σ^2 and ρ be defined as in the previous section, and $\gamma = \sigma_p^2/\sigma^2$, then $\mathrm{Var}(y_{ijkl}) = \sigma^2(1+\gamma)$. Under the alternative hypothesis in Equation 20.3, the power of the test can be obtained as follows:

$$1-\beta \approx 1-\Phi\left(z_{\alpha/2}-\frac{\delta}{\sqrt{\gamma+\frac{1}{n}\left(\rho+\frac{1-\rho}{K}\right)}}\right)+\Phi\left(-z_{\alpha/2}-\frac{\delta}{\sqrt{\gamma+\frac{1}{n}\left(\rho+\frac{1-\rho}{K}\right)}}\right), \quad (20.6)$$

where $\delta = d/\sigma$. To achieve the desired power of $1-\beta$ at the α level of significance, the sample size needed per treatment is

$$n = \frac{(z_{\alpha/2}+z_\beta)^2}{\delta^2-\gamma(z_{\alpha/2}+z_\beta)^2}\left(\rho+\frac{1-\rho}{K}\right). \quad (20.7)$$

Remarks

Let n_{old} be the sample size with $K=1$ (i.e., there is single measure for each subject). Then, we have $n = \rho n_{old} + (1-\rho)n_{old}/K$. Thus, the sample size (with recording replicates) required for achieving the desired power is a weighted average of n_{old} and n_{old}/K. Note that this relationship holds under both a parallel and a crossover design. Table 20.1 provides sample sizes required under a chosen design (either parallel or crossover) for achieving the same power with single recording ($K=1$), three recording replicates ($K=3$), and five recording replicates ($K=5$).

Note that if ρ closes to 0, then these repeated measures can be treated as independent replicates. As it can be seen from the above, if $\rho \approx 0$, then $n \approx n_{old}/K$. In other words, sample size is indeed reduced when the correlation coefficient between recording replicates is close to 0 (in this case, the recording replicates are almost independent). Table 20.2 shows the sample size reduction for different values of ρ under the parallel design. However, in practice, ρ is expected to be close to 1. In this case, we have $n \approx n_{old}$. In other words, there is not much gain for considering recording replicates in the study.

In practice, it is of interest to know whether the use of a crossover design can further reduce the sample size when other parameters such as d, σ^2, and ρ remain the same. Comparing formulas (20.5) and (20.7), we conclude that the sample size reduction by using a crossover design depends upon the parameter $\gamma = \sigma^2_p/\sigma^2$, which is a measure

TABLE 20.1

Sample Size for Achieving the Same Power with K Recording Replicates

ρ	K		
	1	3	5
1.0	n	$1.00n$	$1.00n$
0.9	n	$0.93n$	$0.92n$
0.8	n	$0.86n$	$0.84n$
0.7	n	$0.80n$	$0.76n$
0.6	n	$0.73n$	$0.68n$
0.5	n	$0.66n$	$0.60n$
0.4	n	$0.60n$	$0.52n$
0.3	n	$0.53n$	$0.44n$
0.2	n	$0.46n$	$0.36n$
0.1	n	$0.40n$	$0.28n$
0.0	n	$0.33n$	$0.20n$

TABLE 20.2

Sample Sizes Required under a Parallel-Group Design

	Power = 80%					Power = 90%				
	ρ					ρ				
(K, δ)	0.2	0.4	0.6	0.8	1.0	0.2	0.4	0.6	0.8	1.0
(3, 0.3)	81	105	128	151	174	109	140	171	202	233
(3, 0.4)	46	59	72	85	98	61	79	96	114	131
(3, 0.5)	29	38	46	54	63	39	50	64	73	84
(5, 0.3)	63	91	119	147	174	84	121	159	196	233
(5, 0.4)	35	51	67	82	98	47	68	89	110	131
(5, 0.5)	23	33	43	53	63	30	44	57	71	84

of the relative magnitude of period variability with respect to the within-period subject marginal variability. Let $\theta = \gamma/(z_{\alpha/2} + z_\beta)^2$, then by Equations 20.5 and 20.7, the sample size n_{cross} under the crossover design and the sample size n_{parallel} under the parallel-group design satisfy $n_{\text{cross}} = n_{\text{parallel}}/2(1-\theta)$. When the random period effect is negligible, that is, $\gamma \approx 0$ and hence $\theta \approx 0$, then we have $n_{\text{cross}} = n_{\text{parallel}}/2$. This indicates that the use of a crossover design could further reduce the sample size by half as compared to a parallel-group design when the random period effect is negligible (based on the comparison of the above formula and the formula given in Equation 20.5. However, when the random period effect is not small, the use of a crossover design may not result in sample size reduction. Table 20.3 shows the sample size under different values of γ. It is seen that the possibility of sample size reduction under a crossover design depends upon whether the carryover effect of the QT intervals could be avoided. As a result, it is suggested that a

TABLE 20.3

Sample Sizes Required under a Crossover Design with $\rho = 0.8$

	Power = 80%				
	γ				
(K, δ)	0.000	0.001	0.002	0.003	0.004
(3, 0.3)	76	83	92	102	116
(3, 0.4)	43	45	47	50	53
(3, 0.5)	27	28	29	30	31
(5, 0.3)	73	80	89	99	113
(5, 0.4)	41	43	46	48	51
(5, 0.5)	26	27	28	29	30
	Power = 90%				
	γ				
(K, δ)	0.000	0.001	0.002	0.003	0.004
(3, 0.3)	101	115	132	156	190
(3, 0.4)	57	61	66	71	77
(3, 0.5)	36	38	40	42	44
(5, 0.3)	98	111	128	151	184
(5, 0.4)	55	59	64	69	75
(5, 0.5)	35	37	39	40	42

sufficient length of washout be applied between dosing periods to wear off the residual (or carryover) effect from one dosing period to another. For a fixed sample size, the possibility of power increase by crossover design also depends on parameter γ.

20.1.3 Extension

In the previous section, we consider models without covariates. In practice, additional information such as some PK responses, for example, area under the blood or plasma concentration time curve (AUC) and the maximum concentration (C_{max}), which are known to be correlated to the QT intervals may be available. In this case, models (20.1) and (20.2) are necessarily modified to include the PK responses as covariates for a more accurate and reliable assessment of power and sample size calculation (Cheng and Shao, 2007).

Parallel-Group Design
After the inclusion of the PK response as covariate, model (20.1) becomes

$$y_{ijk} = \mu_i + \eta x_{ij} + e_{ij} + \varepsilon_{ijk},$$

where x_{ij} is the PK response for subject j. The least square estimate of η is given by

$$\hat{\eta} = \frac{\sum_{i=1}^{2}\sum_{j=1}^{2}(\bar{y}_{ij.} - \bar{y}_{i..})(x_{ij} - \bar{x}_{i.})}{\sum_{i=1}^{2}\sum_{j=1}^{2}(x_{ij} - \bar{x}_{i.})^2}.$$

Then, $(\bar{y}_{1..} - \bar{y}_{2..}) - \hat{\eta}(\bar{x}_{1.} - \bar{x}_{2.})$ is an unbiased estimator of $\mu_1 - \mu_2$ with variance

$$\left[\frac{(\bar{x}_{1.} - \bar{x}_{2.})^2}{\sum_{ij}(x_{ij} - \bar{x}_{i.})^2/n} + 2\left(\rho + \frac{1-\rho}{K}\right)\right]\frac{\sigma^2}{n},$$

which can be approximated by

$$\left[\frac{(v_1 - v_2)^2}{\tau_1^2 + \tau_2^2} + 2\left(\rho + \frac{1-\rho}{K}\right)\right]\frac{\sigma^2}{n},$$

where $v_i = \lim_{n\to\infty}\bar{x}_{i.}$, and $\tau_i^2 = \lim_{n\to\infty}\sum_{j=1}^{n}(x_{ij} - \bar{x}_{i.})^2/n$. Similarly, to achieve the desired power of $1 - \beta$ at the α level of significance, the sample size needed per treatment group is given by

$$n = \frac{(z_{\alpha/2} + z_\beta)^2}{\delta^2}\left[\frac{(v_1 - v_2)^2}{\tau_1^2 + \tau_2^2} + 2\left(\rho + \frac{1-\rho}{K}\right)\right]. \tag{20.8}$$

In practice, v_i and τ_i^2 are estimated by the corresponding sample mean and sample variance from the pilot data. Note that if there are no covariates or the PK responses are balanced across treatments (i.e., $v_1 = v_2$), then formula (20.8) reduces to Equation 20.5.

Crossover Design

After taking the PK response into consideration as a covariate, model (20.2) becomes

$$y_{ijkl} = \mu_i + \eta x_{ijl} + \beta_{il} + e_{ijl} + \varepsilon_{ijkl}.$$

Then, $(\bar{y}_{1...} - \bar{y}_{2...}) - \hat{\eta}(\bar{x}_{1..} - \bar{x}_{2..})$ is an unbiased estimator of $\mu_1 - \mu_2$ with variance

$$\left[\gamma + \left(\frac{(\bar{x}_{1..} - \bar{x}_{2..})^2}{\sum_{ijl} (x_{ijl} - \bar{x}_{i..})^2 / n} + 1 \right) \left(\rho + \frac{1-\rho}{K} \right) \sigma^2 \right],$$

which can be approximated by

$$\left[\gamma + \left(\frac{(v_1 - v_2)^2}{\tau_1^2 + \tau_2^2} + 1 \right) \right] \left(\rho + \frac{1-\rho}{K} \right) \sigma^2,$$

where $v_i = \lim_{n \to \infty} \bar{x}_{i..}$ and $\tau_i^2 = \lim_{n \to \infty} \Sigma_{jl} (x_{ijl} - \bar{x}_{i..})^2 / n$. To achieve the desired power of $1 - \beta$ at the α level of significance, the sample size required per treatment group is

$$n = \frac{(z_{\alpha/2} + z_\beta)^2}{\delta^2 - \gamma(z_{\alpha/2} + z_\beta)^2} \left[\frac{(v_1 - v_2)^2}{\tau_1^2 + \tau_2^2} + 1 \right] \left(\rho + \frac{1-\rho}{K} \right). \tag{20.9}$$

When there are no covariates or PK responses satisfy $v_1 = v_2$, formula (20.9) reduces to Equation 20.7.

Formulas (20.8) and (20.9) indicate that under either a parallel-group or a crossover design, a larger sample size is required to achieve the same power if the covariate information is to be incorporated.

20.1.4 Remarks

Under a parallel-group design, the possibility of whether the sample size can be reduced depends upon the parameter ρ, the correlation between the QT recording replicates. As indicated earlier, when ρ closes to 0, these recording repeats can be viewed as (almost) independent replicates. As a result, $n \approx n_{old}/K$. When ρ is close to 1, we have $n \approx n_{old}$. Thus, there is not much gain for considering recording replicates in the study. On the other hand, assuming that all other parameters remain the same, the possibility of further reducing the sample size by a crossover design depends upon the parameter γ, which is a measure of the magnitude of the relative period effect.

When analyzing QT intervals with recording replicates, we may consider change from baseline. It is, however, not clear which baseline should be used when there are also recording replicates at baseline. Strieter et al. (2003) proposed the use of the so-called time-matched change from baseline, which is defined as measurement at a time point on the postbaseline day minus measurement at same time point on the baseline. The statistical properties of this approach, however, are not clear. In practice, it may be of interest to investigate relative merits and disadvantages among the approaches using (i) the most recent recording replicates, (ii) the mean recording replicates, or (iii) time-matched recording replicates as the baseline. This requires further research.

20.2 Propensity Analysis in Nonrandomized Studies

As indicated in Yue (2007), the use of propensity analysis in nonrandomized trials has received much attention, especially in the area of medical device clinical studies. In nonrandomized study, patients are not randomly assigned to treatment groups with equal probability. Instead, the probability of assignment varies from patient to patient depending on the patient's baseline covariates. This often results in noncomparable treatment groups due to imbalance of the baseline covariates and consequently invalidates the standard methods commonly employed in data analysis. To overcome this problem, Yue (2007) recommends the method of propensity score developed by Rosenbaum and Rubin (1983, 1984) be used.

In her review article, Yue (2007) described some limitations for using propensity score. For example, propensity score method can only adjust for observed covariates and not for unobserved ones. As a result, it is suggested that a sensitivity analysis be conducted for possible hidden bias. In addition, Yue (2007) also posted several statistical and regulatory issues for propensity analysis in nonrandomized trials, including sample size calculation. In this discussion, our emphasis will be placed on the issue of sample size calculation in the context of propensity scores. We propose a procedure for sample size calculation based on weighted Mantel–Haenszel (WMH) test with different weights across score subclasses.

In the next section, a proposed WMH test, the corresponding formula for sample size calculation, and a formula for sample size calculation ignoring strata are briefly described. Subsequent subsections summarize the results of several simulation studies conducted for the evaluation of the performance of the proposed test in the context of propensity analysis. A brief concluding remark is given in Section 20.6.5.

20.2.1 Weighted Mantel–Haenszel Test

Suppose that the propensity score analysis defines J strata. Let n denote the total sample size, and n_j the sample size in stratum j ($\sum_{j=1}^{J} n_j = n$). The data on each subject comprise the response variable $x = 1$ for response and 0 for no response; j and k for the stratum and treatment group, respectively, to which the subject is assigned ($1 \leq j \leq J; k = 1, 2$). We assume that group 1 is the control. Frequency data in stratum j can be described as follows:

	Group		
Response	1	2	Total
Yes	x_{j11}	x_{j12}	x_{j1}
No	x_{j21}	x_{j22}	x_{j2}
Total	n_{j1}	n_{j2}	n_j

Let $O_j = x_{j11}$, $E_j = n_{j1} x_{j1} / n_j$, and

$$V_j = \frac{n_{j1} n_{j2} x_{j1} x_{j2}}{n_j^2 (n_j - 1)}.$$

Then, the WMH test is given by

$$T = \frac{\sum_{j=1}^{J} \hat{w}_j (O_j - E_j)}{\sqrt{\sum_{j=1}^{J} \hat{w}_j^2 V_j}},$$

where the weights $\hat{w}_j$ converges to a constant w_j as $n \to \infty$. The weights are $\hat{w}_j = 1$ for the original Mantel–Haenszel (MH) test and $\hat{w}_j = \hat{q}_j = x_{j2}/n_j$ for the statistic proposed by Gart (1985).

Let $a_j = n_j/n$ denote the allocation proportion for stratum j ($\Sigma_{j=1}^{J} a_j = 1$), and $b_{jk} = n_{jk}/n_j$ denote the allocation proportion for group k within stratum j ($b_{j1} + b_{j2} = 1$). Let p_{jk} denote the response probability for group k in stratum j and $q_{jk} = 1 - p_{jk}$. Under H_0: $p_{j1} = p_{j2}$, $1 \le j \le J$, T is approximately $N(0, 1)$. The optimal weights maximizing the power depend on the allocation proportions $\{(a_j, b_{1j}, b_{2j}), j = 1, \dots, J\}$ and effect sizes $(p_{j1} - p_{j2}, 1, \dots, J)$ under H_1.

20.2.2 Power and Sample Size

In order to calculate the power of WMH, we have to derive the asymptotic distribution of $\Sigma_{j=1}^{J} \hat{w}_j (O_j - E_j)$ and the limit of $\Sigma_{j=1}^{J} \hat{w}_j^2 V_j$ under H_1. We assume that the success probabilities $(p_{jk}, 1 \le j \le J, j = 1, 2)$ satisfy $p_{j2} q_{j1}/(p_{j1} q_{j2}) = \phi$ for $\phi \ne 1$ under H_1. Note that a constant odds ratio across strata holds if there exists no interaction between treatment and the propensity score when the binary response is regressed on the treatment indicator and the propensity score using a logistic regression. The following derivations are based on H_1. It can be verified that

$$O_j - E_j = \frac{n_{j1} n_{j2}}{n_j} (\hat{p}_{j1} - \hat{p}_{j2})$$

$$= \frac{n_{j1} n_{j2}}{n_j} (\hat{p}_{j1} - p_{j1} - \hat{p}_{j2} + p_{j2}) + \frac{n_{j1} n_{j2}}{n_j} (p_{j1} - p_{j2})$$

$$= n a_j b_{j1} b_{j2} (\hat{p}_{j1} - p_{j1} - \hat{p}_{j2} + p_{j2}) + n a_j b_{j1} b_{j2} (p_{j1} - p_{j2}).$$

Thus, under H_1, $\Sigma_{j=1}^{J} \hat{w}_j (O_j - E_j)$ is approximately normal with mean $n\delta$ and variance $n\sigma_1^2$, where

$$\delta = \sum_{j=1}^{J} w_j a_j b_{j1} b_{j2} (p_{j1} - p_{j2})$$

$$= (1 - \phi) \sum_{j=1}^{J} w_j a_j b_{j1} b_{j2} \frac{p_{j1} q_{j1}}{q_{j1} + \phi p_{j1}}$$

and

$$\sigma_1^2 = n^{-1} \sum_{j=1}^{J} w_j^2 \frac{n_{j1}^2 n_{j2}^2}{n_j^2} \left(\frac{p_{j1} q_{j1}}{n_{j1}} + \frac{p_{j2} q_{j2}}{n_{j2}} \right)$$

$$= \sum_{j=1}^{J} w_j^2 a_j b_{j1} b_{j2} (b_{j2} p_{j1} q_{j1} + b_{j1} p_{j2} q_{j2}).$$

Also, under H_1, we have

$$\sum_{j=1}^{J} w_j^2 V_j = n\sigma_0^2 + o_p(n),$$

where

$$\sigma_0^2 = \sum_{j=1}^{J} w_j^2 a_j b_{j1} b_{j2} (b_{j1} p_{j1} + b_{j2} p_{j2})(b_{j1} q_{j1} + b_{j2} q_{j2}).$$

Hence, the power of WMH is given as

$$1 - \beta = P(\,|T| > z_{1-\alpha/2}|\,H_1)$$
$$= P\left(\frac{\sigma_1}{\sigma_0} Z + \sqrt{n}\frac{|\delta|}{\sigma_0} > z_{1-\alpha/2}\right)$$
$$= \Phi\left(\frac{\sigma_0}{\sigma_1} z_{1-\alpha/2} - \sqrt{n}\frac{|\delta|}{\sigma_1}\right),$$

where Z is a standard normal random variable and $\Phi(z) = P(Z > z)$. Thus, the sample size required for achieving a desired power of $1 - \beta$ can be obtained as

$$n = \frac{(\sigma_0 z_{1-\alpha/2} + \sigma_1 z_{1-\beta})^2}{\delta^2}. \qquad (20.10)$$

Following the steps as described in Chow, Shao, and Wang (2003c), the sample size calculation for the WMH test can be carried out as follows:

1. Specify the input variables.
   - Type I and II error probabilities, (α, β).
   - Success probabilities for group 1 $p_{11}, \ldots, p_{J1}$, and the odds ratio ϕ under H_1. Note that $p_{j2} = \phi p_{j1}/(q_{j1} + \phi p_{j1})$.
   - Incidence rates for the strata, $(a_j, j = 1, \ldots, J)$. (Yue proposes to use $a_j \approx 1/J$.)
   - Allocation probability for group 1 within each stratum, $(b_{j1}, j = 1, \ldots, J)$.
2. Calculate n by

$$n = \frac{(\sigma_0 z_{1-\alpha/2} + \sigma_1 z_{1-\beta})^2}{\delta^2},$$

   where

$$\delta = \sum_{j=1}^{J} a_j b_{j1} b_{j2}(p_{j1} - p_{j2}),$$

$$\sigma_0^2 = \sum_{j=1}^{J} a_j b_{j1} b_{j2} (b_{j1} p_{j1} + b_{j2} p_{j2})(b_{j1} q_{j1} + b_{j2} q_{j2})$$

$$\sigma_1^2 = \sum_{j=1}^{J} a_j b_{j1} b_{j2} (b_{j2} p_{j1} q_{j1} + b_{j1} p_{j2} q_{j2})$$

Sample Size Calculation Ignoring Strata

We consider ignoring strata and combining data across J strata. Let $n_k = \Sigma_{j=1}^{J} n_{jk}$ denote the sample size in group k. Ignoring strata, we may estimate the response probabilities by $\hat{p}_k = n_k^{-1} \Sigma_{j=1}^{J} x_{j1k}$ for group k and $\hat{p} = n^{-1} \Sigma_{j=1}^{J} \Sigma_{k=1}^{J} x_{j1k}$ for the pooled data. The WHM ignoring strata reduces to

$$\tilde{T} = \frac{\hat{p}_1 - \hat{p}_2}{\sqrt{\hat{p}\hat{q}\left(n_1^{-1} + n_2^{-1}\right)}},$$

where $\hat{q} = 1 - \hat{p}$.

Noting that $n_{jk} = n a_j b_{jk}$, we have

$$E(\hat{p}_k) \equiv p_k = \frac{\sum_{j=1}^{J} a_j b_{jk} p_{jk}}{\sum_{j=1}^{J} a_j b_{jk}} \tag{20.11}$$

and $E(\hat{p}) \equiv p = \Sigma_{j=1}^{J} a_j (b_{j1} p_{j1} + b_{j2} p_{j2})$. So, under H_1, $\hat{p}_1 - \hat{p}_2$ is approximately normal with mean $\tilde{\delta} = p_1 - p_2$ and variance $n^{-1} \tilde{\sigma}_1^2$, where $\tilde{\sigma}_1^2 = p_1 q_1 / b_1 + p_2 q_2 / b_2$, $q_k = 1 - p_k$ and $b_k = \Sigma_{j=1}^{J} a_j b_{jk}$. Also, under H_1, $\hat{p}\hat{q}(n_1^{-1} + n_2^{-1}) = n^{-1} \tilde{\sigma}_0^2 + o_p(n^{-1})$, where $\tilde{\sigma}_0^2 = pq(b_1^{-1} + b_2^{-1})$. Hence, the sample size ignoring strata is given as

$$\tilde{n} = \frac{(\tilde{\sigma}_0 z_{1-\alpha/2} + \tilde{\sigma}_1 z_{1-\beta})^2}{(p_1 - p_2)^2}. \tag{20.12}$$

Analysis based on propensity score is to adjust for possible unbalanced baseline covariates between groups. Under balanced baseline covariates (or propensity score), we have $b_{11} = \cdots = b_{J1}$, and, from Equation 20.11, $p_1 = p_2$ when $H_0: p_{j1} = p_{j2}$, $1 \leq j \leq J$ is true. Hence, under balanced covariates, the test statistic $\tilde{T}$ ignoring strata will be valid too. However, by not adjusting for the covariates (or, propensity), it will have a lower power than the stratified test statistic T; see Nam (1998) for Gart's test statistic. On the other hand, if the distribution of covariates is unbalanced, we have $p_1 \neq p_2$ even under H_0, and the test statistic $\tilde{T}$ ignoring strata will not be valid.

Remarks

For nonrandomized trials, the sponsors usually estimate sample size in the same way as they do in randomized trials. As a result, the U.S. FDA usually gives a warning and requests sample size justification (increase) based on the consideration of the degree of overlap in the propensity score distribution. When there exists an imbalance in covariate distribution between arms, a sample size calculation ignoring strata is definitely biased. The use of different weights will have an impact on statistical power but will not affect the consistency of the proposed WMH test. Note that the sample size formula by Nam (1998) based on the test statistic proposed by Gart (1985) is a special case of the sample size given in Equation 20.10, where $\hat{w}_j = \hat{q}_j$ and $w_j = 1 - b_{j1}p_{j1} - b_{j2}p_{j2}$.

20.2.3 Simulations

Suppose that we want to compare the probability of treatment success between control ($k = 1$) and new ($k = 2$) device. We consider partitioning the combined data into $J = 5$ strata based on propensity score, and the allocation proportions are projected as $(a_1, a_2, a_3, a_4, a_5) = (0.15, 0.15, 0.2, 0.25, 0.25)$ and $(b_{11}, b_{21}, b_{31}, b_{41}, b_{51}) = (0.4, 0.4, 0.5, 0.6, 0.6)$. Also, suppose that the response probabilities for control device are given as $(p_{11}, p_{21}, p_{31}, p_{41}, p_{51}) = (0.5, 0.6, 0.7, 0.8, 0.9)$, and we want to calculate the sample size required for a power of $1 - \beta = 0.8$ to detect an odds ratio of $\phi = 2$ using two-sided $\alpha = 0.05$. For $\phi = 2$, the response probabilities for the new device are given as $(p_{12}, p_{22}, p_{32}, p_{42}, p_{52}) = (0.6667, 0.7500, 0.8235, 0.8889, 0.9474)$. Under these settings, we need $n = 447$ for MH.

In order to evaluate the performance of the sample size formula, we conduct simulations. In each simulation, $n = 447$ binary observations are generated under the parameters (allocation proportions and the response probabilities) for sample size calculation. MH test with $\alpha = 0.05$ is applied to each simulation sample. Empirical power is calculated as the proportion of the simulation samples that reject H_0 out of $N = 10,000$ simulations. The empirical power is obtained as 0.7978, which is very close to the nominal $1 - \beta = 0.8$.

If we ignore the strata, we have $\tilde{p}_1 = 0.7519$ and $\tilde{p}_2 = 0.8197$ by Equation 20.11 and the odds ratio is only $\tilde{\phi} = 1.5004$, which is much smaller than $\phi = 2$. For $(\alpha, 1 - \beta) = (0.05, 0.8)$, we need $\tilde{n} = 1151$ by Equation 20.12. With $n = 422$, $\tilde{T}$ with $\alpha = 0.05$ rejected H_0 for only 41.4% of simulation samples.

The performance of the test statistics, T and $\tilde{T}$, are evaluated by generating simulation samples of size $n = 447$ under

$$H_0 : (p_{11}, p_{21}, p_{31}, p_{41}, p_{51}) = (p_{12}, p_{22}, p_{32}, p_{42}, p_{52}) = (0.1, 0.3, 0.5, 0.7, 0.9).$$

Other parameters are specified at the same values as above. For $\alpha = 0.05$, the empirical type I error is obtained as 0.0481 for T with MH scores and 0.1852 for $\tilde{T}$. While the empirical type I error of the MH stratified test is close to the nominal $\alpha = 0.05$, the unstratified test is severely inflated. Under this H_0, we have $\tilde{p}_1 = 0.7953$ and $\tilde{p}_1 = 0.7606$ ($\tilde{\phi} = 0.8181$), which are unequal due to the unbalanced covariate distribution between groups.

Now, let us consider a balanced allocation case, $b_{11} = \cdots = b_{j1} = 0.3$, with all the other parameter values being the same as in the above simulations. Under above H_1: $\phi = 2$, we need $n = 499$ for T and $\tilde{n} = 542$ for $\tilde{T}$. Note that the unstratified test $\tilde{T}$ requires a slightly larger sample size due to loss of efficiency. From $N = 10,000$ simulation samples of size $n = 499$, we obtained an empirical power of 0.799 for T and only 0.770 for $\tilde{T}$. From similar simulations under H_0, we obtained an empirical type I error of 0.0470 for T with MH scores and 0.0494 for $\tilde{T}$. Note that both tests control the type I error very accurately in this case. Under H_0, we have $\tilde{p}_1 = \tilde{p}_2 = 0.78$ ($\tilde{\phi} = 1$).

20.2.4 Concluding Remarks

Sample size calculation plays an important role in clinical research when designing a new medical study. Inadequate sample size could have a significant impact on the accuracy and reliability of the evaluation of treatment effect, especially in medical device clinical studies, which are often conducted in a nonrandomized fashion (although the method of propensity score may have been employed to achieve balance in baseline covariates). We propose a unified sample size formula for the WMH tests based on large-sample assumption. We found through simulations that our sample size formula accurately maintains the power. When the distribution of the covariates is unbalanced between groups, an analysis ignoring the strata could be severely biased.

20.3 ANOVA with Repeated Measures

In clinical research, it is not uncommon to have multiple assessments in a parallel-group clinical trial. The purpose of such a design is to evaluate the performance of clinical efficacy and/or safety over time during the entire study course. Clinical data collected from such a design are usually analyzed by means of the so-called ANOVA with repeated measures. In this section, formulas for sample size calculation under the design with multiple assessments are derived using the method of ANOVA with repeated measures.

20.3.1 Statistical Model

Let B_{ij} and H_{ijt} be the illness score at baseline and the illness score at time t for the jth subject in the ith treatment group, $t = 1, \ldots, m_{ij}$, $j = 1, \ldots, n_i$, and $i = 1, \ldots, k$. Then, $Y_{ijt} = H_{ijt} - B_{ij}$, the illness scores adjusted for baseline, can be described by the following statistical model (see, e.g., Longford, 1993; Chow and Liu, 2000):

$$Y_{ijt} = \mu + a_i + S_{ij} + b_{ij}t + e_{ijt},$$

where α_i is the fixed effect for the ith treatment ($\sum_{i=1}^{k} \alpha_i = 0$), S_{ij} is the random effect due to the jth subject in the ith treatment group, b_{ij} is the coefficient of the jth subject in the ith treatment group, and e_{ijt} is the random error in observing Y_{ijt}. In the above model, it is assumed that (i) S_{ij}'s are independently distributed as $N(0, \sigma_S^2)$, (ii) b_{ij}'s are independently distributed as $N(0, \sigma_\beta^2)$, and (iii) e_{ijt}'s are independently distributed as $N(0, \sigma^2)$. For any given subject j within treatment i, Y_{ijt} can be described by a regression line with slope b_{ij} conditioned on S_{ij}, that is,

$$Y_{ijt} = \mu_{ij} + b_{ij}t + e_{ijt}, t = 1,\ldots,m_{ij},$$

where $\mu_{ij} = \mu + \alpha_i + S_{ij}$. When conditioned on S_{ij}, that is, it is considered as fixed, unbiased estimators of the coefficient can be obtained by the method of ordinary least squares (OLS), which are given by

$$\hat{\mu}_{ij} = \frac{\sum_{i=1}^{k} Y_i \sum_{i=1}^{k} t_i^2 - \sum_{i=1}^{k} t_i \sum_{i=1}^{k} Y_i t_i}{m_{ij} \sum_{i=1}^{k} t_i^2 - (\sum_{i=1}^{k} t_i)^2},$$

$$\hat{b}_{ij} = \frac{m_{ij} \sum_{i=1}^{k} Y_i t_i - \sum_{i=1}^{k} Y_i \sum_{i=1}^{k} t_i}{m_{ij} \sum_{i=1}^{k} t_i^2 - (\sum_{i=1}^{k} t_i)^2}.$$

The sum of squared error for the jth subject in the ith treatment group, denoted by SSE_{ij} is distributed as $\sigma^2 \chi^2_{m_{ij}-2}$. Hence, an unbiased estimator of σ^2, the variation due to the repeated measurements, is given by

$$\hat{\sigma}^2 = \frac{\sum_{i=1}^{k} \sum_{j=1}^{n_i} SSE_{ij}}{\sum_{i=1}^{k} \sum_{j=1}^{n_i} (m_{ij} - 2)}.$$

Since data from subjects are independent, $\sum_{i=1}^{k}\sum_{j=1}^{n_i} SSE_{ij}$ is distributed as $\sigma^2 \chi^2_{n^*}$, where $n^* = \sum_{i=1}^{k}\sum_{j=1}^{n_i}(m_{ij} - 2)$. Conditioning on S_{ij} and b_{ij}, we have

$$\hat{\mu}_{ij} \sim N\left(\mu_{ij}, \frac{\sigma^2 \sum_{i=1}^{k} t_i^2}{m_{ij} \sum_{i=1}^{k} t_i^2 - (\sum_{i=1}^{k} t_i)^2}\right),$$

$$\hat{b}_{ij} \sim N\left(b_{ij}, \frac{\sigma^2 m_{ij}}{m_{ij} \sum_{i=1}^{k} t_i^2 - (\sum_{i=1}^{k} t_i)^2}\right).$$

Thus, unconditionally,

$$\hat{\mu}_{ij} \sim N\left(\mu + \alpha_i, \sigma_S^2 + \sigma^2 \frac{\sum_{i=1}^{k} t_i^2}{m_{ij} \sum_{i=1}^{k} t_i^2 - (\sum_{i=1}^{k} t_i)^2}\right),$$

$$\hat{b}_{ij} \sim N\left(\beta_{ij}, \sigma_\beta^2 + \sigma^2 \frac{m_{ij}}{m_{ij} \sum_{i=1}^{k} t_i^2 - (\sum_{i=1}^{k} t_i)^2}\right).$$

When there is an equal number of repeated measurements for each subject, that is, $m_{ij} = m$ for all i, j, the intersubject variation, σ^2_S, can be estimated by

$$\hat{\sigma}_S^2 = \frac{\sum_{i=1}^{k} \sum_{j=1}^{n_i} (\mu_{ij} - \mu_{i.})^2}{\sum_{i=1}^{k} (n_i - 1)} - \hat{\sigma}^2 \frac{\sum_{i=1}^{k} t_i^2}{m \sum_{i=1}^{k} t_i^2 - (\sum_{j=1}^{n_i} t_i)^2},$$

where

$$\hat{\mu}_{i.} = \frac{1}{n_i} \sum_{j=1}^{n_i} \hat{\mu}_{ij}.$$

An estimator of the variation of b_{ij} can then be obtained as

$$\hat{\sigma}_\beta^2 = \frac{\sum_{i=1}^{k} \sum_{j=1}^{n_i} (b_{ij} - b_{i.})^2}{\sum_{i=1}^{k} (n_i - 1)} - \hat{\sigma}^2 \frac{m}{m \sum_{i=1}^{k} t_i^2 - (\sum_{j=1}^{n_i} t_i)^2},$$

where

$$\hat{b}_{i.} = \frac{1}{n_i} \sum_{j=1}^{n_i} \hat{b}_{ij}.$$

Similarly, β_i can be estimated by

$$\hat{\beta}_i = \frac{1}{n_i} \sum_{j=1}^{n_i} b_{ij} \sim N(\beta_i, \delta^2),$$

where

$$\delta^2 = \sigma_\beta^2 + \frac{\sigma^2}{n_i} \frac{m}{m \sum_{i=1}^k t_i^2 - (\sum_{j=1}^k t_i)^2}.$$

20.3.2 Hypotheses Testing

Since the objective is to compare the treatment effect on the illness scores, it is of interest to test the following hypotheses:

$$H_0 : \alpha_i = \alpha_{i'} \quad \text{versus} \quad H_a : \alpha_i \neq \alpha_{i'}.$$

The above hypotheses can be tested by using the statistic

$$T_1 = \sqrt{\frac{n_i n_{i'}}{n_i + n_{i'}}} \left(\frac{\hat{\mu}_{i.} - \hat{\mu}_{i'.}}{\sum_{i=1}^k \sum_{j=1}^{n_i} (\hat{\mu}_{ij} - \hat{\mu}_{i.})^2 / \sum_{i=1}^k (n_i - 1)} \right).$$

Under the null hypotheses of no difference in the illness score between two treatment groups, T_1 follows a t-distribution with $\sum_{i=1}^k (n_i - 1)$ degrees of freedom. Hence, we reject the null hypothesis at the α level of significance if

$$|T_1| > t_{\alpha/2, \sum_{i=1}^k (n_i - 1)}.$$

Furthermore, it is also of interest to test the following null hypothesis of equal slopes (i.e., rate of change in illness scores over the repeated measurement):

$$H_0 : \beta_i = \beta_{i'} \quad \text{versus} \quad H_a : \beta_i \neq \beta_{i'}.$$

The above hypotheses can be tested using the following statistic:

$$T_2 = \sqrt{\frac{n_i n_{i'}}{n_i + n_{i'}}} \left(\frac{\hat{b}_{i.} - \hat{b}_{i'.}}{\sum_{i=1}^k \sum_{j=1}^{n_i} (\hat{b}_{ij} - \hat{b}_{i.})^2 / \sum_{i=1}^k (n_i - 1)} \right).$$

Under the null hypotheses, T_2 follows a *t*-distribution with $\Sigma_{i=1}^k(n_i-1)$ degrees of freedom. Hence, we would reject the null hypotheses of no difference in the rate of change in illness scores between treatment groups at the α level of significance if

$$|T_2| > t_{\alpha/2,\Sigma_{i=1}^k(n_i-1)}.$$

20.3.3 Sample Size Calculation

Since the above tests are based on a standard *t*-test, the sample size per treatment group required for the detection of a clinically meaningful difference, Δ, with a desired power of $1-\beta$ at the α level of significance is given by

$$n \geq \frac{2\sigma^{*2}(z_{\alpha/2}+z_\beta)^2}{\Delta^2},$$

where σ^{*2} is the sum of the variance components. When the null hypothesis $H_0: \alpha_i = \alpha_{i'}$ is tested,

$$\sigma^{*2} = \sigma_\alpha^2 = \sigma_S^2 + \frac{\sigma^2 \sum_{i=1}^k t_i^2}{m_{ij}\sum_{i=1}^k t_i^2 - (\sum_{i=1}^k t_i)^2},$$

which can be estimated by

$$\hat{\sigma}_1^2 = \frac{\sum_{i=1}^k \sum_{j=1}^{n_i}(\hat{\mu}_{ij}-\hat{\mu}_{i.})^2}{\sum_{i=1}^k(n_i-1)}.$$

When the null hypothesis $H_0: \beta_i = \beta_{i'}$ is tested,

$$\sigma^{*2} = \sigma_\beta^2 + \sigma^2 \frac{m_{ij}}{m_{ij}\sum_{i=1}^k t_i^2 - (\sum_{i=1}^k t_i)^2},$$

which can be estimated by

$$\hat{\sigma}_2^2 = \frac{\sum_{i=1}^k \sum_{j=1}^{n_i}(\hat{\beta}_{ij}-\hat{\beta}_{i.})^2}{\sum_{i=1}^k(n_i-1)}.$$

20.3.4 An Example

Suppose a sponsor is planning a randomized, parallel-group clinical trial on mice with multiple sclerosis (MS) for the evaluation of a number of doses of an active compound, an active control, and/or a vehicle control. In practice, experimental autoimmune encephalomyelitis (EAE) is usually induced in susceptible animals following a single injection of central nervous system extract emulsified in the adjuvant. The chronic forms of EAE reflect many of the pathophysiologic steps in MS. This similarity has initiated the usage

TABLE 20.4

Clinical Assessment of Induction of EAE in Mice

Stage 0	Normal
Stage 0.5	Partial limp tail
Stage 1	Complete limp tail
Stage 2	Impaired righting reflex
Stage 2.5	Righting reflex is delayed (not weak enough to be stage 3)
Stage 3	Partial hind limb paralysis
Stage 3.5	One leg is completely paralyzed and one leg is partially paralyzed
Stage 4	Complete hind limb paralysis
Stage 4.5	Legs are completely paralyzed and moribund
Stage 5	Death due to EAE

of EAE models for the development of MS therapies. Clinical assessment of the illness for the induction of EAE in mice that are commonly used is given in Table 20.4. Illness scores of mice are recorded between day 10 and day 19 before dosing regardless of the onset or the remission of the attack. Each mouse will receive a dose from day 20 to day 33. The postinoculation performance of each mouse is usually assessed up to day 34. Sample size calculation was performed using the ANOVA model with repeated measures based on the assumptions that

1. The primary study endpoint is the illness scores of mice.
2. The scientifically meaningful difference in illness score is considered to be 1.0 or 1.5 (note that changes of 1.0 and 1.5 over the time period between day 19 and day 34 are equivalent to 0.067 and 0.1 changes in the slope of the illness score curve).
3. The standard deviation for each treatment group is assumed to be same (various standard deviations such as 0.75, 0.80, 0.85, 0.90, 0.95, 1.0, 1.05, 1.1, 1.15, 1.2, or 1.25 are considered).
4. The null hypothesis of interest is that there is no difference in the profile of illness scores, which is characterized between baseline (at dosing) and a specific time point after dosing, among treatment groups.
5. The probability of committing a type II error is 10% or 20%, that is, the power is, respectively, 90% or 80%.
6. The null hypothesis is tested at the 5% level of significance.
7. Bonferroni's adjustment for α significant level for multiple comparisons was considered.

Under the one-way ANOVA model with repeated measures, the formula for the sample size calculation given in the previous subsection can be used.

Table 20.5 provides sample sizes required for three-arm, four-arm, and five-arm studies with α adjustment for multiple comparisons, respectively. For example, a sample size of 45 subjects (i.e., 15 subjects per treatment group) is needed to maintain an 80% power for detection of a decrease in illness score by $\Delta = 1.5$ over the active treatment period between treatment groups when σ^* is 1.25.

TABLE 20.5

Sample Size Calculation with α Adjustment for Multiple Comparisons

Power	σ^*	Sample Size per Arm					
		Three Arms		Four Arms		Five Arms	
		$\Delta = 1.0$	$\Delta = 1.5$	$\Delta = 1.0$	$\Delta = 1.5$	$\Delta = 1.0$	$\Delta = 1.5$
80%	0.75	12	6	14	7	15	7
	0.80	14	6	16	7	17	8
	0.85	16	7	18	8	20	9
	0.90	17	8	20	9	22	10
	0.95	19	9	22	10	25	11
	1.00	21	10	25	11	27	12
	1.05	24	11	27	12	30	14
	1.10	26	12	30	14	33	15
	1.15	28	13	33	15	36	16
	1.20	31	14	35	16	39	17
	1.25	33	15	38	17	42	19
90%	0.75	16	7	18	8	19	9
	0.80	18	8	20	9	22	10
	0.85	20	9	23	10	25	11
	0.90	22	10	25	12	28	13
	0.95	25	11	28	13	31	14
	1.00	28	13	31	14	34	15
	1.05	30	14	34	16	37	17
	1.10	33	15	38	17	41	18
	1.15	36	16	41	19	45	20
	1.20	39	18	45	20	49	22
	1.25	43	19	49	22	53	24

20.4 Quality of Life

In clinical research, it has been a concern that the treatment of disease or survival may not be as important as the improvement of QOL, especially for patients with a chronic or life-threatening disease. Enhancement of life beyond the absence of illness to the enjoyment of life may be considered more important than the extension of life. QOL not only provides information as to how the patients feel about drug therapies, but also appeals to the physician's desire for the best clinical practice. It can be used as a predictor of compliance of the patient. In addition, it may be used to distinguish between therapies that appear to be equally efficacious and equally safe at the stage of marketing strategy planning. The information can be potentially used in advertising for the promotion of the drug therapy. As a result, in addition to the comparison of hazard rates, survival function, or median survival time, QOL is often assessed in survival trials.

QOL is usually assessed by a questionnaire, which may consist of a number of questions (items). We refer to such a questionnaire as a QOL instrument. A QOL instrument is a very subjective tool and is expected to have a large variation. Thus, it is a concern whether the adopted QOL instrument can accurately and reliably quantify patients' QOL. In this section, we provide a comprehensive review of the validation of a QOL instrument, the use of QOL scores, statistical methods for the assessment of QOL, and practical issues that are commonly encountered in clinical trials. For the assessment of QOL, statistical analysis based on subscales, composite scores, and/or overall score are often performed for an easy interpretation. For example, Tandon (1990) applied a global statistic to combine the results of the univariate analysis of each subscale. Olschewski and Schumacher (1990), on the other hand, proposed to use composite scores to reduce the dimensions of QOL. However, owing to the complex correlation structure among subscales, optimal statistical properties may not be obtained. As an alternative, to account for the correlation structure, the following time series model proposed by Chow and Ki (1994) may be useful.

20.4.1 Time Series Model

For a given subscale (or component), let x_{ijt} be the response of the jth subject to the ith question (item) at time t, where $i = 1, \ldots, k$, $j = 1, \ldots, n$, and $t = 1, \ldots, T$. Consider the average score over k questions:

$$Y_{jt} = \bar{x}_{jt} = \frac{1}{k}\sum_{j=1}^{k} x_{ijt}.$$

Since the average scores $y_{j1}, \ldots, y_{jT}$ are correlated, the following autoregressive time series model may be an appropriate statistical model for y_{jt}:

$$y_{jt} = \mu + \Psi(y_{j(t-1)} - \mu) + e_{jt}, \ j = 1, \ldots, n, \ t = 1, \ldots, T,$$

where μ is the overall mean, $|\psi| < 1$ is the autoregressive parameter, and e_{jt} are independent identically distributed random errors with mean 0 and variance σ_e^2. It can be verified that

$$E(e_{jt}y'_{jt}) = 0 \text{ for all } t' < t.$$

The autoregressive parameter ψ can be used to assess the correlation of consecutive responses y_{jt} and $y_{j(t+1)}$. From the above model, it can be shown that the autocorrelation of responses with m lag times is ψ^m, which is negligible when m is large. Based on the observed average scores on the jth subject, $y_{j1}, \ldots, y_{jT}$, we can estimate the overall mean μ and the autoregressive parameter ψ. The OLS estimators of μ and ψ can be approximated by

$$\hat{\mu}_j = \bar{y}_{j.} = \frac{1}{T}\sum_{t=1}^{T} y_{jt}$$

and

$$\hat{\psi}_j = \frac{\sum_{t=2}^{T} (y_{jt} - \bar{y}_{j.})(y_{j(t-1)} - \bar{y}_{j.})}{\sum_{t=2}^{T} (y_{jt} - \bar{y}_{j.})^2},$$

which are the sample mean and sample autocorrelation of consecutive observations. Under the above model, it can be verified that $\hat{\mu}_j$ is unbiased and that the variance of $\hat{\mu}_j$ is given by

$$\text{Var}(\bar{y}_{j.}) = \frac{\gamma_{j0}}{T}\left[1 + 2\sum_{t=1}^{T-1}\frac{T-t}{T}\psi^t\right],$$

where $\gamma_{j0} = \text{Var}(y_{jt})$. The estimated variance of $\hat{\mu}_j$ can be obtained by replacing ψ with $\hat{\psi}_j$ and γ_{j0} with

$$c_{j0} = \sum_{t=1}^{T}\frac{(y_{jt} - \bar{y}_{j.})^2}{T-1}.$$

Suppose that the n subjects are from the same population with the same variability and autocorrelation. The QOL measurements of these subjects can be used to estimate the mean average scores μ. An intuitive estimator of μ is the sample mean

$$\hat{\mu} = \bar{y}_{..} = \frac{1}{n}\sum_{j=1}^{n}\bar{y}_{j.}.$$

Under the time series model, the estimated variance of $\hat{\mu}$ is given by

$$s^2(\bar{y}_{..}) = \frac{c_0}{nT}\left[1 + 2\sum_{t=1}^{T-1}\frac{T-t}{T}\hat{\psi}^t\right],$$

where

$$c_0 = \frac{1}{n(T-1)}\sum_{j=1}^{n}\left[\sum_{t=1}^{T}(y_{jt} - \bar{y}_{j.})^2\right]$$

and

$$\hat{\psi} = \frac{1}{n}\sum_{j=1}^{n}\hat{\psi}_j.$$

An approximate $(1-\alpha)100\%$ confidence interval for μ has limits

$$\bar{y}_{..} \pm z_{1-\alpha/2}s(\bar{y}_{..}),$$

where $z_{1-\alpha/2}$ is the $(1-\alpha/2)$th quantile of a standard normal distribution.

Under the time series model, the method of confidence interval approach described above can be used to assess the difference in QOL between treatments. Note that the assumption that all the QOL measurements over time are independent is a special case of the above model with $\psi = 0$. In practice, it is suggested that the above time series model be used to account for the possible positive correlation between measurements over the time period under study.

20.4.2 Sample Size Calculation

Under the time series model, Chow and Ki (1996) derived some useful formulas for the determination of sample size based on normal approximation. For a fixed precision index $1 - \alpha$, to ensure a reasonable high-power index δ for detecting a meaningful difference ε, the sample size per treatment group should not be less than

$$n_\delta = \frac{c(z_{1-1/2\alpha} + z_\delta)^2}{\varepsilon^2} \quad \text{for} \quad \delta > 0.5,$$

where

$$c = \frac{\gamma_y}{T}\left(1 + 2\sum_{t=1}^{T-1}\frac{T-t}{T}\psi_y^t\right) + \frac{\gamma_u}{T}\left(1 + 2\sum_{t=1}^{T-1}\frac{T-t}{T}\psi_u^t\right).$$

For a fixed precision index $1 - \alpha$, if the acceptable limit for detecting an equivalence between two treatment means is $(-\Delta, \Delta)$, to ensure a reasonable high power ϕ for detecting an equivalence when the true difference in treatment means is less than a small constant η, the sample size for each treatment group should be at least

$$n_\phi = \frac{c}{(\Delta - \eta)^2}(z_{1/2+1/2\phi} + z_{1-1/2\alpha})^2.$$

If both treatment groups are assumed to have the same variability and autocorrelation coefficient, the constant c can be simplified as

$$c = \frac{2\gamma}{T}\left(1 + 2\sum_{t=1}^{T-1}\frac{T-t}{T}\psi^t\right).$$

When $n = \max(n_\phi, n_\delta)$, it ensures that the QOL instrument will have precision index $1 - \alpha$ and power of no less than δ and ϕ in detecting a difference and an equivalence, respectively. It, however, should be noted that the required sample size is proportional to the variability of the average scores considered. The higher the variability, the larger the sample size that would be required. Note that the above formulas can also be applied to many clinical-based research studies with time-correlated outcome measurements, for example, 24-hour monitoring of blood pressure, heart rates, hormone levels, and body temperature.

20.4.3 An Example

To illustrate the use of the above sample size formulas, consider QOL assessment in two independent groups A and B. Suppose a QOL instrument containing 11 questions is to be administered to subjects at week 4, 8, 12, and 16. Denote the mean of QOL score of the subjects in group A and B by y_{it} and u_{jt}, respectively. We assume that y_{it} and u_{jt} have distributions that follow the time series model described in the previous section with common variance $\gamma = 0.5$ square unit and have moderate autocorrelation between scores at consecutive time points, say $\psi = 0.5$. For a fixed 95% precision index, 87 subjects per group will provide a 90% power for the detection of a difference of 0.25 unit in means. If the chosen acceptable limits are $(-0.35, 0.35)$, then 108 subjects per group will have a power of 90% that the 95% confidence interval of difference in group means will correctly detect an equivalence with $\eta = 0.1$ unit. If the sample size is chosen to be 108 per group, it ensures that the power indices for detecting a difference of 0.25 unit or an equivalence are not less than 90%.

20.5 Bridging Studies

In the pharmaceutical industry, the sponsors are often interested in bringing their drug products from one region (e.g., the United States) to another region (e.g., Asian Pacific) to increase the exclusivity of the drug products in the marketplace. However, it is a concern whether the clinical results can be extrapolated from the target patient population in one region to a similar but different patient population in a new region due to a possible difference in ethnic factors. The ICH recommends that a bridging study may be conducted to extrapolate the clinical results between regions. However, little or no information regarding the criterion for determining whether a bridging study is necessary based on the evaluation of the complete clinical data package is provided by the ICH. Furthermore, no criterion on the assessment of the similarity of clinical results between regions is given. In this section, we propose the use of a sensitivity index as a possible criterion for regulatory authorities in the new region to evaluate whether a bridging clinical study should be conducted and the sample size of such a bridging clinical study.

20.5.1 Sensitivity Index

Suppose that a randomized, parallel-group, placebo-controlled clinical trial is conducted for the evaluation of a test compound as compared to a placebo control in the original region. The study protocol calls for a total of n subjects with the disease under study. These $n = n_1 + n_2$ subjects are randomly assigned to receive either the test compound or a placebo control. Let x_{ij} be the response observed from the jth subject in the ith treatment group, where $j = 1, \ldots, n_i$ and $i = 1, 2$. Assume that x_{ij}'s are independent and normally distributed with means μ_i, $i = 1, 2$, and a common variance σ^2. Suppose the hypotheses of interest are

$$H_0 : \mu_1 - \mu_2 = 0 \quad \text{versus} \quad H_a : \mu_1 - \mu_2 \neq 0.$$

Note that the discussion for a one-sided H_a is similar. When σ^2 is unknown, we reject H_0 at the 5% level of significance if

$$|T| > t_{n-2},$$

where $t_{\alpha,n-2}$ is the $100(1-\alpha/2)$th percentile of a t-distribution with $n-2$ degrees of freedom, $n = n_1 + n_2$,

$$T = \frac{\bar{x}_1 - \bar{x}_2}{\sqrt{\frac{(n_1-1)s_1^2 + (n_2-1)s_2^2}{n-2}}\sqrt{\frac{1}{n_1}+\frac{1}{n_2}}},$$ (20.13)

and $\bar{x}_i$ and s_i^2 are the sample mean and variance, respectively, based on the data from the ith treatment group. The power of T is given by

$$p(\theta) = P(|T| > t_{n-2}) = 1 - T_{n-2}(t_{n-2}\,|\,\theta) + T_{n-2}(-t_{n-2}\,|\,\theta),$$ (20.14)

where

$$\theta = \frac{\mu_1 - \mu_2}{\sigma\sqrt{\frac{1}{n_1}+\frac{1}{n_2}}}$$

and $T_{n-2}(\cdot\,|\,\theta)$ denotes the cumulative distribution function of the noncentral t-distribution with $n-2$ degrees of freedom and the noncentrality parameter θ.

Let x be the observed data from the first clinical trial and $T(x)$ be the value of T based on x. Replacing θ in the power function in Equation 20.14 by its estimate $T(x)$, the estimated power can be obtained as follows:

$$\hat{P} = p(T(x)) = 1 - T_{n-2}(t_{n-2}\,|\,T(x)) + T_{n-2}(-t_{n-2}\,|\,T(x)).$$

Note that Shao and Chow (2002) refer to $\hat{P}$ as the reproducibility probability for the second clinical trial with the same patient population. However, a different and more sensible way of defining a reproducibility probability is to define reproducibility probability as the posterior mean of $p(\theta)$, that is,

$$\tilde{P} = P(|T(y)| > t_{n-2}\,|\,x) = \int p(\theta)\pi(\theta\,|\,x)d\theta,$$

where y denotes the data set from the second trial and $\pi(\theta|x)$ is the posterior density of θ, given x. When the noninformative prior $\pi(\mu_1, \mu_2, \sigma^2) = \sigma^{-2}$ is used, Shao and Chow (2002) showed that

$$\tilde{P} = E_{\delta,u}\left[1 - T_{n-2}\left(t_{n-2}\left|\frac{\delta}{u}\right.\right) - T_{n-2}\left(-t_{n-2}\left|\frac{\delta}{u}\right.\right)\right],$$

where $E_{\delta,u}$ is the expectation with respect to δ and u, u^{-2} has the gamma distribution with the shape parameter $(n-2)/2$ and the scale parameter $2/(n-2)$, and given u, δ has the normal distribution $N(T(x), u^2)$.

When the test compound is applied to a similar but different patient population in the new region, it is expected that the mean and variance of the response would be different.

Chow, Shao, and Hu (2002a) proposed the following concept of sensitivity index for the evaluation of the change in patient population due to ethnic differences. Suppose that in the second clinical trial conducted in the new region, the population mean difference is changed to $\mu_1 - \mu_2 + \varepsilon$ and the population variance is changed to $C^2\sigma^2$, where $C > 0$. If $|\mu_1 - \mu_2|/\sigma$ is the signal-to-noise ratio for the population difference in the original region, then the signal-to-noise ratio for the population difference in the new region is

$$\frac{|\mu_1 - \mu_2 + \varepsilon|}{C\sigma} = \frac{|\Delta(\mu_1 - \mu_2)|}{\sigma},$$

where

$$\Delta = \frac{1 + \varepsilon/(\mu_1 - \mu_2)}{C} \tag{20.15}$$

is a measure of change in the signal-to-noise ratio for the population difference, which is the sensitivity index of population differences between regions. For most practical problems, $|\varepsilon| < |\mu_1 - \mu_2|$ and, thus, $\Delta > 0$. By Equation 20.14, the power for the second trial conducted in the new region is $p(\Delta\theta)$.

As indicated by Chow, Shao, and Hu (2002), there are two advantages of using Δ as a sensitivity index, instead of ε (changes in mean) and C (changes in standard deviation). First, the result is easier to interpret when there is only one index. Second, the reproducibility probability is a strictly decreasing function of Δ, whereas an increased population variance (or a decreased population difference) may or may not result in a decrease in the reproducibility probability.

If Δ is known, then the reproducibility probability

$$\hat{P}\Delta = p(\Delta T(x))$$

can be used to assess the probability of generalizability between regions. For the Bayesian approach, the generalizability probability is

$$\tilde{P}_\Delta = E_{\delta, u} \left[1 - T_{n-2} \left(t_{n-2} \left| \frac{\Delta\delta}{u} \right. \right) - T_{n-2} \left(-t_{n-2} \left| \frac{\Delta\delta}{u} \right. \right) \right].$$

In practice, the value of Δ is usually unknown. We may either consider a maximum possible value of $|\Delta|$ or a set of Δ values to carry out a sensitivity analysis (see Table 20.6). For the Bayesian approach, we may also consider the average of $\tilde{P}_\Delta$ over a prior density $\pi(\Delta)$, that is,

$$\tilde{P} = \int \tilde{P}_\Delta \pi(\Delta) d\Delta.$$

Table 20.6 provides a summary of reproducibility probability $\hat{P}_\Delta$ for various sample sizes, respectively. For example, a sample size of 30 will give an 80.5% of reproducibility provided that $\Delta T = 2.92$.

TABLE 20.6

Sensitivity Analysis of Reproducibility Probability $\hat{P}_\Delta$

				n				
ΔT	10	20	30	40	50	60	100	∞
1.96	0.407	0.458	0.473	0.480	0.484	0.487	0.492	0.500
2.02	0.429	0.481	0.496	0.504	0.508	0.511	0.516	0.524
2.08	0.448	0.503	0.519	0.527	0.531	0.534	0.540	0.548
2.14	0.469	0.526	0.542	0.550	0.555	0.557	0.563	0.571
2.20	0.490	0.549	0.565	0.573	0.578	0.581	0.586	0.594
2.26	0.511	0.571	0.588	0.596	0.601	0.604	0.609	0.618
2.32	0.532	0.593	0.610	0.618	0.623	0.626	0.632	0.640
2.38	0.552	0.615	0.632	0.640	0.645	0.648	0.654	0.662
2.44	0.573	0.636	0.654	0.662	0.667	0.670	0.676	0.684
2.50	0.593	0.657	0.675	0.683	0.688	0.691	0.697	0.705
2.56	0.613	0.678	0.695	0.704	0.708	0.711	0.717	0.725
2.62	0.632	0.698	0.715	0.724	0.728	0.731	0.737	0.745
2.68	0.652	0.717	0.735	0.743	0.747	0.750	0.756	0.764
2.74	0.671	0.736	0.753	0.761	0.766	0.769	0.774	0.782
2.80	0.690	0.754	0.771	0.779	0.783	0.786	0.792	0.799
2.86	0.708	0.772	0.788	0.796	0.800	0.803	0.808	0.815
2.92	0.725	0.789	0.805	0.812	0.816	0.819	0.824	0.830
2.98	0.742	0.805	0.820	0.827	0.831	0.834	0.839	0.845
3.04	0.759	0.820	0.835	0.842	0.846	0.848	0.853	0.860
3.10	0.775	0.834	0.849	0.856	0.859	0.862	0.866	0.872
3.16	0.790	0.848	0.862	0.868	0.872	0.874	0.879	0.884
3.22	0.805	0.861	0.874	0.881	0.884	0.886	0.890	0.895
3.28	0.819	0.873	0.886	0.892	0.895	0.897	0.901	0.906
3.34	0.832	0.884	0.897	0.902	0.905	0.907	0.911	0.916
3.40	0.844	0.895	0.907	0.912	0.915	0.917	0.920	0.925
3.46	0.856	0.905	0.916	0.921	0.924	0.925	0.929	0.932
3.52	0.868	0.914	0.925	0.929	0.932	0.933	0.936	0.940
3.58	0.879	0.923	0.933	0.937	0.939	0.941	0.943	0.947
3.64	0.889	0.931	0.940	0.944	0.946	0.947	0.950	0.953
3.70	0.898	0.938	0.946	0.950	0.952	0.953	0.956	0.959
3.76	0.907	0.944	0.952	0.956	0.958	0.959	0.961	0.965
3.82	0.915	0.950	0.958	0.961	0.963	0.964	0.966	0.969
3.88	0.923	0.956	0.963	0.966	0.967	0.968	0.970	0.973
3.94	0.930	0.961	0.967	0.970	0.971	0.972	0.974	0.977

20.5.2 Assessment of Similarity

Criterion for Similarity

Let x be a clinical response of interest in the original region. Here, x could be either the response of the primary study endpoint from a test compound under investigation or the difference of responses of the primary study endpoint between a test drug and a control (e.g., a standard therapy or placebo). Let y be similar to x but is a response in a clinical bridging study conducted in the new region. Using the criterion for the assessment of population and individual bioequivalence (FDA, 2001), Chow, Shao, and Hu (2002) proposed the following measure of the similarity between x and y:

$$\theta = \frac{E(x-y)^2 - E(x-x')^2}{E(x-x')^2/2}, \tag{20.16}$$

where x' is an independent replicate of x and y, x, and x' are assumed to be independent. Note that θ in Equation 20.16 assesses not only the difference between the population means $E(x)$ and $E(y)$ but also the population variation of x and y through a function of mean squared differences. Also, θ is a relative measure, that is, the mean squared difference between x and y is compared with the mean squared difference between x and x'. It is related to the so-called *population difference ratio* (PDR), that is,

$$\text{PDR} = \sqrt{\frac{\theta}{2} + 1},$$

where

$$\text{PDR} = \sqrt{\frac{E(x-y)^2}{E(x-x')^2}}.$$

In assessing population bioequivalence (or individual bioequivalence), the similarity measure θ is compared with a population bioequivalence (or individual bioequivalence) limit θ_U set by the FDA. For example, with log-transformed responses, FDA (2001) suggests

$$\theta_U = \frac{(\log 1.25)^2 + \varepsilon}{\sigma_0^2},$$

where $\sigma_0^2 > 0$ and $\varepsilon \geq 0$ are two constants given in the FDA guidance, which depend upon the variability of the drug product.

Since a small value of θ indicates that the difference between x and y is small (relative to the difference between x and x'), the similarity between the new region and the original region can be claimed if and only if $\theta < \theta_U$, where θ_U is a similarity limit. Thus, the problem of assessing similarity becomes a hypothesis testing problem with hypotheses

$$H_0 : \theta \geq \theta_U \quad \text{versus} \quad H_a : \theta < \theta_U. \tag{20.17}$$

Let $k = 0$ indicate the original region and $k = 1$ indicate the new region. Suppose that there are m_k study centers and n_k responses in each center for a given variable of interest. Here, we first consider the balanced case where centers in a given region have the same number of observations. Let z_{ijk} be the ith observation from the jth center of region k, b_{jk} be the between-center random effect, and e_{ijk} be the within-center measurement error. Assume that

$$z_{ijk} = \mu_k + b_{jk} + e_{ijk}, \quad i = 1, \ldots, n_k, \quad j = 1, \ldots, m_k, \quad k = 0, 1, \tag{20.18}$$

where μ_k is the population mean in region k, $b_{jk} \sim N(0, \sigma_{Bk}^2)$, $e_{ijk} \sim N(0, \sigma_{Wk}^2)$, and b_{jk}'s and e_{ijk}'s are independent. Under model (20.18), the parameter θ in Equation 20.16 becomes

$$\theta = \frac{(\mu_0 - \mu_1)^2 + \sigma_{T1}^2 - \sigma_{T0}^2}{\sigma_{T0}^2},$$

where $\sigma_{Tk}^2 = \sigma_{Bk}^2 + \sigma_{Wk}^2$ is the total variance (between-center variance plus within-center variance) in region k. The hypotheses in Equation 20.17 are equivalent to

$$H_0 : \varsigma \geq 0 \quad \text{versus} \quad H_a : \varsigma < 0, \tag{20.19}$$

where

$$\varsigma = (\mu_0 - \mu_1)^2 + \sigma_{T1}^2 - (1 + \theta_U)\sigma_{T0}^2. \tag{20.20}$$

Statistical Methods

A statistical test of significance level 5% can be obtained by using a 95% upper confidence bound $\hat{\varsigma}_U$ for ς in Equation 20.20, that is, we reject H_0 in Equation 20.19 if and only if $\hat{\varsigma}_U < 0$.

Under model (20.18), an unbiased estimator of the population mean in region k is

$$\bar{z}_k \sim N\left(\mu_k, \frac{\sigma_{Bk}^2}{m_k} + \frac{\sigma_{Wk}^2}{N_k}\right),$$

where $N_k = m_k n_k$ and $\bar{z}_k$ is the average of z_{ijk}'s over j and i for a fixed k. To construct a 95% upper confidence bound for ς in Equation 20.20 using the approach in Hyslop, Hsuan, and Holder (2000), which is proposed for testing individual bioequivalence, we need to find independent, unbiased, and chi-square distributed estimators of σ_{Tk}^2, $k = 0$, 1. These estimators, however, are not available when $n_k > 1$. Note that

$$\sigma_{Tk}^2 = \sigma_{Bk}^2 + n_k^{-1}\sigma_{Wk}^2 + \left(1 - n_k^{-1}\right)\sigma_{Wk}^2, \quad k = 0, 1;$$

$\sigma_{Bk}^2 + n_k^{-1}\sigma_{Wk}^2$ can be unbiasedly estimated by

$$s_{Bk}^2 = \frac{1}{m_k - 1}\sum_{j=1}^{m_k}(\bar{z}_{jk} - \bar{z}_k)^2 \sim \frac{(\sigma_{Bk}^2 + n_k^{-1}\sigma_{Wk}^2)\chi_{m_k - 1}^2}{m_k - 1},$$

where $\bar{z}_{jk}$ is the average of z_{ijk}'s over i and χ_l^2 denotes a random variable having the chi-square distribution with l degrees of freedom; σ_{Wk}^2 can be estimated by

$$s_{Wk}^2 = \frac{1}{N_k - m_k}\sum_{j=1}^{m_k}\sum_{i=1}^{n_k}(z_{ijk} - \bar{z}_{jk})^2 \sim \frac{\sigma_{Wk}^2 \chi_{N_k - m_k}^2}{N_k - m_k},$$

and $\bar{z}_k, s_{Bk}^2, s_{Wk}^2$, $k = 0, 1$, are independent. Thus, an approximate 95% upper confidence bound for ζ in Equation 20.20 is

$$\hat{\varsigma}_U = (\bar{z}_0 - \bar{z}_1)^2 + s_{B1}^2 + \left(1 - n_k^{-1}\right)s_{W1}^2 - (1 + \theta_U)\left[s_{B0}^2 + (1 - n_0^{-1})s_{W0}^2\right] + \sqrt{U},$$

where U is the sum of the following five quantities:

$$\left[\left(|\bar{z}_0 - \bar{z}_1| + 1.645\sqrt{\frac{s_{B0}^2}{m_0} + \frac{s_{B1}^2}{m_1}}\right)^2 - (\bar{z}_0 - \bar{z}_1)^2\right]^2,$$

$$s_{B1}^4\left(\frac{m_1 - 1}{\chi_{0.05;m_1-1}^2} - 1\right)^2,$$

$$\left(1 - n_1^{-1}\right)^2 s_{W1}^4\left(\frac{N_1 - m_1}{\chi_{0.05;N_1-m_1}^2} - 1\right)^2,$$

$$(1 + \theta_U)^2 s_{B0}^4\left(\frac{m_0 - 1}{\chi_{0.95;m_0-1}^2} - 1\right)^2,$$

$$(1 + \theta_U)^2\left(1 - n_0^{-1}\right)^2 s_{W0}^4\left(\frac{N_0 - m_0}{\chi_{0.95;N_0-m_0}^2} - 1\right)^2,$$

and $\chi_{a;l}^2$ is the $100a$th percentile of the chi-square distribution with l degrees of freedom.

Thus, the null hypothesis H_0 in Equation 20.19 can be rejected at approximately 5% significance level if $\hat{\varsigma}_U < 0$. Similar to population bioequivalence testing (FDA, 2001), we conclude that the similarity between two regions (in terms of the given variable of interest) if and only if H_0 in Equation 20.19 is rejected and the observed mean difference $\bar{z}_0 - \bar{z}_1$ is within the limits of ± 0.233.

Consider now the situation where centers contain different numbers of observations in a given region, that is, the jth center in region k contains n_{jk} observations. We have the following recommendations:

1. If all n_{jk}'s are large, then the previously described procedure is still approximately valid, provided that N_k is defined to be $n_{1k} + \cdots + n_{mkk}$ and $\left(1 - n_k^{-1}\right)s_{Wk}^2$ is replaced by

$$\tilde{s}_{Wk}^2 = \frac{1}{N_k - m_k}\sum_{j=1}^{m_k}\sum_{i=1}^{n_{jk}}(z_{ijk} - \bar{z}_{jk})^2.$$

2. If n_{jk} are not very different (e.g., unequal sample sizes among centers are caused by missing values due to reasons that are not related to the variable of interest), then we may apply the Satterthwaite method (Johnson and Kotz, 1972) as follows. First, replace s_{Wk}^2 with $\tilde{s}_{Wk}^2$. Second, replace n_k with

$$n_{0k} = \frac{1}{m_k - 1} \left(N_k - \frac{1}{N_k} \sum_{j=1}^{m_k} n_{jk}^2 \right).$$

Third, replace s_{Bk}^2 with

$$\tilde{s}_{Bk}^2 = \frac{1}{(m_k - 1)n_{0k}} \sum_{j=1}^{m_k} n_{jk} (\bar{z}_{jk} - \bar{z}_k)^2.$$

Then, approximately,

$$\tilde{s}_{Bk}^2 \sim \frac{\left(\sigma_{Bk}^2 + n_{0k}^{-1} \sigma_{Wk}^2 \right) \chi_{d_k}^2}{m_k - 1}$$

with

$$d_k = \frac{(m_k - 1)\left(\hat{\rho}_k^{-1} + n_{0k} \right)^2}{v_k},$$

where

$$\hat{\rho}_k = \frac{\tilde{s}_{Bk}^2}{\tilde{s}_{Wk}^2} - \frac{1}{n_{0k}}$$

and

$$v_k = \frac{1}{m_k - 1} \left[\sum_{j=1}^{m_k} n_{jk}^2 - \frac{m_k}{N_k} \sum_{j=1}^{m_k} n_{jk}^3 + \frac{1}{N_k^2} \left(\sum_{j=1}^{m_k} n_{jk}^2 \right)^2 - n_{0k}^2 \right].$$

Finally, replace $m_k - 1$ with d_k.

Sample Size Calculation

Chow, Shao, and Hu (2002) proposed a procedure for the determining sample sizes m_1 and n_1 in the new region to achieve a desired power for establishment similarity between regions. The procedure is briefly outlined below. We first assume that sample sizes m_0 and n_0 in the original region have already been determined. Let $\psi = (\mu_0 - \mu_1, \sigma_{B1}^2, \sigma_{B0}^2, \sigma_{W1}^2, \sigma_{W0}^2)$ be the vector of unknown parameters and let U be given in the definition of $\hat{\zeta}_U$ and U_β be the same as U but with 5% and 95% replaced by $1 - \beta$ and β, respectively, where β is a given desired power of the similarity test. Let $\tilde{U}$ and $\tilde{U}_\beta$ be U and U_β, respectively, with $(\bar{z}_0 - \bar{z}_k, s_{BT}^2, s_{BR}^2, s_{WT}^2, s_{WR}^2)$ replaced by $\tilde{\psi}$, an initial guessing value for which the value of ζ (denoted by $\hat{\zeta}$) is negative. Then, the sample sizes m_1 and n_1 should be chosen so that

$$\hat{\varsigma} + \sqrt{\hat{U}} + \sqrt{\hat{U}_\beta} \leq 0 \tag{20.21}$$

holds. We may start with some initial values of m_1 and n_1 and gradually increase them until Equation 20.21 holds.

20.5.3 Remarks

In the assessment of sensitivity, regulatory guidance/requirement for determining whether a clinical bridging study is critical as to (i) whether a bridging study is recommended for providing substantial evidence in the new region based on the clinical results observed in the original region and (ii) what sample size of the clinical bridging study is needed for extrapolating the clinical results to the new region with the desired reproducibility probability. It is suggested that the therapeutic window and intrasubject variability of the test compound must be taken into consideration when choosing the criterion and setting regulatory requirements. A study of historical data related to the test compound is strongly recommended before a regulatory requirement is implemented.

The criterion for the assessment of similarity proposed by Chow, Shao, and Hu (2002) accounts for the average and variability of the response of the primary study endpoint. This aggregated criterion, however, suffers the following disadvantages: (i) the effects of individual components cannot be separated and (ii) the difference in averages may be offset by the reduction in variability (Chow, 1999). As a result, it is suggested that an additional constraint be placed on the difference in means between two regions (FDA, 2001). An alternative to this aggregated criterion is to consider a set of disaggregate criteria on average and variability of the primary study endpoint (Chow and Liu, 2000). This approach, however, is not favored by the regulatory agency due to some practical considerations. For example, we will never be able to claim the new region is similar to the original region if the variability of the response of the primary endpoint is significantly lower than that in the original region.

The FDA recommends the use of PDR for the selection of θ_U. In practice, the determination of the maximum allowable PDR should depend upon the therapeutic window and intrasubject variability to reflect the variability from drug product to drug product and from patient population to patient population.

20.6 Vaccine Clinical Trials

Similar to clinical development of drug products, there are four phases of clinical trials in vaccine development. Phase I trials are referred to early studies with human subjects. The purpose of phase I trials is to explore the safety and immunogenicity of multiple dose levels of the vaccine under investigation. Phase I trials are usually on small scales. Phase II trials are to assess the safety, immunogenicity, and early efficacy of selected dose levels of the vaccine. Phase III trials, which are usually large in scale, are to confirm the efficacy of the vaccine in the target population. Phase IV trials are usually conducted for collecting additional information regarding long-term safety, immunogenicity, or efficacy of the vaccine to fulfill regulatory requirements and/or marketing objectives after regulatory approval of the vaccine.

20.6.1 Reduction in Disease Incidence

As indicated in Chan, Wang, and Heyse (2003), one of the most critical steps in the evaluation of a new vaccine is in assessing the protective efficacy of the vaccine against the target disease. An efficacy trial is often conducted to evaluate whether the vaccine can prevent the disease or reduce the incidence of the disease in the target population. For this purpose, prospective, randomized, placebo-controlled trials are usually conducted. Subjects who meet the inclusion/exclusion criteria are randomly assigned to receive either the test vaccine (T) or placebo control (C). Let p_T and p_C be the true disease incidence rates of the n_T vaccinees and n_C controls randomized in the trial, respectively. Thus, the relative reduction in disease incidence for subjects in the vaccine group as compared to the control groups is given by

$$\pi = \frac{p_C - p_T}{p_C} = 1 - \frac{p_T}{p_C} = 1 - R.$$

In most vaccine clinical trials, π has been widely used and is accepted as a primary measure of vaccine efficacy. Note that a vaccine is considered 100% efficacious (i.e., $\pi = 1$) if it prevents the disease completely (i.e., $P_T = 0$). On the other hand, it has no efficacy (i.e., $\pi = 0$) if $p_T = p_C$. Let x_T and x_C be the number of observed diseases for treatment and control groups, respectively. It follows that the natural estimators for p_T and p_C are given by

$$\hat{p}_T = \frac{x_T}{n_T} \quad \text{and} \quad \hat{p}_C = \frac{x_C}{n_C}.$$

Let $\beta = p_T/p_C$, which can be estimated by

$$\hat{\beta} = \frac{\hat{p}_T}{\hat{p}_C}.$$

By Taylor's expansion and the central limit theorem (CLT), $\log(\hat{\beta})$ is asymptotically distributed as a normal random variable with mean $\log(\beta)$ and variance given by

$$\sigma^2 = \frac{1 - p_T}{n_T p_T} + \frac{1 - p_C}{n_C p_C}. \tag{20.22}$$

For a given confidence level of $1-\alpha$, a $(1-\alpha)$ confidence interval of $\log(\beta)$ is given by

$$(\log(\hat{\beta}) - z_{\alpha/2}\hat{\sigma}, \ \log(\hat{\beta}) + z_{\alpha/2}\hat{\sigma}),$$

where $\hat{\sigma}$ is obtained according to Equation 20.22 by replacing p_T and p_C by $\hat{p}_T$ and $\hat{p}_C$, respectively. In practice, sample size is usually determined by specifying the half length (d) of the confidence interval of β. Assuming that $n = n_T = n_C$, it follows that

$$z_{\alpha/2}\sqrt{\frac{1 - p_T}{n p_T} + \frac{1 - p_C}{n p_C}} = d.$$

This leads to

$$n = \frac{z_{\alpha/2}^2}{d^2}\left(\frac{1-p_T}{p_T} + \frac{1-p_C}{p_C}\right).$$

An Example

An investigator is interested in obtaining a 95% confidence interval for π where it is expected that $p_T = 0.01$ and $p_C = 0.02$. It is desirable to have a confidence interval in log-scale with half length of 0.20 ($d = 0.20$). The sample size needed is then given by

$$n = \left(\frac{1.96}{0.2}\right)^2\left(\frac{1-0.01}{0.01} + \frac{1-0.02}{0.02}\right) \approx 14214.$$

20.6.2 Evaluation of Vaccine Efficacy with Extremely Low Disease Incidence

In many cases, the disease incidence rate is extremely low. In this case, a much larger scale of study is required to demonstrate vaccine efficacy as described in the previous subsection. For sufficiently large sample sizes and small incidence rates, the numbers of cases in the vaccine groups and the control groups approximately have the Poisson distribution with rate parameters $\lambda_T(\approx n_T p_T)$ and $\lambda_C(\approx n_C p_C)$, respectively. As a result, the number of cases in the vaccine group given the total number of cases (denoted by S) is distributed as a binomial random variable with parameter θ, that is, $b(S, \theta)$, where

$$\theta = \frac{\lambda_T}{\lambda_C + \lambda_T} = \frac{n_T p_T}{n_T p_T + n_C p_C} = \frac{R}{R+u} = \frac{1-\pi}{1-\pi+u}$$

and $u = n_C/n_T$. Since θ is a decreasing function in π, testing hypotheses that

$$H_0: \pi \leq \pi_0 \quad \text{versus} \quad H_a: \pi > \pi_0$$

is equivalent to testing the following hypotheses:

$$H_0: \theta \geq \theta_0 \quad \text{versus} \quad H_a: \theta < \theta_0,$$

where

$$\theta_0 = \frac{1-\pi_0}{1-\pi_0+u}.$$

Let x_T and x_C be the number of the observed diseases for the treatment and control, respectively. A natural estimator for θ is given by $\hat{\theta} = x_T/(x_T + x_C)$. The test statistic is given by

$$T = \frac{\sqrt{x_T + x_C}\,(\hat{\theta} - \theta_0)}{\sqrt{\theta_0(1-\theta_0)}}.$$

Under the null hypothesis, T is asymptotically distributed as a standard normal random variable. Hence, we reject the null hypothesis at α level of significance if $T > z_\alpha$. Under the alternative hypothesis, the power of the above test can be approximated by

$$1 - \Phi\left(\frac{z_\alpha\sqrt{\theta_0(1-\theta_0)} + \sqrt{x_T + x_C}(\theta_0 - \theta)}{\sqrt{\theta(1-\theta)}}\right).$$

In order to achieve a desired power $1 - \beta$, the total number of diseases needed can be obtained by solving

$$\frac{z_\alpha\sqrt{\theta_0(1-\theta_0)} + \sqrt{x_T + x_C}(\theta_0 - \theta)}{\sqrt{\theta(1-\theta)}} = -z_\beta.$$

This leads to

$$x_T + x_C = \frac{[z_\alpha\sqrt{\theta_0(1-\theta_0)} + z_\beta\sqrt{\theta(1-\theta)}]^2}{(\theta_0 - \theta)^2}.$$

Under the assumption that $n = n_T = n_C$, it follows that

$$n = \frac{[z_\alpha\sqrt{\theta_0(1-\theta_0)} + z_\beta\sqrt{\theta(1-\theta)}]^2}{(p_T + p_C)(\theta - \theta_0)^2}.$$

An Example

Suppose the investigator is interested in conducting a two-arm parallel trial with equal sample size ($u = 1$) to compare a study vaccine with a control in terms of controlling the disease rates. It is expected that the disease rate for the treatment group is 0.001 and the disease rate for the control group is 0.002. The hypotheses of interest are given by

$$H_0 : \theta \leq 0.5 \quad \text{versus} \quad H_a : \theta > 0.5.$$

Hence, $\theta_0 = 0.5$. It can be obtained that

$$\theta = \frac{0.001}{0.001 + 0.002} = \frac{1}{3}.$$

As a result, the sample size needed in order to achieve an 80% ($\beta = 0.20$) power at the 5% ($\alpha = 0.05$) level of significance is given by

$$n = \frac{[1.64\sqrt{0.5(1-0.5)} + 0.84\sqrt{1/3(1-1/3)}]^2}{(0.001 + 0.002)(1/3 - 1/2)^2} \approx 17,837.$$

Thus, 17,837 subjects per arm is needed in order to detect such a difference with an 80% power.

20.6.3 Relative Vaccine Efficacy

In vaccine trials, when the control is a licensed vaccine (an active control), the relative efficacy π can be evaluated through the relative risk (i.e., $R = P_T/P_C$) based on the relationship $\pi = 1 - R$. If the absolute efficacy of the control (i.e., π_C) has been established, one can estimate the absolute efficacy of the test vaccine by

$$\pi_T = 1 - R(1 - \pi_C).$$

For a comparative vaccine trial, it is often designed as a noninferiority trial by testing the following hypotheses:

$$H_0 : R \geq R_0 \quad \text{versus} \quad H_a : R < R_0,$$

where $R_0 > 1$ is a prespecified noninferiority margin or a threshold for relative risk. In practice, the hypotheses regarding relative risk are most often performed based on log-scale. In other words, instead of testing the above hypotheses, we usually consider the following hypotheses:

$$H_0 : \log(R) \geq \log(R_0) \quad \text{versus} \quad H_a : \log(R) < \log(R_0).$$

As it can be seen, this becomes the two-sample problem for relative risk, which has been discussed in Section 4.6. Hence, the detailed discussion is omitted.

20.6.4 Composite Efficacy Measure

As indicated by Chang et al. (1994), in addition to the prevention of the disease infection, a test vaccine may also reduce the severity of the target disease as well. As a result, it is suggested that a composite efficacy measure be considered to account for both incidence and severity of the disease when evaluating the efficacy of the test vaccine. Chang et al. (1994) proposed the so-called burden-of-illness composite efficacy measure.

Suppose n_T subjects were assigned to receive treatment while n_C subjects were assigned to receive control (placebo). Let x_T and x_C be the number of cases observed in the treatment and control groups, respectively. Without loss of generality, we assume the first x_T subjects in the treatment group and x_C subjects in the control group experienced the events. Let s_{ij}, $i = T, C; j = 1, \ldots, x_i$ be the severity score associated with the jth case in the ith treatment group. For a fixed $i = T$ or C, it is assumed that s_{ij} are independent and identically distributed random variables with mean μ_i and variance σ_i^2. Let p_i be the true event rate of the ith treatment group. The hypotheses of interest is given by

$$H_0 : p_T = p_C \quad \text{and} \quad \mu_T = \mu_C \quad \text{versus} \quad H_a : p_T \neq p_C \quad \text{or} \quad \mu_T \neq \mu_C.$$

Let

$$\bar{s}_i = \frac{1}{n_i} \sum_{j=1}^{x_i} s_{ij},$$

$$\bar{x} = \frac{n_T \bar{s}_T + n_C \bar{s}_C}{n_T + n_C},$$

$$s_i^2 = \frac{1}{n_i - 1} \sum_{j=1}^{n_i} (s_{ij} - \bar{s}_i)^2.$$

The test statistic is given by

$$T = \frac{\bar{s}_T - \bar{s}_C}{\sqrt{\bar{x}^2 \hat{p}(1-\hat{p})(1/n_T + 1/n_C) + \hat{p}(s_0^2/n_T + s_1^2/n_C)}}.$$

Under the null hypothesis, Chang et al. (1994) showed that T is asymptotically distributed as a standard normal random variable. Hence, we would reject the null hypothesis if $|T| > z_{\alpha/2}$. Assume that $n = n_T = n_C$ and under the alternative hypothesis, it can be shown that

$$\bar{x} \to_{a.s.} (\mu_T + \mu_C)/2 = \mu_*,$$

$$\hat{p} \to_{a.s.} (p_T + p_C)/2 = p_*.$$

Without loss of generality, we assume $p_T \mu_T > p_C \mu_C$ under the alternative hypothesis. Thus, the power of the above test can be approximated by

$$1 - \Phi\left[\frac{z_{\alpha/2}\sqrt{2\mu_*^2(p_*(1-p_*)) + 2p_*\left(\sigma_T^2 + \sigma_C^2\right)} - \sqrt{n}(\mu_T p_T - \mu_R p_R)}{\sqrt{p_T\left(\sigma_T^2 + \mu_T^2(1-p_T)\right) + p_R\left(\sigma_R^2 + \mu_R^2(1-p_R)\right)}} \right].$$

Hence, the sample size needed in order to achieve a desired power of $1-\beta$ can be obtained by solving

$$\frac{z_{\alpha/2}\sqrt{2\mu_*^2 p_*(1-p_*) + 2p_*\left(\sigma_T^2 + \sigma_C^2\right)} - \sqrt{n}(\mu_T p_T - \mu_R p_R)}{\sqrt{p_T\left(\sigma_T^2 + \mu_T^2(1-p_T)\right) + p_R\left(\sigma_R^2 + \mu_R^2(1-p_R)\right)}} = -z_\beta.$$

This leads to

$$n = \frac{1}{(\mu_T p_T - \mu_R p_R)^2}\left[z_{\alpha/2}\sqrt{2\mu_*^2 p_*(1-p_*) + 2p_*\left(\sigma_T^2 + \sigma_C^2\right)} \right.$$
$$\left. + z_\beta\sqrt{p_T\left(\sigma_T^2 + \mu_T^2(1-p_T)\right) + p_R\left(\sigma_R^2 + \mu_R^2(1-p_R)\right)} \right]^2.$$

It should be noted that the above formula is slightly different from the one derived by Chang et al. (1994), which is incorrect.

An Example
Consider a clinical trial with $\mu_T = 0.20$, $\mu_C = 0.30$, $p_T = 0.10$, $p_R = 0.20$, and $\sigma_T^2 = \sigma_C^2 = 0.15$. The sample size needed in order to have an 80% ($\beta = 0.20$) power for detecting such a difference at the 5% ($\alpha = 0.05$) level of significance is given by

$$n = \frac{1}{(0.2 \times 0.1 - 0.3 \times 0.2)^2}\left[1.96\sqrt{2 \times 0.25^2 \times 0.15 \times 0.85 + 2 \times 0.15 \times 0.3} \right.$$
$$\left. + 0.84\sqrt{0.1(0.15^2 + 0.2 \times 0.9) + 0.2(0.15^2 + 0.3 \times 0.8)} \right]^2$$
$$\approx 468.$$

As a result, 468 subjects per treatment group are required for achieving an 80% power for detecting such a difference in the burden-of-illness score.

20.6.5 Remarks

In the previous subsections, the procedures for sample size calculation in vaccine clinical trials were discussed based on a primary efficacy study endpoint using a parametric approach. Durham et al. (1998) considered a nonparametric survival method to estimate the long-term efficacy of a cholera vaccine in the presence of warning protection. For the evaluation of long-term vaccine efficacy, as indicated by Chan, Wang, and Heyse (2003), the analysis of time to event may be useful for determining whether breakthrough rates among vaccines change over time. However, it should be noted that sample size calculation may be different depending upon the study objectives, the hypotheses of interest, and the corresponding appropriate statistical tests.

Clinical development for vaccine has recently received much attention from both regulatory agencies such as the U.S. FDA and the pharmaceutical industry. For example, Ellenberg and Dixon (1994) discussed some important statistical issues of vaccine trials (related to HIV vaccine trials). O'Neill (1988) and Chan and Bohida (1998) gave asymptotic and exact formulas for sample size and power calculations for vaccine efficacy studies, respectively. Chan, Wang, and Heyse (2003) provided a comprehensive review of vaccine clinical trials and statistical issues that are commonly encountered in vaccine clinical trials.

Bibliography

Abbas, I., Rovira, J., Casanovas, J., and Greenfield, T. 2008. Optimal design of clinical trials with computer simulation based on results of earlier trials, illustrated with a lipodystrophy trial in HIV patients. *Journal of Biomedical Informatics*, 41(6), 1053–1061.

Adcock, C.J. 1988. A Bayesian approach to calculating sample size. *Statistician*, 37, 433–439.

Alizadeh, A.A. and Staudt, L.M. 2000. Genomic-scale gene expression profiling of normal and malignant immune cells. *Current Opinion in Immunology*, 12(2), 219–225.

Anderson, S. and Hauck, W.W. 1990. Considerations of individual bioequivalence. *Journal of Pharmacokinetics and Biopharmaceutics*, 8, 259–273.

Armitage, P. 1955. Tests for linear trends in proportions and frequencies. *Biometrics*, 11, 375–386.

Armitage, P. and Berry, G. 1987. *Statistical Methods in Medical Research*. Blackwell Scientific Publications, London, UK.

Armitage, P., McPherson, C.K., and Rowe, B.C. 1969. Repeated significance tests on accumulating data. *Journal of Royal Statistical Society A*, 132, 235–244.

Arnold, B.F., Hogan, D.R., Colford, J.M., and Hubbard, A.E. 2011. Simulation methods to estimate design power: An overview for applied research. *BMC Medical Research Methodology*, 11(1), 94.

Babb, J.S. and Rogatko, A. 2004. Bayesian methods for cancer phase I clinical trials. In *Advances in Clinical Trial Biostatistics*. Ed. Geller, N.L., Marcel Dekker, Inc, New York, NY, pp. 1–40.

Bailar, J.C. and Mosteller, F. 1986. *Medical Uses of Statistics*. Massachusetts Medical Society, Waltham, MA.

Barry, M.J., Fowler, F.J. Jr., O'Leary, M.P., Bruskewitz, R.C., Holtgrewe, H.L., Mebust, W.K., and Cockett, A.T. 1992. The American Urological Association symptom index for benign prostatic hyperplasia. *Journal of Urology*, 148, 1549–1557.

Bauer, P. 1989. Multistage testing with adaptive designs (with discussion). *Biometrie und Informatik in Medizin und Biologie*, 20, 130–148.

Bauer, P. and Kohne, K. 1994. Evaluation of experiments with adaptive interim analyses. *Biometrics*, 50, 1029–1041.

Bauer, P. and Rohmel, J. 1995. An adaptive method for establishing a dose–response relationship. *Statistics in Medicine*, 14, 1595–1607.

Becker, M.P. and Balagtas, C.C. 1993. Marginal modeling of binary cross-over data. *Biometrics*, 49, 997–1009.

Belyakov, I.M., Ahlers, J.D., Nabel, G.J. et al. 2008. Generation of functionally active HIV-1 specific CD8+ CTL in intestinal mucosa following mucosal, systemic or mixed prime-boost immunization. *Virology*, 381, 106–115.

Benjamini, Y. and Hochberg, Y. 1995. Controlling the false discovery rate: A practical and powerful approach to multiple testing. *Journal of the Royal Statistical Society B*, 57(1), 289–300.

Benjamini, Y. and Yekutieli, D. 2001. The control of the false discovery rate in multiple testing under dependency. *Annals of Statistics*, 29(4), 1165–1188.

Berger, R.L. and Hsu, J.C. 1996. Bioequivalence trials, intersection-union tests and equivalence confidence sets. *Statistical Science*, 11, 283–302.

Berger, J.O., Boukai, B., and Wang, Y. 1997. Unified frequentist and Bayesian testing of a precision hypothesis. *Statistical Science*, 12, 133–160.

Berry, D.A. 1990. *Statistical Methodology in the Pharmaceutical Science*. Marcel Dekker, Inc., New York, NY.

Biswas, N., Chan, I.S.F., and Ghosh, K. 2000. Equivalence trials: Statistical issues. Lecture Note for the One-Day Short Course at Joint Statistical Meetings, Indianapolis, IN.

Black, M.A. and Doerge, R.W. 2002. Calculation of the minimum number of replicate spots required for detection of significant gene expression fold change in microarray experiments. *Bioinformatics*, 18(12), 1609–1616.

Blackwelder, W.C. 1982. Proving the null hypothesis in clinical trials. *Controlled Clinical Trials*, 3, 345–353.

Böhning, D. 1998. Zero-inflated Poisson models and C.A.MAN: A tutorial collection of evidence. *Biometrical Journal*, 40, 833–843.

Böhning, D., Dietz, E., Schlattmann, P., Mendonca, L., and Kirchner, U. 1999. The zero-inflated Poisson model and the decayed, missing and filled teeth index in dental epidemiology. *Journal of the Royal Statistical Society: Series A*, 162, 195–209.

Boyle, P. and Parkin, D.M. 1991. Statistical methods for registries. In *Cancer Registration. Principles and Methods*. Eds. Jensen, O.M., Parkin, D.M., MacLennan, R., Muir, C.S., and Skeet, R.G., IARC Scientific, Lyon, France, pp. 126–158.

Bretz, F. and Hothorn, L.A. 2002. Detecting dose-response using contrasts: Asymptotic power and sample size determination for binary data. *Statistics in Medicine*, 21, 3325–3335.

Breslow, N.E. and Day, N.E. 1980. *Statistical Methods in Cancer Research, Vol. 1: The Analysis of Case-Control Studies*. Oxford University Press, New York, NY.

Breslow, N.E. and Day, N.E. 1987. *Statistical Methods in Cancer Research, Vol. 2: The Analysis of Cohort Studies*. Oxford University Press, New York, NY.

Bryant, J. and Day, R. 1995. Incorporating toxicity considerations into the design of two-stage phase II clinical trials. *Biometrics*, 51, 1372–1383.

Buncher, C.R. and Tsay, J.Y. 1994. *Statistics in Pharmaceutical Industry*, 2nd Edition. Marcel Dekker, Inc., New York, NY.

Burton, A., Altman, D.G., Royston, P., and Holder, R.L. 2006. The design of simulation studies in medical statistics. *Statistics in Medicine*, 25(24), 4279–4292.

Buyse, M.E., Staquet, M.J., and Sylvester, R.J. 1984. *Cancer Clinical Trials: Methods and Practice*. Oxford Medical Publications, New York, NY.

Byar, D.P. 1988. The design of cancer prevention trials. In *Cancer Clinical Trials*. Eds. Scheurlen, H., Kay, R., and Baum, M., Springer, Berlin Heidelberg, pp. 34–48.

Campbell, M.J., Donner, A., and Klar, N. 2007. Developments in cluster randomized trials. *Statistics in Medicine*, 26, 2–19.

Campbell, M.K., Mollison, J., and Grimshaw, J.M. 2001. Cluster trials in implementation research: Estimation of intracluster correlation coefficients and sample size. *Statistics in Medicine*, 20, 391–399.

Capizzi, T. and Zhang, J. 1996. Testing the hypothesis that matters for multiple primary endpoints. *Drug Information Journal*, 30, 349–356.

CAST. 1989. Cardiac Arrhythmia Supression Trial. Preliminary report: Effect of encainide and flecainide on mortality in a randomized trial of arrhythmia supression after myocardial infarction. *New England Journal of Medicine*, 321, 406–412.

Chan, I.S.F. and Bohida, N.R. 1998. Exact power and sample size for vaccine efficacy studies. *Communications in Statistics, A*, 27, 1305–1322.

Chan, I.S.F., Wang, W., and Heyse, J.F. 2003. Vaccine clinical trials. In *Encyclopedia of Biopharmaceutical Statistics*, 2nd Edition. Ed. Chow, S.C., Marcel Dekker, Inc., New York, NY, pp. 1305–1322.

Chang, M.N. 1989. Confidence intervals for a normal mean following group sequential test. *Biometrics*, 45, 247–254.

Chang, M.N. 2004. Power and sample size calculations for dose response studies. Dose–Response Trial Design Workshop, November 15, 2004, Philadelphia, PA.

Chang, M. 2007. Adaptive design method based on sum of p-values. *Statistics in Medicine*, 26, 2772–2784.

Chang, M. and Chow, S.C. 2005. A hybrid Bayesian adaptive design for dose response trials. *Journal of Biopharmaceutical Statistics*, 15, 677–691.

Chang, M. and Chow, S.C. 2006. Power and sample size for dose response studies. In *Dose Finding in Drug Development*. Ed. Ting, N., Springer, New York, NY, 220–241.

Chang, M.N. and O'Brien, P.C. 1986. Confidence interval following group sequential test. *Controlled Clinical Trials*, 7, 18–26.

Chang, M.N., Guess, H.A., and Heyse, J.F. 1994. Reduction in burden of illness: A review efficacy measure for prevention trials. *Statistics in Medicine*, 13, 1807–1814.

Chang, M.N., Wieand, H.S., and Chang, V.T. 1989. The bias of the sample proportion following a group sequential phase II trial. *Statistics in Medicine*, 8, 563–570.

Channon, E.J. 2000. Equivalence testing in dose–reponse study. *Drug Information Journal*, 34, 551–562.

Chen, J.J., Tsong, Y., and Kang, S. 2000. Tests for equivalence or non-inferiority between two proportions. *Drug Information Journal*, 34.

Chen, K.W., Li, G., and Chow, S.C. 1997. A note on sample size determination for bioequivalence studies with higher-order crossover designs. *Journal of Pharmacokinetics and Biopharmaceutics*, 25, 753–765.

Chen, M.L. 1997a. Individual bioequivalence: A regulatory update. *Journal of Biopharmaceutical Statistics*, 7, 5–11.

Chen, T.T. 1997b. Optimal three-stage designs for phase II cancer clinical trials. *Statistics in Medicine*, 16, 2701–2711.

Chen, T.T. and Ng, T.H. 1998. Optimal flexible designs in phase II clinical trials. *Statistics in Medicine*, 17, 2301–2312.

Cheng, B. and Chow, S.C. 2015. Statistical inference for a multiple-stage transitional seamless trials designs with different study objectives and endpoints. Under revision.

Cheng, B., Chow, S.C., and Wang, H. 2006. Test for departure from dose linearity under a crossover design: A slope approach. Unpublished manuscript.

Cheng, B. and Shao, J. 2007. Exact tests for negligible interaction in two-way linear analysis of variance/covariance. *Statistica Sinica*, 17, 1441–1456.

Chinchilli, V.M. and Esinhart, J.D. 1996. Design and analysis of intrasubject variability in cross-over experiments. *Statistics in Medicine*, 15, 1619–1634.

Chow, S.C. 1999. Individual bioequivalence: A review of FDA draft guidance. *Drug Information Journal*, 33, 435–444.

Chow, S.C. 2000. *Encyclopedia of Biopharmaceutical Statistics*. Marcel Dekker, Inc., New York, NY.

Chow, S.C. 2011. *Controversial Issues in Clinical Trials*. Chapman & Hall/CRC, Taylor & Francis, New York, NY.

Chow, S.C. 2015. Challenging issues in assessing analytical similarity in biosimilar. *Biosimilars*, 5, 33–39.

Chow, S.C. and Chang, M. 2011. *Adaptive Design Methods in Clinical Trials*, 2nd Edition. Chapman & Hall/CRC, Taylor & Francis, New York, NY.

Chow, S.C., Chang, M., and Pong, A. 2005. Statistical consideration of adaptive methods in clinical development. *Journal of Biopharmaceutical Statistics*, 15, 575–591.

Chow, S.C. and Chiu, S.T. 2013. Sample size and data monitoring for clinical trials with extremely low incidence rate. *Therapeutic Innovation & Regulatory Science*, 47, 438–446.

Chow, S.C. and Ki, F.Y.C. 1994. On statistical characteristics of quality of life assessment. *Journal of Biopharmaceutical Statistics*, 4, 1–17.

Chow, S.C. and Ki, F.Y.C. 1996. Statistical issues in quality of life assessment. *Journal of Biopharmaceutical Statistics*, 6, 37–48.

Chow, S.C. and Liu, J.P. 1992. *Design and Analysis of Bioavailability and Bioequivalence Studies*. Marcel Dekker, Inc., New York, NY.

Chow, S.C. and Liu, J.P. 1995. *Statistical Design and Analysis in Pharmaceutical Science*. Marcel Dekker, Inc., New York, NY.

Chow, S.C. and Liu, J.P. 1998. *Design and Analysis of Clinical Trials*. John Wiley & Sons, New York, NY.

Chow, S.C. and Liu, J.P. 2000. *Design and Analysis of Bioavailability and Bioequivalence Studies*, 2nd Edition. Marcel Dekker, Inc., New York, NY.

Chow, S.C. and Liu, J.P. 2003. *Design and Analysis of Clinical Trials*, 2nd Edition. John Wiley & Sons, New York, NY.

Chow, S.C. and Liu, J.P. 2004. Design and Analysis of Clinical Trials: Concepts and Methodologies, Third Edition. John Wiley & Sons, New York, NY.

Chow, S.C. and Liu, J.P. 2013. *Design and Analysis of Clinical Trials—Revised and Expanded*, 3rd Edition. John Wiley & Sons, New York, NY.

Chow, S.C., Lu, Q.S., and Tse, S.K. 2007a. Statistical analysis for two-stage seamless design with different study endpoints. *Journal of Biopharmaceutical Statistics*, 17, 1163–1176.

Chow, S.C. and Shao, J. 1988. A new procedure for the estimation of variance components. *Statistics and Probability Letters*, 6, 349–355.

Chow, S.C. and Shao, J. 2002. *Statistics in Drug Research: Methodology and Recent Development*. Marcel Dekker, Inc., New York, NY.

Chow, S.C. and Shao, J. 2004. Analysis of clinical data with breached blindness. *Statistics in Medicine*, 23, 1185–1193.

Chow, S.C., Shao, J., and Hu, O.Y.P. 2002a. Assessing sensitivity and similarity in bridging studies. *Journal of Biopharmaceutical Statistics*, 12, 269–285.

Chow, S.C., Shao, J., and Wang, H. 2002b. Individual bioequivalence testing under 2×3 crossover designs. *Statistics in Medicine*, 21, 629–648.

Chow, S.C., Shao, J., and Wang, H. 2003a. Statistical tests for population bioequivalence. *Statistica Sinica*, 13, 539–554.

Chow, S.C., Shao, J., and Wang, H. 2003b. *In vitro* bioequivalence testing. *Statistics in Medicine*, 22, 55–68.

Chow, S.C., Shao, J., and Wang, H. 2003c. *Sample Size Calculations in Clinical Reserach*. Marcel Dekker, New York, NY.

Chow, S.C., Shao, J., and Wang, H. 2008. *Sample Size Calculation in Clinical Research*, 2nd Edition. Taylor & Francis, New York, NY.

Chow, S.C., Song, F.Y., and Bai, H. 2016. Analytical similarity assessment in biosimilar studies. *AAPS Journal*, doi: 10.1208/s12248-016-9882-5.

Chow, S.C. and Tse, S.K. 1990. A related problem in bioavailability/bioequivalence studies: Estimation of intra-subject variability with a common CV. *Biometrical Journal*, 32, 597–607.

Chow, S.C. and Tse, S.K. 1991. On variance estimation in assay validation. *Statistics in Medicine*, 10, 1543–1553.

Chow, S.C. and Tu, Y.H. 2008. On two-stage seamless adaptive design in clinical trials. *Journal of the Formosan Medical Association*, 107, S52–S60.

Chow, S.C. and Wang, H. 2001. On sample size calculation in bioequivalence trials. *Journal of Pharmacokinetics and Pharmacodynamics*, 28, 155–169.

Christl, L. 2015. Overview of regulatory pathway and FDA's guidance for development and approval of biosimilar products in US. Presented at FDA ODAC Meeting, January 7, 2015, Silver Spring, MD.

Chuang-Stein, C. and Agresti, A. 1997. A review of tests for detecting a monotone dose–response relationship with ordinal response data. *Statistics in Medicine*, 16, 2599–2618.

Cochran, W.G. 1954. Some methods for stengthening the common chi-square tests. *Biometrics*, 10, 417–451.

Colton, T. 1974. *Statistics in Medicine*. Little, Brown and Company, Boston, MA.

Conaway, M.R. and Petroni, G.R. 1996. Designs for phase II trials allowing for a trade-off between response and toxicity. *Biometrics*, 52, 1375–1386.

Cornfield, J. 1956. A statistical problem arising from retrospective studies. *Proceedings of the Third Berkeley Symposium on Mathematical Statistics and Probability*, Ed. Neyman, J., 4, 135–148.

Cornfield, J. 1978. Randomization by group: A formal analysis. *American Journal of Epidemiology*, 108, 100–102.

Cox, D.R. 1952. A note of the sequential estimation of means. *Proceedings of the Cambridge Philosophical Society*, 48, 447–450.

Cox, D.R. 1972. Regression models and life tables (with discussion). *Journal of Royal Statistical Society B*, 74, 187–220.

Cox, D.R. and Hinkley, D.V. 1974. *Theoretical Statistics*. Chapman & Hall, London.

Cui, L., Hung, H.M.J., and Wang, S.J. 1999. Modification of sample size in group sequential clinical trials. *Biometrics*, 55, 853–857.

CPMP. 1997. Points to consider: The assessment of the potential for QT interval prolongation by non-cardiovascular products.

Crawford, E.D., Eisenberg, M.A., Mcleod, D.G., Spaulding, J.T., Benson, R., Dorr, A., Blumenstein, B.A., Davis, M.A., and Goodman, P.J. 1989. A controlled trial of Leuprolide with and without flutamide in prostatic carcinoma. *New England Journal of Medicine*, 321, 419–424.

Crowley, J. 2001. *Handbook of Statistics in Clinical Oncology*. Marcel Dekker, Inc., New York, NY.

CTriSoft Intl. 2002. *Clinical Trial Design with ExpDesign Studio*, www.ctrisoft.net., CTriSoft Intl., Lexington, MA.

Cui, X. and Churchill, G.A. 2003. How many mice and how many arrays? Replication in mouse cDNA microarray experiments. In *Methods of Microarray Data Analysis II*. Eds. Johnson, K.F. and Lin, S.M., Kluwer Academic Publishers, Norwell, MA, pp. 139–154.

DeMets, D. and Lan, K.K.G. 1994. Interim analysis: The alpha spending function approach. *Statistics in Medicine*, 13, 1341–1352.

Dietz, K. and Böhning, D. 2000. On estimation of the Poisson parameter in zero-modified Poisson models. *Computational Statistics and Data Analysis*, 34, 441–459.

Dixon, D.O. and Simon, R. 1988. Sample size consideration for studies comparing survial curves using historical controls. *Journal of Clinical Epidemiology*, 41, 1209–1213.

Donald, A. and Donner, A. 1987. Adjustment to the Mantel–Haenszel chi-square statistic and odds ratio variance estimator when the data are clustered. *Statistics in Medicine*, 6, 491–499.

Dong, X., Tsong, Y., and Wang, Y.T. 2017. Adjustment for unbalanced sample size for analytical biosimilar equivalence assessment. *Journal of Biopharmaceutical Statistics*, 27(2), 220–232.

Donner, A. 1992. Sample size requirements for stratified cluster randomization designs. *Statistics in Medicine*, 11, 743–750.

Donner, A., Birkett, N., and Buck, C. 1981. Randomization by cluster sample size requirements and analysis. *American Journal of Epidemiology*, 114, 906–914.

Donner, A. and Klar, N. 2000. *Design and Analysis of Cluster Randomization Trials in Health Research*. Arnold Publishers Limited, London, UK.

Dubey, S.D. 1991. Some thoughts on the one-sided and two-sided tests. *Journal of Biopharmaceutical Statistics*, 1, 139–150.

Dudoit, S., Schaffer, J.P., and Boldrick, J.C. 2003. Multiple hypothesis testing in microarray experiments. *Statistical Science*, 18, 71–103.

Dudoit, S., Yang, Y.H., Callow, M.J., and Speed, T.P. 2002. Statistical methods for identifying differentially expressed genes in replicated cDNA microarray experiments. *Statistica Sinica*, 12, 111–139.

Dunnett, C.W. and Gent, M. 1977. Significance testing to establish equivalence between treatments with special reference to data in the form of 2×2 tables. *Biometrics*, 33, 593–602.

Durham, S.D., Flourney, N., and Li, W. 1998. A sequential design for maximizing the probability of a favorable response. *Canadian Journal of Statistics*, 26, 479–495.

EAST. 2000. Software for the design and interim monitoring of group sequential clinical trials. CYTEL Software Corporation.

Eldridge, S.M., Ukoumunne, O.C., and Carlin, J.B. 2009. The intra-cluster correlation coefficient in cluster randomized trials: A review of definitions. *International Statistical Review*, 77, 378–394.

Ellenberg, S.S. and Dixon, D.O. 1994. Statistical issues in designing clinical trials of AIDS treatments and vaccines. *Journal of Statistical Planning and Inference*, 42, 123–135.

EMEA. 2002. Point to consider on methodological issues in confirmatory clinical trials with flexible design and analysis plan. The European Agency for the Evaluation of Medicinal Products Evaluation of Medicines for Human Use. CPMP/EWP/2459/02, London, UK.

Emrich, L.J. 1989. Required duration and power determinations for historically controlled studies of survival times. *Statistics in Medicine*, 8, 153–160.

Ensign, L.G., Gehan, E.A., Kamen, D.S., and Thall, P.F. 1994. An optimal three-stage design for phase II clinical trials. *Statistics in Medicine*, 13, 1727–1736.

Esinhart, J.D. and Chinchilli, V.M. 1994. Extension to the use of tolerance intervals for assessment of individual bioequivalence. *Journal of Biopharmaceutical Statistics*, 4, 39–52.

Esteban, M. 2009. Attenuated poxvirus vectors MVA and NYVAC as promising vaccine candidates against HIV/AIDS. *Human Vaccines*, 5, 1–5.

Excler, J.L., Tomaras, G.D., and Russell, N.D. 2013. Novel directions in HIV-1 vaccines revealed from clinical trials. *Current Opinion in HIV and AIDS*, 8(5), 421–431.

Fairweather, W.R. 1994. Statisticians, the FDA and a time of transition. Presented at Pharmaceutical Manufacturers Association Education and Research Institute Training Course in Non-Clinical Statistics, Georgetown University Conference Center, February 6–8, 1994, Washington, DC.

Farrington, C.P. and Manning, G. 1990. Test statistics and sample size formulae for comparative binomial trials with null hypothesis of nonzero risk difference or non-unity relative risk. *Statistics in Medicine*, 9, 1447–1454.

FDA. 1988. *Guideline for the Format and Content of the Clinical and Statistical Section of New Drug Application*. U.S. Food and Drug Administration, Rockville, MD.

FDA. 1992. *Guidance on Statistical Procedures for Bioequivalence Studies Using a Standard Two-Treatment Crossover Design*. Office of Generic Drugs, Center for Drug Evaluation and Research, Food and Drug Administration, Rockville, MD.

FDA. 1977. *Guidelines for the Clinical Evaluation of Anti-Infective Drugs (Systemic)*. U.S. Food and Drug Administration, Rockville, MD.

FDA. 1998. *Guidance for Industry on Providing Clinical Evidence of Effectiveness for Human Drug and Biological Products*. Food and Drug Administration, Rockville, MD.

FDA. 1999a. *Average, Population, and Individual Approaches to Establishing Bioequivalence*. Center for Drug Evaluation and Research, Food and Drug Administration, Rockville, MD.

FDA. 1999b. *Guidance for Industry on Bioavailability and Bioequivalence Studies for Nasal Aerosols and Nasal Sprays for Local Action*. Center for Drug Evaluation and Research, Food and Drug Administration, Rockville, MD.

FDA. 2000. *Guidance for Industry: Bioavailability and Bioequivalence Studies for Orally Administered Drug Products: General Considerations*. Center for Drug Evaluation and Research, Food and Drug Administration, Rockville, MD.

FDA. 2001. *Guidance for Industry on Statistical Approaches to Establishing Bioequivalence*. Center for Drug Evaluation and Research, Food and Drug Administration, Rockville, MD.

FDA. 2002. *Guidance for Industry Immiotoxicology Eevaluation of Investigational New Drugs*. Center for Drug Evaluation and Research, the United States Food and Drug Administration, Rockville, MD.

FDA. 2010a. *Draft Guidance for Industry—Adaptive Design Clinical Trials for Drugs and Biologics*. http://www.fda.gov/downloads/Drugs/.../Guidances/ucm201790.pdf

FDA. 2010b. *Guidance for Industry—Non-Inferiority Clinical Trials*. The United States Food and Drug Administration, Rockville, MD.

FDA. 2015. *Guidance on Scientific Considerations in Demonstrating Biosimilarity to a Reference Product*. The United States Food and Drug Administration, Silver Spring, MD.

FDA/TPD. 2003. Preliminary concept paper: The clinical evaluation of QT/QTc interval prolongation and proarrhythmic potential for non-arrhythmic drug products. Released on November 15, 2005. Revised on February 6, 2003.

Feigl, P., Blumestein, B., Thompson, I., Crowley, J., Wolf, M., Kramer, B.S., Coltman, C.A. Jr., Brawley, O.W., and Ford, L.G. 1995. Design of the prostate cancer prevention trial (PCPT). *Controlled Clinical Trials*, 16, 150–163.

Feinstein, A.R. 1977. *Clinical Biostatistics*. The Mosby Company, St. Louis, MO.

Feiveson, A.H. 2002. Power by simulation. *The Stata Journal*, 2(2), 107–124.

Feller, W. 1943. On a general class of 'contagious distributions. *Annals of Mathematical Statistics*, 12, 389–400.

Fine, G.D. 1997. A formula for determing sample size to study dose–response. *Drug Information Journal*, 31, 911–916.

Fisher, L. 1990. *Biostatistics: Methodology for the Health Sciences*. John Wiley & Sons, New York, NY.

Fleiss, J.L. 1986. *The Design and Analysis of Clinical Experiments.* John Wiley & Sons, New York, NY.

Fleming, T.R. 1990. Evaluation of active control trials in AIDS. *Journal of Acquired Immune Deficiency Syndromes,* 3(Suppl. 2), 582–587.

Fleming, T.R. and Harrington, D.R. 1991. *Counting Process and Survival Analysis.* John Wiley & Sons, New York, NY.

Fleiss, J.L. 1981. *Statistical Methods for Rates and Proportions,* 2nd Edition. Wiley, New York.

Freedman, L.S. 1982. Tables of the number of patients required in clinical trials using logrank test. *Statistics in Medicine,* 1, 121–129.

Freedman, L.S., Green, S.B., and Byar, D.P. 1990. Assessing the gain in efficiency due to matching in a community intervention study. *Statistics in Medicine,* 9, 943–952.

Frick, M.H., Elo, O., Haapa, K. et al. 1987. Helsikini heart study: Primary-prevention trial with gemfibrozil in middle-aged men with dyslipidemia. *New England Journal of Medicine,* 317, 1237–1245.

Friede, T. and Schmidli, H. 2010. Blinded sample size reestimation with negative binomial counts in superiority and non-inferiority trials. *Methods of Information in Medicine,* 49, 618–624.

Friedman, L.M., Furberg, C.D., and DeMets, D.L. 1981. *Fundamentals of Clinical Trials.* John Wiley & Sons, New York, NY.

Gadbury, G.L., Page, G.P., Edwards, J., Kayo, T., Prolla, T.A., Weindruch, R., Permana, P.A., Mountz, J.D., and Allison, D.B. 2004. Power and sample size estimation in high dimensional biology. *Statistical Methods in Medical Research,* 13, 325–338.

Gail, M.H. 1985. Applicability of sample size calculations based on a comparison of proportions for use with the logrank test. *Controlled Clinical Trials,* 6, 112–119.

Gail, M.H. and Simon, R. 1985. Testing for qualitative interactions between treatment effects and patient subjects. *Biometrics,* 71, 431–444.

Gart, J.J. 1985. Approximate tests and interval estimation of the common relative risk in the combination of 2 × 2 tables. *Biometrika,* 72, 673–677.

Gasprini, M. and Eisele, J. 2000. A curve-free method for phase I clinical trials. *Biometrics,* 56, 609–615.

Gastañaga, V.M., McLaren, C.E., and Delfino, R.J. 2006. Power calculations for generalized linear models in observational longitudinal studies: A simulation approach in SAS. *Computer Methods and Programs in Biomedicine,* 84(1), 27–33.

Gastwirth, J.L. 1985. The use of maximum efficiency robust tests in combining contingency tables and survival analysis. *Journal of American Statistical Association,* 80, 380–384.

Ge, Y., Dudoit, S., and Speed, T.P. 2003. Resampling-based multiple testing for microarray data analysis. *TEST,* 12(1), 1–44.

Geller, N.L. 1994. Discussion of interim analysis: The alpha spending function approach. *Statistics in Medicine,* 13, 1353–1356.

Genovese, C. and Wasserman, L. 2002. Operating characteristics and extensions of the false discovery rate procedure. *Journal of the Royal Statistical Society, Series B,* 64(3), 499–517.

George, S.L. and Desu, M.M. 1973. Planning the size and duration of a clinical trial studying the time to some critical event. *Journal of Chronic Diseases,* 27, 15–24.

Gilbert, G.S. 1992. *Drug Safety Assessment in Clinical Trials.* Marcel Dekker, Inc., New York, NY.

Glantz, S.A. 1987. *Primer of Biostatistics,* 2nd Edition. McGraw-Hill, Inc., New York, NY.

Golub, T.R., Slonim, D.K., Tamayo, P. et al. 1999. Molecular classification of cancer: Class discovery and class prediction by gene expression monitoring. *Science,* 286(15), 531–537.

Gould, A.L. 1995. Planning and revision the sample size for a trial. *Statistics in Medicine,* 14, 1039–1051.

Graybill, F.A. and Deal, R.B. 1959. Combining unbiased estimators. *Biometrics,* 15, 543–550.

Graybill, F. and Wang, C.M. 1980. Confidence intervals on nonnegative linear combinations of variances. *Journal of American Statistical Association,* 75, 869–873.

Green, S.J. and Dahlberg, S. 1992. Planned versus attained designs in phase II clinical trials. *Statistics in Medicine,* 11, 853–862.

Guimaraes, P. and Palesch, Y. 2007. Power and sample size simulations for randomized play-the-winner rules. *Contemporary Clinical Trials,* 28(4), 487–499.

Gupta, P.L., Gupta, R.C., and Tripathi, R.C. 1996. Analysis of zero-adjusted count data. *Computation Statistics & Data Analysis,* 23, 207–218.

Gupta, S.B., Jacobson, L.P., Margolick, J.B. et al. 2007. Estimating the benefit of an HIV-1 vaccine that reduces viral load set point. *The Journal of Infectious Diseases*, 4, 546–550.

GUSTO. 1993. An investigational randomized trial comparing four thrombolytic strategies for acute myocardial infarction. The GUSTO Investigators. *The New England Journal of Medicine*, 329, 673–682.

Haaland, P.D. 1989. *Experimental Design in Biotechnology*. Marcel Dekker, Inc., New York, NY.

Haidar, S.H., Davit, B., Chen, M.L. et al. 2008. Bioequivalence approaches for highly variable drugs and drug products. *Pharmaceutical Research*, 25, 237–241.

Hamasaki, T., Isomura, T., Baba, M. et al. 2000. Statistical approaches to detecting dose–response relationship. *Drug Information Journal*, 34, 579–590.

Harris, E.K. and Albert, A. 1990. *Survivorship Analysis for Clinical Studies*. Marcel Dekker, Inc., New York, NY.

Haybittle, J.L. 1971. Repeated assessment of results in clinical trials of cancer treatment. *British Journal of Radiology*, 44, 793–797.

Hayes, R.J. and Bennett, S. 1999. Simple sample size calculation for cluster-randomized trials. *International Journal of Epidemiology*, 28, 319–326.

Hayes, R. and Moulton, L. 2009. *Cluster Randomised Trials*. Chapman & Hall/CRC Press, Boca Raton, FL.

Haynes, B.F. and Shattock, R.J. 2008. Critical issues in mucosal immunity for HIV-1 vaccine development. *Journal of Allergy and Clinical Immunology*, 122, 3–9.

Heart Special Project Committee. 1988. Organization, review and administration of cooperative studies (Greenberg report): A report from the Heart Special Project Committee to the National Advisory Council. *Controlled Clinical Trials*, 9, 137–148.

Himabindu, G., Liu, K., Pariser, A., and Pazdu, R. 2012. Rare cancer trial design: Lessons from FDA approvals. Published Online First on June 20, 2012, doi: 10.1158/1078-0432.CCR-12-113.

Hochberg, Y. 1988. A sharper Bonferroni procedure for multiple tests of significance. *Biometrika*, 75, 800–802.

Hochberg, Y. and Tamhane, A.C. 1987. *Multiple Comparison Procedure*. John Wiley & Sons, New York, NY.

Hollander, M. and Wolfe, D.A. 1973. *Nonparametric Statistical Methods*. John Wiley & Sons, New York, NY.

Holm, S. 1979. A simple sequentially rejective multiple test procedure. *Scandinavian Journal of Statistics*, 6, 65–70.

Hommel, G., Lindig, V., and Faldum, A. 2005. Two-stage adaptive designs with correlated test statistics. *Journal of Biopharmaceutical Statistics*, 15, 613–623.

Hothorn, L.A. 2000. Evaluation of animal carcinogenicity studies: Cochran-Armitage trend test vs. multiple contrast tests. *Biometrical Journal*, 42, 553–567.

Howe, W.G. 1974. Approximate confidence limits on the mean of $X + Y$ where X and Y are two tabled independent random variables. *Journal of American Statistical Association*, 69, 789–794.

Hsieh, F.Y. 1988. Sample size formulas for intervention studies with the cluster as unit of randomization. *Statistics in Medicine*, 7, 1195–1202.

Hughes, M.D. 1993. Stopping guidelines for clinical trials with multiple treatments. *Statistics in Medicine*, 12, 901–913.

Hughes, M.D. and Pocock, S.J. 1988. Stopping rules and estimation problems in clinical trias. *Statistics in Medicine*, 7, 1231–1242.

Huque, M.F. and Dubey, S. 1990. Design and analysis for therapeutic equivalence clinical trials with binary clinical endpoints. *Proceedings of Biopharmaceutical Section of the American Statistical Association*, Alexandria, VA, pp. 91–97.

Hussey, M.A. and Hughes, J.P. 2007. Design and analysis of stepped wedge cluster randomized trials. *Contemporary Clinical Trials*, 28(2), 182–191.

Hyslop, T., Hsuan, F., and Holder, D.J. 2000. A small sample confidence interval approach to assess individual bioequivalence. *Statistics in Medicine*, 19, 2885–2897.

ICH. 1998a. Ethnic factors in the acceptability of foreign clinical data. *Tripartite International Conference on Harmonization Guideline*, E5.

ICH. 1998b. Statistical principles for clinical trials. *Tripartite International Conference on Harmonization Guideline, E9.*

ICH. 1999. Choice of control group in clinical trials. *Tripartite International Conference on Harmonization Guideline, E10.*

ICH. 2003. *ICH E14 Guidance on the Clinical Evaluation of QT/QTc Interval Prolongation and Proarrhythmic Potential for Non-Antiarrhythmic Drugs.*

Jansakul, N. and Hinde, J.P. 2002. Score tests for zero-inflated Poisson models. *Computational Statistics and Data Analysis*, 40, 75–96.

Jennison, C. and Turnbull, B. 1989. Interim analysis: The repeated confidence interval approach (with discussion). *Journal of Royal Statistical Society, B*, 51, 305–361.

Jennison, C. and Turnbull, B.W. 1990. Statistical approaches to interim monitoring of medical trials: A review and cormmentary. *Statistics in Medicine*, 5, 299–317.

Jennison, C. and Turnbull, B. 1993. Sequential equivalence testing and repeated confidence intervals, with application to normal and binary response. *Biometrics*, 49, 31–34.

Jennison, C. and Turnbull, B. 2000. *Group Sequential Methods with Applications to Clinical Trials.* Chapman & Hall, New York, NY.

Jennison, C. and Turnbull, B.W. 2006. Adaptive and nonadaptive group sequential tests. *Biometrika*, 93, 1–21.

Johnson, N.L. and Kotz, S. 1970. *Continuous Univariate Distributions-2.* John Wiley, New York.

Johnson, N.L. and Kotz, S. 1972. *Distribution in Statistics.* Houghton Mifflin Company, Boston, MA.

Johnson, N.L., Kotz, S., and Kemp, A. 1992. *Univariate Discrete Distributions*, 2nd Edition. John Wiley, New York.

Jones, B. and Kenward, M.G. 1989. *Design and Analysis of Cross-Over Trials.* Chapman & Hall, New York, NY.

Joseph, L. and Belisle, P. 1997. Bayesian sample size determination for normal means and differences between normal means. *Statistician*, 44, 209–226.

Joseph, L., Wolfson, D.B., and Berger, R.D. 1995. Sample size calculations for binomial proportions via highest posterior density intervals (with discussion). *Journal of the Royal Statistical Society, Series D (The Statistician)*, 44, 143–154.

Jung, S.H. 2005. Sample size for FDR-control in microarray data analysis. *Bioinformatics*, 21, 3097–3104.

Jung, S.H., Bang, H., and Young, S.S. 2005. Sample size calculation for multiple testing in microarray data analysis. *Biostatistics*, 6(1), 157–169.

Jung, S.H., Chow, S.C., and Chi, E.M. 2007. On sample size calculation based on propensity analysis in non-randomized trials. *Journal of Biopharmaceutical Statistics*, 17, 35–41.

Keene, O.N., Jones, M.R., Lane, P.W., and Anderson, J. 2007. Analysis of exacerbation rates in asthma and chronic obstructive pulmonary disease: Example from the TRISTAN study. *Pharmaceutical Statistics*, 6, 89–97.

Kessler, D.A. 1989. The regulation of investigational drugs. *New England Journal of Medicine*, 320, 281–288.

Kessler, D.A. and Feiden, K.L. 1995. Faster evaluation of vital drugs. *Scientific American*, 272, 48–54.

Khatri, C.G. and Shah, K.R. 1974. Estimation of location of parameters from two linear models under normality. *Communications in Statistics—Theory and Methods*, 3, 647–663.

Kim, J.H., Perks-Ngarm, S., Excler, J.-L., and Michael, N.L. 2010. HIV vaccines: Lessons learned and the way forward. *Current Opinion in HIV and AIDS*, 5(5), 428–434.

Kim, K. 1989. Point estimation following group sequential tests. *Biometrics*, 45, 613–617.

Kim, K. and DeMets, D.L. 1987. Confidence intervals following group sequential tests in clinical trials. *Biometrics*, 43, 857–864.

Kruskal, W.H. and Wallis, W.A. 1952. Use of ranks in one-criterion variance analysis. *Journal of American Statistical Association*, 47, 583–621.

Lachin, J.M. 1981. Introduction to sample size determination and power analysis for clinical trials. *Controlled Clinical Trials*, 2, 93–113.

Lachin, J.M. and Foulkes, M.A. 1986. Evaluation of sample size and power for analysis of survival with allowance for nonuniform patient entry, losses to follow-up, noncompliance, and stratification. *Biometrics*, 42, 507–519.

Lakatos, E. 1986. Sample size determination in clinical trials with time-dependent rates of losses and noncompliance. *Controlled Clinical Trials*, 7, 189–199.

Lakatos, E. 1988. Sample sizes based on the log-rank statistic in complex clinical trials. *Biometrics*, 44, 229–241.

Lan, K.K.G. and DeMets, D.L. 1983. Discrete sequential boundaries for clinical trials. *Biometrika*, 70, 659–663.

Lan, K.K.G. and Wittes, J. 1988. The B-value: A tool of monitoring data. *Biometrics*, 44, 579–585.

Landau, S. and Stahl, D. 2013. Sample size and power calculations for medical studies by simulation when closed form expressions are not available. *Statistical Methods in Medical Research*, 22(3), 324–345.

Landis, J.R., Heyman, E.R., and Koch, G.G. 1976. Average partial association in three-way contingency tables: A review and discussion of alternative tests. *International Statistical Review*, 46, 237–254.

Lasdon, L.S. and Waren, A.D. 1979. Generalized reduced gradient software for linearly and non-linearly constrained problems. In *Design and Implementation of Optimization Software*. Ed. Greenberg, H., Sijthoff and Noordhoff, Pubs, The Netherlands, 363–397.

Laubscher, N.F. 1960. Normalizing the noncentral t and F distribution. *Annals of Mathematical Statistics*, 31, 1105–1112.

Lee, A.H., Wang, K., and Yau, K.K.W. 2001. Analysis of zero-inflated Poisson data incorporating extent of exposure. *Biometrical Journal*, 43, 963–975.

Lee, M.L.T. and Whitmore, G.A. 2002. Power and sample size for DNA microarray studies. *Statistics in Medicine*, 21, 3543–3570.

Lee, S.J. and Zelen, M. 2000. Clinical trials and sample size considerations: Another prespective. *Statistical Science*, 15, 95–110.

Lee, Y., Shao, J., and Chow, S.C. 2004. The modified large sample confidence intervals for linear combinations of variance components: Extension, theory, and application. *Journal of the American Statistical Association*, 99, 467–478.

Lee, Y., Shao, J., Chow, S.C., and Wang, H. 2002. Test for inter-subject and total variabilities under crossover design. *Journal of Biopharmaceutical Statistics*, 12, 503–534.

Lehmacher, W. and Wassmer, G. 1999. Adaptive sample size calculations in group sequential trials. *Biometrics*, 55, 1286–1290.

Li, C.S., Lu, J.C., Park, J., Kim, K., Brinkley, P.A., and Peterson, J.P. 1999. Multivariate zero-inflated Poisson models and their applications. *Technometrics*, 41, 29–38.

Li, G., Shih, W.C.J., and Wang, Y.N. 2005. Two-stage adaptive design for clinical trials with survival data. *Journal of Biopharmaceutical Statistics*, 15, 707–718.

Lin, Y. and Shih, W.J. 2001. Statistical properties of the traditional algorithm-based designs for phase I cancer clinical trials. *Biostatistics*, 2, 203–215.

Lindley, D.V. 1997. The choice of sample size. *Statistician*, 44, 167–171.

Liu, G. and Liang, K.Y. 1997. Sample size calculations for studies with correlated observations. 53(3), 937–947.

Liu, J.P. and Chow, S.C. 1992. Sample size determination for the two one-sided tests procedure in bioequivalence. *Journal of Pharmacokinetics and Biopharmaceutics*, 20, 101–104.

Liu, K.J. 2001. Letter to the editor: A flexible design for multiple armed screening trials. *Statistics in Medicine*, 20, 1051–1060.

Liu, Q. 1998. An order-directed score test for trend in ordered 2xK tables. *Biometrics*, 54, 1147–1154.

Liu, Q. and Chi, G.Y.H. 2001. On sample size and inference for two-stage adaptive designs. *Biometrics*, 57, 72–177.

Liu, Q., Proschan, M.A., and Pledger, G.W. 2002. A unified theory of two-stage adaptive designs. *Journal of American Statistical Association*, 97, 1034–1041.

Longford, N.T. 1993a. Regression analysis of multilevel data with measurement error. *British Journal of Mathematical and Statistical Psychology*, 46, 301–312.

Longford, N.T. 1993b. *Random Coefficient Models*. Oxford University Press Inc., New York, NY.

Lu, Q., Chow, S.C., and Tse, S.K. 2014. On two-stage adaptive design with count data from different study durations under Weibull distribution. *Drug Designing*, 3(114), doi: 10.4172/2169-0138.1000114.

Lu, Q., Tse, S.K., Chow, S.C., Chi, Y., and Yang, L.Y. 2009. Sample size estimation based on event data for a two-stage survival adaptive trial with different durations. *Journal of Biopharmaceutical Statistics*, 19, 311–323.

Lu, Q., Tse, S.K., and Chow, S.C. 2010. Analysis of time-to-event data under a two-stage survival adaptive design in clinical trials. *Journal of Biopharmaceutical Statistics*, 20, 705–719.

Lu, Q.S., Tse, S.K., Chow, S.C., and Lin, M. 2012. Analysis of time-to-event data with non-uniform patient entry and loss to follow-up under a two-stage seamless adaptive design with Weibull distribution. *Journal of Biopharmaceutical Statistics*, 22, 773–784.

Maca, J., Bhattacharya, S., Dragalin, V., Gallo, P., and Krams, M. 2006. Adaptive seamless phase II/III designs: Background, operational aspects, and examples. *Drug Information Journal*, 40, 463–474.

Mak, T.K. 1988. Analysing intraclass correlation for dichotomous variables. *Applied Statistics*, 37, 344–352.

Malik, M. and Camm, A.J. 2001. Evaluation of drug-induced QT interval prolongation. *Drug Safety*, 24, 323–351.

Mantel, N. and Haenzsel, W. 1959. Statistical aspects of the analysis of data from retrospective studies of disease. *Journal of National Cancer Institute*, 22, 719–748.

Margolies, M.E. 1994. Regulations of combination products. *Applied Clinical Trials*, 3, 50–65.

Marcus, R., Peritz, E., and Gabriel, K.R. 1976. On closed testing procedures with special reference to ordered analysis of variance. *Biometrika*, 63, 655–660.

Mattapallil, J.J., Douek, D.C., Buckler-White, A. et al. 2006. Vaccination preserves CD4 memory T cells during acute simian immunodeficiency virus challenge. *The Journal of Experimental Medicine*, 203, 1533–1541.

McCullagh, P. and Nelder, J.A. 1983. Quasi-likelihood functions. *Annals of Statistics*, 11, 59–67.

Mdege, N.D., Man, M.S., Taylor, C.A., and Torgerson, D.J. 2011. Systematic review of stepped wedge cluster randomized trials shows that design is particularly used to evaluate interventions during routine implementation. *Journal of Clinical Epidemiology*, 64(9), 936–948.

Mehrotra, D.V. and Railkar, R. 2000. Minimum risk weights for comparing treatments in stratified binomial trials. *Statistics in Medicine*, 19, 811–825.

Meier, P. 1953. Variance of a weighted mean. *Biometrics*, 9, 59–73.

Meier, P. 1989. The biggest public health experiment ever, the 1954 field trial of the Salk poliomyelitis vaccine. In *Statistics: A Guide to the Unknown*, 3rd Edition. Eds. Tanur, J.M., Mosteller, F., and Kruskal, W.H., Wadsworth, Belmont, CA, pp. 3–14.

Meinert, C.L. 1986. *Clinical Trials: Design, Conduct, and Analysis*. Oxford University Press, New York, NY.

Miettinen, O. and Nurminen, M. 1985. Comparative analysis of two rates. *Statistics in Medicine*, 4, 213–226.

Mike, V. and Stanley, K.E. 1982. *Statistics in Medical Research: Methods and Issues, with Applications in Cancer Research*. John Wiley & Sons, New York, NY.

Miller, R.G. Jr. 1997. *Beyond ANOVA, Basics of Applied Statistics*, 2nd Edition. Springer-Verlag, New York, NY.

Moineddin, R., Matheson, F.I., and Glazier, R.H. 2007. A simulation study of sample size for multilevel logistic regression models. *BMC Medical Research Methodology*, 7(1), 34.

Moses, L.E. 1992. Statistical concepts fundamental to investigations. In *Medical Uses of Statistics*. Eds. Bailar, J.C. and Mosteller, F., New England Journal of Medicine Books, Boston, MA, pp. 5–26.

Moss, A.J. 1993. Measurement of the QT interval and the risk associated with QT interval prolongation. *American Journal of Cardiology*, 72, 23B–25B.

Müller, H.H. and Schafer, H. 2001. Adaptive group sequential designs for clinical trials: Combining the advantages of adaptive and of classical group sequential approaches. *Biometrics*, 57, 886–891.

Müller, P., Parmigiani, G., Robert, C., and Rousseau, J. 2004. Optimal sample size for multiple testing: The case of gene expression microarrays. *Journal of the American Statistical Association*, 99, 990–1001.

Murray, D.M. 1998. *Design and Analysis of Group-Randomized Trials (Vol. 29)*. Oxford University Press, New York, USA.

Mutter, G.L., Baak, J.P.A., Fitzgerald, J.T., Gray, R., Neuberg, D., Kust, G.A., Gentleman, R., Gallans, S.R., Wei, L.J., and Wilcox, M. 2001. Global express changes of constitutive and hormonally regulated genes during endometrial neoplastic transformation. *Gynecologic Oncology*, 83, 177–185.

Nam, J.M. 1987. A simple approximation for calculating sample sizes for detecting linear trend in proportions. *Biometrics*, 43, 701–705.

Nam, J.M. 1998. Power and sample size for stratified prospective studies using the score method for testign relative risk. *Biometrics*, 54, 331–336.

Neuhauser, M. and Hothorn, L. 1999. An exact Cochran-Armitage test for trend when dose–response shapes are a priori unknown. *Computational Statistics and Data Analysis*, 30, 403–412.

Neyman, J. 1939. On a new class of contagious distributions applicable in entomology and bacteriology. *Annals of Mathematical Statistics*, 10, 35–57.

O'Brien, P.C. and Fleming, T.R. 1979. A multiple testing procedure for clinical trials. *Biometrics*, 35, 549–556.

Olkin, I. 1995. Meta-analysis: Reconciling the results of independent studies. *Statistics in Medicine*, 14, 457–472.

Olschewski, M. and Schumacher, M. 1990. Statistical analysis of quality of life data in cancer clinical trials. *Statistics in Medicine*, 9, 749–763.

O'Neill, R.T. 1988. Assessment of safety. In *Biopharmaceutical Statistics for Drug Development*. Ed. Peace, K., Marcel Dekker, Inc., New York, NY, pp. 543–604.

O'Quigley, J., Pepe, M., and Fisher, L. 1990. Continual reassessment method: A practical design for phase I clinical trial in cancer. *Biometrics*, 46, 33–48.

O'Quigley, J. and Shen, L. 1996. Continual reassessment method: A likelihood approach. *Biometrics*, 52, 673–684.

Orloff, J., Douglas, F., Pinheiro, J. et al. 2009. The future of drug development: Advancing clinical trial design. *Nature Reviews Drug Discovery*, 8(12), 949–957.

Pagana, K.D. and Pagana, T.J. 1998. *Manual of Dignostic and Laboratory Tests*. Mosby, Inc., St. Louis, MO.

Pan, W. 2002. A comparative review of statistical methods for discovering differentially expressed genes in replicated microarray experiments. *Bioinformatics*, 18(4), 546–554.

Pan, W., Lin, J., and Le, C.T. 2002. How many replicated of arrays are required to detect gene expression changes in microarray experiments? A mixture model approach. *Genome Biology*, 3(5), 1–10.

Patulin Clinical Trials Committee (of the Medical Research Council). 1944. Clinical trial of Patulin in the common cold. *Lancet*, 2, 373–375.

Pawitan, Y. and Hallstrom, A. 1990. Statistical interim monitoring of the cardiac arrhythmia suppression trial. *Statistics in Medicine*, 9, 1081–1090.

Peace, K.E. 1987. *Biopharmaceutical Statistics for Drug Development*. Marcel Dekker, Inc., New York, NY.

Peace, K.E. 1990. *Statistical Issues in Drug Research and Development*. Marcel Dekker, Inc., New York, NY.

Peto, R., Pike, M.C., Armitage, P., Breslow, N.E., Cox, D.R., Howard, S.V., Mantel, N., McPherson, K., Peto, J., and Smith, P.G. 1976. Design and analysis of randomized clinical trials requiring prolonged observation of each patient. *British Journal of Cancer*, 34, 585–612.

Petricciani, J.C. 1981. An overview of FDA, IRBs and regulations. *IRB*, 3, 1.

Pham-Gia, T.G. 1997. On Bayesian analysis, Bayesian decision theory and the sample size problem. *Statistician*, 46, 139–144.

Phillips, K.F. 1990. Power of the two one-sided tests procedure in bioequivalence. *Journal of Pharmacokinetics and Biopharmaceutics*, 18, 137–144.

PhRMA. 2003. *Investigating drug-induced QT and QTc prolongation in the clinic: Statistical design and analysis considerations*. Report from the Pharmaceutical Research and Manufacturers of America QT Statistics Expert Team, August 14, 2003.

PHSRG. 1989. Steering Committee of the Physician's Health Study Research Group. Final report of the aspirin component of the ongoing physicians' health study. *New England Journal of Medicine*, 321, 129–135.

PMA. 1993. PMA Biostatistics and Medical Ad Hoc Committee on Interim Analysis. Interim analysis in the pharmaceutical industry. *Controlled Clinical Trials*, 14, 160–173.

Pocock, S.J. 1977. Group sequential methods in the design and analysis of clinical trials. *Biometrika*, 64, 191–199.

Pocock, S.J. 1983. *Clinical Trials: A Practical Approach.* John Wiley & Sons, New York, NY.

Pocock, S.J. and Hughes, M.D. 1989. Practical problems in interim analyses with particular regard to estimation. *Controlled Clinical Trials,* 10, S209–S221.

Podgor, M.J., Gastwirth, J.L., and Mehta, C.R. 1996. Efficiency robust tests of independence in contingency tables with ordered classifications. *Statistics in Medicine,* 15, 2095–2105.

Portier, C. and Hoel, D. 1984. Type I error of trend tests in proportions and design of cancer screens. *Communications in Statistics,* 13, 1–14.

Posch, M. and Bauer, P. 2000. Interim analysis and sample size reassessment. *Biometrics,* 56, 1170–1176.

Press, W.H., Teukolsky, S.A., Vetterling, W.T., and Flannery, B.P. 1996. *Numerical Recipes in Fortran 90.* Cambridge University Press, New York, NY.

Proschan, M.A., Follmann, D.A., and Geller, N.L. 1994. Monitoring multi-armed trials. *Statistics in Medicine,* 13, 1441–1452.

Proschan, M.A. and Hunsberger, S.A. 1995. Designed extension of studies based on conditional power. *Biometrics,* 51, 1315–1324.

Proschan, M.A., Lan, G.K.K., and Wittes, J.T. 2006. *Statistical Monitoring of Clinical Trials: A Unified Approach.* Springer Science+Business Media, LLC, New York, NY.

Quan, H. and Shih, W.J. 1996. Assessing reproducibility by the within-subject coefficient of variation with random effects models. *Biometrics,* 52, 1195–1203.

Rerks-Ngarm, S., Pitisuttithum, P., Nitayaphan, S. et al. 2009. Vaccination with ALVAC and AIDSVAX to Prevent HIV-1 Infection in Thailand. *New England Journal of Medicine,* 361, 1–12.

Reynolds, R., Lambert, P.C., and Burton, P.R. 2008. Analysis, power and design of antimicrobial resistance surveillance studies, taking account of inter-centre variation and turnover. *Journal of Antimicrobial Chemotherapy,* 62(Suppl. 2), ii29–ii39.

Rosenbaum, P.R. and Rubin, D.B. 1983. The central role of the propensity score in observational studies for causal effects. *Biometrika,* 70, 41–55.

Rosenbaum, P.R. and Rubin, D.B. 1984. Reducing bias in observational studies using subclassification on the propensity score. *Journal of American Statistical Association,* 95, 749–759.

Rosenberger, W. and Lachin, J. 2003. *Randomization in Clinical Trials.* John Wiley & Sons, New York, NY.

Ruberg, S.J. 1995a. Dose response studies: I. Some design considerations. *Journal of Biopharmaceutical Statistics,* 5, 1–14.

Ruberg, S.J. 1995b. Dose response studies: II. Analysis and interpretation. *Journal of Biopharmaceutical Statistics,* 5, 15–42.

Rubinstein, L.V., Gail, M.H., and Santner, T.J. 1981. Planning the duration of a comparative clinical trial with loss to follow-up and a period of continued observation. *Journal of Chronic Diseases,* 34, 469–479.

Sander, C. 2000. Genomic medicine and the future of health care. *Science,* 287(5460), 1977–1978.

Sargent, D. and Goldberg, R. 2001. A flexible design for multiple armed screening trials. *Statistics in Medicine,* 20, 1051–1060.

Schall, R. and Luus, H.G. 1993. On population and individual bioequivalence. *Statistics in Medicine,* 12, 1109–1124.

Schoenfeld, D. 1981. The asymptotic properties of nonparametric tests for comparing survival distributions. *Biometrika,* 68, 316–319.

Schoenfeld, D.A. 1983. Sample-size formula for the proportional-hazards regression model. *Biometrics,* 499–503.

Schuirmann, D.J. 1987. A comparison of the two one-sided tests procedure and the power approach for assessing the equivalence of average bioequivalence. *Journal of Pharmacokinetics and Biopharmaceutics,* 15, 657–680.

Self, S. and Mauritsen, R. 1988. Power/sample size calculations for generalized linear models. *Biometrics,* 44, 79–86.

Self, S., Mauritsen, R., and Ohara, J. 1992. Power calculations for likelihood ratio tests in generalized linear models. *Biometrics,* 48, 31–39.

Self, S., Prentice, R., Iverson, D. et al. 1988. Statistical design of the women's health trial. *Controlled Clinical Trials,* 9, 119–136.

SERC. 1993. EGRET SIZ: Sample size and power for nonlinear regression models. Reference Manual, Version 1. Statistics and Epidemiology Research Corporation.

Serfling, R.J. 1980. *Approximation Theorems of Mathematical Statistics*. John Wiley & Sons, New York.

Shaffer, J.P. 2002. Multiplicity, directional (Type III) errors, and the null hypothesis. *Psychological Methods*, 7, 356–369.

Shao, J. and Chow, S.C. 1990. Test for treatment effect based on binary data with random sample sizes. *Australian Journal of Statistics*, 32, 53–70.

Shao, J. and Chow, S.C. 2002. Reproducibility probability in clinical trials. *Statistics in Medicine*, 21, 1727–1742.

Shao, J. and Tu, D. 1999. *The Jackknife and Bootstrap*. Springer-Verlag, New York, NY.

Shapiro, S.H. and Louis, T.A. 1983. *Clinical Trials, Issues and Approaches*. Marcel Dekker, Inc, New York, NY.

Sheiner, L.B. 1992. Bioequivalence revisited. *Statistics in Medicine*, 11, 1777–1788.

Shih, J.H. 1995. Sample size calculation for complex clinical trials with survival endpoints. *Controlled Clinical Trials*, 16, 395–407.

Shih, W.J. 1993. Sample size re-estimation for triple blind clinical trials. *Drug Information Journal*, 27, 761–764.

Shih, W.J. and Zhao, P.L. 1997. Design for sample size re-estimation with interim data for double-blind clinical trials with binary outcomes. *Statistics in Medicine*, 16, 1913–1923.

Shipley, M.J., Smith, P.G. and Dramaix, M. 1989. Calculation of power for matched pair studies when randomization is by group. *International Journal of Epidemiology*, 18, 457–461.

Shirley, E. 1977. A non-parametric equivalent of William' test for contrasting increasing dose levels of treatment. *Biometrics*, 33, 386–389.

Simon, R. 1989. Optimal two-stage designs for phase II clinical trials. *Controlled Clinical Trials*, 10, 1–10.

Simon, R., Radmacher, M.D., and Dobbin, K. 2002. Design of studies with DNA microarrays. *Genetic Epidemiology*, 23, 21–36.

Snedecor, G.W. and Cochran, W.G. 1989. *Statistical Methods*, 8th Edition. Iowa State University Press, Ames, IA.

Spiegelhalter, D.J. and Freedman, L.S. 1986. A predictive approach to selecting the size of a clinical trial, based on subjective clinical opinion. *Statistics in Medicine*, 5, 1–13.

Spilker, B. 1991. *Guide to Clinical Trials*. Raven Press, New York, NY.

Spriet, A. and Dupin-Spriet, T. 1992. *Good Practice of Clinical Trials*. Karger, S. Karger, AG, Medical and Scientific Publication, Basel.

Stewart, W. and Ruberg, S.J. 2000. Detecting dose response with contrasts. *Statistics in Medicine*, 19, 913–921.

Storey, J.D. 2002. A direct approach to false discovery rates. *Journal of the Royal Statistical Society B*, 64(1), 479–498.

Storey, J.D. 2003. The positive false discovery rate: A Bayesian interpretation and the q-value. *Annals of Statistics*, 31(6), 2013–2035.

Storey, J.D., Taylor, J.E., and Siegmund, D. 2004. Strong control, conservative point estimation and simultaneous conservative consistency of false discovery rates: A unified approach. *Journal of the Royal Statistical Society, Series B*, 66(1), 187–205.

Storey, J.D. and Tibshirani, R. 2001. *Estimating false discovery rates under dependence, with applications to DNA microarrays*. Technical Report 2001-28, Department of Statistics, Stanford University.

Storey, J.D. and Tibshirani, R. 2003. SAM thresholding and false discovery rates for detecting differential gene expression in DNA microarrays. In *The Analysis of Gene Expression Data: Methods and Software*. Eds. Parmigiani, G., Garrett, E.S., Irizarry, R.A., and Zeger, S.L., Springer, New York, 272–290.

Strieter, D., Wu, W., and Agin, M. 2003. Assessing the effects of replicate ECGs on QT variability in healthy subjects. Presented at Midwest Biopharmaceutical Workshop, May 21, 2003, Muncie, IN.

Stuart, A. 1955. A test for homogeneity of the marginal distributions in a two-way classification. *Biometrika*, 42, 412–416.

Sutton, A.J., Cooper, N.J., Jones, D.R., Lambert, P.C., Thompson, J.R., and Abrams, K.R. 2007. Evidence-based sample size calculations based upon updated meta-analysis. *Statistics in Medicine*, 26(12), 2479–2500.

Sutton, A.J., Donegan, S., Takwoingi, Y., Garner, P., Gamble, C., and Donald, A. 2009. An encouraging assessment of methods to inform priorities for updating systematic reviews. *Journal of Clinical Epidemiology*, 62(3), 241–251.

Tandon, P.K. 1990. Applications of global statistics in analyzing quality of life data. *Statistics in Medicine*, 9, 819–827.

Taylor, D.W. and Bosch, E.G. 1990. CTS: A clinical trials simulator. *Statistics in Medicine*, 9(7), 787–801.

Temple, R. 1993. Trends in pharmaceutical development. *Drug Information Journal*, 27, 355–366.

Temple, R. 2003. Presentation at Drug Information Agency/FDA Workshop, Washington DC.

Tessman, D.K., Gipson, B., and Levins, M. 1994. Cooperative fast-track development: The fludara story. *Applied Clinical Trials*, 3, 55–62.

Thall, P.F., Simon, R.M., and Estey, E.H. 1995. Baysian sequential monitoring designs for single-arm clinical trials with multiple outcomes. *Statistics in Medicine*, 14, 357–379.

Thall, P.F., Simon, R.M., and Estey, E.H. 1996. New statistical strategy for monitoring safety and efficacy in single-arm clinical trials. *Journal of Clinical Oncology*, 14, 296–303.

Thomas, J.G., Olson, J.M., Tapscott, S.J., and Zhao, L.P. 2001. An efficient and robust statistical modeling approach to discover differentially expressed genes using genomic expression profiles. *Genome Research*, 11, 1227–1236.

Thompson, S.G., Pyke, S.D., and Hardy, R.J. 1997. The design and analysis of paired cluster randomized trials: An application of meta-analysis techniques. *Statistics in Medicine*, 16, 2063–2079.

Ting, N., Burdick, R.K., Graybill, F.A., Jeyaratnam, S., and Lu, T.C. 1990. *Journal of Statistical Computation and Simulation*, 35, 135–143.

Todd, S. 2003. An adaptive approach to implementing bivariate group sequential clinical trial designs. *Journal of Biopharmaceutical Statistics*, 13, 605–619.

Troendle, J.F., Korn, E.L., and McShane, L.M. 2004. An example of slow convergence of the bootstrap in high dimensions. *American Statistician*, 58, 25–29.

Tsai, A.A., Rosner, G.L., and Mehta, C.R. 1984. Exact confidence interval following a group sequential test. *Biometrics*, 40, 797–803.

Tse, S.K., Chow, S.C., Lu, Q., and Cosmatos, D. 2009. On statistical tests for homogeneity of two zero-inflated Poisson populations. *Biometrical Journal*, 51, 159–170.

Tsiatis, A.A. and Mehta, C. 2003. On the inefficiency of the adaptive design for monitoring clinical trials. *Biometrika*, 90, 367–378.

Tsiatis, A.A., Rosner, G.L., and Mehta, C.R. 1984. Exact confidence interval for following a group sequential test. *Biometrics*, 40, 797–803.

Tsong, Y. 2015. Analytical similarity assessment. *Presented at Duke-Industry Statistics Symposium*, October 22–23, 2015, Durham, NC.

Tu, D. 1997. Two one-sided tests procedures in establishing therapeutic equivalence with binary clinical endpoints: Fixed sample performances and sample size determination. *Journal of Statistical Computation and Simulation*, 59, 271–290.

Tukey, J.W. and Heyse, J.F. 1985. Testing the statistical certainty of a response to increasing doses of a drug. *Biometrics*, 41, 295–301.

Tygstrup, N., Lachin, J.M., and Juhl, E. 1982. *The Randomized Clinical Trials and Therapeutic Decisions*. Marcel Dekker, Inc., New York, NY.

UNAIDS. 2009. UNAIDS Annual Report 2009: Uniting the World Against AIDS. Geneva, Switzerland.

Van den Broek, J. 1995. A score test for zero-inflation in a Poisson distribution. *Biometrics*, 51, 738–743.

van den Oord, E.J.C.G. and Sullivan, P.F. 2003. A framework for controlling false discovery rates and minimizing the amount of genotyping in gene-finding studies. *Human Heredity*, 56(4), 188–199.

Wang, H. and Chow, S.C. 2002. A practical approach for parallel trials without equal variance assumption. *Statistics in Medicine*, 21, 3137–3151.

Wang, H., Chow, S.C., and Chen, M. 2005. A Bayesian approach on sample size calculation for comparing means. *Journal of Biopharmaceutical Statistics*, 15, 799–807.

Wang, K., Yau, K.K.W., and Lee, A.H. 2002. A zero-inflated Poisson mixed model to analyze diagnosis related groups with majority of same-day hospital stays. *Computer Methods and Programs in Biomedicine*, 68, 195–203.

Wang, S.K. and Tsiatis, A.A. 1987. Approximately optimal one-parameter boundaries for group sequential trials. *Biometrics*, 43, 193–200.

Wang, T.R. and Chow, S.C. 2015. On establishment of equivalence acceptance criterion in analytical similarity assessment. *Presented at Poster Session of the 2015 Duke-Industry Statistics Symposium*, October 22–23, 2015, Durham, NC.

Watkins, D.I., Burton, D.R., Kallas, E.G. et al. 2008. Nonhuman primate models and the failure of the Merck HIV-1 vaccine in humans. *Nature Medicine*, 14, 617–621.

West, M., Blanchette, C., Dressman, H., Huang, E., Ishida, S., Sprang, R., Zuzan, H., Olson, J., Marks, J., and Nevins, J. 2001. Predicting the clinical status of human breast cancer by using gene expression profiles. *Proceedings of the National Academy of Sciences*, 98, 11462–11467.

Westfall, P.H. and Young, S.S. 1989. P-value adjustments for multiple tests in multivariate binomial models. *Journal of the American Statistical Association*, 84, 780–786.

Westfall, P.H. and Young, S.S. 1993. *Resampling-Based Multiple Testing: Examples and Methods for P-Value Adjustment*. Wiley, New York, NY.

Westfall, P.H. and Wolfinger, R.D. 1997. Multiple tests with discrete distributions. *American Statistician*, 51, 3–8.

Westfall, P.H., Zaykin, D.V., and Young, S.S. 2001. Multiple tests for genetic effects in association studies: Methods in Molecular Biology. In *Biostatistical Methods*. Ed. Looney, S., Humana Press, Toloway, NJ, pp. 143–168.

Whitehead, J. 1993. Sample size calculation for ordered categorical data. *Statistics in Medicine*, 12, 2257–2271.

Whitehead, J. 1997. Bayesian decision procedures with application to dose-finding studies. *International Journal of Pharmaceutical Medicine*, 11, 201–208.

Wiens, B.L., Heyse, J.F., and Matthews, H., 1996. Similarity of three treatments, with application to vaccine development. *Proceedings of the Biopharmaceutical Section of the American Statistical Association*, Alexandria, VA, 203–206.

Wilcoxon, F. 1945. Individual comparisons by ranking methods. *Biometrics*, 1, 80–83.

Wilcoxon, F. and Bradley, R.A. 1964. A note on the paper "Two sequential two-sample grouped rank tests with application to screening experiments". *Biometrics*, 20, 892–895.

Williams, D.A. 1971. A test for differences between treatment means when several doses are compared with a zero dose control. *Biometrics*, 27, 103–118.

Williams, D.A. 1972. The comparison of several dose levels with a zero dose control. *Biometrics*, 28, 519–531.

Witte, J.S., Elston, R.C., and Cardon, L.R. 2000. On the relative sample size required for multiple comparisons. *Statistics in Medicine*, 19, 369–372.

Wolfinger, R.D., Gibson, G., Wolfinger, E.D., Bennett, L., Hamadeh, H., Bushel, P., Afshari, C., and Paules, R.S. 2001. Assessing gene significance from cDNA microarray expression data via mixed models. *Journal of Computational Biology*, 8(6), 625–637.

Wooding, W.M. 1993. *Planning Pharmaceutical Clinical Trials: Basic Statistical Principles*. John Wiley & Sons, New York, NY.

Wu, M., Fisher, M., and DeMets, D. 1980. Sample sizes of long-term medical trials with time-dependent noncompliance and event rates. *Controlled Clinical Trials*, 1, 109–121.

Yue, L. 2007. Statistical and regulatory issues with the application of propensity score analysis to non-randomized medical device clinical studies. *Journal of Biopharmaceutical Statistics*, 17, 1–13.

Yuen, H.K., Chow, S.C., and Tse, S.K. 2015. On statistical tests for homogeneity of two zero-inflated Poisson populations. *Journal of Biopharmaceutical Statistics*, 25(1), 44–53.

Zhang, N.F. 2006. The uncertainty associated with the weighted mean of measurement data. *Metrologia*, 43, 195–204.

Zhu, H. 2017. Sample size calculation for comparing two Poisson or negative binomial rates in non-inferiority or equivalence trials. *Statistics in Biopharmaceutical Research*, 9(1), 107–115.

Zhu, H. and Lakkis, H. 2014. Sample size calculation for comparing two negative binomial rates. *Statistics in Medicine*, 33, 376–387.

Index

Printed in the United States
by Baker & Taylor Publisher Services